SECOND EDITION

AGRICULTURAL FUTURES AND OPTIONS

Principles and Strategies

Wayne D. Purcell

Virginia Polytechnic Institute and State University

Stephen R. Koontz

Colorado State University

Prentice Hall
Upper Saddle River, New Jersey 07458

Library of Congress Cataloging-in-Publication Data

Purcell, Wayne D.

Agricultural futures and options : principles and strategies / Wayne D. Purcell, Stephen R. Koontz.—2nd ed.

p. cm.

Includes bibliographical references and index.

ISBN 0-13-779943-8

1. Commodity futures. 2. Options (Finance) 3. Produce trade—Finance. 4. Agriculture—Economic aspects. I. Koontz, Stephen R. II. Title.

HG6046.P87 1999

332.63'28—dc21 98-27294

CIP

Acquisitions Editor: Charles Stewart
Editorial Assistant: Jennifer Stagman
Editorial Production Services: WordCrafters Editorial Services, Inc.
Managing Editor: Mary Carnis
Director of Production and Manufacturing: Bruce Johnson
Prepress Manufacturing Buyer: Marc Bove
Marketing Manager: Melissa Bruner
Cover Designer: Miguel Ortiz
Printer/Binder: Courier Westford

Simon & Schuster/A Viacom Company
Upper Saddle River, NJ 07458

The first edition was published in 1991 by Macmillan Publishing Company, a division of Macmillan, Inc.

Printed in the United States of America

10 9 8 7 6 5 4 3 2 1

ISBN 0-13-779943-8

Prentice-Hall International (UK) Limited, *London*
Prentice-Hall of Australia Pty. Limited, *Sydney*
Prentice-Hall Canada, Inc., *Toronto*
Prentice-Hall Hispanoamericana, *Mexico*
Prentice-Hall of India Private Limited, *New Delhi*
Prentice-Hall of Japan, *Tokyo*
Simon & Schuster Asia Pte. Ltd., *Singapore*
Editora Prentice-Hall do Brasil, Ltda., *Rio de Janeiro*

CONTENTS

PREFACE

In a historical context, highly visible trade in commodity futures is still relatively new. During the 1960s and 1970s, price volatility in the agricultural commodities increased significantly. Changes in government policies and increased exposure to a world market brought price uncertainty to the grain and oilseed complex. Volatile prices for feedstuffs, fluctuating demand for beef and pork, and short-run changes in the number of cattle and hogs in feeding programs brought increased price variability to the livestock complex. Exposure to the risk of unpredictable price fluctuations became a major economic problem for agricultural producers.

In the 1980s, still other sources of risk surfaced. Interest rates moved to unprecedented levels and became more volatile. Exchange rates for the major world currencies fluctuated. The U.S. stock market proved vulnerable to the surrounding economic instability and stock prices began to fluctuate, often wildly. The 1985 farm bill legislation paved the way for loan rates often well below market prices, and grain prices became even more volatile as exposure to global market uncertainties was accentuated.

To an economic environment that was increasingly characterized by exposure to price risk came new futures instruments and a revitalization of older instruments. The grain and oilseed futures were more widely used. Futures contracts in live (slaughter) cattle were initiated in 1964, and trade in hog futures followed in 1966. Feeder cattle futures were started in 1971. Interest rate and foreign currency futures were initiated in the 1970s, and trade in stock index futures burst on the scene in 1982. A bit later in the 1980s, trade in options on futures for the agricultural commodities was authorized, and a significant new and flexible price risk management tool was introduced. All the developments were motivated by a common objective: to provide a means of protecting against increasingly dangerous and unpredictable fluctuations in cash prices.

In the 1990s, the trends toward uncertain market conditions increased their speed. The North American Free Trade Agreement and the General Agreement on Tariffs and Trade formalized the moves toward increased world trade. New farm bill legislation, passed in early 1996, brought an end to the long-standing programs of acreage set-asides and deficiency payments. Even the dairy farmer was now facing price risk. The need for effective price risk management escalated.

Potential users of the futures and options markets are sometimes intimidated by change and the introduction of new services. Adaptation and use are often delayed.

Beginners often perceive the markets to be complex and obscure, but they are neither. The primary objective of this book is to provide the beginning student of the markets with the understanding and basic tools needed to use the futures markets and options on futures contracts effectively.

In terms of level, the first edition of this book was written with a third-year student in a college or university program in mind. It was intended also to fit the needs of the second-year student in a two-year program. It was definitely a beginning-level book. The only prerequisites were an interest in the markets and some exposure to basic principles of economics. The student or the private-sector reader who is interested and curious and who has no formal education in economics should have been very comfortable with the book.

The book was designed for a one-semester course or a two-quarter sequence. Each succeeding chapter was designed to take a logical step forward in the education process. Both fundamental (supply–demand) and technical analysis (charts, chart interpretations, etc.) are stressed. The orientation is toward the agricultural sector, but the relatively new financial futures were also covered.

This second edition builds on the basic framework of the first edition. Coverage of options on futures is expanded, with more attention to option premiums. Options on futures are increasingly popular with decision makers, and more detailed coverage is merited. Separate chapters on foreign currencies and on stock market applications have been added. As our exposure to a global market grows, currency exchange risk becomes more important and has direct ramifications for the American farmer. And all the markets—stock, financial, agricultural, currency—are becoming closely intertwined.

Each chapter contributes to the base of understanding on risk management, but Chapter 11 is perhaps the most important chapter in the book. It integrates the fundamental and technical analyses of the markets and demonstrates strategies tied to the financial position of the firm and to the abilities of the firm manager. Chapters 1 through 10 provide the necessary base and Chapter 11 pulls it all together in what is intended to be a unique and creative way.

Chapters 12 and 13 put it all in a proper perspective. The tools that will allow successful use of the futures markets and options are relatively easy to acquire. The perspective and the discipline that are sufficient conditions for effective use of the markets are not that easy to acquire. Throughout this new edition, that thread of thought is present, and a conscious effort is made to deal with the intangibles of the market as well as the more apparent and observable dimensions. An explicit example of this orientation is Chapter 6, which deals with the psychology of the markets and of the trader.

In putting the initial book together some choices had to be made. Not everything can be or should be covered in one book. Many futures instruments including crude oil, gasoline, and lumber are not covered in any detail. The coverage of financial futures and foreign currencies was limited to one chapter, and that reflects another choice that was difficult to make. It is important for the beginner in these markets to recognize the risk of volatile interest rates and the implications to export demand of changing exchange rates. The expanded coverage demonstrates those issues and contributes to the breadth of the book that is so important in a beginning treatment. The new chapter on stock market futures and options applications reflects the growing importance of futures-based risk management in this important market.

References listed at the ends of the chapters were selected because they deal with the concepts in the chapters. There is no attempt, in a beginning-level book, to provide an extensive listing of the literature, and that approach represents another choice that had to be made. In the end, we focused on a coverage that is sufficiently broad to give the reader an accurate impression of the scope of activity in futures, but is sufficiently detailed to give the important working knowledge of futures trade.

All this is important. The futures markets offer tremendous economic potential to the individual decision maker and to sectors of our economy. The individual has the opportunity to manage exposure to price risk. A particular sector, such as the agricultural sector, has the opportunity to transfer the costs of exposure to price risk to the speculator who is outside the sector and who is willing to accept that risk.

We need to move to use of the markets from an understanding of what they can and cannot do with an open mind and from an informed perspective as to what trade in futures and options on futures can offer. Contributing to that perspective has been an equally important objective of the book. As readers, we hope you enjoy reading this edition of the book as much as we have enjoyed preparing it.

Wayne D. Purcell
Stephen R. Koontz

CHAPTER 1

THE BASICS OF COMMODITY FUTURES

146

Trade in commodity futures contracts via the organized exchanges currently seen in the United States goes back to the 1860s. The basic concept is much older. There are records of trade in contractual obligations, similar to the modern day futures contracts, in China and Japan in earlier centuries.

The current widespread and growing interest in commodity futures emerged during the 1970s. Extreme price variability in the grains, oilseeds, fibers, and livestock commodities brought with it a sense of urgency and a need for mechanisms to manage exposure to price risk. Instability in the economy late in the decade and into the early 1980s brought double-digit inflation, a prime interest rate that exceeded 20 percent, and widespread uncertainty. Farm policy moved away from approaches that pegged specific prices for key agricultural commodities and toward a posture that would allow U.S. prices to move with the world market. The instability and uncertainty gave impetus to trade in futures contracts for such diverse items as the agricultural commodities, treasury bills, lumber, foreign currencies, copper, and heating oil.

There is little reason to expect the variability, instability, and related exposure to price risk to disappear. Increasingly, commodities produced and processed in the U.S. are bought, sold, and traded in a world market. That world-market exposure has been formalized during the 1990s by the North American Free Trade Agreement (NAFTA) and the world-level General Agreement on Tariffs and Trade (GATT). That worldwide involvement means the U.S. producer, processor, or handler is exposed to the uncertainties of weather, political unrest, and changing levels of exchange rates throughout the world. U.S. soybean producers and processors experienced a price range from \$4.50 to \$11.00 per bushel during the 1980s. Intrayear price moves in excess of \$3.00 per bushel occurred five times during the decade.

During the 1990s, the price range has been from \$5.14 to \$9.03, with within-year price ranges well above \$2.00 per bushel. In corn, the 1990s have seen a price range in futures from \$2.04 to an extraordinary \$5.54, and the range in cash prices in many domestic market areas has been even wider. With exposure to such price risk in soybeans, corn, and other commodities comes the need to understand trade in commodity futures and to develop the capacity to use futures and options on futures effectively in managing exposure to that price risk. The need is on both sides of the ledger. The record corn prices of 1995 and 1996 brought a profit bonanza to farmers who had crops to sell, but they proved devastating to dairy farmers, livestock producers, and poultry firms who use corn as an important input. Clearly, there are pressing and growing needs for farmers to manage exposure to variable selling prices and equally pressing needs for users of agricultural commodities to manage exposure to variable costs. The financial integrity and economic well-being of the entire agricultural and agribusiness sector is at stake.

The objective of this book is to provide you with a working knowledge of commodity futures and options on futures. Emphasis will be on the agricultural commodities, but the concepts and applications will be relevant for any product or instrument for which futures are traded. Coverage starts at the basic level and then proceeds to the development of tools and techniques that provides all the potential user should need to develop and manage an effective program of price-risk management. Managing exposure to the increasingly volatile cash markets will be critically important to the viability of any business, farm or nonfarm, that is involved in the commodity business as we move toward the year 2000 and beyond.

In this initial chapter, emphasis will be on *what* is being traded, *why* trade in futures contracts exists, and how that trade is conducted. Coverage is very basic and is designed to answer questions the beginner tends to ask such as: What is being traded? Why is there no futures contract for some months? Why does trade in commodity futures protect against cash market price risk, and how does it work? Who trades futures, and why? What risks, if any, does the trader face if the risk of cash-price fluctuations is eliminated? How, if at all, does the futures market influence the cash market, or vice versa? What does trading futures cost? In the final analysis, who benefits from trade in futures?

Typically, as a beginner, you see the entire process as more complex and complicated than it really is. By covering the basics, the chapter removes some of the barriers to understanding and provides a base on which to build. You are encouraged to spend the time and effort needed on this chapter to get the basics down. Don't move forward from this chapter until your basic questions have been answered.

COMMODITY FUTURES: WHAT

A commodity futures contract is a legal instrument calling for the holder of that contract either to deliver or to accept delivery of a commodity on or by some future date. By definition, therefore, a commodity futures contract is what the terminology implies—a contractual obligation. When commodity futures are traded, it is this contractual obligation that is being traded. A trader who sells (goes short)[1] a commodity futures contract has incurred a legal and binding commitment to deliver that commodity, meeting the conditions explained in the contract, on or before a specified date. In trade jargon, "going short" means selling futures contracts. Conversely, a trader who buys (goes long) a commodity futures contract has incurred a legal and

[1]A listing of terms used in trading commodity futures is included as a glossary at the end of the book. As new terms are introduced in the text, each will be explained or illustrated, but you will be able to refer to the Glossary whenever review is needed or when the meaning of the term or concept is not apparent.

binding commitment to accept delivery of that commodity. The points or markets at which delivery can be completed are identified in the futures contract. Appendix 1A to this chapter lists widely traded futures, the exchanges on which they are traded, the size of the contracts offered, and related detail.

It is important for the beginning student of the markets to recognize that futures contracts can also be bought **or** *sold by a trader who has no position in the actual physical* **or** *cash commodity.* In particular, selling a commodity futures contract when you have no actual physical commodity on hand is not "selling something you do not have." An individual, partnership, or any other form of business operation always has a contractual promise or commitment that can be sold. Traders who are not involved with the cash commodity are called speculators and, as will become apparent later, they are essential to the operation of the futures markets.

A bit of perspective is important here. You need to focus on the basic issue involved. The futures market is related to the cash market, but is a separate market. As you move ahead, keep in mind that the reason there is a futures market is that producers, processors, and users have an interest in avoiding exposure to the risk of variable prices. This cash market risk, because there is a futures market, can be transferred to someone else. Keep this important point in mind.

Trade in commodity futures is the buying and selling of contractual obligations calling for future delivery of a specified commodity. Producers, processors, buyers, and others producing, trading, or using the physical commodity will be involved. These traders are called hedgers. But it is not necessary that all futures traders either possess or wish to possess the actual physical commodity. Traders called speculators who neither have nor wish to have a position in the actual commodity provide volume in the futures markets and contribute to the process of discovering a price for later time periods. The objectives of the hedgers and speculators are different, but both types of traders can be and are involved, and both types are important.

How Trade Is Conducted

Figure 1.1 provides a flow diagram of the actual mechanics of trade. Most of the communication required is by phone or developing communication systems such as the Internet. Traders contact their broker and place an order to buy or sell. The order is transmitted to the floor of the exchange in Chicago, New York, Kansas City, or some other city in the U.S. or around the world in which a futures exchange is located. The order is time stamped and passed on to a floor broker representing the brokerage firm that received the original order. Orders are usually still filled by an open outcry auction system. The floor broker "asks" a certain price if it is a sell order or "bids" a certain price if it is a buy order. If other brokers, representing some other client(s), have orders that allow them to buy or sell at the price the first broker is asking or bidding, a trade is completed. Each broker records the price and who bought or sold, and the trade is time-stamped and entered into the records of the brokerage firms involved.

All this sounds complicated, but it is not. What we have is just a process that keeps a record of when the order from the broker is received. This is the "time-stamp" feature. If there are orders to buy or sell at the same price, the orders must be filled or completed in chronological order based on the time they are received by the repre-

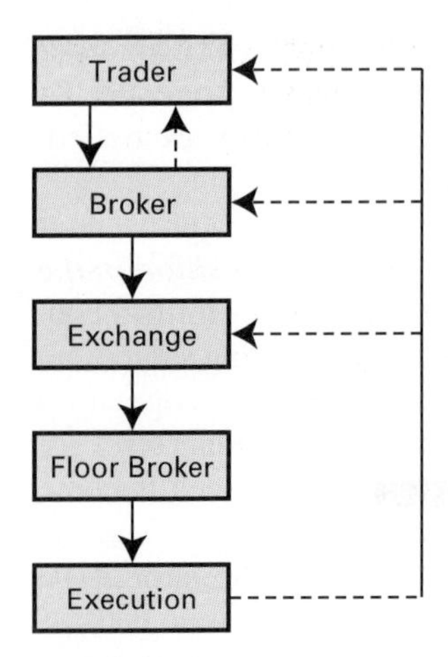

FIGURE 1.1
Flow Diagram for the Placing and Execution of an Order to Buy or Sell Commodity Futures

sentative in the trading pits. The "open outcry" feature is required. Trade is via a competitive auction, and the floor broker or other trader must "cry out" the willingness to buy or sell at a particular price. Hand signals are used to complement the verbal bids or offers in the often noisy trading pits. It is simply an auction where the buyers and sellers deal directly with each other.[2]

Figure 1.1 suggests there is a reverse flow of information to the trader through the broker or directly by mail. What the trader gets after the order is filled will vary across brokerage firms but will typically include

1. The date of the transaction;
2. A description of the transaction;
3. The price at which the transaction occurred;
4. The profit or loss from the trade if the transaction offset a position established earlier;
5. Commission charges if the transaction offset a position established earlier;
6. The account balance prior to this transaction; and
7. The account balance after the transaction.

The commission charges are simply the brokerage firm's charges for performing the trading service. The term *offset* indicates that the order eliminates a previously established position. For example, if corn futures were bought 10 days earlier, an order to sell the same quantity of corn futures contracts would offset *or* cancel the initial purchase.

If the order establishes an initial position in the futures, usually no commission charges are involved immediately. Commissions for trade in futures are charged after

[2]Responding to changing needs, some of the major exchanges are exploring and moving to trade via electronic networks. Such systems allow 24-hour access and trading from users around the world and will involve a departure from the traditional "open outcry" trading techniques. During the 1990s, there has been a tendency to move toward electronic and often computerized systems to handle the volume of trade needed and to make the overseeing and regulatory functions of regulatory agencies such as the Commodity Futures Trading Commission (CFTC) and the Security Exchange Commission (SEC) easier and more effective.

buy–sell or sell–buy transactions are completed (after a "round turn" has been completed). A round turn is a buy and later sale, or a sale and later buy, which eliminates any position or obligation in the markets. Buying 10 live cattle futures contracts on October 1 and selling 10 live cattle futures contracts on December 1 would complete a round turn, for example. The trader will now have no obligation in the futures market and commissions will be charged by the brokerage firm that handled the business. Some brokerage firms, you will find, "split" their commission charges on options. They charge part of the fee when the position is established, the rest when the position is closed out.

It is also after the round turn is completed that a profit or loss[3] will be added to, or subtracted from, the beginning account balance. You are encouraged to pause at this point and think about how selling at $3.10 and buying back at $2.85 can generate a profit. Also, think about how completion of a round turn eliminates any position in the markets.

Keep it simple for now. Selling futures means you have incurred an obligation to deliver a commodity on or by a later date. But the futures market is not intended to be a market that completes physical delivery. You will almost always want to offset the obligation to deliver by buying back an equal number of futures contracts before they mature. You will find this meets your objective of transferring the risk of price fluctuations to someone else and you can do it all with "paper" transactions. That is what the futures market is supposed to do.

Later chapters will address how to place an order with the broker, which type of order to use in a particular instance, and related detail. At this point, *it is sufficient for you to just understand that trade occurs via an open outcry and competitive auction process*. Since the prices that evolve from the auction process are so widely watched and widely used, it is useful to pause for a moment to reflect on how important it is that the futures market discover prices that accurately reflect the underlying supply–demand relationships.

Orders to buy or sell are filled on the floor of the futures exchanges by an open outcry and competitive auction process. The trader is represented by a broker whose job is to seek "fills" of the trader's orders. The resulting prices are highly visible and are widely used by decision makers as price expectations. You can avoid price risk in the cash market by buying and selling futures contracts in this related but separate market.

The Accounting Process

To be allowed to trade futures contracts, the trader must complete necessary forms supplied by the brokerage firm and deposit required "margin money." The forms essentially transfer the risk of, and responsibility for, any losses from the brokerage firm to the trader and are standard in format and content. The concept of margin money requires more explanation.

[3]A profit is earned if, for example, corn futures are bought at $3.10 and sold at $3.20. Conversely, a loss would be incurred if the futures were bought at $3.10, the market does not go up, and they later have to be sold at $2.85. Similarly, selling initially at $3.10 and buying back later at $2.85 would earn a profit. There will be profits in the futures trades if you can meet the old adage of "buy low and sell high," or "sell high and buy low."

There are two types of margin requirements. First, the *initial margin* sets out the minimum monies traders must have on deposit with their broker to trade one futures contract. Among characteristics of the initial margin requirements are the following:

1. The minimum initial margin is set by the exchanges on which the particular contract is traded. For example, the Chicago Board of Trade (CBOT) offers a soybean futures contract of 5,000 bushels of soybeans of a specified quality. The CBOT sets the minimum initial margin requirement for the soybean contract.
2. Initial margins have historically been set at 5 percent or less of the face value of the contract. A 5,000-bushel soybean contract at $8 per bushel has a value of $40,000. Applying a 5 percent rule, initial margin requirements per contract would be expected to be around $2,000. The margins can and do vary over time with a tendency for the CBOT or other exchange to raise the requirements when prices are volatile.
3. Brokerage firms can legally charge more than the minimum margin set by the exchange. They are not allowed to charge less if that particular exchange establishes minimum margin levels for a particular commodity.

The second type of margin is the *maintenance margin*. It specifies the minimum level at which the account must be maintained and becomes a threshold or trigger point to signal a "margin call" when the position the trader has established is losing money. The margin call requests additional money if the positions are to be kept in place. As a general rule of thumb, the maintenance margin requirement will be about two-thirds of the initial margin.

To illustrate the use of margins, let's assume we have speculators who hear about drought damage to the nation's corn crop and decide they want to trade corn futures. A broker is contacted, the necessary account forms are completed, and a speculator is told the margin requirements for a 5,000-bushel corn contract on the CBOT are $1,200 initial and $800 maintenance per contract.[4] Our speculator puts in an order to buy December corn futures at $3.50, and the order to buy at $3.50 is filled on July 2.[5]

Table 1.1 provides a chronological record of what could happen if the expectation that corn prices will go higher turns out to be wrong in the short run. It uses a single 5,000-bushel futures contract to illustrate.

On July 5, the December corn futures are at $3.40 at the close of trade—$.10 per bushel (bu.) below the $3.50 level at which the futures were bought. The trader has suffered an account loss of $500 (5,000 times $.10 per bu.). The account balance per contract is pulled down through the $800 maintenance level ($1,200 – $500 = $700) and a margin call of $500 would be issued to cover the losses and restore the $1,200 account balance. If the margin call is not answered and money is not sent to the broker within a prescribed time limit (three to four business days with many brokerage firms, but check with your broker), the trader's position can be liquidated, and the trader will have to absorb any losses.

[4]Many brokerage firms will require a minimum account balance and/or a certain level of net worth before they will open an account. This discussion of margins assumes any such beginning requirements have already been satisfied. Such up-front requirements are designed to ensure the trader will be able to handle any losses that might be incurred and will vary significantly across brokerage firms. You should check around for the best deal.

[5]The July 2 date was picked for illustrative purposes. Any date could have been used between the time trade in December futures contracts is started (often as early as August of the previous year) and the maturity date of the December contract during the third week of December.

TABLE 1.1 Accounting for Margins and Margin Calls for a Long Position in December Corn, 50,000-Bushel Contract

Date	Price ($ per bu.)	Action	Margin Action	Balance ($)
		Initial margin = $1,200		
		Maintenance margin = $800		
July 2	$3.50	Buy December corn futures @ $3.50.		$1,200
July 3	3.46			1,000
July 4	Holiday			
July 5	3.40		$500 call	1,200
July 6	3.33			850
July 9	3.28		$600 call	1,200
July 10	3.31			1,350
July 11	3.38			1,700
July 12	3.40			1,800
July 13	3.47			2,150
July 16	3.56			2,600
July 17	3.66			3,100
July 18	3.70			3,300
July 19	3.71			3,350
July 20	3.75			3,550
.	.		.	.
.	.		.	.
.	.		.	.
Sept. 21	$3.90	Sell December corn futures @ $3.90		$4,300

The chronological record in Table 1.1 shows increasing margin requirements as prices decline and, after the market reverses and prices move higher, an accumulation of surplus in the account that could be withdrawn by the trader.[6] *You should spend some time regenerating the margin calls shown in the table.* It is important to understand, for example, why the price decline from $3.40 to $3.33 (July 5 to July 6) did not generate a margin call, but a margin call *is* subsequently generated by the $3.28 price on July 9. The $3.28 is $.12 below the July 5 level of $3.40 when the account balance was restored to $1,200, and the $.12 decline brings a $600 dip in the account balance. The $800 maintenance margin is penetrated ($1,200 – 600 = $600) again, and a margin call is issued for $600 to bring the account balance back up to $1,200.

The trader's account is updated daily using a "mark-to-market" approach. That is, the balance of the account is updated daily to reflect the market level in the form of the official settlement price for the futures contract for each day. If you check your newspaper, electronic market wire, or the *Wall Street Journal*, there is often a range of prices within which the market is trading at the close. The range is usually small, and the exchange has to pick a "settlement price" near the middle of that range for accounting and margin calculation purposes.

[6]Most brokerage firms will, if the trader prefers, transfer the surplus funds to an account earning money market rates. If the price levels subsequently decline and more margin funds are needed, the money can be transferred back to the commodity trading account to cover emerging margin needs. This eliminates the opportunity cost of funds being tied up with the brokerage firm and earning nothing when the trader does not have a position in the markets.

Note that traders do not pay the face value of the corn contract ($\$3.50 \times 5{,}000 = \$17{,}500$) when they buy the contract. What traders deposit is the required margin. At the end of the trade, when the contract is sold at $3.90 on September 21, the round turn has been completed. The trader's account is then credited with the profit on the trade of $.40 per bushel and debited for the broker's commission charge, usually around $75 per contract for a round turn, or about $.015 per bushel. It should be noted here that the discussion of margins applies only to trade directly in futures. Trade in the relatively new options on futures either has no margin requirement or somewhat different requirements, depending on whether the user is buying or selling. This issue will be detailed in Chapter 7 on options.

It is after the round turn is completed that commissions are charged and the trader's account is credited (debited) with the profit (loss) from the trade. Prior to the completion of the round turn, the only money transferred is the initial margin, any added margin money going to the brokerage firm, or any excess margin coming from the brokerage firm to the trader. The money sent to the broker to answer margin calls is used by the brokerage firm to meet its margin requirements at the exchange. Selling the futures on September 21 eliminates the commitment to accept delivery of the corn as a buyer of corn futures.

Margins and margin requirements are often confusing to the beginner. Just keep in mind that only margins are required to trade, but that you are responsible as a trader for any losses. That responsibility suggests margin calls will have to be answered to keep the position in futures in place if the market trend moves against the initial position. Keep this basic need in mind and just remember: for a hedger looking to avoid price risk, the interest cost on margin money is just a small business expense.

Months Traded and Why

Futures contracts are not traded for each month. Appendix 1B to this chapter shows the months for which futures are traded for several widely traded agricultural commodities and selected other futures. Before proceeding, it is important that you understand why the futures exchanges establish trade in specific months and usually tend to resist requests to extend trade to each month of the year.

Even before any detailed examination of exactly how trade in futures allows transfer of cash-price risk, it should be clear to you that the markets must offer a high level of liquidity. Cattle feeders seeking protection against price risk on a pen of cattle they have just bought or portfolio managers of banks seeking protection against rising interest rates must have confidence they can buy or sell the needed futures without delays. There must be some trader willing to take the other side of the transaction without a time delay and in a volume adequate to cover cattle feeders' and portfolio managers' needs. That is what is meant by the term "liquidity."[7]

[7]There are two measures of the level of activity in futures markets. One is *open interest*, the total number of contracts that have been bought and sold and are still in place. The second is *trading volume*, the number of contracts traded in a particular day. When open interest and trading volume are relatively high, the liquidity in the market is then typically adequate for effective trading by large and small traders, both hedgers and speculators. These concepts will be covered in more detail in later chapters.

Experience has shown that a necessary condition for liquidity is active involvement by speculators. The speculators enter the futures markets looking for profits. They want to buy low and sell high, or sell high and buy back low. Risk capital is needed to cover margin requirements and to absorb losses when the speculative trade turns out to be a losing trade.

There is, at any point in time, a limited supply of speculative capital. Speculators have many investment alternatives from which to choose, and futures trade is just one of those alternatives. Within trade in futures contracts, there are many different commodities and products available. The speculator interested primarily in agricultural commodities can trade live cattle, beef trimmings, feeder cattle, hogs, pork belly, and milk futures on the Chicago Mercantile Exchange (CME), or soybeans, soybean meal, soybean oil, corn, oats, and wheat futures on the CBOT. Other classes of wheat are traded on the Kansas City and Minneapolis exchanges, and a New York Exchange offers futures trade in cotton and selected dairy products. Orange juice and petroleum products such as crude oil are traded via associations established at the New York Exchange. Coffee, sugar, and cocoa are also traded in New York. If the speculator feels comfortable with the nonagricultural commodities, there are numerous futures contracts in interest rate futures, stock indices, and foreign exchange futures. Add the widely traded precious metal futures—gold, silver, copper, platinum—and it is clear that a great deal of competition exists for the speculative investment dollar within the futures complex.

If trade were conducted in each month for the many futures contracts traded, the markets could be "thin" and liquidity would suffer. When there is only limited trade in a futures contract for a particular month, the filling of orders will be difficult. In other words, there may not be enough liquidity to meet the needs of hedgers interested in eliminating exposure to price risk.

The exchanges therefore tend to go to an alternate month format or select months that best fit the needs of those involved in the cash markets. In feeder cattle, for example, trade in feeder cattle futures is for steers weighing 600–800 pounds that are ready to move into feedlots. Production programs tend to be of two types. Calves carried through the winter months are generally sold as feeder cattle in the spring months, and calves carried through the spring and summer grazing season are typically sold as feeder cattle in the fall months. There is, therefore, a strong seasonal pattern to the production and pricing of feeder cattle, and the months for which futures are traded have been selected to fit the production patterns.

Trade in feeder cattle futures was started in 1971, and for many years, the Chicago Mercantile Exchange listed futures contracts for March, April, and May and then for August, September, October, and November. As suggested, the months were selected to fit the production patterns in the cash markets. In the early 1980s, receiving strong encouragement from producer groups, the exchange initiated trade in a January futures. Trade in the January contract continues to be small, but has been adequate for the exchange to keep the contract listed. It appears the January feeder cattle futures helped to fill the time gap between November and March and attracted attention of users and traders for that reason.[8]

[8]A January contract for live cattle futures, or slaughter cattle futures, was started at the same time. The contract never attracted interest and was later dropped. Apparently, there was less need for a January contract in live cattle when December and February contracts are traded, and the time span between the existing contracts for fed cattle coming out of feedlots was less than that for feeder cattle.

In the corn market, the strong seasonal pattern emerges again. With harvest in October and November, the December contract is positioned to discover the price for the new crop in any particular year. Typically very high levels of open interest and trading volume occur in the December contract, and the liquidity is there to meet the needs of the small farmer who needs to establish hedges, and the large-volume grain elevator who needs to hedge the cash contracts being extended to producers.

The primary motivation behind the selection of months is therefore tied to the nature of the cash market programs and to the need for market liquidity. A related and less frequently mentioned, or less frequently admitted, motivation is to give the speculator the impression that profit opportunities might exist.

Speculative capital tends to move toward areas and toward futures contracts that offer profit potential. In a production year plagued by drought problems, speculators will flock to the corn and soybean futures in search of major profit opportunities. If futures were traded for each month for corn, the speculators tend to see fewer opportunities. Any chance to arbitrage between the cash and futures markets or between different months of the futures market is diminished when futures are traded for each month, and the CBOT is afraid that the speculators will move their capital to other alternatives.

To illustrate the important concept of arbitrage, consider the situation in which September corn futures are at an unusually high premium relative to December futures. Speculators who have data showing the historical relationships may see the chance to profit by selling September futures and buying December futures. This arbitrage activity will tend to help restore a near-normal relationship between September and December and the speculators' trades add liquidity to the market. If there were corn futures for October and November as well (there are none), speculators might see less opportunity for profitable arbitrage activity and take their risk capital to some other opportunity.

Users who do not understand the importance of speculative activity will continue to ask for more futures contracts. Hedgers often want a contract for each month so they can "match" their cash program more closely in futures trades. The exchanges will tend to resist in an effort to protect liquidity and ensure the viability of their contracts. A compromise will always be necessary, but that compromise will seldom if ever involve moving to trade of futures contracts for each month. Of the commodities or products listed in Appendix 1B, crude oil traded on the New York Mercantile Exchange is an exception. With worldwide interest in petroleum as a primary source of energy, there is apparently enough hedging and speculative interest to support trade in each calendar month. You will find the same thing is happening in some interest rate futures where the volume traded is often huge and lack of liquidity is not a concern.

Futures contracts are not traded for each month of the year. The months selected for trade represent efforts by the exchanges to offer trade in those months necessary for those involved in the related cash markets while ensuring liquidity in those months for which trade is conducted. The concerns for liquidity reflect the importance of speculative activity in the markets and you should start to build a perception (correct) that trade in commodity futures would be impossible without the presence and activity of the speculators.

Trade in commodity futures contracts exists for two related reasons. First, the trading of contracts for futures delivery contributes to the process of price discovery. Second, trade in commodity futures contracts provides a mechanism which can be used to reduce or eliminate the risk of fluctuating cash prices faced by those dealing in the physical commodity. This process of transferring cash-price risk is called *hedging*.

Commodity Futures and Price Discovery

In any market in which prices are not set by the sellers on a cost-plus or administered-price basis, buyers and sellers come together or communicate and negotiate prices and related terms of trade. *The process of gathering and interpreting information on supply and demand, formulating an asking (or bid) price, the give and take during the negotiations, and the dynamic adjustments to new information as it becomes available across time is called price discovery.* It is, as implied, an ongoing and continuous process.

The futures markets provide centralized and highly visible trade within which information can be received, interpreted, and incorporated into a discovered price. By definition, the futures market is an anticipatory or forward-pricing market. It is attempting to "discover" what the price of a commodity will be at some time in the future. *The price of a commodity futures contract on a particular day can be meaningfully interpreted as the consensus of those trading on that day as to what the price will be at the future point in time.* That consensus is based on the available information and the consensus will change over time as expectations for supply and demand levels for the future time period change. It is very important that you develop an understanding of this price discovery function of futures markets and why it is so important.

In a storable commodity such as corn, the cash price is directly related to the futures market. The bid price posted at a local elevator is the price of the appropriate futures contract minus an adjustment for location since the corn futures are traded in Chicago at the CBOT. Figure 1.2 provides a plot of the closing price for a recent March corn futures contract and the cash price (Central Illinois) from October 1 until the March futures contract "matures" and trade ceases during the third week of March. The two price series tend to move together in parallel fashion. *It is primarily in the futures market that the price is being discovered, and it is the futures market that is recording and interpreting changes in the available body of information that will influence prices for later time periods.*

You should pause and reflect for a moment. The cash prices shown in Figure 1.2 are prices offered by grain elevators in central Illinois. The futures price is the price *for the March futures* on the same days for which cash prices were recorded *starting with October 1*. It is the daily discovered price for March futures that is the key indicator of price, and the cash price is tied to the futures price. Elevators base their cash bids on the closest futures contract. Chapter 2 will deal in detail with the difference between the cash and futures markets (it is called "basis") and why it might change over time. It is sufficient here to recognize that the task of discovering price for corn, wheat, soybeans, cotton, and other storable commodities is awarded primarily to the futures markets. The cash price is tied directly to the futures price.

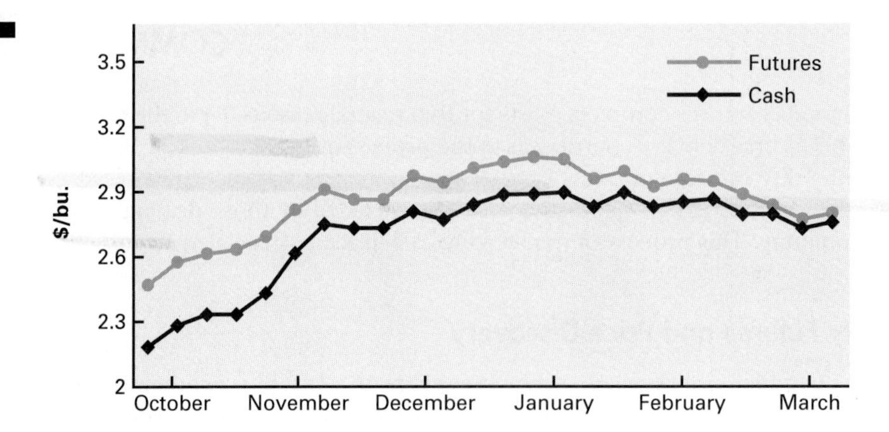

FIGURE 1.2
Cash Corn and March Corn Futures Prices, October to March, Recent Year

The relationship between futures and cash prices is less stable and less apparent in a commodity such as slaughter cattle. In the grains, the difference between December and March corn will typically approximate the cost of storing corn from December to March. For slaughter cattle, there is no storage function. The cash and futures markets are more nearly separable, with each functioning in the presence of the other. Unlike the crops in which supply is fixed once harvest is completed, the supply of cattle can change throughout the year. The futures market has to register this possibility of intrayear supply changes. This becomes apparent if we consider the impact of a drought that forces cattle off pasture and increases slaughter levels, depressing current cash prices. But the unexpected and forced increase in slaughter will influence the supply of cattle in later time periods. Prices for distant futures contracts might move higher, reflecting expectations for decreased cattle slaughter in the later time periods.

The plot in Figure 1.3 relates futures quotes for December live or fed cattle futures to cash prices in Western Kansas from early August until early November. Early in the period, the December futures are above cash, and that relationship persists until early November. As new information emerges about the likely prices in December, the December futures move below cash in early November. Eventually, the two series will tend to converge as the maturity date of the December futures approaches.[9]

The price discovery function of the futures markets receives very little attention from the user and casual observer of the markets, but is arguably more important than the much-discussed risk transfer or hedging function. It was suggested earlier that it is important that you build an appreciation for the importance of this role of the futures markets. Some simple examples should help early understanding, and we will come back to this issue in later chapters.

During the January–April period, many midwestern producers have to make a decision on how many acres to plant in corn and soybeans respectively. *In making*

[9]All futures contracts have a specified "maturity date." The December live cattle futures will mature on December 20 or the business day immediately preceding the 20th day of the contract month. The exchanges can provide detailed information on their futures contracts, which includes maturity or expiration times for the various contracts. This type of information is also available from commodity brokers and in calendar format on electronic systems.

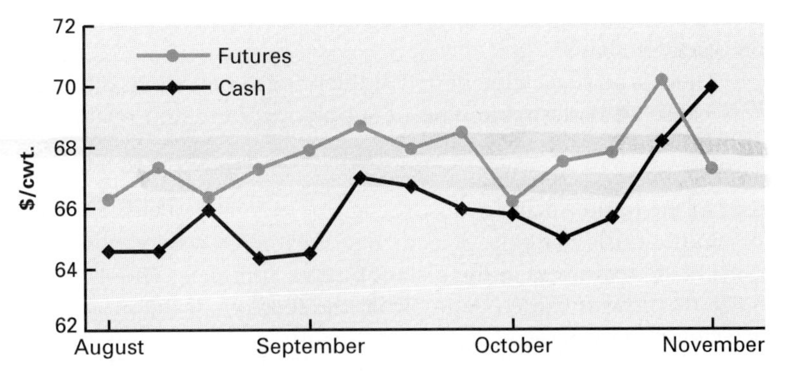

FIGURE 1.3
Cash-Futures Relationships: Western Kansas Cash and December Cattle Futures Prices

that decision, some type of price expectation has to be used. Surveys indicate that producers increasingly watch futures prices and use distant futures prices as a source of price expectations. Keep in mind that what is needed is an expectation for price in the harvest period, which is up to 10 months later. The futures markets are not always accurate predictors of later cash prices, but they are one source of price expectations that are highly visible and available to everyone.

To illustrate, assume it is March 1 and a particular producer is wrestling with the decision on how many acres to plant in corn and in soybeans. Historically, it has taken soybean prices about 2.4 times corn prices to generate equal levels of profitability on a per-acre basis. On March 1, November soybean futures are trading near $7.50 per bushel and December corn futures are trading near $2.50.[10] That gives a price ratio of 3:1, and soybeans look more favorable than corn. The producer reacts to the opportunity and decides to plant more acres in soybeans and decrease acreage in corn compared to recent years.

Many thousands of producers are watching the same set of price expectations and considering making a similar change.[11] As the realization that a widespread switch to soybeans is occurring spreads through the markets, traders of corn and soybean futures start to recognize they have underestimated producers' willingness and ability to switch from corn to soybeans. The price for November soybean futures starts to move down from $7.50, and the discovered price for December corn futures moves above $2.50. At planting time during May, November soybeans are trading at $6.60 and the December corn futures are up to $2.75. The information base has changed and the discovered prices must adjust accordingly. If producers continued to switch from corn to soybeans during the planting period, the November soybean prices could be

[10]As will become apparent in later chapters, it is the November soybean futures and the December corn futures that the producer would use to hedge or forward-price soybeans and corn, respectively. Here, it is sufficient to recognize that the trading levels of those distant futures contracts on April 1 can influence planting decisions and to start building recognition that those planting decisions will influence the prices eventually "discovered" for soybeans and corn later in the year.

[11]During the 1980s and early 1990s, meeting the planting requirements to qualify for farm program subsidies limited this type of response. The early 1996 farm bill legislation allows flexibility in planting decisions, and this type of supply response involving switching of crops will be very important in future years.

driven down to a level that profits are small or negative compared to profits from corn on a per-acre basis.

There is no suggestion here that the price expectations offered in the futures markets cause an undesirable type of supply response and related price volatility. *The futures markets must react to changes in prospective supplies, and individual decision makers must be aware that the markets will react.* This is a basic rule that every user of the futures market must recognize and understand.

In the cattle markets, placements of cattle on feed vary with the price expectations being registered in the distant futures contracts. Those reactions appear to stabilize the flow of cattle coming from the feedlots. If the projected supplies of cattle in a later period are small, traders of the futures contracts for the later period discover a higher price, and feedlot operators respond to a possible profit opportunity by placing more cattle. If projected supplies for the distant period appear to be getting too large, the futures contracts for the distant period reflect that in the form of lower prices, and feedlot operators reduce placements. Over time, the futures markets thus have the capacity to stabilize supplies of fed cattle and stabilize fed cattle prices. If producers overreact, however, and do not recognize the aggregate influence of similar decisions by many producers at the same time, the price levels for future time periods can be significantly influenced.

In the storable commodities, the intermonth price patterns can also influence producers' decisions. In winter wheat, to illustrate, the harvest is in June and early July. The premium shown by March futures over the July harvest-period futures is widely seen as a market-determined price for storage. If the premium is large, producers are encouraged to store wheat. Storage affects the supply that is available to the market both at harvest and later in the crop year as March approaches. The price expectations being registered in the markets have the potential to influence the decisions of those involved in producing and/or storing the commodity.

It is clear that the price expectations being reflected by distant futures can and do influence the final supply of product for the later time period. Those distant price expectations have the potential to change producers' decisions. Futures prices must then adjust to the realization that supplies are being changed. This is all a very logical and legitimate part of the price discovery process. To be an accurate predictor of prices in the future time period, the futures market must correctly anticipate the direction and magnitude of the response of decision makers, and that is a very difficult assignment. *It is important, therefore, that you realize that decision makers who respond to the very visible futures markets as a source of price expectations should follow through and get the prices established.* In other words, the decision maker should proceed to use the futures market, or use cash contracts, to establish price and eliminate exposure to price risk. This need sets the stage for discussion of the second and usually most visible function of trade in futures, the hedging or risk-transfer function.

Trade in commodity futures, in registering the influence of changes in information on supply and demand, provides information that has the potential to influence prices in the cash market. Decision makers must be aware of the possibility of a supply response to changed price expectations and seek protection against the risk of falling prices. It is especially important that the individual decision maker keep in mind that many other producers might be considering the same changes or

adjustments and try to anticipate the price implications of those adjustments. You have to be aware of a micro–macro trap here. Individual (micro) decisions will not change prices, but add them all together (macro) and a major price change might be coming. You should start to build an understanding of how important it is for you to protect yourself against the price changes in the micro–macro trap. We will come back to this issue often.

The Hedging Mechanism

The futures market provides an opportunity for the decision maker to escape much *or* all of the risk of fluctuating cash prices. Hedging, as a means of avoiding exposure to the risk of fluctuating prices, can be effective if two basic requirements about the way cash and futures prices behave are met:

1. Over time, cash and futures prices will respond to the underlying forces of supply and demand in such a way that prices in the two markets tend to move together, and
2. As the maturity date of the futures contract approaches, the cash and futures markets will tend to converge and approach some predictable difference called a basis.

To illustrate why these two requirements are important and to illustrate the concept of basis, it is useful to examine the hedging framework in its simplest form.

Date	Cash Market	Futures Market
October 1	An elevator manager is holding significant inventories of soybeans. He is concerned that a large crop being planted in Brazil will push soybean prices lower and cause the value of his inventory to decline.	No position yet. The manager checks his market news wire, checks with his broker, and learns soybean futures for the following May are trading around $6.40.

The manager considers the situation. He recognizes that by selling an appropriate number of soybean futures contracts (each contract is for 5,000 bushels, CBOT or 1,000 bushels, Mid-America Exchange, or MidAm), his cash inventory will be protected if (1) there is reason to expect the futures market to move down if the cash market falls, and (2) there is reason to expect the two markets to come together in May.

To reset the scene:

Date	Cash Market	Futures Market
October 1	Calculates "break-even" cost of 80,000 bushels of soybeans bought in the cash market plus storage costs and plus a $.10-per-bushel profit margin at $6.10 per bushel.	Deposits margin money with his broker and sells 16 May futures (5,000-bu. contracts) at $6.40.

Date	Cash Market	Futures Market
Early May	Soybeans are sold at $5.70 per bushel.	Buys 16 May futures at $6.00, eliminating any obligation to deliver soybeans.
Net:	$.40 "loss" per bushel.	Profit on the round turn of $.40 per bushel.
	Overall: $.10-per-bushel profit to storage before hedging costs.	

This illustration implies the difference between cash and May futures (the basis) is –$.30 on October 1 and is still –$.30 in early May. Technically, the basis is always defined as cash minus futures. It is easy to see why almost all elevators have a policy calling for hedges on inventory. It will be noted later that there is some risk associated with behavior of the basis, but the key point here deals with the underlying mechanism. *With a stable basis, the $.10 profit to the storage program that is reflected in the $6.10 needed break-even price is realized exactly.* Without the hedge, the program would have incurred an operating loss of $.30 and would not have captured the $.10-per-bushel profit—a result of $.40 per bushel below the hedge result. This is obviously important. Start to think about what would have happened if, when the cash market declined, the May futures stayed the same or increased. There would have been no protection. You should stop and think about the obvious—the hedge will not work if cash and futures do not move together.

A hedge will serve its function of offsetting the risk of fluctuating cash prices if and only if the cash and futures markets move together over time and the cash-futures basis approaches an expected level at the end of the decision period.

This statement is true whether we are looking at cattle, corn, wheat, interest rates, sugar, crude oil, or any other commodity or service for which futures contracts are traded. *The behavior of the cash-futures basis is critical.*

Notice we have ignored the nominal costs of trading the futures (commission costs plus interest on margin money). Be aware also that the hedge would have "protected" the manager if the cash price had increased. The round turn trade in the futures would have lost money in a rising market, offsetting unexpected gains in the cash market. For example, the cash market could have moved up to $6.50, earning a $.40 gain in the cash market. But the May futures would now be up to the $6.80 level (using the –$.30 basis or cash-futures difference again), and would have to be bought back around $6.80. The round turn in futures (selling at $6.40, buying back at $6.80) would lose $.40. The profit to the storage program would still be the $.10-per-bushel margin that was established by the hedge. These points will be illustrated many times later in the book. The purpose here is to illustrate the importance of the two markets moving together and converging. *Before proceeding, you should definitely try other price levels to confirm that, before the costs of the hedge are accounted for, the net profit will always be $.10 per bushel, regardless of the absolute price level, as long as the cash-futures basis in early May is –$.30.*

Whether cash and futures markets move together is a hypothesis that can be tested empirically. Empirical examination shows that (1) the two markets do tend to move together over time, and (2) any risk associated with a variable and unpredictable basis is significantly smaller than the risks associated with unhedged fluctuating cash prices. The risk of unexpected basis behavior will be covered later, but it is *not* large and *it is much smaller for virtually every commodity than is the risk of cash-price fluctuations.*

The assumption regarding behavior of the cash and futures markets can also be supported on the basis of logical reasoning. Both markets are operating within the same economy, are being barraged daily with information from the same economic developments, and are trading the same commodity. The only difference is in time. Cash prices are for today's market; futures prices are for some period in the future. As the maturity date of the futures contract approaches, the two markets are discovering the price for the same commodity *in the same time period*, and the prices would be expected to converge on some predictable basis level. After all, the underlying supply–demand forces are now representing the same time period and represents essentially the same set of information in both the cash and futures markets. Any difference, as the maturity date of the futures contracts approaches, should be a reflection of location. The futures contracts are traded in Chicago, New York, Kansas City, or in some other city, and the specific cash market may be at some distance from those cities.

The hypothesis regarding convergence of the cash and futures markets can thus be confirmed by both empirical observation and logical reasoning. Logic further suggests that the opportunity for arbitrage will ensure that the futures contract will not close or mature above cash prices by more than the costs of delivering the commodity under the provisions of the futures contract. During the maturity month of the futures contract, there is a designated period within which delivery can be completed. If the futures price is above cash by more than delivery costs, any experienced trader can

1. Buy the relatively low-priced cash product;
2. Sell futures contracts and notify the exchange of intent to deliver; and
3. Deliver the product under the provisions of the futures contract at a profit.

Delivery procedures vary across commodities, ranging from shipment of a bulky product such as fed cattle to designated delivery points to the transfer of a certificate identifying grain in an approved warehouse. When futures prices are too high, the arbitrage buying boosts the cash market and the selling forces down the futures. *These actions tend to force futures to converge to a level of cash price plus delivery costs. All that is needed, remember, for the hedge to be effective is that the cash futures basis at the end of the production period approach an expected level. Arbitrage between cash and futures markets by experienced traders will make this happen for you.*

To illustrate further, consider the live cattle futures and assume the position of an experienced trader in fed cattle located on the terminal market at Omaha, one of the designated delivery points. On a day in early June, when delivery under the June live cattle futures is now being allowed, the trader observes

1. Cash cattle of deliverable quality selling at $70.00 per hundredweight;
2. June futures trading at $73.00 per hundredweight; and
3. Available estimates of the costs of delivery under the specifications of the live cattle futures at $1.50 per hundredweight.

The cost estimate is based on past experience, and represents USDA grader time, paperwork costs, pen rental, and any other incidentals associated with live cattle delivery. The trader can buy the cash cattle, sell the June futures, and meet the obligation for the short position in June live cattle futures by delivering the cattle versus buying back the futures. If the cattle can indeed be bought at $70.00 and the futures can be sold at $73.00, the trader can net $1.50 per hundredweight ($3.00 gross margin less the $1.50 delivery costs). Traders' arbitrage actions tend to boost cash prices as they buy cash cattle, push futures down as they sell futures, and move the cash futures basis toward the expected –$1.50 per hundredweight, a basis level that reflects the $1.50-per-hundredweight delivery costs. If many traders (or cattle feeders or other producers) deliver in the futures market, the expected convergence on a basis of –$1.50 per hundredweight is essentially guaranteed by that delivery process, and users can count on the hedge to work.[12]

If futures prices are low relative to cash, another set of actions evolves to force convergence. Holders of long positions in futures contracts can, by declining to sell the futures and complete the round turn, accept delivery of the physical product. This removes buying power from the cash market, as packers, for example, accept delivery of cattle versus buying cash cattle, and forces hedgers and speculators holding short positions in futures to bid up futures trying to complete their round turn. Convergence is again assured.

For some commodities and futures instruments such as feeder cattle, a new lean hog futures starting with the February 1997 contract, and interest rate futures, there is no physical delivery process and a *cash settlement* approach is employed. Any remaining contracts on the maturity date of the futures are "settled" using a widely published cash price or cash-price index. The sometimes difficult physical delivery of the commodity or instrument is avoided, but the economic forces forcing convergence are the same. A producer holding short hedges in feeder cattle, for example, will tend to hold the short positions if the futures are above cash. Holding rather than buying back tends to move the futures down toward the cash-price series (called the National Feeder Steer Price Series), and convergence is assumed. Conversely, if the cash series is above the futures, holders of short positions will hurry to buy them back since being "cash settled" at a higher price would mean losses to the short futures position. The buying action boosts the feeder cattle futures and prompts cash-futures convergence.

There is interest in moving to cash settlement for any commodity or instrument in which (1) the use of a futures contract would be difficult with actual delivery (such as futures for a stock index) and (2) a representative and broad cash-price index can be developed (such as the feeder cattle, with weighted average prices across 12 midwestern states). There are clear advantages to elimination of the physical delivery

[12]The idea is that the *threat of delivery* will ensure that the cash futures basis will approximate the $1.50 differential. The futures market is not intended to be an alternative market that actually handles the cattle. The possibility of delivery is included with the hope that the basis will always converge to an expected level, make the hedges effective, and no deliveries will be required. It is the possibility of delivery that *guarantees* the basis will approach some expected level and keeps the risk associated with basis variability to acceptable levels. Starting with the June 1995 live cattle futures, the buyer of futures who does not offset that position and chooses to accept delivery can request that the delivery be made in certified packing plants in carcass form. This change was intended to improve the delivery process, but the idea of arbitrage is still there and is still relevant.

process, but the need for a cash-price series or index that is broadly based and free from possible manipulation has stopped the move to cash settlement for some commodities such as fed cattle (the live cattle futures).

Figures 1.4 and 1.5 record the paths of cash and futures prices for hogs and for soybeans, respectively, picture the convergence discussed here, and suggest that the process works for all commodities. The forces that cause convergence between cash and futures for hogs and soybeans are the same as those demonstrated for cattle where delivery is still involved. The two figures offer interesting patterns of price movement.

In both cases, the futures are slightly above the cash prices when the futures contract is closed and trading stops. *You should not be concerned about this apparent lack of convergence.* Remember, the cash-price series used in the plots are not for Chicago, and the cash-futures basis must reflect a location adjustment. It must also reflect a cost of delivery as illustrated earlier for cattle. *It does not matter whether the convergence is to a basis of zero or to some nonzero level. As long as convergence to a predictable level is consistent and reliable, the hedge will work.*

Figures 1.4 and 1.5 also demonstrate again the difference in basis patterns for storable versus nonstorable commodities. The basis for the soybeans in Figure 1.5 is fairly stable, especially in November and December. The basis for hogs in Figure 1.4 is not. The early futures prices for hogs in Figure 1.4 suggest that the cash prices of $50–55 per hundredweight will fall into the $40s by December, and cash prices did move lower as December approached. Again, the implication of the different time period in the cash and futures prices for a nonstorable commodity is apparent. The supply of hogs was expected to be higher in December, pushing prices lower, and the December futures contract was registering this expectation earlier in the year.

Later chapters will offer many examples of the hedge process and the mechanics of incorporating a basis allowance. *The need here is for you to understand that there are logical economic forces in the marketplace that ensure that the cash and futures difference will converge on some expected basis level.* The ideal situation, to repeat and for emphasis, would be one in which there are no deliveries under the futures contract because the threat of delivery ensures cash-futures convergence to predictable levels.

Trade in commodity futures provides a hedging mechanism that can be used to eliminate or reduce the risk of cash-price fluctuations. The effectiveness of the hedge will vary with the extent to which the cash

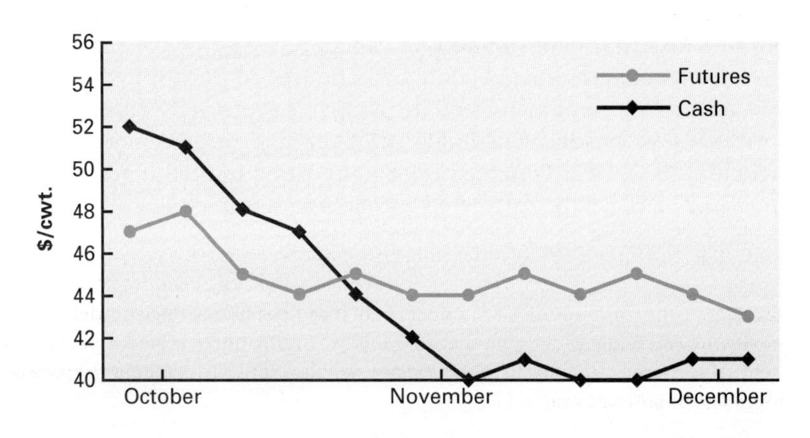

FIGURE 1.4
Convergence of Cash and Futures Prices for Hogs: Omaha Cash and December Futures Prices

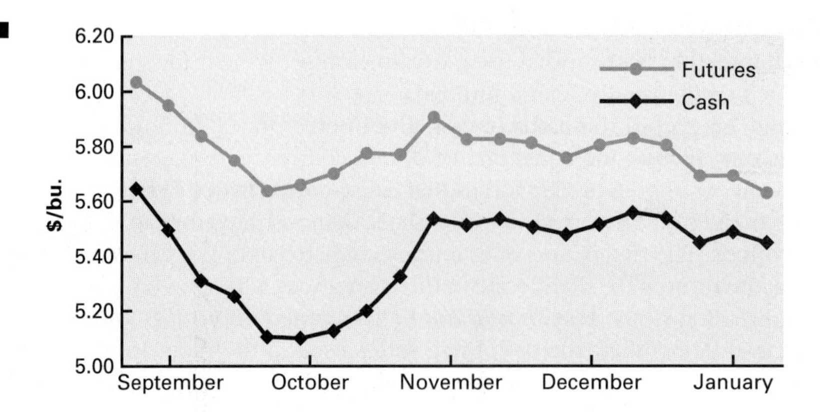

FIGURE 1.5
Convergence of Cash and Futures Prices for Soybeans: Illinois Cash and January Futures Price

and futures move together and converge during the maturity month of the futures contract. In other words, hedging will be effective if the behavior of the cash futures basis is either stable or the basis moves to an expected or predictable level. There are forces in the marketplace to ensure that the necessary convergence to an expected basis level will occur to a workable degree of reliability, and this convergence makes hedging work.

COMMODITY FUTURES: HOW

Trade in commodity futures involves two general categories of traders identified earlier, speculators and hedgers. Thc objectives of the two types of traders are different. Speculators are looking for profits; hedgers are looking for protection against cash-price risk. The two groups may differ in the analytical techniques they employ, and they often differ in the frequency of trades.

Hedgers' Strategies

Consistent with an overall objective of reducing exposure to price risk, hedgers use a variety of approaches to trade in futures. A simple approach is to use the futures market as a means of *hedging or forward-pricing* when costs are more than covered and a profit margin is being offered.

Selling or going short a commodity futures contract is hedging or forward-pricing the underlying commodity, and the hedger is protected against the risk of falling cash prices.[13] The producer or holder of a product must decide, at any point in time, whether the forward-pricing opportunity being offered is acceptable. The expected *forward price* can be calculated as follows:

[13]Users of the commodities are concerned about rising cash prices because their costs would go up. These hedgers are called *long hedgers*. In this initial chapter, the illustrations will stick with the more widely seen *short hedge* for agricultural producers. The long hedge will be explained and explored in later chapters.

Forward Price = Futures Price + Basis

The basis, in this instance, is an adjustment to localize the futures price. It is the expected difference between the futures price and the cash price in the producer's local market where the cash product will be sold. Keep in mind that cash prices in most production areas are below futures prices, so the basis is usually negative. Thus, the forward price on any day is the current futures price for the later period minus the absolute size of the expected closing basis, the basis at the end of the production period. In developing the basic concepts, the costs of the futures trades are ignored for simplicity. The costs, in the form of commissions and interest on margin funds, do not change these basic illustrations, and we will introduce them later in the book.

By comparing the forward price to budgeted production costs, the producer of the product can decide whether the forward price is acceptable. If the answer is yes, futures contracts are sold and held until the end of the period when the product is sold in the cash market and the futures contracts are bought back to complete the round turn in the futures market. This "buying back" eliminates any futures market commitment to deliver the product.

Using the futures market as a forward-pricing mechanism and hedging when the price level is considered acceptable is a simple approach. Deciding what price should be accepted and when to price is the most difficult part of this strategy. The practice of selling a futures contract and then holding that futures position until the cash commodity is sold is called a *conservative* hedging strategy in the trade.

As hedgers gain more experience in trading, they often seek to become effective *selective* hedgers as contrasted to the conservative hedger. The producer seeks to "select" when to have the hedge in place and when it should be removed and put the producer back in the position of being a cash market speculator. The analytical tools needed to provide a base for effective selective hedging strategies will be developed in detail in later chapters. Here, we just need to demonstrate the basic procedures.

Table 1.2 illustrates a simple conservative hedge for corn assuming the producer decides to hedge in June after the crop is up and growing. You should go over this

TABLE 1.2
Demonstration of a Short Hedge for a Corn Grower

Date	Cash Market	Futures Market
June 3	Estimate production cost @ $2.10. Estimate October basis @ –$.30. Forward price = $2.80 – .30 = $2.50 Profit margin = $2.50 – 2.10 – $.40	December @ $2.80.
June 3		Sell December @ $2.80.
October 10	Cash price = $2.40 Net = $.30	Buy back December @ $2.70. Net = $.10
	Realized price = $2.50 Overall profit = $.40*	
Costs = 1 to 1.5 cents per bushel in commission costs plus interest on margin funds for 4 to 5 months.		

*The overall net is exactly equal the projected profit because the final basis is the projected –$.30. With the cash market at $2.40, the December corn futures are $2.70, which results in a final basis of –$.30. The realized "forward price" can always be calculated by adding the cash price and the net from the round turn in futures, or $2.40 + .10 = $2.50.

illustration carefully, making sure you understand each number. The $2.10 cost comes from budgeted costs of production. The estimated October basis of –$.30 must come from historical records of cash prices *in October* compared to December corn futures *in October*. *It is very important to get these details down before proceeding.*

The forward price is a direct result of the –$.30 basis adjustment, and the expected profit margin is then a direct result of the forward price and estimated production costs. On October 10, the cash product is sold in the regular cash market at $2.40—the highest local bid the producer can find. The December futures are bought back at $2.70, exactly $.30 above the cash selling price. The realized price is the expected $2.50 which generates a $.40 per bushel profit.

Table 1.3 repeats the illustration in Table 1.2, but illustrates what happens when the final or closing basis is –$.35, not the expected –$.30 per bushel. The results illustrate what is meant by *basis risk*. The weaker (more negative) basis than was expected reduces the net returns from the hedge by $.05 per bushel. *You should spend whatever time is necessary on Table 1.3 to ensure complete understanding of what is being done and why the results are as shown, and what would happen if the ending basis is still at other levels.* Confirm, for example, that the "overall" will be $.45 if the cash price in October is $2.70 and December futures are bought back at $2.95. Table 1.3 moves beyond the simplistic use of a constant basis in earlier discussion *and this extension is very important. It demonstrates better what the real world often shows.*

Table 1.4 demonstrates a hedge on cattle and assumes the hedge is established by selling October futures in June when the feeder cattle are bought and placed in the feeding pens. The cattle are projected to finish in mid-September. In this example, the cash price moves sharply lower and the hedge protects the producer against major losses. The hedge nets $5.50 per hundredweight before commissions because the closing basis is –$1.50, not the –$2.00 that was incorporated on July 10. If the October futures had been near $61 on September 15 and had to be bought back at that level, the overall would have been the expected $5.00 per hundredweight.

Careful examination of the examples reminds us that whenever the cash futures difference or basis at the end of the program equals the expected basis incorporated

TABLE 1.3
Demonstration of a Short Hedge for a Corn Grower When Final Basis Does Not Equal Projected Basis

Date	Cash Market	Futures Market
June 3	Estimate production cost @ $2.10. Estimate October basis @ –$.30. Forward price = $2.80 – .30 = $2.50 Profit margin = $2.50 – 2.10 – $.40	December @ $2.80.
June 3		Sell December @ $2.80.
October 10	Cash price = $2.40 Net = $.30	Buy back December @ $2.75. Net = $.05
	Realized price = $2.45 Overall profit = $.35*	

Costs = 1 to 1.5 cents per bushel in commission costs plus interest on margin funds for 4 to 5 months.

*The overall net is not equal the projected profit because the final basis is –$.35 rather than the projected –$.30. With the cash market at $2.40, the December corn futures are $2.75, which results in a final basis of –$.35. The realized forward price is $2.40 + .05 = $2.45, not the expected $2.50.

TABLE 1.4
Demonstration of a Short Hedge for a Cattle Feeder

Date	Cash Market	Futures Market
July 10	Estimate cost @ $65.00 per cwt. Estimate Sept. basis @ –$2.00 per cwt. Forward price = $72 – 2.00 = $70.00 Profit margin = $70 – 65 = $5.00	October live cattle @ $72.00
July 10		Sell October @ $72
September 15	Cash price = $59 Net = –$6.00	Buy back October @ $60.50 Net = $11.50
	Overall = $5.50*	

*The overall net is $5.50 per cwt. because the final basis is –$1.50, not the expected or projected –$2.00 basis at the time the hedge was placed. Note that basis risk can help or hurt the final margin.

in the hedge decision, the net before commissions is exactly the expected profit margin regardless of whether prices go up or down. It follows, therefore, *that the hedger is still exposed to the risk that the final basis will not be the expected basis.* Emergence of basis risk, where the final basis is not equal to the expected basis, results in a realized price and a realized profit that will not exactly equal the expected levels. *The realized forward price is the cash price plus the net from the futures trade, and it will equal the forward price only if the closing basis is at the expected level.*

To make sure this is clear, you should go back and work though the results in Table 1.2 if the cash price is $2.40 when the cash corn is sold and the futures have to be bought back at $2.60. The overall or realized profit margin is $.50 and the realized forward price is $2.60. *Make sure you can reproduce results of this type before proceeding.* Fill in the blanks for realized profit and forward price when the October 10 results are:

Cash	December Futures	Overall Profit	Forward Price
$2.10	$2.50		
2.50	2.75		
3.10	3.32		
3.30	3.70		

The results indicate that a basis that is stronger than had been expected at the end of the production period is favorable for the producer. Thus, as noted in Table 1.4, basis risk can be "good" or "bad" for you as a hedger. Keep in mind that as an individual producer, you have no control over either the cash price or the futures price. Producers will have to sell at the cash price that is being offered at harvest and buy back the futures at whatever level the futures market is trading.

The hedge that uses the futures directly therefore guarantees the hedger a specific price and a specific profit margin subject to basis risk. Producers, firms holding an inventory, and short hedgers tend to get upset when the price level goes up and the hedge "protects" them from the benefits of higher prices in the cash market. If the cash cattle prices in Table 1.4 were $79 on September 15 instead of $59, to illustrate, the unhedged or cash speculative program would have received a $14 per hundredweight windfall return. The hedged program will receive around $6, depending on the final basis, and the windfall gain in the cash market will be denied by the hedge. Producers often see the missed opportunity as a loss and may be reluctant to hedge again.[14]

The use of options on futures, a relatively new tool that emerged in the mid-1980s, has the potential to eliminate these *opportunity costs* that come with short hedges in rising markets. The options therefore appear to have the potential to overcome the reluctance of producers to use the markets for fear they will peg prices below those that later develop.

The basic difference when using options is easy to demonstrate. The forward price from the straight hedge using futures was calculated as follows:

Forward Price = Futures + Basis

A *put option* gives the hedger the right, but not the obligation, to a short position in the futures. A floor price is established as follows:

Floor Price = Futures Price + Basis – Premium on Put Option

A particular futures price, called a *strike price*, is selected and adjusted for basis and the premium for that particular strike price. Detail on the use of options will be left until later chapters. Here, you only need to recognize that *there is now available an alternative approach that establishes a price floor for the producer rather than establishing a particular price*. The options thus have the potential to (1) reduce or eliminate the producer's concerns over the opportunity cost of pegging price too low, and (2) eliminate the producer's concerns over financing of margin calls. In using options to establish a price floor, the producer pays the option premium up front *and there is no exposure to margin calls*. The options are an exciting new tool and Chapter 7 is allocated to coverage of the options later in the book.

Both the direct use of the futures as a hedge and options leave the user exposed to basis risk. How big is this risk and how does it compare to the risk in the cash market?

The answer depends on where the producer is located. The corn producer in central Illinois may see very little variation. In Virginia, where corn production is much more susceptible to the vagaries of the weather, the basis can be more volatile. Across the past five years, the corn-producer in the primary corn-producing area in Virginia saw the October basis (cash in October less December futures) average –$.03 per bushel with a range of –$.15 to + $.09. *Obviously, some risk is still associated with*

[14]Actually, what has occurred is what economists call "opportunity costs." The opportunity cost of taking a course of action is what you give up by not taking some other course of action. Here, the opportunity cost of hedging is the $14 per hundredweight gain that would have been received by being a cash market speculator. Producers do tend to think about the situation this way, but you should start to develop the capacity to evaluate the alternatives in terms of risk exposure. In comparing two alternatives, the mean-variance measures of income are widely used in the literature. The importance of the mean or average levels is apparent, but it is the variance—a statistical measure of variability—that becomes critical to many decision makers. Many risk-averse decision makers are willing to take a lower mean level of income over time in order to achieve a less variable income stream. The preferences in terms of mean-variance measures will vary across individual decision makers with their attitude toward risk and their financial ability to carry risk. There will be more emphasis on this in later chapters, but you need to pause and think about your attitude toward risk in the context of these illustrations.

the basis. This topic will be dealt with explicitly in Chapter 2 and will be covered by examples throughout the book. *Basis risk is part of the hedge or option program and you cannot avoid it.*

> **Hedgers are looking for protection against unpredictable moves in cash programs. The most common need is protection against falling prices by the producer. Protection is gained by selling or going short in the futures and relying on the economic forces that ensure that the cash-futures basis will approach the expected level at the end of the program. The relatively new options allow the producer to establish a price floor and eliminate concerns over margin calls and the concern that prices will go sharply higher after prices are set in the futures. Regardless of the tool or technique, the need is for protection against costly moves in cash prices.**

Speculators' Strategies

The strategies adopted by speculators range from being similar to those of the hedger to intraday programs that have little parallel in hedging programs. The trading activity of speculators provides the volume and liquidity necessary for effective hedging programs. *The speculator accepts the price risk the hedger is seeking to transfer.*

Speculative programs range from high-volume intraday trading at one extreme to a trend-following system that calls for a trade every few weeks or few months. Traders operating on the floor of the exchanges or those at outlying points who monitor trade using electronic equipment with real-time (no lag) price quotes are important in "making the market." Hundreds or even thousands of contracts might be traded in a given day with no overnight positions—that is, all positions are canceled or offset before the end of the trading day. These *day traders* or short-term traders are looking for small price moves. They use analytical techniques based on hours or even minutes versus days or weeks and rely on their skill and the knowledgeable placement of buy or sell orders to limit potential losses and to stay on the profitable side of the action.

Speculators in the soybean futures at the Chicago Board of Trade who specialize in intraday trades might buy July soybeans at $7.105 just after trade opens. The reasoning might be as simple as the fact that the $7.105 opening is below the close of $7.14 the previous day, and that the trading range for most days encompasses the previous day's close. The speculator expects to see prices at $7.14 or higher later in the day.

Assume a particular speculator does in fact buy 100,000 bushels of the July soybean futures at $7.105 within a few minutes after the market opens. The lowest price recorded early in the period is $7.10. The speculator can place a *sell-stop order* at $7.09. The sell-stop order will be activated and filled at the first available price if the market touches $7.09 from above. If the market rallies to $7.135, long positions can be offset by selling at a $.03 per-bushel profit. The variable costs of trades will be extremely small for this type of floor trader and the $.03 per-bushel is a $3,000 gross profit on the 100,000 bushels. The same speculator might trade many times during the day, always looking for a potential gain that exceeds the risk exposure.

At the other extreme, speculators who are trend followers pay little or no attention to the intraday gyrations of the market. They trade more like selective hedgers. They seek to follow the age-old adage that directs one to "ride your winners and cut your losers." The objective of the program is to use fundamental and/or technical analysis[15] to isolate the long-term trend in the market and trade with the trend. Stop orders are typically used to protect the traders' equity if their analysis proves wrong and/or to lift the position when the price trend changes relative to expectations. Use of the stop orders is covered in detail in Chapter 4.

Speculators in commodity futures are thus investors looking for profits. With margin requirements set at around 5 percent of the face value of a contract, the financial leverage and the potential to make money with a minimal investment are apparent. But the hoped-for results are not always there. Surveys indicate 75–80 percent of the speculative accounts lose money. This is especially true of the many small accounts being handled by investors with little knowledge of analytical techniques and less awareness of the need for trading discipline.

Trade in the highly visible futures (cattle, hogs, corn, soybeans, wheat, interest rate futures, stock indices, foreign currencies, gold, silver, etc.) is successful because speculative activity is present. When the speculators are not present, the contract may disappear.

During the 1970s, for example, the Chicago Mercantile Exchange attempted to start a contract in broilers. After a few months of limited activity, the contract was eliminated.[16] The Chicago Board of Trade has tried a slaughter cattle contract on several occasions, but the trading volume has never been sufficient. A new exchange was launched in New Orleans to trade rice futures, short-staple cotton futures, and soybeans priced at the New Orleans export point. The exchange did not survive. After the farm bill legislation of early 1996, new dairy futures contracts were introduced in both New York and Chicago. It is not clear whether they will succeed. Not only must the contract have the potential to serve a hedging need in the presence of significant exposure to price risk, it must also attract speculative interest and speculative capital. If the speculator is not present and involved, the market can suffer from the related problems of low trading volume and a lack of liquidity.

Before leaving this brief discussion on the involvement and role of the speculator, an important point about the net gain from trade in futures can be made. It is true that for every dollar made in futures, someone else loses a dollar. Critics are then prone to call it a zero-sum game with no net gains for anyone. But that is not true for an entire economic sector. *The producer or the processor of an agricultural commodity has the chance to transfer the cost of exposure to cash-price risk to speculators who are willing participants and who are typically outside the agricultural sector. There can be, therefore, a net gain to a particular sector of the economy in the form of transferring the costs of exposure to price risk to investors trading as speculators outside that sector.* This is a less widely recognized benefit of trade in futures and extends beyond the obvious potential benefits to the individual hedger.

The key is that the speculator and the hedger need each other. If there is no need for protection against cash price fluctuations, there is no

[15]In simple terms, fundamental analysis deals with the basic supply and demand forces that determine prices. Technical analysis relies on the past history of those prices as a base for anticipating prices in the future. Fundamental analysis is covered in detail in Chapter 3, and technical analysis is covered in Chapters 4 and 5. They are then integrated in later chapters.

[16]Interestingly, a broiler contract was proposed again by the CME in 1990. There was apparently a growing expressed need among integrated producers and users such as fast-food chains for a risk-transfer futures investment. But the contract failed again, at least partly because little or no speculative interest developed.

economic justification for trade in futures and there will be no contract. The speculator, therefore, will have no contract to trade if there is no potential for hedging activity. In reciprocal manner, the hedger needs the speculator. If there is no speculator to accept the risk the hedger wishes to transfer and to provide the much needed liquidity and trading volume, the hedger will be denied access to futures trade as a risk-transfer mechanism and as an aid to price discovery.

FUTURES MARKET REGULATION

Users of the futures markets must have confidence in the exchanges and in the trading process. Keenly aware of the importance of their "image," the exchanges adopt strict self-regulatory rules and requirements. As trade in futures grew in the early 1970s, however, there was a growing perception that the public needed and deserved a regulatory agency to monitor and oversee trade and protect the interests of the trader, small or large, against any type of trade-related abuse.

The *Commodity Futures Trading Commission Act of 1974* created a federal agency, the Commodity Futures Trading Commission (CFTC), that has borne much of the responsibility for overseeing trade in futures contracts for the agricultural commodities. A complete coverage of the CFTC and its organization and activities is not needed in this beginning text, but it is important that you recognize that the futures markets *are* subject to the watchful eye of a federal regulatory agency. The reference by Perry Kaufman at the end of the chapter provides detailed coverage of the CFTC and how it functions.

With the advent of the financial futures and the foreign currency instruments in the 1980s, the scope of exchanges such as the Chicago Mercantile Exchange and the Chicago Board of Trade moved beyond the traditional agricultural commodities. With that expansion also came discussion of what agency should oversee trade in futures. The Security Exchange Commission (SEC), the federal agency that monitors trade in stocks, started to show increased interest in futures trade. Trade in futures for the S&P 500 stock index, for example, created the opportunity to hedge the risk in a stock portfolio. The interrelations between the cash and futures sides of financial instruments were being formalized and there were varying opinions on how the regulatory function should be handled.

The mix of agencies and the charge to each will continue to evolve over time, but that is not the important need here. The need here is to recognize that the futures markets are regulated in an effort to protect the interests of the trading public. These regulations are likely to be even more stringent in the future as trade grows and expands and as the exchanges are increasingly computerized and the surveillance opportunities are enhanced by computerization and advances in technology. This is especially true after an FBI investigation in the late 1980s revealed both violations of rules and questionable practices at some of the major exchanges. Confidence of the user is extremely important in the futures markets and the overseeing and regulatory functions are sure to be enhanced as use of the markets grows and expands.

The futures markets are regulated by federal agencies. The exchanges regulate themselves, recognizing the importance of their image to the trading public. In the future, as the use of the markets grows, regulation

is likely to be more extensive and more stringent. You can be confident that your trade in futures and options is reasonably secure and safe.

SUMMARY

Trade in commodity futures and options on those futures is a complex process. New terms and concepts have been introduced in this chapter, but you should not be concerned if everything is not crystal clear at this point. All that is required before proceeding is an understanding of the basic considerations.

Futures markets exist to assist in the process of price discovery and to provide a mechanism that allows producers, holders, and users of a commodity to transfer price risk. The transfer is accomplished by taking equal and opposite positions in the futures and cash markets. A producer of corn, for example, sells corn futures to gain protection against falling cash corn prices. If the cash futures difference or basis is stable or predictable, the protection will be effective and complete.

Before proceeding, you should review all the figures, tables, and illustrations in this chapter. Make sure you understand the important role that basis plays in determining the success of hedging efforts. Review what *price discovery and price risk transfer* are and review *who trades futures and why*. Focus attention on the important difference between the floor price set by options and the specific price set by hedging. Put yourself in the position of a *speculator and hedger*, respectively, and make sure you understand what each is trying to do and *why each is an integral part of the market*. If there is no mystery left when you complete that review, move on to Chapter 2. You now have established the base upon which an understanding of how to use the markets can be developed.

KEY POINTS

- Buying or selling a futures contract means acceptance of a *legal contractual commitment* to accept delivery of, or deliver, the underlying physical commodity on or by a specific date in the future. In practice, however, very few deliveries are actually made. The obligation is offset by buying back, to illustrate, the same number of contracts that was sold earlier.
- Futures markets contribute to *price discovery* by registering the impact of changing information in a centralized and competitive pricing process. The discovered prices in the futures markets are visible to the public and to all potential users and are *increasingly used as price expectations*.
- *The discovered futures prices for distant time periods have the potential to change future supplies*, and those changes will be registered in the form of price changes as part of the overall price discovery process. The possibility of a *supply response* is especially important in the livestock commodities, even within the year.
- Futures markets provide an opportunity for growers, users, and holders of a physical commodity to *transfer the price risk* associated with their positions to someone else. This process is called *hedging*.
- The success of a hedge depends on behavior of the cash-futures difference or *basis*. To the extent the basis does not move to expected levels, the net result of

the hedge can be changed relative to expectations. This possibility is called *basis risk*.

- *The possibility of delivery and/or arbitrage between cash and futures* for both physical delivery and cash settlement futures instruments *causes the cash and futures markets to converge* and keeps the level of basis risk at acceptable levels—at levels well below the risk of unpredictable moves in most cash markets.
- The relatively new options have the potential to establish *price floors* and *eliminate concerns about margins and fears that an opportunity cost will be incurred from pricing too low* when placing hedges directly in the futures in rising markets.
- Speculators are critical to the success of futures markets. They generate *volume* and *liquidity* and *accept the risk* the hedger wishes to avoid, thus allowing a net gain to a particular sector by transferring the costs of exposure to price risk to someone else.

USEFUL REFERENCES

Stephin C. Blank, Colin A. Carter, and Brian Schmiesing, *Futures and Options Markets: Trading in Commodities and Financials,* Prentice Hall, Upper Saddle River, N.J., 1991. An intermediate-level coverage with some emphasis on the financial markets.

Robert E. Fink and Robert B. Feduniak, *Futures Trading: Concepts and Strategies*, New York Institute of Finance (Prentice-Hall), New York, 1988. A beginning-level book that treats many commodities, options, and financial futures in a readable way.

Frederick F. Horn, *Trading in Commodity Futures*, 2nd ed., New York Institute of Finance, New York, 1984. General treatment across many commodities.

Perry J. Kaufman (ed.), *Handbook of Futures: Commodity, Financial, Stock Index, and Options*, John Wiley & Sons, New York, 1984. The book is a compilation of writings on many topics by various authors. It is a large but useful reference book.

APPENDIX 1A. CONTRACTS, CONTRACT SIZES, EXCHANGES

Instrument	Exchange	Contract Size
Com	CBOT	5,000 bu
Corn	MidAm	1,000 bu
Wheat	CBOT	5,000 bu
Wheat	MidAm	1,000 bu
Wheat	KC	5,000 bu
Wheat	MGE	5,000 bu
Soybeans	CBOT	5,000 bu
Soybeans	MidAm	1,000 bu
Live cattle	CME	40,000 lb
Live cattle	MidAm	20,000 lb
Feeder cattle	CME	50,000 lb
Lean hogs	CME	40,000 lb
Live hogs	MidAm	15,000 lb
Pork bellies	CME	40,000 lb
Soybean meal	CBOT	100 tons
Soybean meal	MidAm	50 tons
Cotton	CTN	50,000 lb
T-bills	IMM	$1,000,000
T-bills	MidAm	$500,000
T-bonds	CBOT	$100,000
T-bonds	MidAm	$50,000
Frozen orange juice	CTN	15,000 lb
Crude oil	NYMEX	1,000 bbl
Lumber	CME	80,000 bd ft

Where:

CBOT refers to the Chicago Board of Trade.

CME refers to the Chicago Mercantile Exchange.

CTN refers to the New York Cotton Exchange.

IMM refers to the International Monetary Market (division of CME).

KC refers to the Kansas City Board of Trade (hard winter wheat).

MGE refers to Minneapolis Grain Exchange (spring wheats).

MidAm refers to the Mid-America Commodity Exchange (in Chicago).

NYMEX refers to the New York Mercantile Exchange.

APPENDIX 1B. MONTHS FOR WHICH FUTURES ARE TRADED BY COMMODITY

Month	Corn	Wheat	Soybeans	Feeder Cattle	Live Cattle	Lean Hogs	Pork Bellies	T-Bills	T-Bonds	Cotton	Crude Oil
Jan.			X	X							X
Feb.					X	X	X				X
Mar.	X	X	X	X			X	X	X	X	X
Apr.				X	X	X					X
May	X	X	X	X			X			X	X
June					X	X		X	X		X
July	X	X	X			X	X			X	X
Aug.			X	X	X	X	X				X
Sept.	X	X	X	X				X	X		X
Oct.				X	X	X				X	X
Nov.			X	X							X
Dec.	X	X			X	X		X	X	X	X

CHAPTER 2

CASH-FUTURES BASIS: THE CONCEPT, USES, ISSUES

The difference between cash and futures prices, defined as basis, was introduced in Chapter 1. It is an important concept. As discussed in Chapter 1, the effectiveness of hedging efforts depends on the behavior of the basis. Basis has other and very important dimensions and applications, however.

Changes in the cash-futures basis can signal developments in the underlying supply–demand fundamentals. Basis thus becomes a barometer of market strength or weakness. The level of the basis is an important part of a decision process in determining whether storage of grain, cotton, or other storable commodities will be profitable. And as was noted briefly in Chapter 1, it is the level of the basis relative to the cost of delivery that becomes important in the delivery process as actions are taken by producers holding short hedges or market arbitrageurs to ensure cash-futures convergence. The level of basis will also determine the type and level of arbitrage activities employed by traders to force convergence in a cash-settled futures instrument. Basis is clearly an important concept, and it deserves more coverage.

CALCULATING AND RECORDING

Information on basis patterns for a particular market area is very valuable. On any specific day, there are potentially as many measures of basis for a particular cash market as there are futures contracts. On November 1, for example, the cash price for corn in a particular market area can be aligned with the December, March, May, July, and September corn futures contracts, all the futures contracts traded for corn. We will deal with the question of which futures contract to use later in the chapter. It is sufficient here to stick with technique. Table 2.1 shows a format that is useful in recording basis.

A time calendar in months runs down the left side of the table. Across the top are the various futures contract months. In each cell is the mean and the range (shown in

TABLE 2.1
Suggested Format for Recording of Cash-Futures Basis for Corn

Calendar Months		Futures Months				
		Mar.	May	July	Sept.	Dec.
January						
February						
April	The mean and range of the basis can be shown for each					
May	combination of calendar months and futures months in $ per bushel.					
June						
July						
August						
September						
October						–$.21 (–.36 to –.16)
November						
December						

parentheses) of the *observed* basis for corn across the past five years.[1] If you want to know the history of the basis in October using the December futures, we can see that it has averaged –$.21 with a range of –$.36 to –$.16 across the five-year period. You need to make sure you understand how these numbers are generated.

The entries in the cell where the calendar month of October intersects with the December futures column come from subtracting the average close of December corn futures during October from the average cash price during October. The calculations might be as implied in Table 2.2. For each of the five Octobers across a five-year period, the average close of the December futures during October would be subtracted from the average cash price for the October months. October is used in the example because it is the normal month in which corn is harvested. The December futures are used since there are no corn futures contracts for October and November and the December futures are the correct choice to be used in hedging a growing corn crop. *You should take care to note it is cash prices in October of past years versus trading levels of December futures during October in those past years that are used to build the basis history.* The worst-case basis of –$.36, recorded in Table 2.1, apparently occurred in year 5. Note this was the year in Table 2.2 when the average cash price was $2.60 and the average close for the December futures during October was $2.96.

The same procedure would be followed for each of the other calendar months to complete a large table of basis data. Clearly, the same thing could be done weekly by using weekly average cash prices and a weekly average of the close in December corn futures or some other futures contract of interest. Weekly data would provide more detail and are seen by many users, especially large farmers, to be worth the extra effort. Keep in mind that the product is bought and sold on a daily basis and the weekly or even daily data will provide better estimates of the basis to be expected than would monthly averages. Whether any increase in accuracy is worth the added effort would be for each user to decide, but with electronic spreadsheets on your computer

[1]The five-year period is selected for illustrative purposes. In practice, the user wants as much data as possible, but should not go back to early years when the basis patterns may have changed due to changing transportation costs, new storage facilities having been built, etc. Five years may be a reasonable choice in many settings.

TABLE 2.2
Procedure for Calculating a Cash-Futures Basis for Corn*

Year (Month)	Cash Price	Futures Prices (Basis)				
		Mar.	May	July	Sept.	Dec.
1 (Jan.)	2.41	2.65 (–.24)	2.72 (–.29)	2.75 (–.34)	2.60 (–.19)	2.50 (–.09)
1 (Feb.)	2.49	2.81 (–.32)	2.90 (–.41)	2.95 (–.46)	2.82 (–.33)	2.75 (–.26)
.						
.						
.						
5 (Oct.)	2.60	2.87 (–.27)	2.90 (–.30)	2.88 (–.28)	2.91 (–.31)	2.96 (–.36)
5 (Nov.)	2.02	2.47 (–.45)	2.50 (–.54)	2.60 (–.58)	2.47 (–.45)	2.40 (–.44)
5 (Dec.)	2.10	2.54 (–.41)	2.65 (–.55)	2.68 (–.58)	2.55 (–.45)	2.40 (–.30)

*All data are in dollars per bushel.

or the opportunity to subscribe to an advisory service that maintains these data sets, it is not difficult.

When it is absolutely necessary to go back and create a historical data set, some time savers are possible. Take a cash quote in the middle of the week, Wednesday for example, and record it. Match that cash quote against the close of the futures contract for a particular futures contract for the same Wednesday.

For example, assume the cash quote on Wednesday, October 4, is $3.10. Align that price with the closing price for December futures on October 4, say $3.45. A basis of –$.35 would then be allocated for week 1 in October for this particular year. Do that for all four weeks in the month and then average the results. You get a useful approximation of the basis pattern across recent years with far less number crunching involved. By scanning the calculations, the best and worst basis can be quickly identified. Recall again that the wide basis, the negative basis that is largest in absolute value, is the worst-case basis. The term "weak" is typically used to describe these basis levels that are negative and large in absolute terms. The producer would always prefer a small negative or even a positive basis, available primarily in deficit-producing areas, because that means a better selling price for the cash corn relative to the futures prices. Such a basis with cash price at or above futures is a "strong" basis.

Information on historical basis patterns is often available from the state extension service. The local grain elevator will have some information, and some commodity brokers keep a record of basis levels for several commodities in market areas involving their clients. Historical data on futures prices are available from the exchanges or from private-sector vendors who sell computerized data sets. At the worst, the decision maker may face the task of going to the library and working in the microfilm or microfiche areas. Procedurally, find a cash quote from the local newspaper for a day in the middle of the week and match that against the closing prices for the futures contracts of interest for the same day. Futures prices are always available on microfilm or other recordings of the *Wall Street Journal*. Once the information is gathered, many decision makers use a computerized spreadsheet program (such as Microsoft Excel) to do the calculations, generate the basis tables, and keep them updated. The computerized versions also allow interyear comparisons and plots across different years that show any difference in the basis patterns. Such comparisons are valuable and electronic aides make them practical.

The discussion to this point implies interest in the historical pattern in the cash futures basis for corn at harvest in October. *As development proceeds, it will be apparent to you that the basis pattern of interest will be a function of the pricing need and the decision situation involved.* A hog producer with hogs scheduled to sell in late May will be interested in the cash hog prices in late May against June hog futures across recent years. There is no futures contract for hogs for the month of May. A wheat producer interested in using basis levels to guide a yes–no decision on storage, a decision to be made in late June, will want information on late June cash prices matched against late June levels of the following May (or March, possibly) futures. For a storage program, you will see that this is an *opening basis*, the basis when you start or "open" the storage program. The producer will also want information on the expected *closing basis* late in the storage period as May 1 approaches. Exactly which basis will be needed will become clear as the various decision situations are presented and illustrated. The way the data are collected and recorded does not vary across commodities and across applications, however.

As illustrated in Chapter 1, it is the estimate of cash futures basis that is used to localize the futures price and to determine the forward price being offered by the futures market. During a particular decision period, once a particular basis level is pulled from historical data, it will not change. This is important, because it gives the producer a means of evaluating the cash-contract bids by the local elevator for later delivery.

To illustrate, assume a corn producer has found that the average cash futures basis for October, in the local market area, has averaged –\$.30 per bushel across the past five years with a worst-case basis of –\$.42. During July, rainfall starts to boost crop prospects and there are reasons to be concerned that prices will be lower at harvest. The producer calls the local elevator and learns the elevator is offering \$2.60 for October delivery. *Is this a good offer, a competitive offer?*

At the time, the producer notes the December corn futures are trading at \$3.00. The cash-contract offer for harvest delivery from the local elevator is reflecting a basis level of –\$.40 (\$3.00 futures versus the \$2.60 cash offer), and the producer is concerned.[2] The –\$.40 is \$. 10 per bushel wider than the average the basis tables show and is within \$.02 of the worst-case basis. Recognizing that he or she will be exposed to basis risk, the producer may well decide to hedge the corn at an expected forward

[2]Elevators tie their offers for cash-forward contracts to the futures market by including a margin that will be reflected in the basis. If a producer accepts the \$2.60 cash offer, the elevator will immediately hedge its position by selling or going short in the futures market to protect against the risk of prices dropping below \$2.60. The important point here is for the reader to recognize the value of basis tables in evaluating the cash bid. *If* the cash bid is reflecting a worst-case basis, *then* the producer might decide it is better to hedge directly in the futures rather than paying the elevator to handle the hedge. Keep in mind that the elevator is facing basis risk and is paying the costs of the futures trades in the cash contract offer. We will come back to this issue in later chapters. The key point here is that most producers who are familiar with trading futures will not be willing to accept a cash contract that allows the elevator to pass all or most of the basis risk to the producer in the form of a low cash bid. Over time, as producers learn to look at the alternatives, the competitive environment in the community improves and the cash bids by the local elevators may start to look better relative to the forward price being offered directly by the futures. If there is competition, the elevator may have to get more efficient or accept a smaller margin.

price of $2.70 ($3.00 futures plus –$.30 average basis) directly in the futures versus accepting the cash-contract offer. After all, the producer reasons, the elevator will hedge the cash-contract positions in the futures—and the producer is not willing to pay as much as $.10 per bushel for the elevator to do the hedging.

A key point here: *Once the producer chooses to use the average basis, the –$.30, it will not change all year.* Some other producer in the same cash market area might choose to use a more conservative estimate of basis—say –$.35—but once that choice is made, there is no reason to change it during the particular growing season. Remember, it is based on historical data. The basis tables will not change until they are updated at the end of the year.

Knowledge of basis levels thus gives you as a decision maker an alternative to cash-contract offers and a means of evaluating these offers. When the cash-contract bid does not look competitive, the corn or other product can be hedged directly in the futures. *Without good basis records and data, these types of informed comparisons would not be possible.*

Basis information is very important. It is worth the effort required to gather historical information on basis, record it, and keep it updated for the five or more most recent years. Basis patterns will be shown to be important for pricing actions for growing crops, for livestock programs, for evaluating prices being offered in cash contracts for later delivery, and for storage decisions for the storable commodities.

LEVEL OF BASIS

Many people equate basis with transportation costs, but it is more complicated than that. It is true, however, that costs of moving the grain or the livestock is a very important component of basis. Corn is worth more at a barge point along the Mississippi than it is 50 miles from the waterway. Slaughter hogs that are being produced near the packing plant are worth more to the plant than they would be if they are a three-hour truck haul away. Transportation costs have increased over time and basis levels have adjusted to reflect those increases. Let's recognize that location and the related transportation costs are key determinants of basis level and look at some other forces that will cause the basis to be weak (more negative) or strong (less negative or even positive).

Forces that cause basis to be weak include the following:

Good weather at harvest

Big crops in competing countries

Big surplus stocks

Inadequate storage capacity on farms

Light participation in the government's price support programs

Shortage of rail cars or barges

The fact that producers need cash immediately

What we find is that anything that forces the producer to sell the cash crop immediately will tend to generate a weak basis at harvest. Good harvest-period weather brings the crop to market in a rush, and strains the capacity of the storage and transportation facilities. The elevator with little or no storage space is not an aggressive buyer. Cash bids decline, and the basis tends to become more negative. Often, the "forced" sales are a local condition and the futures market does not trade lower with the cash market.

Big crops and big stocks put the buyer in the driver's seat. The processors of soybeans, for example, do not need to bid up cash prices because they know there is an abundant supply of product available. There is little or no incentive to bid up cash prices to get control of the physical product to hold in inventory, and the cash price can drift lower relative to futures prices. The threat of delivery by producers and others holding short positions in futures will eventually constrain basis levels, but there is a great deal of latitude left within that process. Almost any set of basis data for a particular market area and time period will have a range of $.10 to $.20 per bushel for corn across five years and an even wider range for soybeans and possibly wheat.

Problems of inadequate storage, a short supply of rail cars, and so forth, will allow the cash market in a particular market area to drop relative to the futures. In the Southeast, for example, there has been an increase in soft red wheat production in recent years. But there is little on-farm storage, and most of the commercial space is dedicated to corn and soybeans. This means the wheat producer in South Carolina, for example, may be able to buy commercial elevator space in Charleston elevators from harvest in June to some time in late summer. As September approaches, the elevator is going to pressure producers to sell wheat to make room for corn. The necessity of selling in a cash market that recognizes it is something of a distress sale puts pressure on *regional* cash prices. The futures market in Chicago will not necessarily dip to lower prices with the South Carolina cash market. The result is a weaker basis in South Carolina because of limited storage capacity.

Producers who historically were not in the government program are denied access to the loan price[3] as a source of cash flow and may have to dump their product on a harvest-period market that really does not need the grain. Cash prices tended to be forced down relative to futures, especially in states where a smaller percentage of the crop base was in the government program. Again, the local or regional basis is weaker due to lower local or regional cash prices. The forces influencing the cash price will not exert the same level of influence on the more nearly national futures price, and the local basis becomes weaker.

If producers need cash to pay off loans or meet other financial obligations, the cash product may be sold regardless of what the market signals are suggesting in terms of storage. Once again, cash prices can be forced lower in states or regions that were not heavily involved in the government programs and had no other means of meeting immediate cash flow needs.

The opposite of the foregoing cases will, of course, tend to make the basis stronger. If farmers have on-farm storage or can rent commercial space at favorable rates, they can delay sales until the harvest-period pressure subsides. If the corn or

[3]The "loan price" was set in farm bill legislation prior to 1996 and was the dollar value per bushel the producer can get at harvest by entering the grain in government storage programs. If prices go up later, the producer paid nominal costs of storage and reclaimed the grain for sale in the cash market. If cash prices never went up enough to justify reclaim, the grain was "forfeited" and the producer kept the cash from the initial program entry. With the 1996 farm legislation, the chances are much smaller that this type of program will be influencing basis patterns in future years.

wheat is eligible for a government loan program, then cash flow needs can be met by entering the product in the government loan program. The crop can be sold in subsequent months or forfeited to the government after the harvest-period pressure on price has diminished.

A small crop or small and tight stocks in a particular area will always help the basis. In deficit-producing areas, processors will want to control the local supplies if the crop is small, and they may bid up local cash prices to attract the producer's interest in selling. A moment's reflection will suggest that an adequate supply of rail cars to reach distant, especially export, markets and financially sound producers precludes having to dump product on a local market that cannot handle it or does not want it. Local cash prices are supported accordingly, and the basis does not weaken significantly.

Before closing this discussion, it should be noted that changing interest rates have affected long-term basis patterns. It costs more to hold $8.00 soybeans when the interest rate is 12 percent rather than 6 percent. A buyer, facing high interest costs on an inventory of cash product, will bid less for the soybeans. But interest rates seldom change enough within a year to threaten the effective use of basis patterns for decision-making purposes.

Location and related transportation costs are the primary reasons for basis levels in a particular market area. But there are other factors like storage capacity, the level of participation in government programs, weather at harvest, and the financial position of producers that will influence the level of the cash-futures basis at a particular point in time in a particular market area.

BASIS AS A BAROMETER

The futures market discovers price for the national market and considers the overall set of supply-demand information. Within this broad framework, the cash prices—and therefore the basis—can be influenced by market forces that generate information on what is going on behind the scenes. *The astute decision maker will benefit from monitoring basis patterns.*

Consider a grain exporting firm that seeks to buy grain, move it into the world market, and meet a targeted per-bushel profit margin. In such a "pass through" operation, the firm derives a bid to corn producers by subtracting an operating margin (including the targeted per-bushel profit) from bids it is receiving in the world market. If the firm needs 25,000 bushels of grain to complete a unit train load of wheat headed toward the New Orleans export point, the firm's cash bids will often be raised to encourage local producers to sell their grain. Bidding up the last 25,000 bushels of grain may be preferable to tolerating interest costs on the entire train of grain for several days. The firm accepts a smaller operating margin than it might like to see, especially on the last 25,000 bushels, and the cash-futures basis improves.

Observing that change in the cash-futures basis can tell the astute producer something about export market activity and the overall level of demand in the marketplace. If the basis is improving because export buyers are raising cash bids to pry the grain loose from producers and country elevators, the producer feels better about holding the grain in storage. This is especially true if the producer is speculating on the possibility of stronger cash prices. The grain is held with an increasing confidence that cash prices will rally. Improving demand in the cash markets and

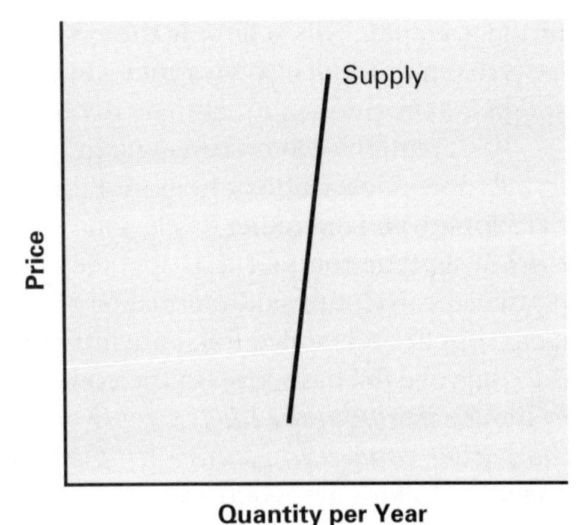

FIGURE 2.1
The Nature of the Supply Function for Grain for the Entire Year

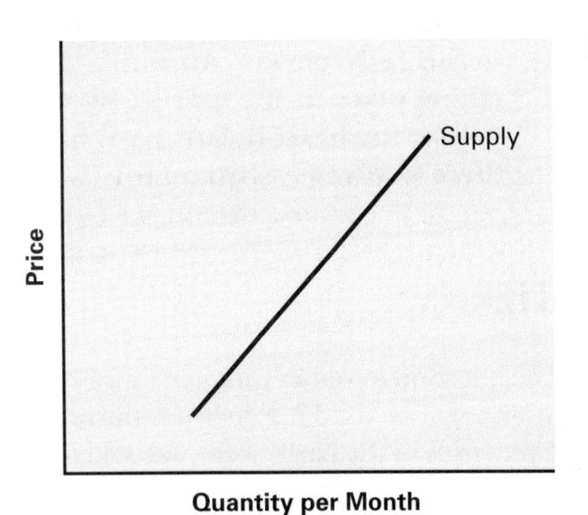

FIGURE 2.2
The Nature of the Supply Function for Grain for One Month

improving basis levels will also tend to boost the futures market. Trading levels of distant futures might move up enough to allow the producer to now forward-price the stored product at a profit even though profitable hedges were not being offered earlier. Basis developments relative to expectations based on historical data and historical basis patterns can become important input to many types of decisions.

The economics of what is happening are demonstrated in Figures 2.1 and 2.2. For the year, once harvest is complete, the supply curve is essentially vertical (Figure 2. 1). The supply curve is therefore very inelastic, which means it has little slope and suggests producers can offer little or no quantity response to a change in price.[4]

In Figure 2.2, the supply curve for a time period as short as one month *does* have some slope, and producers can change the quantity available to the market in response to a price change. Within the year, producers can and will pull grain out of storage and offer it to the export or domestic markets. As the price goes up, more

[4]The quantity available to the U.S. is not totally fixed for the year, of course, if we recognize that we are now involved in a world market. By exporting and importing product, any one country in the world market can change quantity in response to a price change.

product is sold. This selling is the response to the higher cash price. The improved basis that may result also becomes a barometer of what is going on in the marketplace and helps the producer make the decision of whether to sell and when to sell.

Basis patterns can be revealing in livestock as well. Increasingly, meatpackers are trying to schedule cattle or hogs into their plants to avoid costly variability in raw material supplies and operating levels. One way to do this is to offer producers a *basis contract,* a written contract that specifies delivery date(s), quality specifications, and a particular cash-futures differential or a basis level. If the packer is not able to get sufficient supplies scheduled for some future week, one way to attract more hogs or cattle is to improve the basis level in the contract offer. *That change in basis sends a signal to producers about the likely supply of hogs or cattle in the area and the attitude of the packer concerning future supplies.* Improved basis offers suggest the packer is concerned about the adequacy of supply for the future time period and this is very valuable information to the producer who is making decisions on exactly when to sell.

Changes in basis can be a barometer of the level of demand, or supply, in the cash market. An astute decision maker can often infer what is taking place in the market by watching the behavior of the basis. An improving basis means cash bids are moving up relative to futures, and there is always a reason for those improved cash prices.

USES OF BASIS PATTERNS

The question raised in Chapter 1 and earlier in this chapter concerning which futures month will be used for decision purposes is immediately answered when we look at direct uses of the basis. Here, we will look at hedging during a production season and then look at hedging a product held in inventory, the so-called storage hedge. As implied earlier, you will find that the basis data relevant to these two important decisions will be different.

For wheat to be harvested in June, the producer of a growing crop will want to have information on the cash futures basis using the *July* futures. For corn to be harvested in October, the harvest period, cash-futures basis levels using the *December* futures will be of interest. In soybeans to be harvested in October or early November, it will be the *November* futures that will be relevant. *It is always the futures contract that is at or beyond the end of the harvest period that will be used.* You should recall that futures are not traded for every month and review Appendix 1B in Chapter 1.

Let's use corn to illustrate. From early in the year, even before the crop is planted, the producer can monitor the December futures to see what forward price the futures market is offering. Recall the forward price is defined as:

$$FP = FUT + BASIS$$

Where

FP = forward price,

FUT = price for the futures contract to be used, and

$BASIS$ = the expected basis for the period when the crop will be harvested and sold in the cash market.

It is the *harvest-period basis,* using the December futures, that will be used to adjust or localize the quote coming from the Chicago Board of Trade. In early May, the producer might be interested in the cash-futures basis using the May futures as a barometer of the market for corn still being held in storage, but that is *not* the basis needed to estimate and monitor forward-pricing opportunities for the growing crop. For the growing crop, the basis tied to the December futures will be of concern.

So, in early May, producers would come down the left column in Table 2.1 until they reach October (if harvest will be in October) and then go across to the column tagged with December futures. For the rest of the year, since the basis tables are based on historical information, whatever basis level is picked for use in that local market will continue to be used.[5]

The same approach is taken if the enterprise being hedged is cattle or if we are dealing with an agricultural bank in its attempts to gain protection against higher interest rates. *You always pick the futures month beyond the end of the cash program or decision period.* As a rule of thumb, you should not use November soybeans, for example, unless you are reasonably sure the soybeans will be harvested by November 15. During the last few days of trading before the November futures contract matures on the next to last Thursday of the month, basis performance can be quite variable. If there is good reason to expect harvest will be delayed until late November or early December, the producer should move out to the January futures (the next futures month) and hedge the soybeans using the January futures.

For a production hedge, use the futures month just past the projected end of the harvest. That futures contract will be used in calculating historical basis data. The normal contracts are November futures for soybeans, December for corn, October or December for cotton, July for winter wheat, and December for spring wheat.

The storage hedge is different. Here, we will be especially interested in the basis pattern for the end of the storage period, the *closing basis* for the program. In this decision situation, the basis at the beginning of the potential storage period, the *opening basis,* is also important in estimating or projecting the probable basis improvement and the related *profit* potential to the storage program. Thus, there is interest in the opening basis and the closing basis. The opening basis is the current cash price versus the current price for the distant futures at the beginning of the storage period. The closing basis is the cash price versus futures price for that same distant futures at the end of the storage period.

Consider the case in which storage of corn is being considered, with the storage period to run from October 1 to May 1. Table 2.3 records the framework needed to make the storage decision intelligently, and profitably, over time and demonstrates the very important storage hedge.

[5]As suggested earlier, not all decision makers will use the average basis. Some will lean toward the worst-case basis across recent years to be conservative. That way, the chances are smaller that the closing basis in the fall will be worse than the one used to estimate the forward price. Using the worst-case basis thus tends to ensure that any basis surprises that will occur at harvest will be favorable rather than unfavorable. If this is not clear, review the hedging illustrations in Chapter 1 that show a final basis not equal to the expected basis used when the hedge was initially planned and established. Table 1.3 provides an illustration.

TABLE 2.3
Demonstration of the Use of Basis and Basis Patterns in Making the Hold–Sell or Storage Decision

Date	Cash Prices and Basis	Futures Action
October 1	Cash bids @ \$2.40, May futures @ \$2.85, opening basis = –\$.45	
	Cost of holding October 1 to May 1, including interest on the \$2.40 cash corn, estimated @ \$.21.	
	Expected closing cash futures basis on May 1 = –\$.15	
	Expected profit = expected basis improvement less costs of holding = (\$.45 – .15) – .21 = \$.09	
October 1		Sell May corn futures @ \$2.85.
May 4	Sell cash corn @ \$2.50.	Buy back May futures @ \$2.65.
	Net = –\$.11	Net = \$.20
	Overall = \$.09	

Costs = 1–2 cents per bushel commissions plus interest on margin funds.

Note that the opening basis on October 1 *using the May futures is* –\$.45 per bushel. With a relatively wide basis at harvest, interest in storing should improve as the decision maker looks ahead from an October 1 vantage point. After all, a wide basis at harvest means cash prices are relatively low and there is always resistance to selling at what looks like low prices. *But will storage be profitable?*

The answer comes from comparing the projected basis improvement with the estimated costs of holding the product. Note that in Table 2.3, the closing basis near May 1 (from historical tables) is *expected* to be –\$.15. That means an expected basis improvement of \$.30 per bushel (–\$.45 to –\$.15), and the costs of holding the grain are being estimated at \$.21. Those costs include interest on the \$2.40 corn, aeration, shrinkage, handling, and personal property taxes where applicable. Having considered all these costs, it looks as if storage has a good chance of being profitable.

The various parts of Table 2.3 deserve more emphasis before proceeding. *On October 1, the cash bid on October 1 against the October 1 trading level of the May futures sets the opening basis of –\$.45.* No historical tables are needed here, and there are no projections required. The cash bid and the trading level of the May 1 futures are known with certainty on October 1. Producers can simply pick up the phone to check cash bids at local elevators and look in the newspaper, check with a broker, or use the electronic service they subscribe to to get October 1 prices for May futures.

The expected closing basis for May 1, the –\$.15, *does* come from the historical tables. Recalling the format for recorded basis data discussed earlier in the chapter, that –\$.15 could be the average basis across the past five years, it could be the worst-case basis around May 1 across the past five years, or it could be a level within the range of observed basis levels that the decision maker decides to use. *It is important to remember that the decision maker has to select a particular basis level from the historical tables.* As discussed earlier, using the worst-case or weakest basis would help to ensure that the storage program would not be subjected to a basis around May 1 that is weaker (or more negative) than that used in evaluating profitability of the storage program. But there are still no guarantees. If some combination of cash prices and futures levels on May 4 emerges that gives a basis of –$.24, the hedged storage program breaks even.[6] Any basis on May 4 that is weaker than –$.24 would mean the program actually loses money. You must keep in mind that the individual producer is a price taker in both markets. Producers sell the cash product as high as they can, and buy back the futures position as low as possible—but individual decision makers can exert no influence on either of the two markets.

Before proceeding, it is important to spend time on Table 2.3. Be prepared to work through the final results with varying closing basis levels. Confirm, for example, that before commission and interest costs on margin money, a cash price of $2.70 and a May futures price of $2.75 on May 4 would result in an overall profit of $.19 per bushel, not the projected $.09. Confirm also that a cash price of $2.70 and a May futures price of $2.94 on May 4 would result in an overall profit of zero, not the projected $.09 per bushel.

Another way of dealing with the storage decision is to recognize that storage can be profitable when the projected closing basis allows pricing the corn above the break-even near May 1. Using the information from Table 2.3, the forward price for May 1 is

$$\begin{aligned} FP &= \text{Futures} + \text{Basis} \\ &= \$2.85 - .15 \\ &= \$2.70. \end{aligned}$$

With the corn forward-priced at $2.70, the projected break-even is $2.61 ($2.40 cash bid plus the $.21 costs of storage) and the profit potential is again $.09 per bushel. In other words, *the low cash price at harvest ($2.40) and its related weak basis creates* a *situation such that the grain can be forward-priced at a profit subject to basis risk.* The same basis improvement present in Table 2.3 generates the opportunity to forward-price above the break-even cost for May 1.[7]

Across the years, producers have built on-farm storage and then feel the storage has to be used. When prices are high at harvest, they store. When prices are low at harvest, they store. The attitude is that, since prices are usually at their lows for the year during harvest, storage should be profitable. But that approach is wrong. *It is basis and projected basis patterns that should be used in making the storage decision.*

[6]With a basis allowance of –$.15 near May 1, the projected profit is $.09 per bushel. If the *actual* closing basis near May 1 is –$.24, the program will only break even *no matter what the general price level turns out* to be. Any cash futures combination near May 1 that results in a basis weaker than –$.24 would result in a loss. Note that a –$.15 basis allowance suggests a profit of $.09, so a basis when the program is terminated that is $.09 "worse" than –$.15 (i.e., –$.24) will result in a break-even position. Anything worse than –$.24 will mean a loss. You should spend some time on this point and recognize there *is some possibility that a new worst-case basis could develop and turn the $.09 per-bushel profit expectation into a loss.*

[7]You should pause and note that costs of production do not enter into the storage decision. On October 1, all production costs are fixed. It is the cash bid of $2.40 on October 1 that is being bypassed (the opportunity cost, in economic terms) and it is the $2.40 that should be used to "cost" the corn into the storage program.

Research shows that holding any storable product as a cash-market speculator seldom works. Cash prices must increase enough to cover all costs, including shrink and, very important, interest on the value of the inventory. If you have to continue or renew a note with a 12 percent interest rate because you do not sell soybeans at $6.00 at harvest, interest costs alone will be $.06 per bushel per month! If no extension of a note is involved, the money could be earning 8–9 percent in a certificate of deposit, and that rate should be charged as the opportunity cost of holding the soybeans.

The point is, the cash market will seldom rally enough to cover all storage costs based on examination of past price patterns in the Midwest (a surplus production area) and also in the Southeast (a deficit production area). The rule therefore really becomes:

> **Hold and store grain or other storable commodities only when projected improvement in basis allows the commodities to be hedged or forward-priced at a profit.**

This fundamental rule for the storage decision explains the often-heard observation that the manager of a grain elevator does not much care about the *level* of the cash market, that the manager only cares about *basis.* Upon reflection, you would agree. *A well-managed grain elevator will always hedge its inventory, so it is in fact basis and basis patterns that are important.* The elevator will always try to buy cash commodities at harvest-period prices low enough to allow the inventory to be hedged at a profit given their extensive knowledge of historical basis patterns.

When the basis is weak at harvest and the projected basis improvement indicates storage will be profitable, the farmer with on-farm storage can do the same thing as the elevator. Having the storage bins gives the opportunity. Insisting on using the bins at all times as a cash market speculator will typically lose money. Remember, the bins are a fixed cost and fixed costs should never influence the hold–sell or storage decision.

Immediately, then, interest switches to how often the basis patterns do in fact say yes to storage. Figure 2.3 plots corn basis, using the May futures, from October 1 through the end of April for three representative years. The cash price used is the central Illinois price for # 2 yellow corn. With the costs of storing the corn from October 1 to May 1 usually in the $.20 to $.35 per bushel range—and it does differ for $3.00 corn versus $2.00 corn because of interest costs—there are years in which storing and forward pricing makes good money.

Year 2, for example, shows a harvest-period basis of about –$.53 and the basis improves to about –$.05 on May 1. The basis improvement was $.48, and there was good money to be made by storing. Conversely, years 1 and 3 offered no better than a breakeven situation. It is interesting to note that these actual basis patterns move to essentially the same level on May 1 even though the opening basis levels in October were significantly different. All the closing basis levels are in a range of –$.02 to –$.10 per bushel.

Figure 2.3 clearly demonstrates the need for a conscious decision on storage. In years (such as year 2) when the harvest-period basis is wide, storage may offer excellent profits. In other years, the best decision is to sell the cash grain at harvest and not insist on using the storage facilities. Across a number of commodities and across areas of the country, examination reveals the same patterns. *In some years, storage offers excellent profits, but in a majority of the cases, the harvest-period basis is saying "sell the cash."* Those basis-related signals are usually correct.

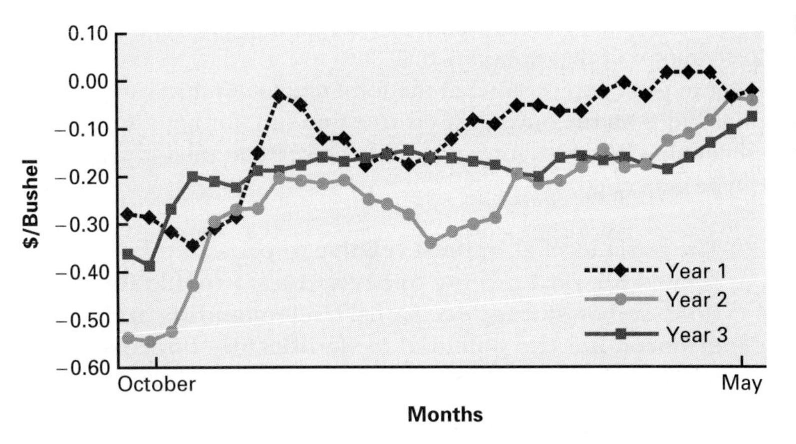

FIGURE 2.3
Basis Patterns for Corn, October to May, for Three Consecutive Years

Returning to the storage rule and restating in terms of the forward-pricing opportunities for emphasis, we can calculate the possible profits to storage as

$$PRO = FP - BE$$

where

PRO = expected profits,

FP = forward price, and

BE = the break-even price (harvest price + storage costs).

The rule then becomes:

Store and forward-price when the forward price offered exceeds the break-even price by a sufficient amount to justify the acceptance of the inherent basis risk.

You should stop and think a bit here about possible management strategies. Look at the basis path in year 2 in Figure 2.3. About all the basis improvement we might have expected based on an average closing basis of –$.05 for the historical data had been realized by mid-December. At that point, the producer should sell the cash product, buy back or offset the short position in May futures, and *terminate the storage program.* The expected basis gain has been realized and the costs of storing the commodity have been cut in half. The program will be much more profitable than anticipated, and you can terminate the program by selling cash and buying back the futures whenever you like. That flexibility is an obvious advantage of storing and forward-pricing in futures versus using a cash contract from the local grain elevator that calls for early May delivery.

The examples clearly show the importance of basis risk. *One reason that cash-forward contracts are so popular is that producers do not want to deal with the basis risk and are willing to pay someone else to handle the basis issues and accept exposure to basis risk for them.* At this point, you should pause and reflect on how much you should be willing to pay for avoiding exposure to basis risk. It will vary across producers, of course, but not many producers would be willing to pay the equivalent of the worst-case basis in the form of a lower price in cash contract offers. Recognition of that notion is built into the wording of the storage rule. There is no point in giving up revenue in the form of a lower-than-can-be-justified cash contract price bid from the buyer. This is true for corn, for hogs, for cotton, or for any other commodity for which futures contracts are traded and for which historical basis tables can be generated.

> **The basis level at harvest relative to projected later basis levels, and the related forward-pricing opportunities, provide the keys to profitable hold–sell or storage decisions. Understanding and using this decision criterion has the potential to significantly improve the profit position of producers who have access to storage capacity.**

BASIS AS A DECISION CRITERION

It is clear from the foregoing discussion that the actual basis level at harvest is the key to the storage decision. But the basis level is not always wide enough to justify storage. What does the decision maker who wants to benefit from expected higher prices do when the harvest-period basis and the projected basis improvement do not support storage?

There are a number of alternatives, including holding the product as a cash market speculator. But we have already argued that being a cash market speculator seldom works. Fortunately, there are alternatives and most involve using the basis as a guide. If the basis is too narrow at harvest to justify storage, producers can look at (1) a basis contract, (2) delayed pricing contracts, and (3) selling the cash and buying the futures.[8]

Basis Contract

The format of a basis contract is illustrated in Table 2.4. The buyer and seller agree on October 10 on a basis level tied to some distant futures contract—here the May corn futures. A negotiated percentage (usually around 75 percent) of the value of the product at harvest is paid to the producer at the time the contract is signed. The elevator gets the grain immediately. Between harvest and a specified ending date, usually May 1 if the May futures are involved, producers must decide when they are ready to close out the program and establish the price. In the table, it is assumed the producer makes that decision on March 21.

Several important features of the basis contract approach deserve emphasis.

1. *Most storage costs to producers are eliminated.* Here, the producer incurs only the interest cost on 25 percent of the harvest-period value. This is the important feature—price levels do not *have* to rally enough to cover all costs of holding the grain for the program to be profitable. If the producer does not have outstanding debt, interest earned on the $1.50 received at harvest might also be credited to the program.

[8]Later, it will be apparent that buying a call option on futures versus a direct position in futures is a possibility. But the procedure is much the same, and we will illustrate here by dealing directly in the futures. Coverage of option strategies is reserved for Chapter 7.

TABLE 2.4 Demonstration of a Basis Contract for a Corn Producer and Elevator

Date	Prices	Action
October 10	Cash = $2.00	Elevator pays a negotiated 75 percent of $2.00 ($1.50) to producer. Agrees to pay producer $.10 under May futures prior to May 1. Producer signs basis contract and transfers corn to elevator.
		Elevator buys May corn futures.
March 21	May futures = $2.41	Producer asks to "settle" and gets $2.31.
		The elevator sells the May corn futures and pays the producer a net price of $2.31.

2. *The elevator or other buyer gets the grain and the producer loses control.* If the elevator runs into financial trouble, the producer is in trouble.
3. *Basis contracts work to the benefit of buyer and seller when the buyer needs the grain and the seller is expecting prices to move higher.* Remember, the way the buyer shakes the grain loose to fill the train, meet processing needs, and so on, is to bid up the cash market and narrow the basis—and that is when the basis contract works. The buyer wants the grain immediately, and the bidding up of the cash grain and the willingness to negotiate a basis contract suggest to the producer that prices might increase.
4. *Very importantly, the producer is still fully exposed to price risk.* It does little good to have a contract for $.10 under May if you stand and watch the May futures go significantly lower in price. *Producers must monitor and protect themselves against major price breaks.* All the techniques to be developed in later chapters to help pinpoint when to hedge or forward-price will be equally appropriate in determining when to close out the basis contract or when to seek price protection.
5. Table 2.4 does not include specifics on futures prices, but *it is pertinent to note that the elevator is accepting the basis risk.* If the basis is wider than the $.10 allowed in the contract, the elevator incurs the costs. In a rising market, to illustrate, the futures position will not provide full protection to the elevator if the basis widens and is –$.20 or –$.25 when the producer decides to close the pricing component of the contract. The value of cash inventories has declined relative to the futures, and any holder of inventories will want to see the cash-futures basis strengthen.

The basis contract is thus an excellent approach when the harvest-period basis does not suggest storage but there is reason to expect prices to move up. The basis contract eliminates most storage costs, and the producer can benefit from even a small increase in price.

Delayed Pricing Contracts

The deferred or delayed pricing contracts or programs work much like the basis contract except the producer gives up title to the grain and has neither basis nor price set. How much, if anything, the producer gets up front varies. In some cases, it is a matter of getting free storage in a commercial facility for a few weeks or months. In other instances, the producer may be allowed an extended period to price that delays the actual sale or pricing into another tax year. One approach is to pick a date or pick a time period and the producer receives the going cash market offer on that future day.

Generally, the deferred pricing schemes are not favorable to the producer. Price level could fall and the basis could widen, so both types of risk are present and are being carried by the producer. We could argue it is better than being a cash market speculator because not all costs of storing are present, but that depends on what up-front benefits accrue to the producer. On the negative side, the possession of the grain has transferred to the elevator, and the producer has lost control in the event of financial difficulty by the elevator.

If a basis contract is not available, producers have a third option when the basis is too strong to suggest storage. They can simply sell the cash product and buy futures for a corresponding number of bushels.

Sell Cash, Buy Futures

Selling cash and buying futures essentially captures the benefits of a strong basis. This approach has several attributes that are worthy of consideration.

1. *All costs of storage for the physical product are eliminated* since the producer gets the full harvest-period cash market value when the cash product is sold.
2. *Basis risk is largely eliminated* since the cash position has been closed out while the basis is favorable.
3. *The producer, holding long positions in the futures, is fully exposed to price risk*. If price levels in futures fall, margin calls will be involved and losses will accumulate.
4. The long position in futures is a speculative position in the futures market.[9]

The approach is not necessarily a bad one even recognizing the possible tax implications. *It is often far better than being a cash market speculator where the producer pays all the storage costs for the privilege of speculating in the cash market.* Speculating in futures can be much cheaper, and any small increase in price levels could move the program to the profitable side of the ledger.

Table 2.5 provides a comparison of several alternatives when the basis is unusually strong (small negative or even positive) at harvest and the prices trend higher through the year. Table 2.6 repeats the comparisons when the price trend is down. It is apparent, upon examination of the tables, that virtually any strategy will be superior to holding the product as a cash market speculator.

The tables are important, but the numbers and the results may not be intuitively obvious to you. To make sure all this is clear, let's work through the results shown in Table 2.5.

[9]There is a popular feeling among producers that so long as they do not take a bigger position in futures than was the cash position they just sold, they are somehow "hedging." This is false. Once the cash position is eliminated, there is no way the "equal and opposite" acid test of what is a hedge can be met. Appendix 2A to this chapter summarizes the implications to the tax position of the producer and provides general guides to distinguish a hedge from a speculative position. You are cautioned to remember that these are general guides and are not legal opinions.

TABLE 2.5 Comparison of Alternatives When the Basis Is Strong at Harvest: Price Trend Is Up

October 1	Cash bids = $2.10 per bushel May futures = $2.30 Cost of storage = $.28
April 30	Cash market = $2.30 May futures = $2.40
Strategies:	
Cash speculator	–$.08 net compared to $2.10 sale
Basis contract*	$.17 net compared to $2.10 sale
Sell cash, buy futures**	$.09 net compared to $2.10 sale

*Assumes settlement $.10 under May futures on April 30 and $.03 per bushel interest cost on the 25 percent of value not paid on October 1.

**Assumes commission costs of $.01 per bushel for the round turn in the futures.

The cash speculator holds $2.10 corn and incurs a cumulative cost of storage of $.28. The break-even price is therefore $2.38, but the corn is sold on April 30 at $2.30—the best cash bid available. The program loses $.08 per bushel.

Using a basis contract eliminates all the costs except the interest on the 25 percent of harvest-period value, the $.525 per bushel not paid on October 1. This illustration thus assumes the basis contract called for 75 percent of the October 1 value to be paid on October 1. At 10 percent for 7 months, the interest cost on the $.525 runs $.03 per bushel. The producer receives $2.30 on May 1 if the contract called for $.10 under May, an improvement of $.20 per bushel over the October 1 cash bid of $2.10. Deduct the $.03 interest costs, and the result is a $.17 net improvement over selling on October 1.

In selling cash and buying the futures, all costs are eliminated except the costs of trading futures. A $.01-per-bushel commission cost is used, and no interest on margin funds is included since the futures account is generating revenue in an upward-trending market. Buying futures at $2.30 and selling them at $2.40 grosses $.10 and nets $.09 after commissions compared to selling October 1.

TABLE 2.6 Comparison of Alternatives When the Basis Is Strong at Harvest: Price Trend Is Down

October 1	Cash bids = $2.10 per bushel May futures = $2.30 Cost of storage = $.28
April 30	Cash market = $1.90 May futures = $2.00
Strategies:	
Cash speculator	–$.48 net compared to $2.10 sale
Basis contract*	–$.23 net compared to $2.10 sale
Sell cash, buy futures**	–$.33 net compared to $2.10 sale

*Assumes settlement $.10 under May futures on April 30 and $.03 per bushel interest cost on the 25 percent of value not paid on October 1.

**Assumes commission costs of $.01 per bushel for the round turn in the futures and $.02 per bushel interest cost on added margin-call funds.

Thus, both the basis-contract and the sell-cash-buy-futures strategies are significantly superior to the cash speculative position. The relatively strong basis on October 1 was discouraging holding the cash grain, and this message is typically correct. If the –$.10 closing basis was representative of the historical levels, only $.10 in basis improvement was potentially available from October 1 to April 30. *The market was saying do not store.*

In Table 2.6, the comparisons are similar. Both the basis-contract and the sell-cash-buy-futures approaches lose less than the cash speculative strategy. Relative to selling on October 1, the basis-contract approach would lose $.23 as futures decline to $2.00 and the price settlement is at $1.90. The $.03 interest cost is still present.

Selling the cash and buying futures is $.33 below selling at harvest. When the futures market declines, margin calls will be involved. An added $.02 per bushel cost is incorporated to reflect the interest costs of the increased margin money. The net is therefore the $2.10 cash sale less the $.30 loss in futures less the $.03 combined commission and margin interest costs, or $1.77. That result is $.33 worse than selling at harvest for $2. 10.

Both Table 2.5 and 2.6 reflect modest movements in price. In a year in which the cash price goes up dramatically, the cash speculative strategy would look more favorable. The storage costs are still present, however, and those costs will pull the net from a cash speculative strategy below the other approaches *unless* the cash market goes up significantly relative to the futures as the basis strengthens in a major way. Such basis developments are not highly likely during a storage period.

The regular storage hedge is not shown in Tables 2.5 and 2.6. When the price trend is up (Table 2.5), the storage hedge would lose $.20. The forward price of $2.20, using a –$.10 basis for April 30, is $.18 below the break-even of $2.38 ($2.10 cash bid plus $.28 storage). There would be an added $.02 in costs of the hedge including interest on small margin calls. When the prices go down, the losses would be around –$.19—the –$.18 that was "locked in" plus $.01 commission costs. The futures position would have earned revenue in the down market so no interest cost on margin money is assessed. *Clearly, if there is reason to expect prices to decline during the storage period, any inventories that are being held should be hedged.*

The basis level, and expected movements in the basis, are key components of successful decisions on storage. By taking advantage of knowledge on how to manage basis, the decision maker can employ more effective overall strategies than just being a cash market speculator. When a profitable storage opportunity is not available, other strategies that are related to basis and basis behavior are often superior to being a cash market speculator.

BASIS CONTRACTS IN PRODUCTION PROGRAMS

By offering a basis contract that looks good to a producer, a grain elevator is helping to ensure it will get grain at harvest. Meatpackers who offer basis contracts to cattle feeders or hog producers are after the same thing—they want to ensure a stable flow of quality product into their facilities. In extending a basis contract in an effort to secure the raw material, the buyer is accepting the basis risk.

If there is no pricing provision in the contract, the buyer will not necessarily have a position in futures. The objective is to secure a supply of raw materials. The basis risk is still present. Assume a grain elevator extends a basis contract specifying a price of $.15 under November futures for soybeans at harvest. The contract ensures the producer will deliver soybeans to the elevator, but the elevator is still exposed to basis risk. *If, at harvest, the basis is wider than the level specified in the basis contract, the buyer will pay a net price that is above the going cash market.* A basis of –$.25 versus the contracted –$.15 in soybeans will mean the buyer is $.10 worse off via the basis contract than if he or she had just waited and bought in the cash market. It is apparent that if the basis is more favorable than the contracted level, the buyer will benefit. The same holds true for the buyer of cattle or hogs via basis contracts.[10]

The motivation, once again, is to tie down a supply of raw material and to schedule it into the elevator or into the processing facility. Buyers will want to accomplish that task without giving up too much in terms of exposure to basis risk. Clearly, buyers would prefer to establish basis levels in the contracts that are weaker (more negative or smaller positive) than they expect the final basis to be. Sellers want the strongest possible basis that can be negotiated. Again, we see the importance of producers having complete information on historical basis levels and basis patterns. If the producer does not have historical basis information, the elevator or other buyer is in a superior bargaining position in negotiating the details of the basis contract.

Keep in mind that the buyer is receiving a benefit in the form of guaranteed supplies. If that benefit is significant—and it certainly can be—then the buyer should be willing to pay a premium in the form of accepting basis risk. If, through superior knowledge of basis patterns, the buyer can secure supplies and transfer much or all of the basis risk to the producer, then the buyer has a double advantage. It is *very* important that the cattle feeder, for example, not accept a basis contract of "$1.00 under June" if the historical data set on basis levels in the packer's market area shows –$1.00 per hundredweight to be a worst-case basis and that the more likely basis is +$.50 per hundredweight.

Buyers use basis contracts to acquire product or raw material and stabilize flows into their facilities. In the process, basis risk is often transferred from the producer to the buyer during the production period. If the basis level in the contract is to the disadvantage of the producer given historical patterns, however, the risk is transferred back to the producer. It is important to have good basis information.

SUMMARY

The cash-futures basis is extremely important to the decision maker. Looking at the behavior of cash prices versus the nearby futures can provide an indication of the strength of demand in the cash market. The expected harvest-period basis allows the

[10]If the producer is also offered a pricing feature that is established at the time the contract is signed, as in the typical basis contracts in storage programs, the buyers will need to have a long position in futures. Basis risk is still transferred to the buyer, however

producer to monitor the forward price being offered by the futures market and gives a means of comparison with cash contract offers by the local elevator for harvest-period delivery.

By extending the basis calculations to distant futures, the producer has the foundation for profitable hold–sell or storage decisions. Storable product should be placed in storage when the projected basis improvement exceeds the costs of carrying the product. If the basis is too small at harvest to suggest storage, the product should be sold. If there are sound reasons to expect higher prices, then basis contracts, a deferred pricing plan, or even selling the cash and buying futures will usually outperform a cash speculative program and reduce the decision maker's risk exposure by reducing or eliminating the costs of storing the physical product.

Individual decision makers who do not have access to historical basis patterns should invest the time and effort to get the data. Knowledge of basis levels and basis patterns is extremely valuable in virtually any decision that involves use of futures markets as a price risk management tool.

KEY POINTS

- *The behavior* of the basis, not the *level of the basis,* is the key to the effectiveness of hedging programs.
- The level *and changes in the level* of the basis using the nearby futures can be a barometer of the strength in demand for the cash product and can also give some indication of the supply of product available to the market.
- *Basis patterns* are the key to effective storage decisions. The rule is to *store only when the projected improvement in the basis exceeds the costs of storing the product.*
- The best storage opportunities occur *when the basis is unusually weak (cash significantly below futures) at harvest.*
- When the basis is too strong at harvest to suggest storage, *the use of basis contracts, deferred pricing plans, or even selling the cash product and buying futures will usually be more effective than holding the product as a cash market speculator.*
- *Basis contracts leave the producer fully exposed to price risk,* and the decision maker should be alert to profitable opportunities to close out the basis contract.
- *Delayed pricing contracts typically leave the producer exposed to both price and basis risk* and are generally not effective marketing techniques for the producer.
- *Selling the cash product and buying futures at harvesttime can be an effective approach,* but it will typically be *viewed as a speculative position in futures* for tax purposes.
- *Historical data on basis levels and patterns are extremely valuable* and put the producer in a much better bargaining position with the buyer of grain or livestock. Developing over time a set of historical data and developing measures of the frequency with which various basis levels are observed *puts the individual producer in a more effective bargaining position in dealing with buyers.*

USEFUL REFERENCES

Wayne D. Purcell, *Managing Price Risk in Ag Community Markets,* John Deere Publishing, 1997. Chapter 4 of this reference provides a very basic treatment of cash-futures basis.

David Rinehimer, "Hedging," in Perry J. Kaufman, ed., *Handbook of Futures Markets: Commodity, Financial, Stock Index, and Options,* John Wiley & Sons, New York, 1984. In developing hedging examples, the author deals with basis patterns, the seasonality of basis, and related topics that would add to the coverage in this chapter.

APPENDIX 2A. SPECULATION AND TAX IMPLICATIONS

The tax treatment of trade in commodity futures is important to the user. If trade is ruled to be speculative in nature, deductions for tax purposes are restricted. Individuals may be able to deduct only $3,000 in a particular tax year, and any added losses must be carried forward to future years. Certain corporate entities may not be able to deduct any losses that are ruled to be speculative in nature.

If the activity is ruled to be legitimate hedging activity, then any losses on the futures side of the hedge program are deductible. In a year in which the market moves up significantly in the futures, the futures side of the transaction can show a large negative result. The decision maker is "covered" since the cash market will be moving up too, but it is clearly important that the trades be treated as hedges for tax purposes so that the negative accumulation in the futures account will be effectively deductible and taxes are paid only on the net profit from the hedged program. For a long hedge, of course, it could be a declining market that accumulates a negative balance in the futures account, and the tax implications are the same. It is important that the decision maker have an understanding of what is and what is not hedging.

There are apparently no clear court rulings on what is and what is not hedging, especially for the selective hedging approach popular with many traders. In this appendix, we offer strictly a layman's interpretation of the guidelines that appear to have been used in past court cases. *You are cautioned to keep in mind that these are not legal opinions and should be used accordingly.*

Criteria that appear to be used in the court rulings and in the opinions of the Internal Revenue Service are the following:

1. The trades should meet the "equal and opposite" test. That is, the futures position should be opposite the position in cash and should not be larger than the position in cash. For example, a corn producer who expects to produce 50,000 bushels of corn should not be short more than 50,000 bushels in corn futures. A cattle feeder who will need 20,000 bushels of corn to complete a feeding program is "short" in the cash market, and should not be "long" more than 20,000 bushels of corn in the futures market. This test of equal and opposite positions is typically the first criterion applied by the IRS and by the courts in legal rulings.
2. The futures positions should be designed to protect the user against disadvantageous moves in the cash market. An elevator holding 500,000 bushels of wheat in storage would be hurt if the price levels fall. Therefore, the elevator would be expected to be short in the futures market because that is the position that would protect against a decline in the cash market. If the elevator places long hedges to protect against rising costs of grain to fill an anticipated sale, the futures transactions should be accounted separately for the long hedges.
3. There should not be frequent trades in futures. The IRS and the courts may disallow trades that meet the equal and opposite test and the test of being positioned to protect against disadvantageous moves in the cash market if there are frequent trades. This is often an issue in the selective hedging programs. At the extreme, day trades (in which positions are placed and removed in the same trading day) are sure to raise serious questions about whether the program is a legitimate hedging program. The selective hedging program likely has a better chance of passing the test if the trades are infrequent and are clearly designed to protect

against major moves in the price levels based on some defensible criterion that is being used to signal change in the direction of price trends.

4. The trades in the futures should not have a profit motivation. Any appearance of trade in futures to earn profits and to help support an otherwise unprofitable business program is likely to receive an adverse reaction from the IRS and from the courts.
5. The futures trades should be an integral and important part of the ongoing daily activity of the business. This tends to be an overall criterion that suggests the business is using the futures property, meeting all the other "tests," and would have difficulty operating successfully without the futures activities.

In general, the decision maker can enhance the probability of a positive ruling in terms of what is hedging versus speculation if there is a clear intent to protect against adverse moves in the cash market, if the equal and opposite test is met, and if there is not evidence of frequent trades and a related profit motivation.

You should recognize that at the time of this writing, we are not aware of any court ruling that clearly establishes criteria to determine what is and is not hedging in a selective hedging program. A program that does not show frequent trades and has a documented and demonstrable reason for lifting and replacing hedges is more likely to receive a positive interpretation given the positions being adopted by the Treasury Department in recent years, but there are no ironclad guarantees that even that type of program will not be viewed as speculative in a particular situation and at a particular point in time.

CHAPTER 3

FUNDAMENTAL ANALYSIS: SUPPLY AND DEMAND

INTRODUCTION

There is a continuing debate over fundamental versus technical analysis of the commodity markets and which should be employed. The debate is not really necessary. Both approaches are important, and the two approaches are in fact complementary. Each approach has inherent strengths and weaknesses. The complementarity of fundamental and technical approaches to analysis of the markets will become apparent as Chapters 3, 4, and 5 are developed, but the essence of the issue can be captured quite easily. The supply–demand fundamentals will ultimately determine price, but the technical dimension of the markets is useful in guiding the timing of actions as the supply–demand balance is being sought via price discovery processes.

It is, as suggested, a tautology that the interaction of supply and demand determines price. In the final analysis, the prices being discovered in the futures market must honor what is happening to the supply and demand relationships and the supply–demand balance. But we must remember that the futures market is attempting to discover the price that will balance supply and demand *for some future time period*. That day-to-day effort to discover the correct price is based on less-than-precise estimates of the levels of supply and demand for that future time period. Across the time span during which the price discovery process is being completed, the market is periodically "shocked" by changes in the information base. It is not surprising, therefore, that the path of discovered prices changes over time.

Analysis of the fundamental supply–demand information does not have to be sophisticated and it does not have to be capable of generating highly accurate predictions of future prices to be effective for the hedger or speculator. What is needed is the capacity to anticipate the direction of price moves and to formulate an impression of the likely price range. *The direction in which price will move will often determine what price-risk management strategy will be employed.* The producer who is a selective hedger is very interested in being able to anticipate the direction of price movement. If the consensus of the supply–demand information seems to be calling for prices to trend higher, the proper position for a selective hedger is that of cash market speculator. If the consensus is for lower prices, then the need is for aggressive placing of short hedges. For the long hedger interested in protection against higher input costs, the correct positions are just reversed, of course, but there is still keen interest in the probable direction of price movements.

Technical analysis will be covered in detail in Chapters 4 and 5, and it will be presented as the key to the timing of actions. But technical analysis will not be effective if there is a naive reliance on some technical indicator that is generating a price that the emerging supply–demand balance essentially guarantees will not develop. The basic point is important and deserves emphasis: *Fundamental analysis is needed to identify price direction and probable price ranges within the decision period. Then, technical analysis will be valuable in guiding the timing of market actions within the price ranges generated by the forces of supply and demand.*

THE SUPPLY–DEMAND FRAMEWORK

Often, we see the supply–demand framework presented in the form of a market equilibrium. A supply function and a demand function are shown and a single equilibrium or market-clearing price is demonstrated. Figure 3.1 demonstrates with the equilibrium price at level P.

In an after the fact context, a single market-clearing price makes sense. After all the changes in information have been registered, there is—conceptually, at least—a single equilibrium price that balances or matches the forces of supply and demand. The price of Choice steers in Omaha for last year was $74.30 per hundredweight, for example. But marketing and pricing decisions cannot be made in an *ex post* context. They have to be made during periods when a great deal of uncertainty exists about the exact level of supply and demand. Decisions are therefore made in the face of high levels of price uncertainty.

Figure 3.2 is a better picture of what is actually happening. Buyers and sellers bring to the marketplace some preconceived expectations of the "true" supply and demand. But access to information differs, the information base is never complete and perfect, and not all buyers and sellers would interpret the same set of information in exactly the same way. What we have, therefore, is a *distribution* of the estimated sup-

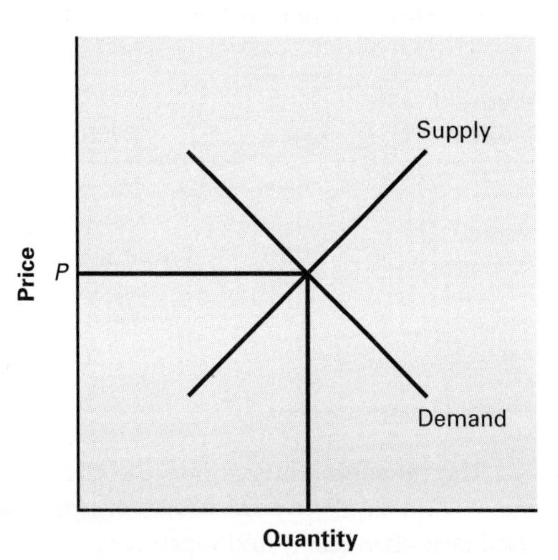

FIGURE 3.1
Demonstration of Supply and Demand and a Single Equilibrium Price

FIGURE 3.2
Range of Prices Due to Varying Estimates of Supply and Demand

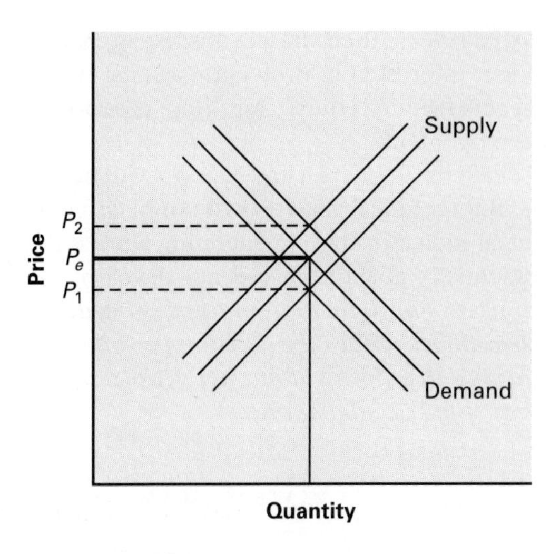

ply and demand curves, with some tendency toward more frequent estimates near the middle of the distribution. In Figure 3.2, actual transaction prices can occur in a range of P_1 to P_2, with some tendency for them to concentrate around P_e, the equilibrium price. The extent to which prices do concentrate around P_e and the size of the range between P_1 and P_2 will depend primarily on how complete and accurate the underlying information is at any point in time and how easy it is to interpret that information.

On a particular day, therefore, the prices in the futures markets for cattle, corn, interest rates, or any other commodity or instrument are being discovered by buyers and sellers who have varying impressions of what the price level should be. During the day, if there is no significant influx of new information to shock the market, a consensus will tend to develop. For that day, the closing or settlement price is the best representation of that consensus and is the best single indication of the price expectation for the future time period.

Figure 3.3 demonstrates. For each trading day, we see the trading range for a futures contract represented by a vertical bar. The horizontal "dash" represents the closing or settlement price. The typical format in daily newspapers, the *Wall Street Journal*, electronic market news wires, and so on, is as follows, using soybeans to illustrate in cents per bushel:

Futures Month	Open	High	Low	Close	Change
May	590.5	594.75	590.0	593.25	+ 2.25
July	605.0	608.5	613.5	607.00	+ 2.25
Aug.	609.0	613.0	608.5	611.25	+ 1.50
Sept.	609.0	613.0	608.0	611.50	+ 2.00
Nov.	614.5	618.0	613.0	615.50	+.50
Jan.	625.0	627.5	624.0	625.25	+.25
Mar.	635.5	637.5	634.0	636.00	–.25

The "change" entry shows the change relative to the closing price for the previous trading day. The terms *close* (or *closing price*) and *settle* (or *settlement price*) are used interchangeably and mean essentially the same thing. If there is a useful distinc-

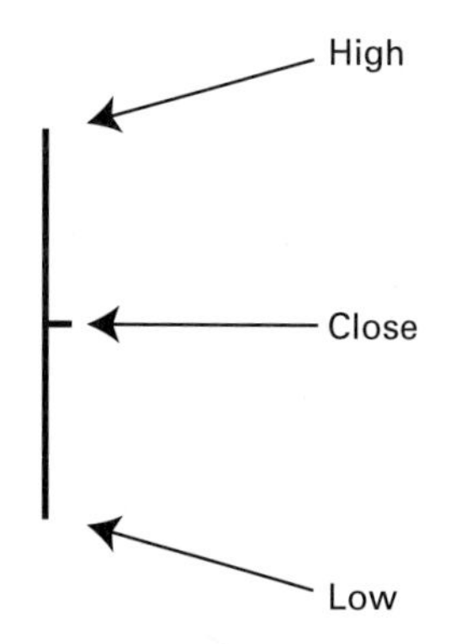

FIGURE 3.3
Daily High, Low, and Closing Prices for Futures

tion, it is that the closing price often shows a price range, and the settlement price is a price in that range designated by the exchanges as the official price for accounting purposes.

During the trading session, the consensus floats in a price range much like the range P_1 to P_2 in Figure 3.2. If new information enters the price discovery process, the range established early in the day may be expanded as the new information is received and incorporated. Across a number of days, the market probes into new higher prices, new lower prices, or both, as information enters the marketplace and is subjected to varying and imprecise interpretations by the traders. The fact that there is a trading range during the trading day is thus demonstrating the same thing that is demonstrated in Figure 3.2. *The information being used by futures traders is not perfect, it varies across traders in terms of access, and will never be interpreted exactly the same way by any two traders.* There is, therefore, a price range within the day and prices can and do move significantly up or down over time as the information on supply and/or demand changes and is interpreted in different ways.

Figure 3.4 shows the bar chart for a recent soybean futures contract to demonstrate changing prices within the year. The purpose here is to demonstrate that prices *do* vary a great deal within the year as the flow of information changes. The discovered prices for 1997 soybeans traded across a fairly wide range. In drought-stricken years, the price range within the year will be much wider. The price-discovery process is not an exact science, and the discovered prices do react to new information and do move on a seasonal basis as harvest approaches or as weather threatens the crop.

It is clearly important for you, as a decision maker, to be able to formulate a usefully accurate estimate of what the price range is likely to be across a decision period. And it is important that you be able to anticipate, with a useful degree of accuracy, the direction and magnitude of the price response to a new "shock" of information on supply and demand. To do that requires a basic understanding of the important economic forces that shift supply and/or demand. That basic understanding will require the decision maker to master a few simple tools of fundamental analysis.

The equilibrium price is the single price that would balance supply and demand and clear the market. But the levels of supply and demand for a future time period are never known with certainty. The prices discovered in the futures market will reflect that uncertainty and will trace out some distribution over time as new information on supply and demand enters the market and prices adjust to reflect the change in the information.

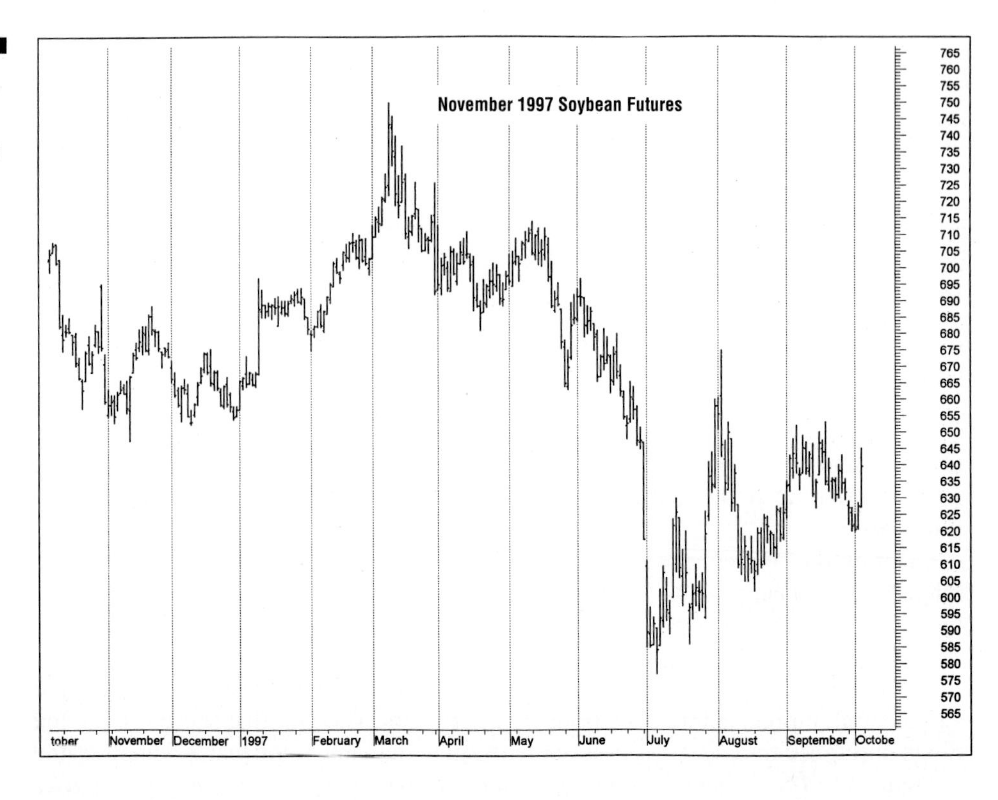

FIGURE 3.4
Demonstration of Changing Futures Prices within a Year for Soybeans

THE SUPPLY SIDE: CROPS

For the important crops, the monitoring of the supply situation should start with the stocks that are carried into the year. Table 3.1 demonstrates, showing the widely used supply–demand balance sheet format for corn.

The ending stocks for one crop year become the beginning stocks for the next crop year. The crop year runs from September 1 to the following August 31. In the table, 426 million bushels are carried forward from the 1995/96 crop year to the 1996/97 crop year. To the beginning stocks, add production and we have the total supply for the 1996/97 crop year which started on September 1, 1996. During the year, that total supply must be used in some way or it ends up in ending stocks and must be carried forward to the next crop year.

Before harvest, the production for the current crop year is, of course, an estimate. The USDA generates those estimates using information on planted acres, estimates of harvested acres, and estimates of yields.

Early in the year, the market is attempting to anticipate both the acreage and yield figures. The government programs prior to 1996 had a set-aside requirement that obviously influenced how many acres were planted. The USDA typically releases a Prospective Plantings report relatively early in the year, in late March in recent years. This report gives the first publicly available information on the acreage that is likely to be planted. Such information gets reflected in the early-year efforts to project the upcoming crop year. The June 12, 1997, estimates shown in Table 3.1 use the available information on planting intentions to generate the estimate of an 8.8-billion-bushel crop for the 1997 growing season.

TABLE 3.1
Supply–Demand Balance: Corn

Category	1995/96	1996/97	1997/98
		Crop Year	
		(million bu.)	
Beginning stocks	1,558	426	909
Production	7,374	9,293	9,840
Imports	16	10	10
Total supply	8,948	9,729	10,759
Feed, residual	4,696	5,325	5,600
Food, seed, ind.	1,598	1,670	1,760
Exports	2,228	1,825	2,050
Total use	8,522	8,820	9,410
Ending stocks	426	909	1,349
Average price	$3.24	$2.70–2.75*	$2.25–2.65*

*Estimate as of June 12,1997.

A bit later in the year, on August 1 in recent years, the planted acreage estimates are refined via surveys of producers, and this information is made available later in August. It is thus midsummer before the planted acreage is known with reasonable accuracy, and yields have to be estimated during this period to allow generation of production estimates. The first yield estimate that is based on survey data comes in the August *Crop Production* report, reflecting conditions of August 1. *At best, the information is imprecise, is subject to sampling errors when the surveys are conducted, and can be radically changed by weather developments.*

Table 3.2 illustrates a calendar of major reports for agricultural commodities during a representative calendar year. The information base can and obviously does change during the year for any and all of the commodities.

TABLE 3.2
Illustrative Calendar of Major Agricultural Commodity Reports

Date	Report
January 7	Poultry Slaughter
January 11	Crop Production
January 14	Crop Production—Annual Grain Stocks, Winter Wheat Seedings, World Agricultural Supply and Demand
January 22	Cattle on Feed, Cold Storage, Livestock Slaughter
January 29	Eggs, Chickens, and Turkeys
February 3	Poultry Slaughter
February 5	Cattle (January 1 Inventory)
February 9	Crop Production, World Agricultural Supply and Demand
February 16	Cattle on Feed
February 22	Livestock Slaughter, Cold Storage
February 24	Eggs, Chickens, and Turkeys
March 3	Poultry Slaughter
March 9	Crop Production, World Agricultural Supply and Demand
March 11	Livestock Slaughter
March 18	Cattle on Feed, Cold Storage—Annual
March 21	Cold Storage

Continues

TABLE 3.2
Continued

Date	Report
March 23	Eggs, Chickens, and Turkeys
March 25	Livestock Slaughter
March 31	Grain Stocks, Prospective Plantings, Hogs and Pigs
April 1	Poultry Slaughter
April 11	Crop Production, World Agricultural Supply and Demand
April 21	Eggs, Chickens, and Turkeys
April 22	Cattle on Feed, Cold Storage, Livestock Slaughter
May 3	Poultry Slaughter
May 10	Crop Production, World Agricultural Supply and Demand
May 16	Cattle on Feed
May 20	Cold Storage, Livestock Slaughter
May 23	Eggs, Chickens, and Turkeys
June 1	Poultry Slaughter
June 9	Crop Production, World Agricultural Supply and Demand
June 17	Cattle on Feed
June 23	Eggs, Chickens, and Turkeys
June 24	Livestock Slaughter
June 30	Grain Stocks, Hogs and Pigs
July 1	Poultry Slaughter
July 12	Crop Production, World Agricultural Supply and Demand
July 22	Cattle on Feed, Cold Storage, Livestock Slaughter
July 25	Eggs, Chickens, and Turkeys
July 29	Cattle (July 1 Inventory)
August 2	Poultry Slaughter
August 11	Crop Production, World Agricultural Supply and Demand
August 15	Cattle on Feed
August 19	Livestock Slaughter
August 22	Cold Storage
August 24	Eggs, Chickens, and Turkeys
September 2	Poultry Slaughter
September 12	Crop Production, World Agricultural Supply and Demand
September 16	Cattle on Feed
September 23	Cold Storage, Eggs, Chickens, and Turkeys, Livestock Slaughter
September 30	Grain Stocks, Hogs and Pigs
October 3	Poultry Slaughter
October 12	Crop Production, World Agricultural Supply and Demand
October 21	Cattle on Feed, Livestock Slaughter, Cold Storage
October 24	Eggs, Chickens, and Turkeys
November 2	Poultry Slaughter
November 9	Crop Production, World Agricultural Supply and Demand
November 18	Cattle on Feed
November 21	Cold Storage
November 23	Eggs, Chickens, and Turkeys
November 28	Livestock Slaughter
December 2	Poultry Slaughter
December 12	Crop Production, World Agricultural Supply and Demand
December 16	Cattle on Feed
December 21	Cold Storage
December 22	Eggs, Chickens, and Turkeys, Livestock Slaughter

Source: USDA, Economic Research Service, National Agricultural Statistics Service, World Agricultural Outlook Board.

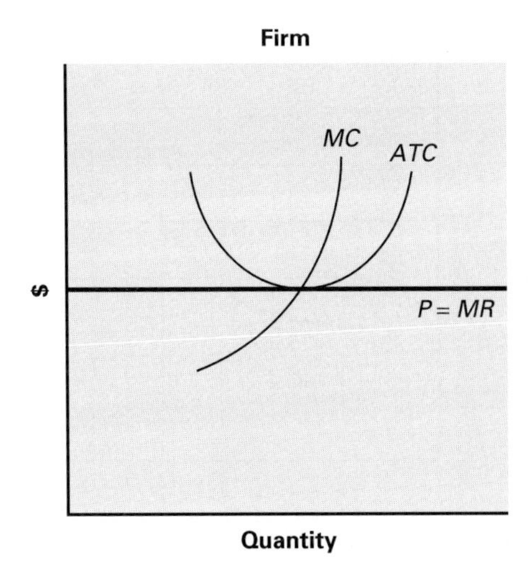

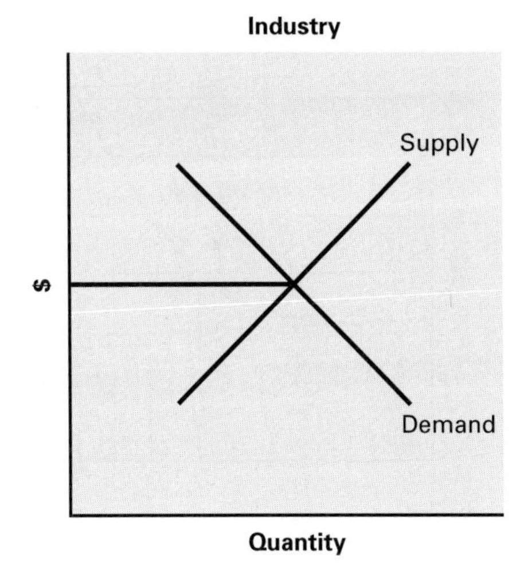

FIGURE 3.5
Profit Maximization for Individuals at the Industry Price

In general, therefore, the futures market reacts to changes in the available information on supply and demand throughout the year. But the process is more complex than it first appears, especially early in the year. Before the crops are planted, the futures market must anticipate how decision makers will react to the available price expectations given the current government programs and other factors that could influence decisions on what crops to plant. In the livestock commodities, the need is to anticipate how many cattle will be placed on feed and to anticipate how producers will react to a particular economic environment in deciding to expand or contract the breeding herd in hogs. To understand that process, we have to start with coverage of the economics of how producers decide.

In the context of basic economics, it is easy to demonstrate how producers' decisions have to be made. Figure 3.5 shows the situation facing the individual decision maker at the producer level in agriculture. In an industry that approaches the textbook conditions for pure competition,[1] the individual decision maker has to react to expectations for the industry-determined price.

To maximize profit, the individual decision maker operates at the point at which marginal revenue (*MR*) equals marginal cost (*MC*) of production.[2] Since the demand curve facing the individual producer is completely elastic and is a horizontal line, the

[1]The product being produced is essentially homogeneous, there are no noneconomic barriers to entry, and there are many producers, each producer too small to exert significant influence on price. In some commodities, there are significant economies of size and/or initial investment requirements, but these do not block entry for the well-financed firm. The concept of pure competition is covered in most beginning economic texts, but the coverage seldom extends to the issue of implications to specific decision situations. The coverage here will help you in your attempts to grasp just how important this issue is to the individual producer.

[2]This type of marginal analysis is widely employed in basic economics. What the $MR = MC$ criterion says, in lay terms, is that the firm will expand output as long as what it gets back from an added bushel or hundredweight (the *MR*) exceeds what it costs to produce and offer that added unit (the *MC*). As output expands, *MC* will tend to move up as the physical capacity of the plant or the operation is stretched. By increasing production up to the point that the two marginal flows are equal, the firm maximizes its profits for that decision period.

FIGURE 3.6
Generation of the Supply Curve for the Individual Farm Firm

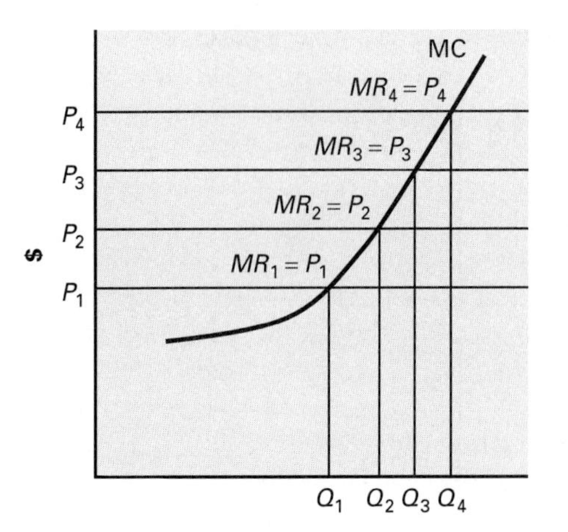

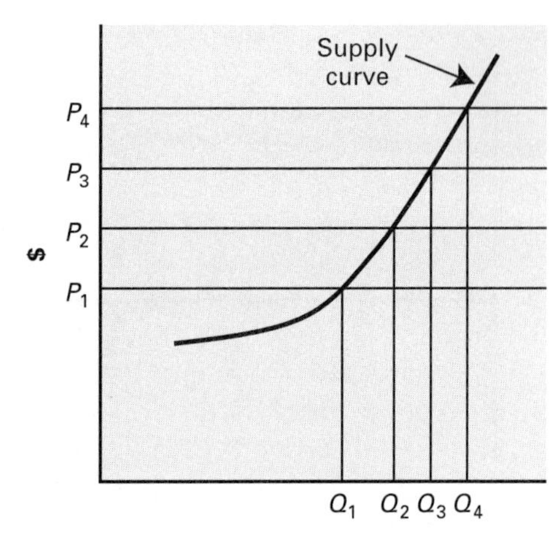

quantity offered will change as the price (which is *MR* to the firm) changes at the industry level.[3] If you visualize a large number of industry-determined prices and think of the $MR = MC$ profit maximizing criterion for each, a schedule of the quantities that would be offered at alternative prices by an individual firm is generated. For a given level of technology, for given cost levels of the variable inputs, and for a given level of prices for the other commodities that could be produced, that schedule becomes the supply curve for the individual firm. That supply curve or schedule supply shows the quantities the firm would offer at each alternative price.

Figure 3.6 illustrates this generation of a supply curve for the individual farm firm as the industry-wide price and therefore *MR* to the firm changes. For a given level of technology and the related cost structure, the manager of the individual firm will adjust the level of the variable input employed and change output so that $MC = MR$ at the various levels of the industry-determined price. If the industry-determined price drops below average variable cost (*AVC*) and remains there, the firm will eventually cease operations and be forced out of business.

The supply curve for the firm slopes up and to the right. It increases at an increasing rate, reflecting the basic economic fact that it is difficult to change production processes in the short run. As prices move higher, it is even more difficult to respond, and the short-run supply response of the typical firm is therefore very inelastic. Over time, it is easier to respond, and the curve will not be so steep. *The futures market has to handle all this*. If hog prices are high enough to elicit expansion, the futures market has to understand that prices for slaughter hogs will increase in the short run as the expansion is launched by withholding gilts (unbred females) from slaughter, but will decrease some 9 to 12 months later as the expanded numbers of pigs reach slaughter weights. The futures market is expected to offer prices that reflect that complex set of decisions by many thousands of producers.

The futures market thus discovers a price for the distant time period that gets brought into the firm's decision process as a price expectation. *That "price expecta-*

[3]In lay terms, again, this means the individual firm will sell whatever level of output they seek to sell at the industry-determined price. How much or how little the firm sells will have no impact on the price on that particular day; that is, the "price line" facing the individual producer is flat or horizontal at a particular industry-determined price. The individual producer is a price taker, not a price setter.

tion" is the expected industry-level price, discovered in the futures market, that the producer pulls down to the local market level and uses in decisions on level of production. As producers go through the decision process and employ the marginal analysis just discussed to varying degrees of sophistication, the aggregate supply of the particular crop for the U.S. is generated. Clearly, the same process is employed across all crops or products, and resources will be allocated or reallocated to the areas in which the price expectation is best given the costs of production.[4]

This discussion has used some economic jargon, and it is important that there be no confusion tied to terminology. What is being said, and what Figure 3.5 shows, is that the individual producer will respond to increased price expectations and will offer increased output at the higher prices. That pattern of responses is what generates the supply curve on the right-hand side of Figure 3.6. The response will not be the same across producers because their production programs and costs differ and because they will differ in attitudes and management abilities. Nonetheless, there is a basic consistency that spans differences in response. *Producers do have to use some type of price expectation and some type of estimate of costs to decide how much to produce, and the summation of all those decisions gives us the total quantity that will be available to the market at alternative price expectations.*

Having explained briefly how individual producers decide, it is then productive to deal with what is sometimes called the *micro–macro paradox* in production agriculture. This phenomenon is critically important to the futures market as it attempts to discover price, and brings much of the price variability to the markets. It is, therefore, important that the user of the futures markets understand what is involved.

At the "micro" level, what the individual producer does will not exert a significant influence on price. Remember, as an individual producer, you are a price taker. But the "macro" or aggregate impact, if many producers make the same adjustments, can be devastating. As was discussed in Chapter 1, it is important that the individual producer keep in mind that when soybeans look more profitable than corn, given the existing price expectations, other producers are looking at the same situation. A widespread response to the same initial set of price expectations can generate a big supply response and a major price change in the opposite direction. You need to keep this in mind. Remember: *It is the job of fundamental analysis to help identify the direction and probable range of price movement.* This possibility of an overreaction by many small producers may be tough to identify and accurately predict, but it is all part of the process. This uncertainty establishes the need for futures markets!

Before proceeding, you should again stop and review. The intent here is to simply document that producers *do* respond to price expectations. One important source

[4]The criterion

$$\frac{MVPx_{1(Y_1)}}{Px_1} = \frac{MVPx_{1(Y_2)}}{Px_1} = \frac{MVPx_{2(Y_1)}}{Px_2} = \ldots \frac{MVPx_{i(Y_j)}}{Px_i} = 1$$

allocates the i inputs to the j products so that profits are maximized to the entire firm. The symbolism *MVP* refers to the *marginal value product* of the particular input for a particular crop. If there are no capital restrictions so that all the ratios are equal to 1, then for each crop, the *MVP* (a measure of marginal revenue) is equal the cost of the input (a measure of marginal cost). This complex-looking "equation" may help you to understand how the various inputs (the X_i) are allocated across various crop possibilities (the Y_j). All it really says is that resources are allocated across alternatives such that the return to the last dollar spent in each alternative use is equal.

of price expectations is the futures market. If the distant futures quotes for corn, soybeans, hogs, cattle, and so on, are employed as price expectations, then the decision processes of the many small producers will bring an aggregate or macro response to the price expectations. Many producers have never heard of marginal revenue and marginal cost, but they all go through a similar mental process that involves revenue and cost flows as they make decisions on level of production and on redirecting their efforts and resources. All the complicated-looking developments in this chapter reveal is that at the industry-determined price or industry-determined price expectation (which can be the futures price), producers will offer a supply that varies with the efficiency and related costs of their operation. And, most important, producers *will* respond to changes in price expectations.

Table 3.3 may help drive the point home. Soybean-planted acreage has never been influenced directly by set-aside requirements in government programs. It is clear that acreage, and therefore production, tends to surge after years in which price was relatively high. Strong prices in the late 1970s brought rapid increases in acreage and the record 71.6 million acres in the 1979–80 crop year. Production nearly doubled relative to the mid-1970s. The drought-related increase in price in 1983–84 brought a rebound to 67.8 million acres in 1984–85. With prices dropping to and below the $5.00 level, acreage then trended below 60 million acres. The 1988 drought brought $7.42 prices and acreage jumped to 60.7 million acres in 1989–90, but was back below 58 million acres in 1990. The flood-ravaged crops of 1993 brought another set of adjustments, and the huge acreage (70.9 million acres) in 1997 came after near-

TABLE 3.3
Planted Acreage, Production, and Prices for Soybeans, 1975–76 to 1997–98 Crop Years

Crop Year	Planted Acreage (million acres)	Production (million bushels)	Average Farmer Price ($ per bushel)
1975–76	54.6	1,547	4.92
1976–77	50.2	1,288	6.81
1977–78	58.8	1,762	5.88
1978–79	64.4	1,870	6.66
1979–80	71.6	2,268	6.28
1980–81	70.0	1,792	7.57
1981–82	67.8	2,000	6.04
1982–83	70.9	2,190	5.69
1983–84	63.8	1,636	7.81
1984–85	67.8	1,861	5.78
1985–86	63.1	2,099	5.05
1986–87	60.4	1,940	4.78
1987–88	58.2	1,938	5.88
1988–89	58.8	1,549	7.42
1989–90	60.7	1,927	5.70
1990–91	57.8	1,926	5.74
1991–92	59.2	1,987	5.58
1992–93	59.2	2,190	5.56
1993–94	60.1	1,871	6.40
1994–95	61.7	2,517	5.48
1995–96	62.6	2,177	6.72
1996–97	64.2	2,382	7.35
1997–98	70.9	2,727	6.20–6.80*

*Estimate as of March 12, 1998.

record prices in 1995 and 1996. *In soybeans and in other crops, producers respond to the presence of higher prices and to the expectation of high prices.*

Producers must decide how much to produce, and they are assumed to act so as to maximize profits. But the behavioral reaction of individual producers is impossible to predict accurately in terms of magnitude. As a result, estimates of the total supply vary prior to and during a crop year. Add the weather and related yield uncertainty and it is clear why the prices being discovered in the futures markets will have to change and adjust during the year.

Having stressed the importance of monitoring developments on the supply side, it is important that you understand that the process is not impossible. The USDA releases supply–demand reports throughout the year. Table 3.2 lists these reports as the *World Agricultural Supply and Demand Estimates*. The reports are available by subscription, and Appendix 3A provides a broad listing of the available reports and how they can be ordered. Appendix 3A also shows an Internet address at which the reports can be accessed.

Private advisory services are available by subscription to assist the user in keeping up with developments and in interpreting what they mean. The extension services at most land grant universities offer advisory letters by mail, by electronic networks, and by satellite TV presentations.

The market news wires play a particularly important role in this process. Examples are Commodity News Service, Reuters, Globalink offered by Profession Farmers of America, and DTN. Costs range from $30 per month to $300–400 depending on the services requested. Transmissions range from FM band to satellite, which requires a small dish-type receiver.

Most of the wire services offer a survey of analysts' expectations for important USDA reports prior to the release of the reports. These surveys are especially interesting to the beginner because they help clarify what constitutes a "shock" to the markets and they also help to clarify just what constitutes new information to the futures markets. To illustrate, assume a monthly grain stocks report shows the following as a percent of the previous year:

Corn	95%
Wheat	95%
Soybeans	94%

The casual observer would then expect to see corn futures, for example, go sharply higher. After all, the stocks are down 5 percent! This is often the interpretation given by the newspapers and other media coverage. Prices are expected to increase, and talk of what it will mean to food costs to consumers is almost sure to follow. But this is all wrong and overly naive. The fact that the numbers are down is not the important point.

What matters is what the report says relative to prereport expectations. Coming into the report, the traders in corn futures are employing a base of information, a set of expectations, in terms of what the stocks are. Let's assume the prereport survey suggested that the discovered prices for corn just prior to the report were based on this set of expectations relative to year-ago numbers:

	Average	Range
Corn Stocks	92%	89.5–94.0%

After the report, the corn futures will almost assuredly trade lower, not higher. The top end of the range of estimates on the stocks (94%) relative to year-ago levels is below the report number (95%). *This report is a surprise and will be a shock to the market.* Because it was not correctly anticipated, the report has a great deal of informational value. There is a basic rule here: It is not the numbers in the reports, but the numbers relative to prereport expectations that will influence the markets.

It is important to monitor the supply side of the markets. That monitoring is not difficult given the many reports offered publicly by the USDA and by state extension services. Private advisory services also assist in this process, and an electronic market news wire is available to virtually everyone at a nominal to modest cost. The release of prereport estimates by professional analysts helps clarify why some reports elicit major price responses and some reports do not.

THE DEMAND SIDE: CROPS

The demand or "disappearance" components are also demonstrated in Table 3.1. Depending on the crop, the type of domestic usage will vary, but the export volume is always the big unknown and is the toughest component of demand to predict.

In corn, for example, the domestic feed usage is very important, but this number is not impossible to predict with reasonable accuracy. We know how many cattle, hogs, chickens, and so forth, we have on January 1, and that gives a base upon which to estimate feed usage during the year. The number of cattle on feed or the number of hogs kept for breeding can change within the year, of course, and that brings a degree of imprecision to the estimates of feed usage.

The USDA has developed models to predict feed usage of corn and total feedgrains. Decision makers can reap the benefits of those analytical efforts by monitoring the supply–demand reports that are released periodically throughout the crop year. The same reports that bring the basic supply-side information also provide estimates of the demand or usage levels throughout the year.

The export side brings much of the uncertainty. Table 3.4 records the quantity of corn, wheat, and soybeans exported since the late 1970s. It is clear that both the quantity exported and the exports as a percent of production vary considerably over time. For wheat, for example, the percentage of production that is exported has ranged from 37.7 to 78.3 percent. Within the year, much the same thing can happen. The estimates of exports within the year will vary significantly, reflecting developing crop conditions in other producing countries, changes in the level of the U.S. dollar which affects the costs of U.S.-produced grain, economic conditions in buying countries, and many other factors. In other important crops, such as soybeans and corn, the relative importance of exports varies, but export demand still tends to be the volatile component.

For all the crops, exports are often the most variable of the "disappearance" components. The level of exports varies with crop conditions in other countries, the trading level of the U.S. dollar against other curren-

TABLE 3.4
Exports, Production, Exports/Production for Corn, Wheat, and Soybeans, 1977–78 to 1997–98 Crop Years

	Exports			Production			Exports/Production		
Crop Year	**Corn**	**Wheat**	**Soy**	**Corn**	**Wheat**	**Soy**	**Corn**	**Wheat**	**Soy**
	(million bushels)			(million bushels)			(%)		
1977–78	1,948	1,124	700	6,425	2,036	1,762	30.3	55.2	39.7
78–79	2,133	1,194	739	7,078	1,776	1,870	30.1	67.2	39.5
79–80	2,433	1,375	875	7,939	2,134	2,268	30.6	64.4	38.6
80–81	2,355	1,514	724	6,645	2,374	1,792	35.4	63.8	40.4
81–82	1,967	1,773	929	8,202	2,799	2,000	24.0	63.3	46.5
82–83	1,870	1,509	905	8,235	2,765	2,190	22.7	54.6	41.3
83–84	1,835	1,429	740	4,166	2,420	1,567	44.0	59.0	47.2
84–85	1,865	1,424	598	7,674	2,595	1,861	24.3	54.9	32.1
85–86	1,241	915	741	8,877	2,425	2,099	14.0	37.7	35.3
86–87	1,504	1,004	757	8,250	2,092	1,940	18.2	48.0	39.0
87–88	1,725	1,592	785	7,064	2,105	1,905	24.1	75.8	41.4
88–89	1,650	1,450	550	4,352	1,810	1,472	41.1	78.3	34.0
89–90	2,275	1,300	590	7,527	2,036	1,927	30.2	63.9	30.6
90–91	1,725	1,068	557	7,934	2,736	1,926	21.7	39.0	28.9
91–92	1,584	1,280	684	7,475	1,981	1,987	21.2	64.6	34.4
92–93	1,663	1,354	770	9,477	2,467	2,190	17.5	54.9	35.2
93–94	1,328	1,228	589	6,336	2,396	1,871	21.0	51.3	31.5
94–95	2,177	1,188	838	10,103	2,321	2,517	21.5	51.2	33.3
95–96	2,228	1,241	851	7,374	2,183	2,177	30.2	56.8	39.1
96–97	1,795	1,001	882	9,293	2,285	2,382	19.3	43.8	37.0
*97–98	1,625	1,075	950	9,366	2,527	2,727	17.3	42.5	34.8

*Estimate as of March 12, 1998.

cies, the presence or absence of government programs to subsidize exports, and other sources of uncertainty. The uncertainty on the demand side adds to the uncertainty on the supply side and makes efforts to discover the correct equilibrium or market-clearing prices for future time periods difficult.

ENDING STOCKS

The *ending stocks* figure is perhaps the single most important entry in the supply–demand tables. It measures the surplus or leftover stocks that must be carried forward to the next crop year.

Figures 3.7, 3.8, and 3.9 show plots of prices and ending stocks as a percent of total usage for the same crop year for corn, wheat, and soybeans. An algebraic function has been fitted to the data through the 1997–98 crop year. This simple approach becomes a useful framework in generating an initial estimate of what the price level for the upcoming crop year will be. The procedures involved in fitting the functions are presented in Appendix 3B to this chapter, and the algebraic models for each crop

FIGURE 3.7
Corn Price versus Ending Stocks as Percent of Use, 1975–1998

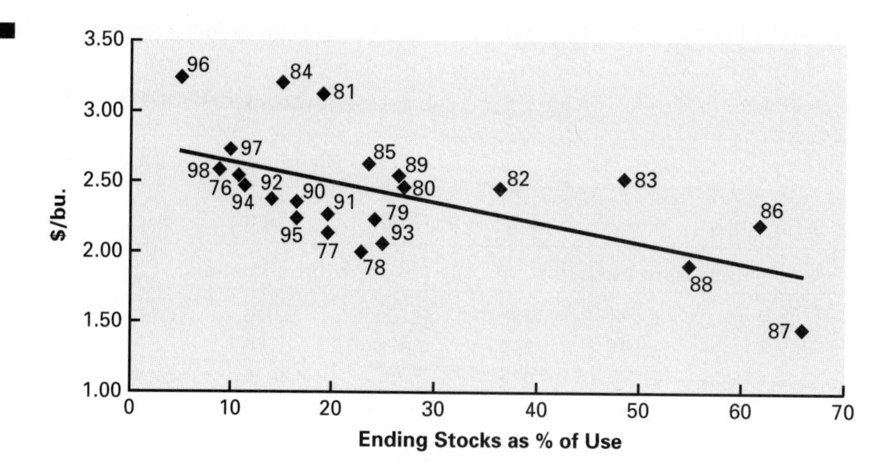

FIGURE 3.8
Wheat Price versus Ending Stocks as Percent of Use, 1975–1998

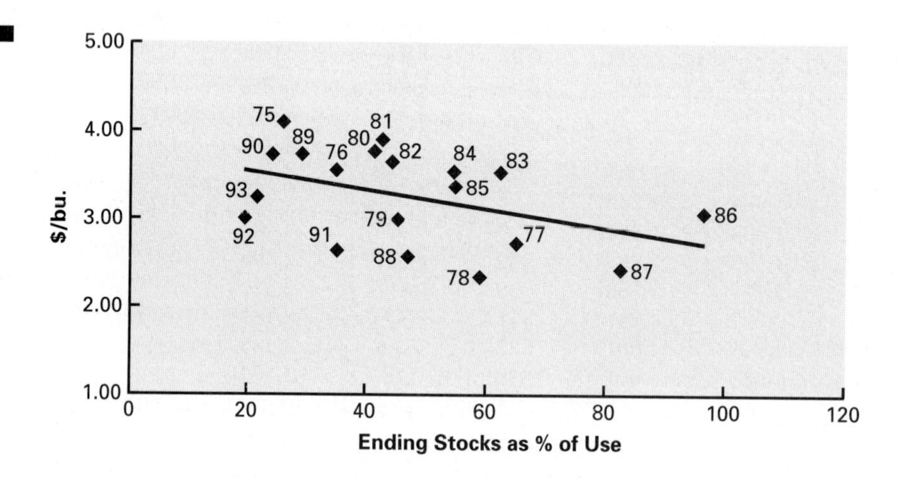

FIGURE 3.9
Soybean Price versus Ending Stocks as Percent of Use, 1975–1998

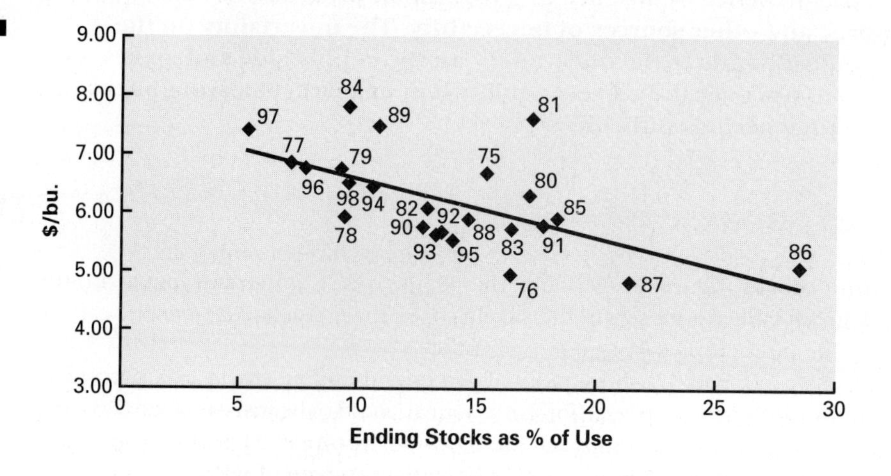

are also provided. Added detail on how the algebraic models can be used most effectively is also included. If you would prefer not to deal with the math you need not worry—there are simplistic approaches that will still be very effective; they will be discussed in this section.

As estimates of *total usage and ending stocks* are generated during the crop year, and prior to the planting season, it is possible to generate an initial estimate of price for the year. Using corn to illustrate, the steps are as follows:

1. Calculate ending stocks as a percent of usage and locate the point on the horizontal axis.
2. Move vertically up to the fitted curve, and then extend a horizontal line to the vertical axis.
3. Read off an estimate of the price on the vertical axis given the fitted relationship between price and ending stocks as a percent of usage.

One possible modification of steps 1 to 3 is to *force* the line to go through the point for the most recent complete crop year. In other words, just sketch a curve parallel to the fitted line and *draw it through the observation for the most recent complete crop year*. Then, use the estimates of ending stocks as a percent of use for the current year and generate an estimate of price from your new curve. This modification is demonstrated in detail in the appendix using the algebraic functions, and essentially shifts the curve to make sure it "fits" the most recent observation. Over time, it will be important to continue reestimating the curve to make sure it is representative of recent years. This is especially important when the relationship between price and ending stocks as a percent of use appears to be changing as is the case in recent years.

The result is a useful beginning estimate of the average price for the year, and it offers important perspective to the decision maker. It makes little sense, for example, to sit and wait for a chance to forward-price corn at $3.50 if your initial estimate suggests an average price for the crop year of $2.60. Sure, there can be and will be lots of variation around your initial estimate over time, *but the approach gives you a good idea of the general price level that will be observed during the year if no major shocks to the information base, especially to the crop production estimates, emerge.*

To the user of the futures, this ability to formulate an educated estimate of probable price levels is important. Consider the corn producer, to illustrate, who is trying during March to decide (1) whether to hedge corn, and (2) how much to hedge as the December corn futures approach contract highs in the mid-$2.60s. The producer works through the price-ending stocks either graphically or using the algebraic equation (see Appendix 3B) and generates a producer-level average cash price for the crop year of $2.70. There is now a reason to expect higher futures prices since the cash-futures basis at harvest is negative in the producing areas, and that expectation can be brought into the hedging decision. Clearly, such basic fundamental analysis is important to an effective hedging program.

A moment's reflection shows why the ending stock figures are so important. They are the residual after accounting for total supply and all the components that make up total demand or usage. In the context of Figure 3.2, the relationship between ending stocks and price attempts to capture the impact of estimates of both supply and demand and to generate a price estimate from an approach that is simple and easy to use.

There is no attempt here to reproduce the analytical developments that are provided in detail in books focusing on agricultural price analysis or elementary econometrics. Rather, the approach is to identify the key issues and discuss how they have impact. The USDA's sophisticated analytical models are used to develop the estimates in each supply–demand report, and these reports provide the information base that drives the futures markets. The appendices provide detail on how to get the reports, and references on analytical procedures are shown at the end of the chapter. The simple two-dimensional graphs are very revealing and will help you generate a useful forecast of price, a forecast that can be updated during the year as the USDA periodically releases supply–demand reports.

Analysis of the relationship between ending stocks and price attempts to capture the impact of estimates of both supply and demand. During the year, as estimates of supply and demand are changed, the ending-stocks framework can be used to generate updated estimates of average price for the year. This simple procedure helps to determine the probable direction of price movement on a year-to-year basis, and helps to establish a price range within which price variations are likely to occur. It provides important input to the user of the futures markets.

THE SUPPLY SIDE: LIVESTOCK

In the livestock sector, supply for the year is directly related to the inventory at the beginning of the year. Within the year, the supply response is limited to what can be changed within the time framework of one year.

Table 3.5 shows January 1 inventory numbers for all cattle and the beef cow herd. Figure 3.10 provides a scatter plot of beef production against the inventory numbers from the table, the "total cattle" numbers.

The relationship in Figure 3.10 is not perfect, but there *is* a positive relationship. Deviations from the linear relationship that has been fitted to the data occur primarily due to cyclical developments, percentage of cattle being fed, and changes in cattle type. During 1973–75, the herd was being expanded by holding back heifers. Consequently, production was less in those years relative to the January 1 inventories, and

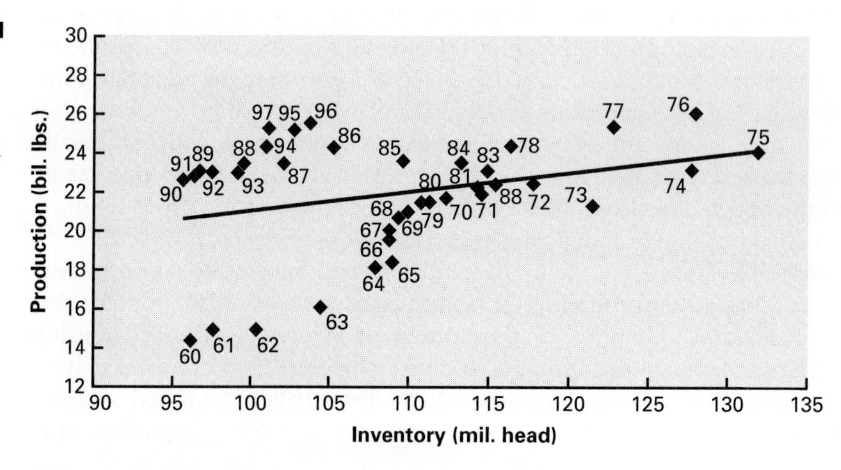

FIGURE 3.10
Cattle Inventory Numbers and Beef Production, 1960–1997

TABLE 3.5
Total Cattle Inventory and the Beef Cow Herd, U.S., 1950–1998

Year	Total Cattle Numbers	Beef Cow Herd
	(1,000 head)	
1950	77,963	16,743
1955	95,592	25,659
1960	96,236	26,344
1965	109,000	34,238
1970	112,369	36,689
1971	114,578	37,878
1972	117,862	38,810
1973	121,539	40,932
1974	127,788	43,182
1975	132,028	45,712
1976	127,980	43,901
1977	122,810	41,443
1978	116,375	38,738
1979	110,864	37,062
1980	111,242	37,107
1981	114,351	38,773
1982	115,444	39,230
1983	115,001	37,940
1984	113,360	37,494
1985	109,582	35,393
1986	105,378	33,633
1987	102,118	33,945
1988	99,622	33,183
1989	96,740	32,488
1990	95,816	32,454
1991	96,393	32,520
1992	97,556	33,007
1993	99,176	33,365
1994	100,988	34,650
1995	102,775	35,156
1996	103,487	35,228
1997	101,460	34,271
1998	99,501	33,683

the estimate of production, generated by the algebraic model fitted to the data, will be too large. Deviations on the other side of the expectations occur when the herd was still being liquidated in the 1985–87 period and a high percentage of the cattle were being fed. Also, the increased use of crossbreeding programs with the larger breeds has increased production per head. The model that was fitted to the data is shown in Appendix 3B, but there is no need to get deeply involved in the mathematics. *The need is to recognize that the January 1 inventory will be a major factor in determining beef production for the year.*

The same approach can be employed in hogs, sheep, and poultry. What is brought into the year will set the stage for production within the year. Figure 3.11 shows a scatter plot between December 1 inventories and commercial production during the following calendar year for the pork sector. When the breeding herd is being expanded as it was in the early 1970s, production looks unusually low. When the herd is being liquidated as it was in 1984 through 1987, production looks unusually high. It is also

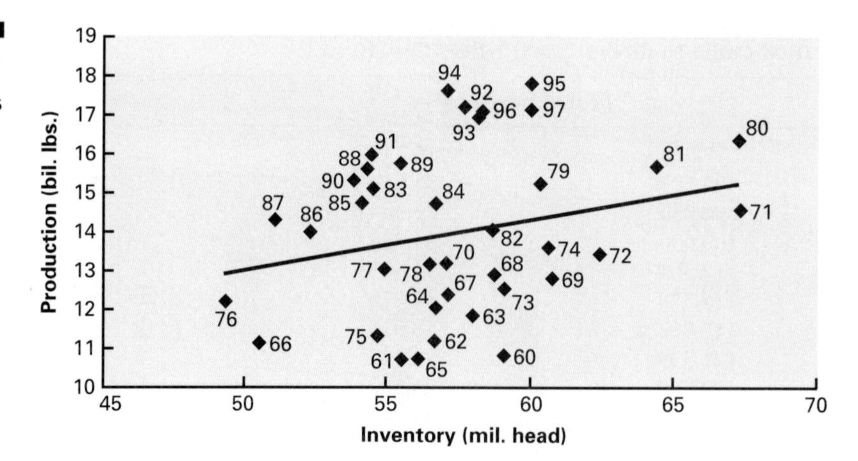

FIGURE 3.11
December 1 (of Previous Year) Hog Numbers and Pork Production, 1960–1997

apparent that production for a given herd size is increasing in recent years, reflecting the increase in pigs per sow and the increases in production efficiency. Some well-managed programs are now producing well over 20 slaughter hogs per sow per year. Over two litters per year are being produced on average, and the average litter size and number of pigs saved are both increasing. Compare the points for 1979 and 1988, for example. Production levels in the two years are comparable, but the inventory for 1979 was 60 million head compared to approximately 54 million head for 1988. Production increases in recent years are more modest but are still recognizable.

Beginning inventories for the year will be an important determinant of production in the year. In recent years, production levels for a given cattle or hog inventory are being increased by technological advances in production and by more effective management.

Seasonal Patterns: Cattle

Within the year, there can be significant variations in production tied to producers' decisions. In cattle, the key is the number of cattle placed on feed.

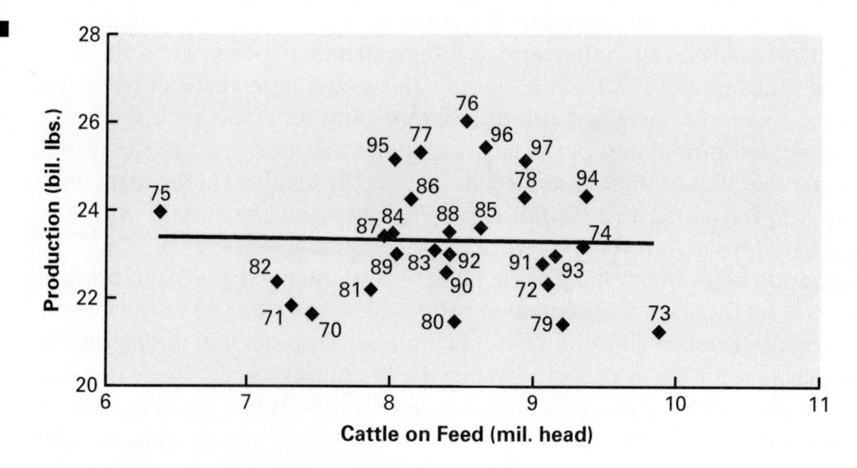

FIGURE 3.12
Beef Production as a Function of January 1 Cattle-on-Feed Numbers, 1970–1997

TABLE 3.6 Content and Format of April 1997 Monthly 7-State* Cattle-on-Feed Report

	Number			1997 as % of	
	1995	1996	1997	1995	1996
	(1,000 head)			(percent)	
On feed, Mar. 1	8,227	8,152	8,769	107	108
Placed on feed during Mar.	1,681	1,666	1,694	101	102
Fed cattle marketed during Mar.	1,513	1,476	1,497	99	101
Other disappearance during Mar.	67	56	62	93	111
On feed, Apr. 1	8,328	8,286	8,904	107	107
Number on feed by class, Apr. 1					
Steers and steer calves	5,530	5,375	5,417	98	101
Heifers and heifer calves	2,762	2,877	3,431	124	119
Cows and bulls	36	34	56	156	165
Number on feed by weight groups, Mar. 1**					
Less than 600 lbs.		262	321		
600–699 lbs.		544	538		
700–799 lbs.		772	721		
800 Plus		370	386		
Total		1,948	1,966		

*The 7 states include AZ, CA, CO, IA, KS, NE, and TX; all numbers are 1,000+ capacity feedlots.

**All states, 1,000+ capacity feedlots.

Source: *Cattle on Feed*, USDA-NASS, April 1997.

Steers and heifers coming out of the feedlots produce carcasses with heavier weights than cow or nonfed slaughter of other types. Figure 3.12 shows a scatter plot of the relationship between the number of cattle on feed on January 1 and beef production. The statistically weak and negative relationship is a bit surprising at first glance, but it essentially confirms the significant increase in production per head in recent years. The high levels of production in 1976–78 reflect the rapid rate of herd liquidation during that period. In the 1979–1981 period, there was a short-lived turn to herd building, and the decreased slaughter of cows and nonfed heifers pulls production below expected levels compared to the fitted relationship.

Table 3.6 shows the format for the 7-state monthly cattle on feed report. The reports are closely watched and widely employed by traders in live cattle and feeder cattle futures.

A 13-state report historically provided information on the number of cattle on feed by weight groupings.[5] By applying an average daily gain of 2.5 to 3.5 pounds per day, depending on cattle type and the season of the year, it was possible to project how many fed cattle would be coming to market in a future quarter or even a future month. The USDA provides frequent estimates of beef production for future calendar quarters in its *Livestock, Dairy, and Poultry Situation and Outlook Report.* Relatively sophisticated analytical models are employed in the forecasts and decision makers can take

[5]In the early 1990s, the weight groupings were dropped from the 13-state reports. Responding to industry concerns, weight groups were later reinstated and are now shown in the monthly 7-state report, which covers feedlots above 1,000-head capacity.

advantage of that expertise by subscribing to the reports. Estimates of production are released periodically to electronic market news services for immediate access by decision makers, eliminating the time lag involved in waiting on the written reports. Decision makers can thus take advantage of the USDA's publicly available forecasts and need not try to do the projections personally unless there is a reason to believe more accurate estimates can be generated.

Table 3.7 shows beef production for 32 recent quarters compared to the USDA estimate from two quarters earlier. The estimates are reasonably accurate with a tendency to underestimate production. It is possible that the USDA models have not yet captured the increased production per head from a genetically improved herd in recent years.

The large errors can often be explained, and those explanations will help to drive home the importance of the micro-macro paradox in production agriculture and of information shocks to the market. Clearly, a reason is needed when the models miss by as much as 5 to 7 percent. A micro–macro trap is set for you when this happens, and you need to be alert to it.

In 1989, cattle feeders were very optimistic on prices in the first quarter. Prices started to decline, and the feedlots held the cattle, waiting for prices to recover. Average slaughter weights started to climb dramatically. Corn costs were going down, making it cheaper to continue feeding the cattle. By the end of the second quarter and into the third quarter, the overfed cattle had to be sold, and beef production jumped. Prices of choice steers on the Omaha market declined sharply relative to expectations, and the USDA estimated cattle being sold in the early summer months were losing over $100 per head. *The macro or aggregate impact of individual decisions to hold the cattle for a hoped-for price recovery was devastating.*

In early 1996, a major drought was prompting accelerated slaughter of cattle in Texas, and corn prices were increasing rapidly on the way to record highs in the summer months. Production surged early in the year, but by the fourth quarter, actual production was 5.19 percent below USDA forecasts. The record corn prices and the early-year weak selling prices for cattle prompted cattle feeders to cut back on production plans. When forecasting only two quarters into the future, it is difficult to correctly anticipate all these risk factors.

After the slaughter numbers are projected using the inventory and cattle-on-feed data, average weights become very important. Traders monitor these data daily, and market news and market information systems record daily average weights in some live cattle markets and report average slaughter weights.

Changes in weights become very important determinants of price for a number of reasons. Obviously, a significant increase in average weights increases the tonnage of beef, and that moves the short-run supply curve for the beef sector to the right. If the own-price demand elasticity for beef is around –0.65,[6] suggesting an elasticity for live cattle at the producer level of around –0.5, each 1 percent increase in tonnage will cause a 2 percent decrease in cattle prices, other factors being equal. But the impact on the distribution of grades for cattle coming out of the feedlot and the desirability of the cattle is perhaps even more important than the price pressure from the increased tonnage.

[6]Kuo S. Hang and Richard C. Haidacher, "An Assessment of Price and Income Effects on Changes in Meat Consumption," in Reuben C. Buse, ed., *The Economics of Demand*, University of Wisconsin, October 1989.

TABLE 3.7 USDA Beef Production Forecasts for Two Quarters in the Future Compared to Actual Production, 1989–1996

Quarter (Year)	Actual Beef Production	USDA Prediction	Prediction Error	Percent Prediction Error
	(million lbs.)		(lbs.)	(%)
1 (1989)	5,529	5,475	–54	–0.98%
2	5,777	5,400	–377	–6.53%
3	5,892	5,475	–417	–7.08%
4	5,775	5,500	–275	–4.76%
1 (1990)	5,508	5,450	–58	–1.05%
2	5,736	5,775	39	0.68%
3	5,823	6,050	227	3.90%
4	5,567	5,675	108	1.94%
1 (1991)	5,385	5,500	115	2.14%
2	5,693	5,725	32	0.56%
3	6,013	6,000	–13	–0.22%
4	5,709	5,775	66	1.16%
1 (1992)	5,597	5,450	–147	–2.63%
2	5,726	5,900	174	3.04%
3	5,991	6,100	109	1.82%
4	5,654	5,725	71	1.26%
1 (1993)	5,357	5,500	143	2.67%
2	5,690	5,825	135	2.37%
3	6,076	6,125	49	0.81%
4	5,819	5,800	–19	–0.33%
1 (1994)	5,745	5,675	–70	–1.22%
2	6,042	5,925	–117	–1.94%
3	6,377	6,225	–152	–2.38%
4	6,114	5,900	–214	–3.50%
1 (1995)	5,877	5,950	73	1.24%
2	6,312	6,100	–212	–3.36%
3	6,602	6,400	–202	–3.06%
4	6,252	6,225	–27	–0.43%
1 (1996)	6,303	6,125	–178	–2.82%
2	6,642	6,425	–217	–3.27%
3	6,390	6,700	310	4.85%
4	6,084	6,400	316	5.19%

When cattle get held in the feedlots for longer than normal periods for any reason, an increased percentage starts to move into the yield grade 4 category.[7] Estimates of the difference in weight of lean cuts range up to the USDA's 4.6 percent difference per yield grade, but the price impact can sometimes be much greater than even the 4.6 percent differentials would suggest.

[7]The yield grades for cattle range from 1 to 5 with yield grade 1 showing the highest ratio of lean cuts to total carcass weight, grade 5 the lowest. Over 95 percent of cattle coming from the feedlots fall in the 2–4 range, with yield grade 3 having become the "par" grade or norm in cash market trade in cattle. Yield grade 4 cattle face sharp price discounts.

The market for yield grade 4 cattle is narrow, and the demand appears to be very inelastic. Any significant increase in yield grade 4 cattle in the flow of cattle to market can drive the price for yield grade 4 carcasses sharply lower. Prices for yield 4 Choice carcasses have dropped as much as $20 per hundredweight below the price of yield grade 3 Choice carcasses. That $20 difference clearly exceeds the estimated differences due to different yields of lean cuts.

Employed as a visible indicator that cattle are "backing up" in the feedlots, a surge in the percent of yield 4 cattle strengthens the packers' bargaining position. They know that the cattle are getting too heavy in the lots, that feed conversion efficiency starts to deteriorate rapidly as the cattle get heavy, and that feedlots will be forced to sell the cattle within a matter of days. Since they keep a close watch on the showlists of cattle at all the major feedlots, the packer buyers know when they are in the driver's seat, and they will attempt to take advantage of the situation by buying as low as possible and improving their profit margins.

The period from quarter 4 of 1984 through quarter 3 of 1985 provides an excellent case study. Table 3.8 records average weights plus carcass and live cattle prices by months during the period. The live prices are for Choice slaughter steers at Omaha, and the carcass prices are for yield grade 3 Choice steer carcasses in the central U.S. market area. Total beef production is also shown, so it is easy to recognize a price impact over and above what we would have expected due solely to a change in production. There is a combination of increased supply, a shift in the balance of market power to the packer related to the overfed cattle, and a move to a "bearish" sentiment that feeds on itself as the prices dip lower.

The dramatic price reactions shown in Table 3.8 appear to be inconsistent with the basic supply–demand framework, but they are not. That framework has to be extended to include the behavioral dimension of the markets. *This type of reaction (typically an overreaction) generates transaction prices in the extremes of the price distribution shown back in Figure 3.2 and extends the price range within which transactions are seen to occur.* From the viewpoint of the trader in futures, especially the hedger, it is important to get the price moves prompted by a short-run holdback of cattle (or other short-run shocks to the supply side) in proper perspective.

The moves *are* typically short-run in nature, and the moves do often tend to run too far before correcting back to some intermediate price level. Decision makers must guard against getting caught up in the emotion of the markets on supply-prompted moves in price that are destined to be short-run in nature. The biggest mistake, and the one commonly seen, is to panic and set short hedges on the cattle, or whatever is experiencing the price dip, down near the low prices. The futures market will correct at least part of the price move. During 1985, the cash market improved dramatically during October as the average weights started to stabilize and eventually moved back toward normal. In most instances, the futures market will anticipate the stabilization, and the futures market will start its correction before the cash prices start to improve. In 1985, the futures markets moved sharply higher during September, anticipating the improved situation prior to the better cash prices in October.

The 1985 experience was largely repeated in 1994. Choice slaughter steer prices were above $75 in March, but were down to the $63 level by June. Average steer slaughter weights increased by 4 percent during the period, pushing an unusually large percentage of the choice cattle into the yield grade 4 category. The USDA estimated losses to cattle feeders ranging from $11.67 to $16.66 per hundredweight during May to October of 1994—up to $180 per head or more for 1100-lb steers. This was a case of the *micro–macro trap* at its worst.

TABLE 3.8 Average Weights, Prices for Choice Steers and Choice Carcass Beef, November 1984–October 1985

Month	Carcass Weights (year earlier) (lbs.)	Total Production (year earlier) (million lbs.)	Average Price ($ per cwt.)	
			Live	Carcass
November 1984	718 (720)	1924 (1935)	64.29	99.08
December 1984	712 (703)	1830 (1965)	65.32	101.22
January 1985	707 (689)	2066 (1914)	64.35	99.50
February 1985	710 (691)	1768 (1859)	62.80	97.42
March 1985	728 (693)	1858 (1937)	58.58	89.52
April 1985	724 (689)	1936 (1776)	58.72	89.20
May 1985	728 (693)	2089 (2060)	58.58	89.52
June 1985	728 (694)	1898 (1984)	56.69	88.48
July 1985	728 (694)	2059 (1936)	53.26	82.22
August 1985	734 (698)	2123 (2112)	51.94	80.02
September 1985	739 (698)	1985 (1904)	51.94	80.02
October 1985	739 (712)	2108 (2182)	58.02	91.11

Source: Livestock and Poultry: Situation and Outlook Report, ERS, USDA.

What we saw in 1985 and again in 1994 is not unusual in the livestock markets. Each producer or cattle feeder is too small to exert influence on price, but the same reaction by many producers can and will move the price. We have a micro–macro paradox, a *micro–macro trap*, paralleling that covered earlier for the crops and in the discussion of the USDA beef production statistics in which the individual firms (micro level) get hurt by the aggregate (macro level) actions of their peers. *Since no obvious way exists to eliminate the possibility of such developments, it is important to try to ease their influence*. You should try to counter the tendency to follow the crowd, and industry leaders and analysts should keep constant reminders in front of producers that the feedlots must stay current and not get caught holding the cattle. And, of course, the futures markets are available to the hedger to get protection against just such a catastrophic development.[8]

Seasonal price patterns in cattle emerge from forage-based production programs that tend to focus sales in the fall months and from increases or decreases in placements of cattle on feed. Cattle-on-feed reports are available to subscribers and the USDA incorporates the impact of the cattle-on-feed reports in its estimates of quarterly beef production. There is significant potential for price variation within the year and for given levels of January inventories, and potential hedgers must stay informed and be aware of developing changes in the supply–demand balance. You must be aware of the micro–macro trap.

[8]It is interesting to think about the impact of hedging during such periods. If the cattle that are ready to be sold during the price break are hedged, the seller is interested primarily in the performance of the basis and not the absolute level of either cash or futures prices. It would therefore be easy to build an argument that having a larger percentage of the cattle hedged would help to guard against holding cattle to excessive weights. There is no strong incentive to hold the cattle and hope the price will come back up if they are hedged.

Seasonal Patterns: Hogs

The supply side on hogs is more difficult for most market analysts to handle than for cattle. Dramatic price moves after the release of quarterly *Hogs and Pigs* reports by the USDA are the norm rather than the exception. Both cash and futures prices frequently show major price reactions to the reports. The futures market is particularly vulnerable since it is often discovering prices for a future time period using a base of information that turns out to be in error when the reports are released. It is important to remember that the reports are often the surprises to the market that cause major price adjustments.

There are two possible and related reasons for the large postreport price moves. First, the futures market could be performing poorly in its assigned task of price discovery and not be "efficient." Second, the information base being employed by the futures market could be deficient in some important respect.

The concept of market efficiency and the way it gets measured will not be covered in detail here. The article by Purcell and Hudson, listed at the end of the chapter, discusses the concept and provides additional references for the reader who wishes to pursue this interesting issue.

In very simple terms, the market is considered efficient if all the publicly available information on supply and demand is being incorporated and registered in the discovered price. The hypothesis to be tested usually revolves around the notion that since the information tends to hit the market in a random fashion, the efficient market will generate a price path such that day-to-day price changes are independent of each other.

Day-to-day changes in price following the release of *Hogs and Pigs* reports are often *not* independent but are highly correlated. The futures market sometimes has to move to a new price plane that requires a $4- to $5-per-hundredweight total change to fully reflect the new information. A series of *related* day-to-day price changes will be necessary to complete the needed price adjustment. The *limit moves* on the lean hog futures ($2 per hundredweight from the previous day's close) dictate that several days will be required to make such an adjustment.

The June 1997 lean hog futures demonstrates (Figure 3.13). The close on December 27, 1996 was $75.22. After the close, the report showed little of the widely anticipated expansion—and expectations of supply for mid-1997 had to be adjusted. The closes for the next two days were $77.22 and $79.22, respectively, and surely the limit-up move to $77.22 the day after the report was related to the subsequent limit-up move to $79.22! *This is a most uncertain market.*

The problem is both the frequency and content of publicly released reports dealing with the hog sector. A typical quarterly report is shown in Table 3.9. Total numbers, hogs kept for breeding, market hogs, farrowings, and farrowing intentions are provided. The market hog category is divided into weight groupings, but most analysts consider the accuracy of the weight groupings a bit suspect. The major categories have a sampling error of at least 2–3 percent in either direction, and the sampling error in the weight groupings may be even larger.

There are always some analysts and market observers who question the accuracy of the reports, but there is no doubt that the hog futures market responds to the reports. Prior to the release of the report, the major market news services conduct a survey of a number of market analysts and release an average and a range of the prerelease estimates. *It is not unusual to see the actual numbers in the Hogs and Pigs reports fall completely outside the range of estimates, and that type of surprising*

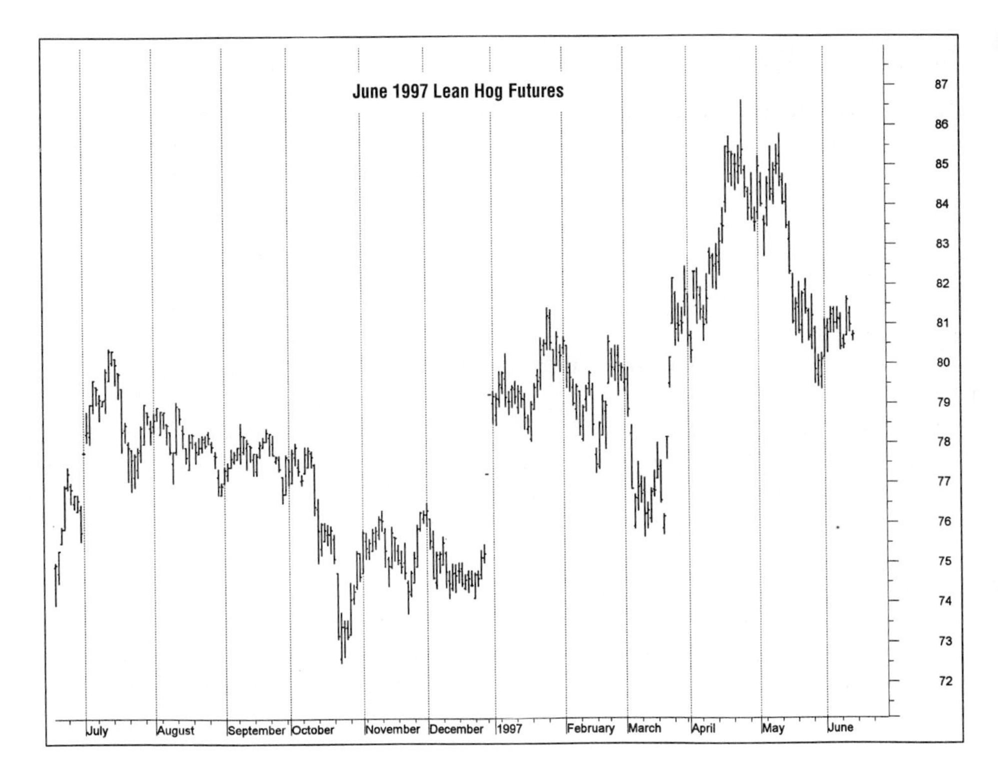

FIGURE 3.13
Futures Chart for June 1997 Lean Hogs

report will always elicit a major price adjustment. The quarterly reports are the primary supply-side information base used by the market in the price-discovery process, and no publicly or privately released information provides a useful alternative or more detailed information.

Abstracting from the question of the accuracy of the reports, the major concerns are the frequency of the releases and the detail provided. The bulk of the slaughter volume in hogs is barrows and gilts, but the slaughter data are not disaggregated into barrows and gilts. The difficulty, then, is being able to track what is happening in the breeding herd. The only indication of female slaughter is sow slaughter, and this data series can be influenced by producer holding of gilts for breeding and for herd expansion. If sow slaughter increases, to illustrate the problem, is it due to a net liquidation of the herd or is it merely replacing aging sows with gilts to renovate the herd? If gilts are being retained at a rate exceeding the normal replacement requirements, *herd expansion can be occurring during periods in which sow slaughter appears to be high* as a percent of total slaughter.

Figure 3.14 relates sow slaughter to the breeding herd with a two-quarter time lag across a recent time period. That is, the breeding herd (in the December–February quarter, for example) is a function of sow slaughter as a percent of total slaughter two quarters earlier, (the June–August quarter, for example). There is clearly substantial variability in the relationship, suggesting that analysts will in fact have problems projecting the breeding herd when we do not know what is happening to gilt slaughter and, related, whether gilts are being retained for the breeding herd. Any attempt to model the relationship shown in Figure 3.14 would give very poor results. The relationship appears to be essentially random.

TABLE 3.9
Content and Format of the December 1996 Hogs and Pigs Report

	1994	1995 (1,000 head)	1996	1996 as % of 1994 (percent)	1996 as % of 1995
March 1 inventory					
All hogs and pigs	57,350	58,465	56,340	98	96
Kept for breeding	7,210	6,998	6,765	94	97
Market	50,140	51,467	49,575	99	96
Market hogs and pigs					
Under 60 pounds	18,780	19,251	18,790	100	98
60–119 pounds	12,190	12,498	11,980	98	96
120–179 pounds	10,430	10,594	10,095	97	95
180 pounds and over	8,740	9,124	8,710	100	95
June 1 inventory					
All hogs and pigs	60,715	59,560	57,200	94	96
Kept for breeding	7,565	7,180	6,870	91	96
Market	53,150	52,380	50,330	95	96
Market hogs and pigs					
Under 60 pounds	22,125	21,270	20,265	92	95
60–119 pounds	13,145	13,060	12,700	97	97
120–179 pounds	9,825	9,865	9,800	100	99
180 pounds and over	8,055	8,185	7,565	94	92
September 1 inventory					
All hogs and pigs	62,320	60,540	58,200	93	96
Kept for breeding	7,415	6,898	6,770	91	98
Market	54,905	53,642	51,430	94	96
Market hogs and pigs					
Under 60 pounds	20,790	20,235	19,330	93	96
60–119 pounds	13,960	13,532	12,800	92	95
120–179 pounds	11,170	10,985	10,600	95	96
180 pounds and over	8,985	8,890	8,700	97	98
December 1 inventory					
All hogs and pigs	59,990	58,264	56,171	94	96
Kept for breeding	7,060	6,839	6,663	94	97
Market	52,930	51,425	49,507	94	96
Market hogs and pigs					
Under 60 pounds	19,556	18,881	18,411	94	98
60–119 pounds	13,087	12,808	12,239	94	96
120–179 pounds	10,941	10,702	10,313	94	96
180 pounds and over	9,346	9,034	8,544	91	95

Source: *Hogs and Pigs*, NASS, USDA, December 27, 1996.

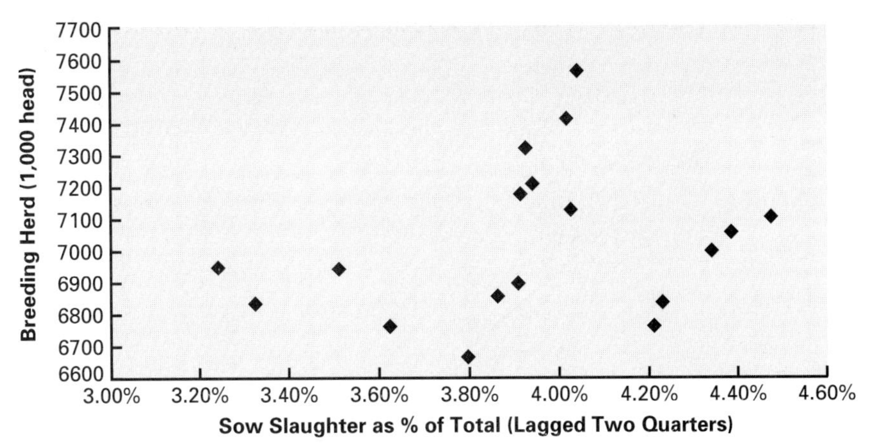

FIGURE 3.14
Breeding Herd as a Function of Sow Slaughter/Total Lagged Two Quarters

If you accept the essence of this discussion, you have to conclude that more information is needed. The immediate question is: Why is this information not provided? Information will be collected when its marginal return (*MR*) is greater than the marginal cost (*MC*) of collecting the data. It appears that neither the public sector (USDA) nor the private sector (brokerage firms, advisory services) feels the *MR* of either more frequent reports or more detailed data in the current quarterly reports exceeds the *MC* of the added information. If we look at the often dramatic moves following release of the quarterly reports, it is difficult to argue that the *MR* is low. Volatile changes in supply and the extreme price changes that come with them are costly not only to producers and to potential investors, but also to society in the form of variable product supplies and variable prices. But the information base is not improving, and there is an important message for you here: It is difficult to predict accurately the supply of hogs that will be available to the market in a future time period given available data. The live hog futures market will have difficulty discovering the correct price because of the lack of information, and will continue to be characterized by major postreport adjustments in price.

In this area, the user of the futures markets will need to prepare for a volatile market. The USDA provides quarterly predictions of pork production, per capita consumption, and prices in its *Livestock, Dairy, and Poultry Situation and Outlook* reports. But they will have some of the same difficulties private analysts experience, and the percentage errors in their predictions tend to be larger than those shown earlier for beef. In terms of impact on strategies, the supply-side problem in hogs suggests the hedger should be aggressive in taking profitable prices whenever the futures market offers them.

Supply fluctuations in hogs can be significant within the year. With only quarterly reports to track producers' decisions on herd expansion or contraction, the supply of hogs in quarters 3 and 4 can be influenced by decisions in quarters 1 and 2, and these decisions are very difficult to anticipate correctly. It appears the often volatile price moves in the live hog futures markets after the release of the quarterly reports could be due to a limited information base, and are not necessarily evidence of inefficiencies in the futures markets.

FIGURE 3.15
Typical Downward-Sloping Demand Curve

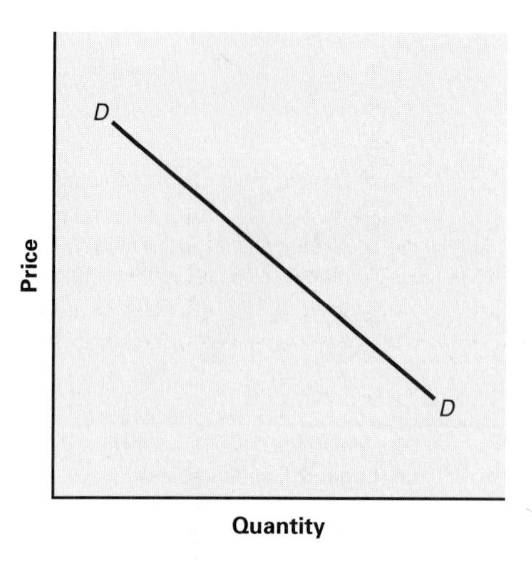

THE DEMAND SIDE: LIVESTOCK AND POULTRY

In dealing with the supply issues for livestock, poultry was not included. The poultry industry is vertically integrated with no obvious exposure to market-determined prices. The level of price-risk exposure has not justified trade in broiler or other poultry-related futures contracts in recent years. On the demand side, however, poultry is a major competitor for market share with pork and beef and must be included.

Technically, demand is a schedule of the quantities that will be taken by consumers at alternative prices. In the livestock and meats, there has been much confusion and misunderstanding about what demand is and is not, and what constitutes a change in demand.

Figure 3.15 illustrates a typical downward-sloping demand curve. Any price–quantity combination that falls on the curve is on the same level of demand. It is only when the entire curve (the entire schedule) changes that demand has changed. Per capita supply, and therefore per capita consumption, can change significantly but that does not mean demand has changed. It is consistent with the law of demand discussed in most beginning economics texts that consumers, at any point in time, will take an increase in supply only at lower prices. *To draw conclusions about what has happened to the level of demand we have to look at both quantity and price.* Scatter plots for beef, pork, and broilers illustrate the issue.

Figure 3.16 shows deflated prices of Choice beef at retail plotted against per capita consumption. The years are identified in the body of the figure. The prices are deflated using the Consumer Price Index (CPI, 1982–84 = 100) to remove the influence of overall price inflation and to allow legitimate year-to-year comparisons.[9]

[9]The price series is divided by the CPI to remove the influence of overall price inflation. This process of "deflating" the price series converts them to a common denominator in dollar terms and ensures that year-to-year price comparisons are not being distorted by overall price inflation.

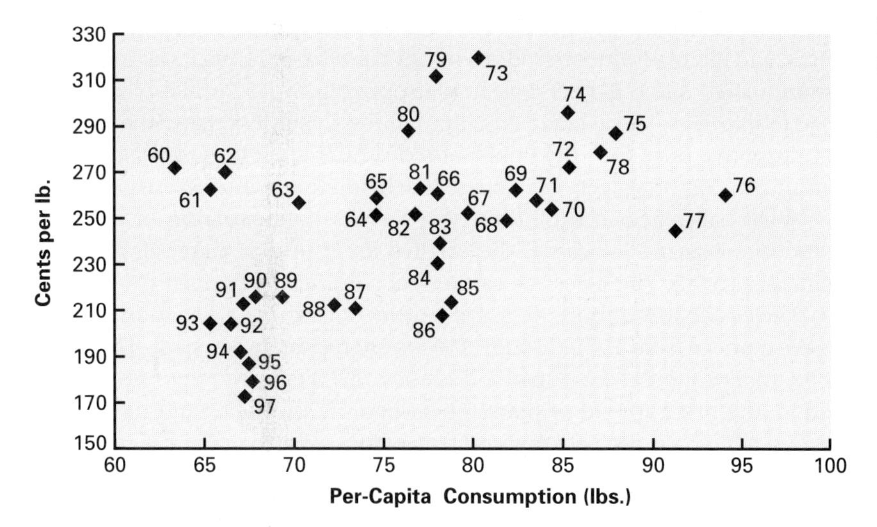

FIGURE 3.16
Per Capita Consumption and Deflated Retail Prices for Beef (CPI, 1982–84=100), 1960–1997

It is easy to find year-to-year changes that suggest demand was increasing. From 1971 to 1972, for example, an increased per capita quantity was taken at significant increases in the inflation-adjusted price. That change indicates demand has increased. We do not know why demand increased, but it is clear that 1972 was on a higher level of demand than was the case in 1971. You should visualize negatively sloping demand curves similar to that shown in Figure 3.16 through each of the points for 1971 and 1972. The price–quantity coordinate for 1972 cannot be on a negatively sloping demand curve that passed through the price–quantity coordinate for 1971.

This type of simple analysis proves very revealing. If we start with 1979, a most negative pattern starts to emerge for beef. Each year, from 1979 through 1986, a reduction in the inflation-adjusted price was required to move essentially a constant per capita supply into consumption. From 1979 through 1986, *a price reduction of over 30 percent in inflation-adjusted prices was required to keep the consumer at the beef counter to buy and consume a largely constant per capita quantity.*

Earlier, in Table 3.5, the total U.S. cattle herd was presented. From 1975 through January 1 of 1990, there was a net liquidation of over 32 million head. It is clear at this point that the liquidation could have been forced at least in part by decreases in demand for beef. *The decision maker trying to anticipate prices for beef and make intelligent use of the futures markets must take the possibility of significant changes in demand into account.*

The essence of the problem is revealed in Table 3.10. Inflation-adjusted (Deflated Retail Price) prices for Choice beef at retail decreased significantly during the 1980s, and the nominal (Retail Price) price was not able to increase during that period.[10] The consumer simply would not pay higher prices. With middleman margins expanding during the period by over 25 percent, the result was severe pressure on calf and stocker

[10]The term *nominal* is used to refer to price, income, or other economic series before they are adjusted for the influence of general price inflation. As implied earlier, the term *deflated* (sometimes called "real") is used to refer to price or income series that have been adjusted for general price inflation.

cattle prices at the producer level. Prices were simply too low to keep the cattle herd intact, and many producers and resources were forced to exit the industry. From 1975 through 1990, the U.S. beef cow herd dropped from 45.7 million head to 33.1 million head (Table 3.5). The average cow herd in the U.S. is less than 50 head. Using 50 head for illustrative purposes, the liquidation of 12.6 million beef cows involves the equivalent of 252,000 producers of average size being forced out of business.

What happened in the industry is a vivid demonstration of derived demand at work. The demand for cattle, the original input or raw material, is derived from the demand at retail. The price for cattle is, accordingly, a derived price.

The concept is important and is worthy of further explanation. In Table 3.10, the nominal prices from 1979 through 1986 were relatively constant. The range was $2.26 to $2.42. For illustrative purposes, assume the retail price *was* constant. During the period, the price spread or margin between the producer and the retailer increased over 20 percent as the packers' and processors' costs went up with overall price inflation. The derived price at the producer level had to go down. The situation can be demonstrated as follows:

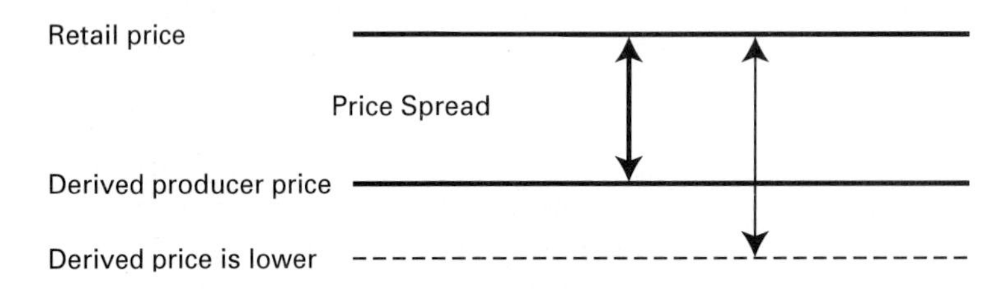

As the price spread expanded, with a largely constant retail price, the producer-level price had to go down (to the dashed line) unless the packer/processor sector increased efficiencies enough to offset the pressures. The increased efficiencies during the period were not enough to eliminate all the pressure at the producer level, and as the derived prices were forced lower, producers were forced out of business. Steer calf prices that had averaged $93.10 per hundredweight in Kansas City in 1979 were in the high $60s during 1982–85 and averaged $69.67 in 1986, and were pushed periodically below $50 during the record-high corn prices in 1996. Budgets show the average total cost of producing calves is $85 to $95 per hundredweight for the typical producer.

It appears the situation started to stabilize in the early 1990s. We could argue, from an examination of Figure 3.16, that the 1993 price–quantity coordinate is near the demand curve that passed through the 1995 price-quantity coordinate. The 1996 and 1997 data suggest the level of demand decreased again, however. It may take several years for the long-term problems to disappear and the demand side of the price equation to start to look more positive. *During the past 15 years, the demand side has been a major cause of price moves to the downside, and it will not suffice for the hedger or the speculator to look only at the supply numbers and implicitly assume demand is constant.*

This latter point is extremely important. Prior to the 1980s, "price analysis" in the cattle markets was heavily supply-side oriented. Changes in cattle on feed and projected supply changes were converted to price changes, with the implicit assumption that demand was constant. But changing lifestyles and related changes in preference patterns changed all that in the 1980s. Trying to just count the supply-side numbers during the 1980s and trade accordingly was difficult for speculators, and the failure to account for decreases in demand left many would-be short hedgers on the sidelines—and hurting financially. Techniques for analyzing short-run changes in demand will be

TABLE 3.10
Per Capita Consumption and Price of Choice Beef at Retail, Actual and Deflated (CPI, 1982–84 = 100), 1970–1997

Year	Per Capita Consumption	Retail Price	Deflated Retail Price
	(lbs.)	(cents/lb.)	
1960	63.3	80.2	270.9
1961	65.4	78.4	262.2
1962	66.1	81.7	269.6
1963	70.2	78.5	256.5
1964	74.7	77.8	251.0
1965	74.6	81.4	258.4
1966	78.1	84.6	260.3
1967	79.8	84.1	251.8
1968	82.0	86.6	248.9
1969	82.5	96.2	262.1
1970	84.4	98.6	254.1
1971	83.7	104.3	257.5
1972	85.5	113.8	272.2
1973	80.5	142.1	320.0
1974	85.4	146.3	296.8
1975	88.0	154.8	287.7
1976	94.2	148.2	260.5
1977	91.4	148.4	244.9
1978	87.2	181.9	279.0
1979	78.0	226.3	311.7
1980	76.4	237.6	288.3
1981	77.1	238.7	262.6
1982	76.8	242.5	251.3
1983	78.2	238.1	239.1
1984	78.1	239.6	231.3
1985	78.8	228.6	212.7
1986	78.4	226.8	206.9
1987	73.4	238.4	209.9
1988	72.3	250.3	211.6
1989	69.3	265.7	214.3
1990	67.8	281.0	214.5
1991	67.2	288.3	212.0
1992	66.4	284.6	203.3
1993	65.4	293.4	203.1
1994	67.0	282.9	190.9
1995	67.4	284.3	186.6
1996	67.6	279.6	178.2
1997	67.2	280.0	174.5

presented later in the chapter and emphasized. *It could be forcefully argued that the dominant shortcoming of the ability of the futures market to discover the correct prices for cattle quickly and efficiently in the 1980s and 1990s was the inability to recognize and account for decreases in demand.*

Figure 3.17 provides a scatter plot for pork. The pattern for pork suggests problems starting in 1980, but the 1987 coordinate looked positive compared to 1986. Note the increase in per capita supplies moving at a significantly higher inflation-adjusted price in 1987 compared to 1986. Demand appears to have increased in 1988 before faltering again in 1989. The 1990s has been largely a period of consoli-

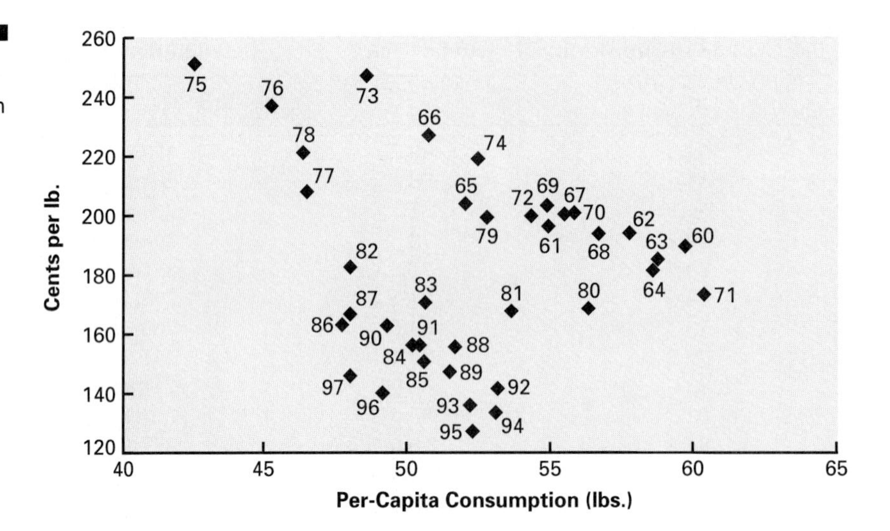

FIGURE 3.17
Per Capita Consumption and Deflated Retail Prices for Pork (CPI, 1982–84=100), 1960–1997

dation, with no compelling evidence to date that demand for pork has started to increase.

To illustrate the importance to producers, let's look at the 1986–88 period. Per capita supplies of pork were up 1.0 percent in 1987 compared to 1986. Using a retail level demand elasticity of –0.67, that would suggest the inflation-adjusted prices would be down 1.5 percent. But pork prices were actually up 2.0 percent. If the middlemen's margins were constant, that swing from a 1.5 percent decrease to a 2.0 percent increase in pork prices would make a big difference in derived hog prices. Prices at the producer level fluctuate more in percentage terms than the changes at retail, so hog prices in 1987 were 5–6 percent higher than they would have been if demand had been constant. Instead of averaging $50.88 at Omaha, the price would have been $48.46. For a 240-pound slaughter hog, that translates to $5.81 per hog—and a big difference for producers.

The other side of the coin is present. Per capita consumption of pork for 1990 and 1996 was essentially constant, but inflation adjusted prices were 162.3 and 140.2 for 1990 and 1996, a 13.6 percent decline. Overall, demand is decreasing in the 1990s, and the obvious demand problems are an important reason why slaughter hog prices dipped below $30 in the fourth quarter of 1994 and were back down to the $30 level in the first quarter of 1998. (It *is* true that much of the price weakness in early 1998 can be traced to the supply side with pork production running 12 percent above 1997 levels.)

Any problems may have been less dramatic in pork than in beef, but they *were and are* important. Table 3.11 provides data paralleling that for beef in Table 3.10. Retail prices were under pressure, and the pressure was relieved primarily in the form of lower prices for hogs to producers prior to 1987. Many producers, especially those that were smaller and less efficient, were driven out of business. At the national level, hog numbers recorded a post-1970 peak at 67.3 million head in 1979, dropped to 50.9 million in 1986, were at 53.8 million head on December 1, 1989, and stood at 59.9 million on December 1, 1997. Supply–demand dynamics are always important in prices received by producers.

The plot for broilers in Figure 3.18 is revealing. Prior to the early 1980s, cost-reducing technology allowed the industry to offer more product at lower and lower

TABLE 3.11
Per-Capita Consumption and Price of Pork at Retail, Actual and Deflated (CPI, 1982–84 = 100), 1970–1997

Year	Per-Capita Consumption (lbs.)	Retail Price	Deflated Retail Price
		(cents/lb.)	
1970	55.9	78.0	201.0
1971	60.4	70.3	173.6
1972	54.4	83.2	199.0
1973	48.7	109.2	245.9
1974	52.6	107.8	218.7
1975	42.6	134.6	250.2
1976	45.3	134.0	235.5
1977	46.6	125.4	206.9
1978	46.4	143.6	220.2
1979	52.9	144.1	198.5
1980	56.4	139.4	169.2
1981	53.7	152.4	167.7
1982	48.1	175.4	181.8
1983	50.7	169.8	170.5
1984	50.3	162.0	156.4
1985	50.7	162.0	150.7
1986	47.8	178.4	162.8
1987	48.1	188.4	165.8
1988	51.7	183.4	155.0
1989	51.6	182.9	147.5
1990	49.4	212.6	162.3
1991	50.5	211.9	155.8
1992	53.2	198.0	141.4
1993	52.3	197.6	136.8
1994	53.2	198.0	133.6
1995	52.4	194.8	127.8
1996	49.2	220.0	140.2
1997	48.0	231.5	144.2

inflation-adjusted prices. But during the 1980s, and paralleling what were apparently preference-related problems in red meats, the demand for broilers started to increase. There is clear indication of increases in demand from 1983 to 1984, from 1985 to 1986, and again from 1987 to 1988 and from 1988 to 1989. Note the ability to sell increased per capita supplies at higher inflation-adjusted prices. It appears that during the 1980s, poultry was starting to be seen in a more positive light by consumers. Further processing and new product development may have been the primary catalysts. Since 1989, the pattern has been one of increased per capita supplies at roughly constant deflated prices.

The result has been a larger market share for poultry as per capita supplies and therefore per capita consumption have continued to increase. Figure 3.19 provides a plot of per capita consumption for beef, pork, and broilers through 1997.

The surging acceptance of poultry by consumers in the 1980s and 1990s brought new competition for beef and pork. At a minimum, anyone using the cattle and hog futures must keep an eye on developments in poultry. If that sector is expecting a 10 percent increase in production, there will be some pressure on poultry prices in spite

FIGURE 3.18
Per Capita Consumption and Deflated Retail Prices for Broilers (CPI, 1982–84=100), 1960–1997

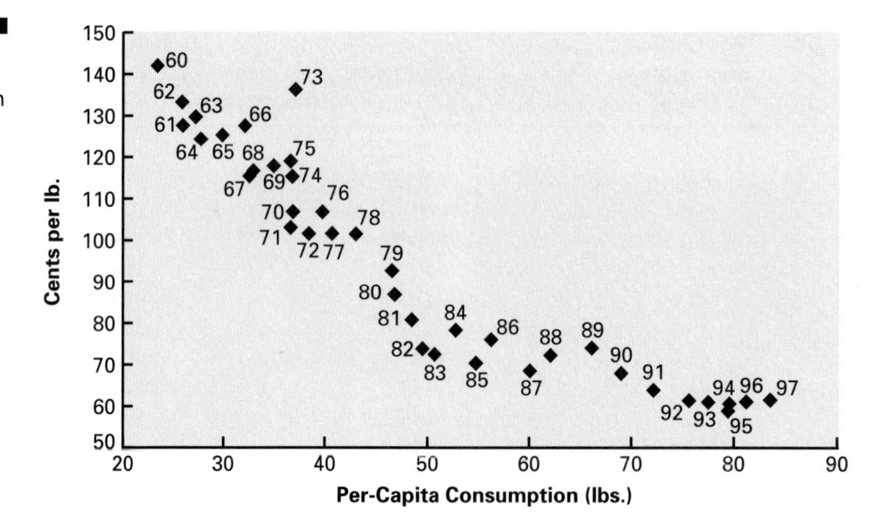

of strong demand and lower poultry prices will decrease demand for pork and beef. The USDA offers production and price projections for broilers and turkeys in the *Livestock, Dairy, and Poultry Situation and Outlook* reports.

Demand for both beef and pork declined significantly during the 1980s and 1990s. Lower derived prices at the producer level forced producers out of business, especially in beef. Across the same time period, the demand for broilers was stable to increasing. The longer run result of the developments in demand has been an increased market share for broilers versus beef and pork. When demand is not stable, analysis of shifts in demand must be incorporated in the fundamental analysis of the markets.

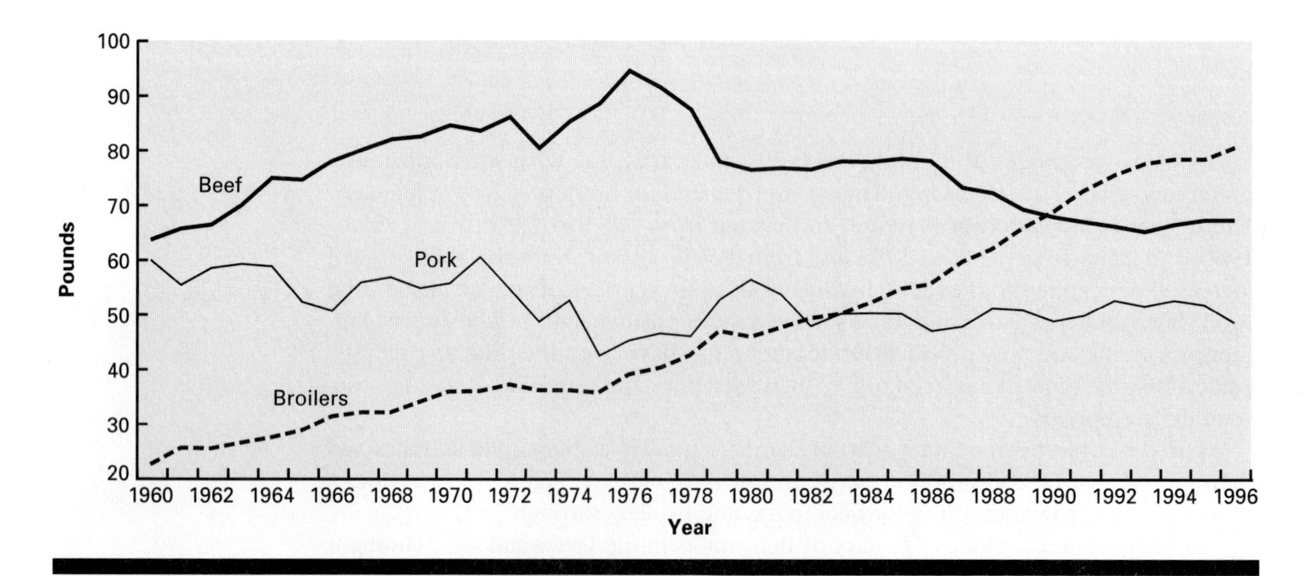

FIGURE 3.19
Per Capita Consumption of Beef, Pork, and Broilers, 1960–1997

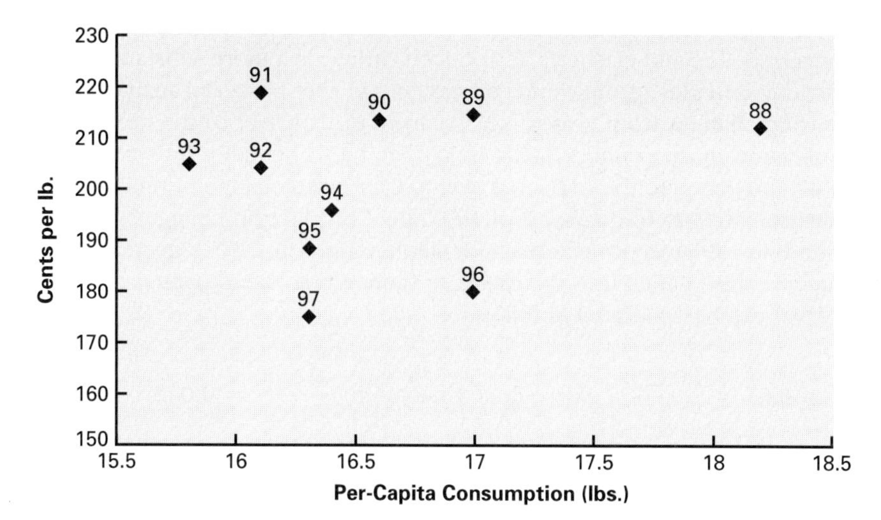

FIGURE 3.20 Quarter 4 Retail Deflated Beef Price (1982–84=100), Per-Capita Consumption, 1988–1997

MONITORING SHORT-RUN DEMAND

Obviously, demand for meats changed during the 1980s. To document changes in demand is not sufficient to estimate the importance of all this to the futures markets and to marketing strategies, however. Recognizing the relevancy of demand analysis should be no problem. What is happening on the demand side is critical. The futures trader, whether a hedger or a speculator, needs a way to get a short-term look at the underlying demand surface. Figure 3.20 offers one approach.

To remove the seasonality from the data, quarterly price-quantity combinations for the same quarter are plotted for the last 10 years. The pattern in recent years for Choice beef looks much like the yearly plots, but that is not surprising.

As a quarter approaches, it is useful to plug in the current retail price, estimate the per capita supply (remember that the USDA projects quarterly supplies *and* per capita consumption in its *Livestock, Dairy, and Poultry Situation and Outlook* reports), extrapolate the current level of inflation in the CPI index, deflate the retail price, and plot the price–quantity coordinate that emerges for the upcoming quarter with the data from recent years. Once a price estimate is generated, the retail price can be reinflated, and estimated farm-to-retail price spreads can be used to generate a live animal price.

An estimate of cattle, hog, or broiler prices generated using such a procedure is a current and useful place to start as a pricing strategy is laid out. As an example, *it makes little sense to sit and wait on a cattle price that would require the retail price series to move to new highs, the price spreads to move to record small levels, or some combination of the two.* Keep in mind that the function of the fundamental analysis is to discover the general price level that should emerge and to identify the probable direction of year-to-year price changes. And keep in mind that the USDA and many private consulting firms provide projections that are based on sophisticated modeling techniques. What we are doing here is making sure you understand the basic economic relationships that are involved and what the USDA and other entities are attempting to model. We need an idea of price direction for the upcoming production or decision period.

A second and widely used approach to monitoring short-run demand employs the concept of demand elasticity, introduced earlier, in a more substantive way. Demand elasticity provides a simple and convenient but very powerful analytical framework to measure changes from year to year or from one quarter to the next. To illustrate its application in determining what is happening to demand, let's use a coefficient of −0.67 and generate the expected year-to-year price change in retail beef prices from quarter 2 of 1996 to quarter 2 of 1997. Beef production changed from 6.642 billion pounds in 1996 to a projected 6.450 billion pounds in 1997, a decrease of 2.8 percent. Plugging this change into the elasticity framework, the expected price for quarter 2 of 1997 can be generated as follows:

$$-0.67 = \frac{-0.028}{X} \quad X = \text{Price Change} = \frac{-0.028}{-0.67} = 0.042$$

Price for quarter 2 of 1997 would be expected to be up 4.2 percent from quarter 2 of 1996 *if the level of demand in 1997 was the same as in 1996*. Since elasticity is a *property* of a demand curve, the framework estimates the price change in response to a given or predicted quantity change, assuming that the level of demand is constant, and assuming therefore that the demand curve has not shifted. The arithmetic would suggest a price of $2.89 for quarter 2 of 1997, up from $2.77 in quarter 2 of 1996.

In application then, the elasticity framework can be used in several ways. It is used here to get an initial impression of whether the level of demand appears to be changing. In the example, the price of Choice beef at retail was $2.77 for quarter 2 of 1996. If the level of demand had been the same, the price in quarter 2 of 1997 should have been $2.89 (1.042 × $2.77). The price for quarter 2 of 1997 was actually $2.79, suggesting that the level of demand during late 1997 was down compared to the level of demand in late 1996. *This is very valuable information*. This information can then be factored into the outlook for early 1997, and the process can be repeated over time to give useful indications of what is happening to the level of demand. *In recent years, the demand for beef and pork have changed enough to make the demand side an important determinant of price levels and of short-term price changes.*

You should be aware that inflation-adjusted or deflated prices are usually employed in the elasticity framework. When the time period is quite brief, such as a quarter-to-quarter or even a year-to-year change, as we show here, using the nominal or observed retail prices would give useful results unless the rate of overall price inflation is unusually high.

Procedurally, the USDA's forecast of beef production for the next quarter can be used to calculate a quarter-to-quarter percentage change. That percentage change is then used in the elasticity framework (coefficient = −0.67) to predict the next quarterly price. For example, the USDA estimated beef production in quarter 2 of 1997 at 6.419 billion pounds. Assume they are forecasting that production in quarter 3 will be up to 6.595 billion pounds given the number of cattle scheduled to come out of the feedlots. This is a 2.75 percent increase. To calculate the expected change in price, use

$$-0.67 = \frac{-0.0275}{X} \quad X = -0.0410$$

where X is the expected change in price from quarter 2, 1997, to quarter 3, 1997. This result can be applied to the quarter 2 price to predict a quarter 3 price. Errors, especially if price predictions are consistently too high, suggest:

1. The impact of declining retail-level demand,
2. The reluctance of retailers to change beef prices in the short run until they are sure some significant change in the supply demand picture has changed, and
3. The elasticity framework will be more effective on year-to-year changes when the time period is long enough to get over retailers' resistance to price changes and for all the supply-side changes to work their way through the system.

Remember, the elasticity framework assumes demand is constant. *When the framework consistently overpredicts price, that is indirect evidence that demand is in fact declining. There is, therefore, much information to be gleaned from the analysis.*

If the framework is used directly at the live animal level, research results show the demand elasticity coefficient to be around –0.5 for hogs and for cattle. If pork production, for example, is projected to be up 8 percent in 1997 compared to 1996, prices for live hogs would be expected to be down a whopping 16 percent! *Clearly, this type of information is useful to the decision maker in establishing a price range within which hog or cattle futures might be expected to trade during coming time periods and in determining the likely direction of price trends during the future time period.*

A few restrictions deserve emphasis.

1. The elasticity coefficients used here are consistent with research results, but elasticity coefficients can change over time.
2. The elasticity framework assumes the level of demand is constant and examines the price implications of a change in supply or a move along the demand curve. If price estimates are higher or lower than those observed, that departure may be evidence that the level of demand has shifted, but you will not necessarily know *why* demand has changed. The short-run analysis of beef prices, for example, suggested demand for beef was declining but it did not reveal why the demand problems were present.

Short-run analysis of demand can be conducted by direct application of the demand elasticity framework. This approach can be very useful in establishing the general price level around which prices would be expected to develop during future time periods and in identifying the probable direction of period-to-period price changes.

SUMMARY

Price is established over time by the interaction of supply and demand. *Fundamental analysis*, involving analysis of the interactions of supply and demand forces, is important to help determine the probable direction of price trends and to provide an idea of the general range within which price will be discovered in the cash and futures markets.

Commodity futures prices are discovered within a supply–demand framework. The observed variability in those prices occurs because information on the levels of supply and demand is imperfect and is subjected to analysis with varying degrees of sophistication.

The supply–demand reports that are released by the USDA for grains, oilseeds, cotton, and other commodities are important in monitoring the fundamental supply–demand picture during the year. Discovered prices change as the estimates of the components of supply or the components of demand change due to changes in crop conditions in the U.S. or around the world and due to macroeconomic changes such as changes in the trading level of the U.S. dollar. The relationship between price and ending stocks, the residual component of the supply–demand table, is *an especially useful place to start in anticipating overall price patterns for the coming crop year*. The data and the projections of supply and demand components are provided by the USDA, and you need only interpret these data and add relatively simple graphical analyses to generate useful price projections.

In livestock, the beginning inventory levels will exert a major influence on probable price levels for the year. Within the year, supply can be adjusted to varying degrees across beef, pork, and poultry, but will not depart in a dramatic way from the general level set by the beginning inventories. Publicly available reports allow the decision maker to monitor possible intrayear changes in supply.

In cattle, the *cattle-on-feed reports* are important indicators of possible changes in intrayear beef supplies. In hogs, the quarterly *hogs and pigs reports* are essentially the only source of information, and some analysts would argue they are both too infrequent and lacking in detail. The USDA employs the cattle and hog data to project estimates of beef and pork production and the *Livestock, Dairy, and Poultry Situation and Outlook* reports are available by subscription.

Across recent years, changes in demand for beef and pork appear to have been increasingly important influences on prices. The evidence suggests significant decreases in demand for the red meats, especially beef from 1979 or 1980 to date. During the same period, there were periodic increases in the demand for poultry. Those *changes in demand appear to have prompted significant changes in inventories and related changes in production* for red meats and poultry. The *demand elasticity framework* provides a useful analytical approach to estimating period-to-period changes and in determining whether the level of demand has changed. If there is evidence that the level of demand is changing, that information can be integrated into the decision process in terms of probable price departures from the price generated from the elasticity framework.

High levels of sophistication and detailed quantitative analysis are not necessary for you to be able to conduct effective fundamental analysis. The need is for recognition and understanding of the supply–demand forces at work in the marketplace. *If the direction of changes in the balance of supply and demand and therefore direction of price during the coming production period or decision period can be accurately anticipated, the individual decision maker has a big advantage in developing an effective hedging plan.*

KEY POINTS

- *Fundamental and technical analysis* of the markets can be *complementary*.
- The *price-discovery process* generates highly variable prices over time because the information base on supply–demand levels is *imprecise* and is subjected to interpretation that *varies in the level of sophistication.*
- The *World Agricultural Supply and Demand Estimates* are the most important periodic releases on supply and demand information for the grains, oilseeds, and other storable commodities.

- The *ending stocks* are perhaps the *most important single entry* in the supply–demand reports for the grains and oilseeds. The relationship between *price* and *ending stocks* is a simple but powerful framework for *generating estimates of expected price levels*.
- In livestock, the *beginning inventories* set the *general supply* for the year. Unless *demand* changes significantly within the year, the general price level will be directly related to beginning inventories.
- Supplies of beef and pork can and do *vary within the year* due to *short-run supply responses* by producers as the number of livestock placed on feed varies. *Changing weights* of cattle and hogs can also be important determinants of intrayear price levels.
- Periodic estimates and forecasts of production, prices, and per capita consumption of meats are released in the USDA's *Livestock, Dairy, and Poultry Situation and Outlook* reports. These reports are *available to individual decision makers* and assist in establishing a price outlook and a price range.
- During the 1980s and 1990s, significant *decreases in demand* for the *red meats* were a primary force in industry changes toward smaller inventories and smaller levels of total production. The market share for poultry increased during the period as per capita supplies of the red meats, especially beef, declined.
- The *demand elasticity framework* is very useful in *predicting period-to-period price changes* and in *determining whether the level of demand has changed*.
- The need is for *understanding* of the *basic forces of supply and demand* and the related ability to *anticipate the direction of price trends*. Highly sophisticated quantitative analysis is not essential.

USEFUL REFERENCES

Jack D. Schwager, *Fundamental Analysis*, John Wiley and Sons, New York, 1997. A massive 639-page effort that explores fundamental analysis in detail.

Dale C. Dahl and Jerome W. Hammond, *Market and Price Analysis: The Agricultural Industries*, McGraw-Hill, New York, 1977. This is one of several very useful references on price analysis for agricultural commodities.

Wayne D. Purcell and Michael A. Hudson, "The Economic Roles and Implications of Trade in Livestock Futures," in Anne Peck, ed., *Futures Markets: Regulatory Issues*, American Enterprise Institutes for Public Policy Research, Washington, D.C., 1985. The authors discuss the price-discovery functions of livestock futures and deal with the issue of inadequate information in the livestock futures markets.

Walter Spilka, Jr., "The USDA Crop and Livestock Information System" in *Handbook of Futures Markets: Commodity, Financial, Stock Index, and Options,* Perry J. Kaufman, ed., John Wiley and Sons, New York, 1991. This reference describes the types of reports released by the USDA and explains the nature and use of the data in the reports. It will be a very useful reference for the beginner who seeks to explore fundamental analysis of commodity prices.

APPENDIX 3A. USDA INFORMATION SERVICES

Presented in this appendix is a list of some reports and periodicals that are available from the USDA Economic Research Service (ERS) and National Agricultural Statistics Service (NASS). Some reports are briefly described and subscription information is provided. The following are ERS Research and Analysis reports:

Agricultural Outlook. ERS subscription. 10 issues. Stock #ERS-AGO. $50.00. Main source for USDA's farm and food price forecasts; short-term outlook for all major areas of the agricultural economy; long-term issue analyses of US agricultural policy, trade forecasts, export-market development, food safety, the environment, farm financial institutions. Includes data on individual commodities, the general economy, US farm trade, farm income, production expenses, input use, prices received and paid, per-capita food consumption, etc.

Farm Business Economics Report. Annual report. 236 pp. September 1997. Stock # ECI-1996. $21.00.

Combines the information from three former separate publications of the series Economic Indicators of the Farm Sector: 1) National Financial Summary, 2) State Financial Summary, and 3) Costs of Production, Major Field Crops and Livestock and Dairy. Includes national and state farm income estimates, farm sector balance sheet, government payments, farm sector debt, and costs of production by commodity. The farm sector remained financially strong in 1995, even though farm sector income was lower than in 1994. NOTE: This publication was formerly called *Economic Indicators of the Farm Sector*.

Foreign Agricultural Trade of the United States (FATUS)/U.S. Agricultural Trade Update. 1998 subscription includes 12 issues of Agricultural Trade Update, plus two annual FATUS supplements. Stock #ERS-FAT. $41.00. Updates the quantity and value of U.S. farm exports and imports, plus price trends. Concise articles analyze specific aspects of the export/import picture. Keeps readers abreast of how U.S. trade stacks up in a global market.

FoodReview. Subscription. 3 issues. Stock # ERS-NFR. $21.00. Featuring the latest data and analyses, FoodReview explores the rapidly changing U.S. food system. Trends in food consumption, food assistance, nutrition, food product development, food safety, and food product trade are analyzed in depth for those who manage, monitor, or depend on the food system. Also includes key indicators of the food sector and updates on Federal policies and programs affecting food.

Rural Development Perspectives. Subscription. 3 issues. Stock # ERS-RDP. $19.00. Non-technical articles on the results of new rural research and what those results mean. Shows the practical application of research in rural banking, aging, housing, the non-metro labor force, poverty, and the effect of farm policies on rural areas. Besides feature articles, each issue also brings you: Rural Indicators—geographic snapshots of trends affecting rural communities; Book Reviews—critical appraisals to keep you abreast of new thinking and theories on rural and small town topics; and Announcements—brief summaries of newly published research on rural areas.

ERS-NASS Products and Services Catalog. This free catalog describes the latest in ERS research reports. It is designed to help you keep up to date in all areas related to food, the farm, the rural economy, foreign trade, and the environment.

Following is a list of ERS *Situation and Outlook* reports and their stock numbers and prices:

Title/number of issues	Stock #	Price
Agricultural Income & Finance (4)	AIS	$27.00
Agriculture & Trade Regionals (4)	WRS	$34.00
Fruit & Tree Nuts (3)	FTS	$27.00
Outlook for U.S. Agricultural Exports (4)	AES	$24.00
Livestock, Dairy, Poultry Outlook (6)	LDP-M	$32.00
Aquaculture (2)	LDP-AQS	$21.00
Sugar & Sweetener (2)	SSS	$22.00
Tobacco (2)	TBS	$22.00
Vegetables and Specialties (3)	VGS	$27.00

This is a list of NASS Statistical Data, stock numbers, and prices:

Title/number of issues	Stock #	Price
Agricultural Chemical Usage (3)	PCU	$35.00
Agricultural Prices (12)	PAP	$61.00
Broiler Hatchery (54)	PBH	$98.00
Catfish Processing (16)	PCF	$43.00
Cattle (14)	PCT	$36.00
Chickens and Eggs (15)	PEC	$45.00
Cold Storage (14)	PCS	$45.00
Cotton Ginnings (14)	PCG	$45.00
Crop Production (17)	PCP	$61.00
Crop Progress (36)	PCR	$74.00
Dairy Products (13)	PDP	$41.00
Egg Products (12)	PEP	$36.00
Farm Labor (4)	PFL	$24.00
Grain Stocks (4)	PGS	$24.00
Hogs and Pigs (4)	PHP	$24.00
Hop Stocks (2)	PHS	$20.00
Livestock Slaughter (13)	PLS	$45.00
Milk Production (13)	PMP	$33.00
Noncitrus Fruits & Nuts (2)	PNF	$24.00
Peanut Stocks & Processing (12)	PPS	$36.00
Potatoes (7)	PPO	$29.00
Poultry Slaughter (12)	PPY	$36.00
Rice Stocks (4)	PRS	$24.00
Sheep and Goats (5)	PGG	$27.00
Turkey Hatchery (16)	PTH	$36.00
Vegetables (6)	PVG	$28.00

Also available from the World Agricultural Outlook Board (WAOB) is *World Agricultural Supply and Demand Estimates*, 12 issues for $40, stock number WASDE. This can be ordered through ERS as well.

The *ERS-NASS Products and Services Catalog* is a valuable source of information about all products and services offered by the USDA and is free of charge.

ERS Ordering Information:

By Phone: 1-800-999-6779

By Fax: (703) 321-8547

By Mail: 5285 Port Royal Road
Springfield, VA 22161

Some reports and periodicals are available on the Internet. Cornell University's Mann Library offers many USDA reports at the following site:

http://www.mannlib.cornell.edu/usda/

Other reports and periodicals as well as electronic data may be downloaded from either the ERS website:

http://www.econ.ag.gov/

the NASS website:

http://www.usda.gov/nass/

or the World Agricultural Outlook Board website:

http://www.usda.gov/oce/waob/waob.htm

Private services are, of course, available on a fee basis. The list would include Professional Farmers of America, Top Farmer, Brock and Associates, Doanes, the Helming Group, Sparks Commodities, and others. Services and costs vary. Many will advertise in

Futures Magazine
P. O. Box 850765
Braintree, MA 02185-9801
1-888-898-5514
Fax: 1-781-848-6450
($39.00 per year)

APPENDIX 3B. MODELS AND APPLICATIONS

A brief explanation of procedure and the fitted algebraic models for the various figures in this chapter are presented in this appendix. In some instances, you may find it more convenient to use the algebraic models than to use the graphs directly. An example would be the price-ending stocks relationships presented in Figures 3.7, 3.8, and 3.9. The application, use, and interpretation of the model results are shown for corn, soybeans, and wheat. This appendix will be most valuable to the reader with some prior exposure to simple statistical models. It is presented as a supplement and is not essential to the succeeding chapters for readers with no statistical background.

Figures 3.7, 3.8, and 3.9

Algebraic models were fitted to the date plotted in Figures 3.7 through 3.9. In each case, a negative relationship exists between price and ending stocks as a percent of total use for the same crop year. Such a relationship would be expected. When stocks are small relative to usage or needs, price will tend to increase to ration usage.

It is also apparent that there is considerable variability around the curves. This too would be expected. A model that expresses price as a function of a measure of ending stocks is a very simple model of a complex set of relationships. Nonetheless, the models serve their intended purpose very well and provide a useful initial estimate of price for the upcoming crop year.

The algebraic model for corn (Figure 3.7) is a simple linear model. A quadratic term, ES^2, was tried as an explanatory or independent variable, but it was not statistically significant. The final model took the following form:

$$PRICE = 2.79 - 0.0139\ (ES)$$

where:

$PRICE$ = Average price to farmers by crop year ($ per bushel), and
ES = Ending stocks as a percent of total use for the same crop year

Statistical properties of the model were

N = 23
(R^2) = .347
$F(1, 21)$ = 11.16

For 1996–97, the most recent complete crop year, ending stocks were 883 million bushels, and total use for the crop year was 8.849 billion bushels. The variable ES would therefore be 9.98. Generating a price estimate for the 1996–97 crop year, we get

$$PRICE = 2.79 - 0.0139\ (9.98) = \$2.65$$

The price is below the observed 1996–97 price of $2.71, suggesting the simple linear model is forecasting slightly too low in the most recent years. One possible adjustment, mentioned in the text, is to force the model to fit the price-stock relationship for the most recent year. Graphically, what we need is to "move" the curve so that it will go through a price of $2.71 in 1996–97.

The forecast error was $.06 ($2.71 minus 2.65). By adding the $.06 to the intercept term (the 2.79), the model is adjusted to fit the 1996–97 scenario and will be

$$PRICE = 2.85 - 0.0139\ (ES)$$

This changes the level of the model, but leaves the relationship between price and ending stocks (the –0.0139 coefficient on ES) unchanged.

Using the revised analytical model and the latest (February 1998) estimates of ending stocks and total use for the 1997–98 crop year (949 million bushels and 9.310 billion bushels), the predicted price for the 1997–98 crop year is

$$PRICE = 2.85 - 0.0139\ (10.2) = \$2.71$$

The 10.2 is, of course, the latest possible (February 11, 1998) estimate of ES when ending stocks are being estimated at 949 million bushels and the total use for the 1997–98 crop year is being estimated at a very large 9.310 billion bushels. This same adjustment process can be used for soybeans or wheat if the model is missing the most recent yearly price by a significant amount. It is demonstrated for corn, but will not be repeated for wheat or for soybeans.

At the time the February 11 USDA estimates were released, the December 1998 corn futures were near $2.85. Since the model forecast is for farm-level cash prices, the results suggest corn futures could trade slightly higher since Midwest basis levels are closer to –$.20 than the –$.14 implied by a $2.85 futures price and a $2.71 cash price. *To the corn producer trying to decide whether to forward price in the $2.60s or carry the risk and look for higher prices, this is very important information.* The producer will feel more comfortable waiting to price, or starting a modest program of pricing, with the expectation of adding more protection later at higher prices. To the user of corn interested in protection against higher prices, the analysis suggests the need for protection. To the speculator, the results suggest a strategy that enters the market from the long side on any price dip.

A caution is in order at this point. Since the model is a linear model, care should be used in applying it to levels of ES that are outside the range of *ES* in the data set used to fit the model. In other words, if stocks relative to total use move to record high or record low levels, the model should be used with care.

The model for soybeans (Figure 3.9) is also a simple linear model. The quadratic component (ES^2) was not statistically significant when included in the model. The final model was as follows:

$$PRICE = 7.53 - 0.0982\ (ES)$$

Statistical properties of the model were:

$N = 24$
$(R^2) = .370$
$F(1,22) = 12.90$

In applying this model, the same caution is needed that was discussed for the corn model. There is a real danger in applying the model to any measure of ES that is outside the range of ES used in estimating the model. It is a linear model, and does not have the quadratic component to bring curvature to the function and to block extreme estimates of PRICE for unusually small or large values of ES.

The latest estimates of ending stocks and total use for the 1997–1998 crop year are 245 million and 2.619 billion bushels, respectively. The variable ES is therefore 9.35.

$$\begin{aligned} PRICE &= 7.53 - 0.0982\ (9.35) \\ &= \$6.61 \end{aligned}$$

During early 1998, the November soybean futures traded in a range of $6.40 to $6.85, with the more recent observations in the $6.50 area. Since harvest period basis levels in most producing areas would be –$.30 to –$.50, the futures market is pricing 1998 soybeans below the model prediction. The hedger or speculator should expect price rallies unless some new and negative information shock hits the market. Strategies should reflect the likelihood that November 1998 soybeans will not exceed $7.00 given the model prediction, and short hedges should be placed aggressively on any price rallies toward $6.75 to $7.00.

In wheat, the model does not show a statistically significant quadratic term and is a linear model as shown by the plot in Figure 3.8. The model is:

$$PRICE = 3.96 - 0.0141\ (ES)$$

Statistical properties of the model were:

$$\begin{aligned} N &= 24 \\ (R^2) &= .269 \\ F(2,22) &= 8.11 \end{aligned}$$

The latest estimates of ending stocks and total use for the 1997–98 crop year (which ends May 31, 1998) are 581 million and 2.386 billion bushels respectively. The variable ES is therefore 24.4. The estimated price would be:

$$\begin{aligned} PRICE &= 3.96 - 0.0141\ (24.4) \\ &= \$3.62 \end{aligned}$$

The $3.62 estimate is above the recent $3.35–3.65 trading levels of the July Chicago wheat futures. This demonstrates the problems that emerge when the variables approach extreme levels. The 581 million bushels in ending stocks is the smallest in an historical context.

In this instance, the user draws two conclusions that help in developing perspective:

1. Wheat prices are likely to be volatile due to the small buffer stocks and the market is vulnerable to any surprise in the weather or surprise development in the export arena.
2. If there is a reason to expect the futures market to move from current levels, the expectation would be for an increase given the model prediction for cash prices.

This provides useful information for the producer. If the cash-futures basis is weak during the upcoming June–July harvest, holding wheat in storage as a cash market speculator has a better chance of being profitable than in normal years. Conversely, if basis is more favorable (as it could be, given the relatively small stocks) the producer should pursue a basis contract, a deferred pricing plan, or even sell cash and buy futures with an expectation for upward trending prices. The speculator in wheat futures should be looking for opportunities to buy or go long on price dips.

Figure 3.10

The model for Figure 3.10, showing the relationship between beef production and January 1 inventories, is

$$BEEFPR = 11.49 + 0.0952 \text{ INV}$$

where

$BEEFPR$ = beef production during the calendar year in billions of pounds, and
INV = January 1 total cattle inventories in millions of head.

An inventory of 100 million head would generate, to illustrate,

$$11.49 + 0.0952 = 21.01 \text{ billion pounds.}$$

This estimate would be too low given production levels of recent years, but application of the relationship is informative. In recent years, with a relatively high percentage of the inventory in feedlots and in the presence of increased production per head of inventory due to technological advancements in breeding, the historical relationship is no longer totally representative. For a given herd size, we can expect even more production during the year than has historically been the case. The R^2 for the model was only .092, indicating that inventories explain only 9 percent of the variation in beef production across the years shown.

Figure 3.11

Figure 3.11 shows a similar plot for pork production relative to December 1 (for previous year) hog inventories. As was the case with beef, it is clear that the production per sow is increasing in recent years. The actual observations are well above the fitted line in recent years. The equation employed to plot the fitted line was of the same general form as the equation used for beef in Figure 3.10 and took the form

$$PORKPR = 6.57 + 0.1294 \text{ (INV)}.$$

Here, the R^2 was only 0.06, indicating many other factors are influencing pork production.

Figure 3.12

Examination of Figure 3.12 suggests increasing production per head in recent years *and* suggests that the January 1 cattle-on-feed numbers are not effective predictors of beef production during the year. Not only is the relationship negative, but the fit is not very effective. The (R^2) for the equation underlying the line shown in the figure is only 0.003, and the equation is not provided due to the poor statistical properties. More detailed and intrayear analyses will clearly be required to explain the variations in beef production in recent years. At a simplistic and beginning level, the relationship shown earlier between January 1 inventory numbers and beef production would be much more effective. What this relationship *does* reveal is how much production can change due to changes in cattle-on-feed numbers *within the year*. It thus supports the need for effective intrayear monitoring of placement patterns in the feedlots as a major determinant of changes in beef production.

CHAPTER 4

TECHNICAL ANALYSIS: THE BAR CHART

INTRODUCTION

The fundamental approach to analysis of the markets was developed in Chapter 3. In this chapter, the technical approach is examined. The fundamental approach establishes the probable direction of price trend and the general price range within which the markets are expected to trade. *The technical dimension guides the timing of pricing decisions within that price range.* To repeat an earlier observation, both approaches are important and are complementary.

Chapter 4 will deal extensively with the bar chart, the most widely used means of monitoring and analyzing price movement in the futures markets. Alternatives to the bar chart are presented in Chapter 5.

TECHNICAL ANALYSIS IN PERSPECTIVE

To gain a full appreciation of the importance of technical analysis, consider the position of a farmer who is considering hedging a growing crop. Costs of production have been carefully budgeted, and the futures market has traded up to a level that will allow a reasonable profit. The farmer acts and sells futures to hedge a significant percentage of projected production.

There is nothing wrong with locking in a price that returns a profit, via a short hedge, but there can be significant *opportunity costs* if the futures market continues to trade higher. The producer is denied the benefits of any still higher prices that develop. By comparison, consider the position of another farmer who monitors the forward prices being offered by the futures market and recognizes that a profit is being offered. But the second farmer can "read" the futures price charts and sees that the market is in a strong uptrend—with still higher prices likely to come. Selling futures in this situation is likely to bring significant margin calls, and with it comes the loss of an opportunity to set prices at still higher levels. Herein lies the much-discussed opportunity cost to the hedger who pays no attention to the technical dimensions of the market. *Prices may be pegged at levels that turn out to be much lower than might have been possible because there is no recognition that prices are trending higher within a price range that is still consistent with the underlying supply–demand fundamentals.*

There is a parallel that might be even more important. Consider the situation in which the pricing opportunities offered by the futures markets are never high enough to cover costs. The producer who waits to price on a "target price" basis, with the target above budgeted costs, may be caught without any price protection if the markets fail to reach the target level and then move sharply lower. The producer who monitors the technical dimensions of the markets sees an important sell signal develop on the charts and moves aggressively to get protection to prevent financially ruinous losses even though the price established by the hedge is below costs.

The technical aspects of the market are much like a road map. They can give direction, and they can keep the decision maker from getting hurt by a lack of perspective in terms of where the market has been and where it might be going. Coverage in this chapter is dedicated to providing the potential user of the futures markets with a "road map" and contributing to the much-needed perspective in terms of where the markets have been and where they are likely to go. The techniques are simple and easy to use and offer tremendous potential for the serious student of the markets. They are not a panacea, but they *will* help in efforts to effectively manage exposure to price risk.

Technical analysis brings an additional set of tools to the decision maker. Disciplined use of the technical dimensions of the market can help the producer avoid the frustrations of margin calls and the significant opportunity costs in a pricing program and can provide a "safety net" against potentially ruinous price moves. The ability to analyze the charts is important to decisions on whether protection should be established and to the correct timing of pricing actions.

THE BAR CHART

There are numerous ways to record, in a technical context, the actions of the futures markets. The most widely used is the *bar chart.* Figure 4.1 illustrates.

On the vertical axis, the bar chart shows a price scale. On the horizontal axis, the chart shows a time calendar typically spanning six months or longer. The chart shows the price action for each trading day. The vertical bar shows the trading range (the high price to the low price) and the horizontal "tic" shows the closing or settlement price. The charts can be updated daily. Most major newspapers show the high, low, and closing price for the important agricultural commodities and financial instruments. A number of national-level firms offer electronic services via FM or satellite transmission that bring the price information directly to the user and offer automatic updating of the bar charts.

A number of widely used bar chart patterns can be used to interpret and predict the price action. Most of the techniques had their origin in the stock markets and have been adapted to the commodity futures markets. As the discussion of the various charting patterns develops, references will be made to the reliability of the various

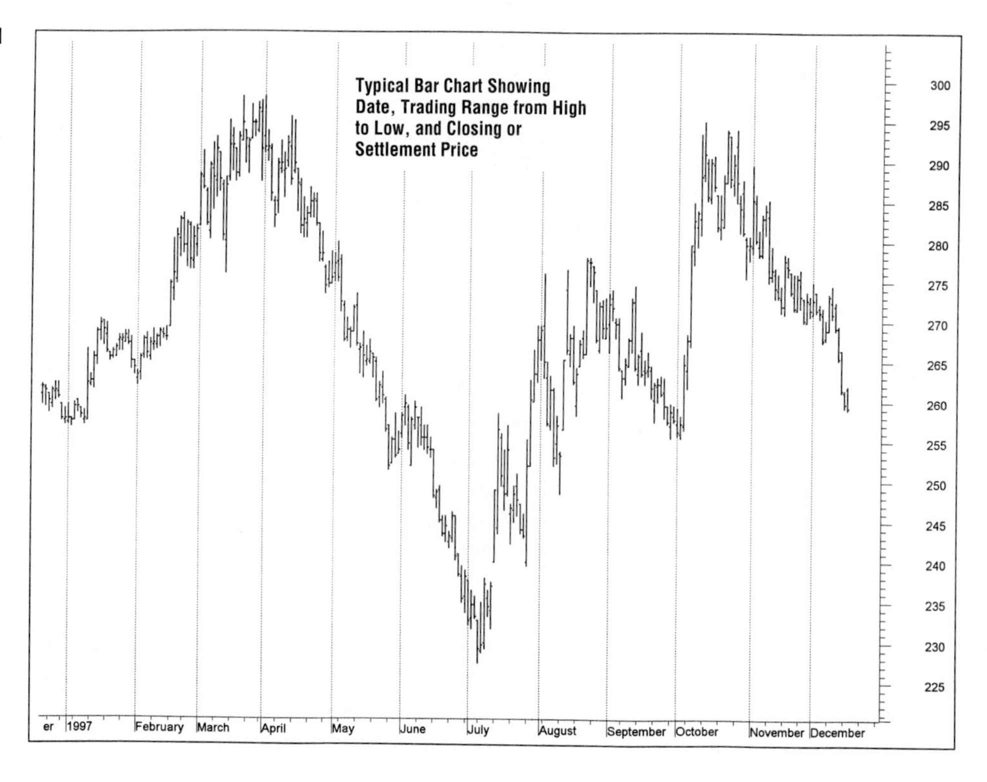

FIGURE 4.1
A Typical Bar Chart Format for a Commodity Futures Contract

"signals" being generated. Specific levels of reliability are difficult to document. There have been attempts by some analysts and researchers to go back through historical charts and generate frequency counts on the number of times a major price move down, for example, followed a "sell signal" that is identified on the charts. In this chapter, calling a particular chart formation a reliable signal or offering more specific indications of reliability is based on awareness of efforts of other analysts and on 25 years of working with and observing the markets.

Trend Lines

Perhaps the most important single tool in technical analysis of the price charts is the simple trend line. Figure 4.2 illustrates with a sketch for an upward-trending market.

When the underlying supply–demand balance is strong enough to generate higher prices, the chart will show an uptrend. In the context of technical analysis, this can be shown by connecting two or more lows in the daily price ranges. Preferably, these lows will be 10 or more trading days apart and the trend line will not be extremely steep.[1] As long as the supply–demand situation is interpreted as sufficiently strong to justify higher prices, the uptrend will remain intact. It is when the

[1]On most commercial charts, the line should not be much steeper than 45 degrees. Obviously, this is a function of the price scale on the chart. The idea is to avoid trying to use trend lines on the short-term, day-to-day variability that characterizes the commodity markets and to isolate the major or more nearly long-term trends that have developed or are developing.

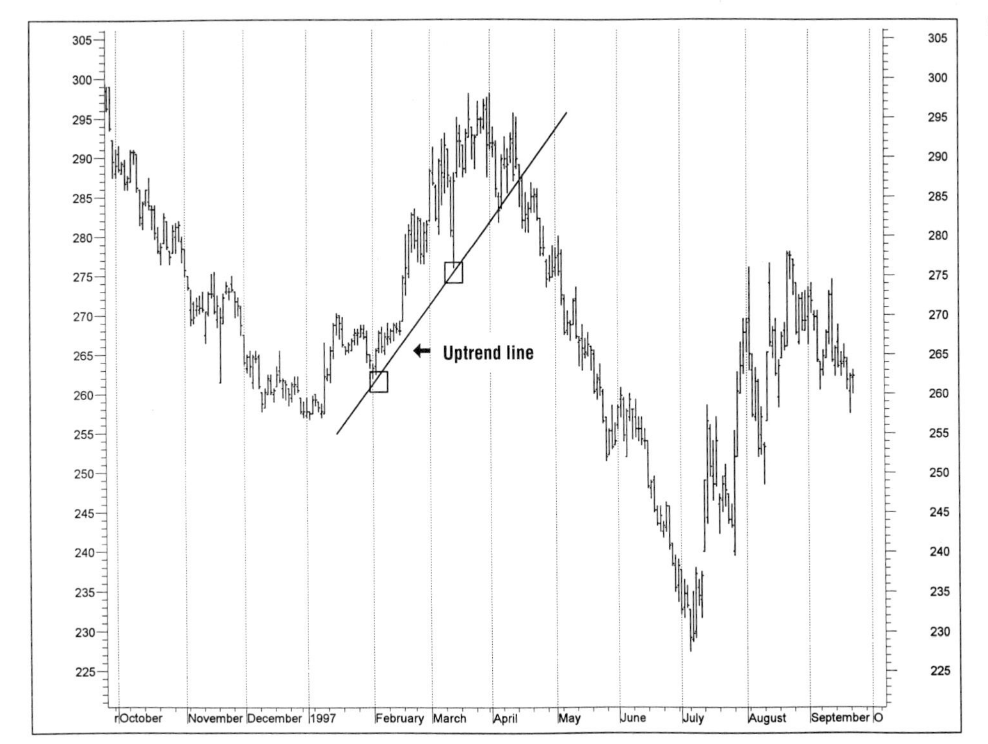

FIGURE 4.2
Illustration of an Uptrend Line Drawn across Two Daily Price Lows

market falters and a close below the trend line is observed that the chart patterns are signaling a change in price direction. The technical pattern is registering an emerging consensus of the supply–demand balance that higher prices can no longer be justified at this point in time.

The sell signal[2] generated by the close below the trend line becomes, to an extent, a self-fulfilling prophecy. All technical analysts, hedgers and speculators, see the same sell signal and are inclined to establish short positions in the market. The concerted action tends to turn the price direction down, at least temporarily.

This does *not* mean that the technicians who trade, based on chart patterns, dominate the markets. *The close below the uptrend line occurs because the consensus of traders' interpretations of the supply–demand balance, based on the available information, is that still higher prices cannot be justified.* If the supply–demand balance is strong enough to justify higher prices, a sell signal generated by the close below the trend line will be ignored and higher prices will emerge. But that scenario is unusual. *An emerging consensus that prices will not move higher is what gener-*

[2]Exactly how and when the market should be sold in response to the sell signal will be covered in detail later in the chapter. There are specific types of orders that can be placed with the broker to sell or "go short" when the daily price range penetrates the trend line or when the close or settlement price is below the trend line. The various orders will be explained as the discussion progresses. The need here is simply to recognize that the close below the trend line is signaling a reversal of the price trend. But it is time for you to start getting familiar with the types of orders explained and demonstrated in Appendix 4A. Refer back to this appendix if needed as the discussions in this chapter progress.

FIGURE 4.3
Illustration of a Downtrend Line Drawn across Two Daily Price Highs

ates the close below the trend line. Prices are therefore likely to move lower in search of a more nearly correct market-clearing or equilibrium price given the available information base.

A parallel exists in downward-trending markets. Downtrend lines are formed by connecting two highs, and a close above the trend line is widely seen as a buy signal. The consensus of the traders is that still lower prices are not justified and the direction of the price trend turns to positive and prices start to move up. Figure 4.3 demonstrates a downtrend line. *The close above the line is a widely recognized buy signal.*

How the trend lines are employed in a market plan depends on whether the hedger is a selective hedger or a conservative hedger. In earlier chapters, the two approaches were introduced and discussed. Selective hedgers will opt to *select* the periods in which they are hedged. A break of an uptrend line will signal the placing of a short hedge. Later, if a downtrend line is broken and a buy signal is generated, the short hedge will be lifted. *Producers of a crop or livestock or holders of inventories are thus selecting periods when they want price protection versus periods when they are willing to be cash market speculators.* They are attempting to *manage* their exposure to cash-price risk.[3] For the long

[3]The preference for such a strategy will vary across producers. Some will seek the occasional high cash prices that accrue to cash market speculators because they are risk takers and are motivated to go after those high prices. Such producers may never use futures or cash contracts to establish prices. At the other extreme, some producers will prefer to establish the hedge and keep it in place because they are not interested in trying to manage selective hedging programs. The selective hedger falls between the two extremes.

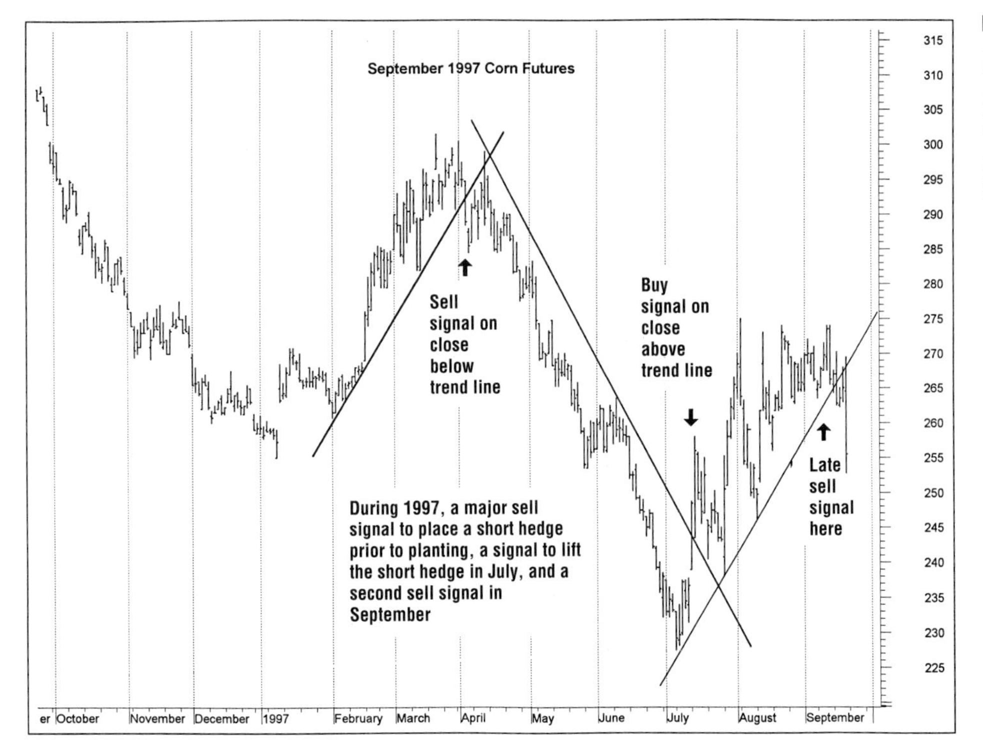

FIGURE 4.4
Illustration of Trends and Related Chart Signals in a Selective Hedging Program

hedger,[4] of course, the actions are reversed. The hedge is placed on buy signals and lifted on sell signals generated by closes that penetrate the trend lines on the charts.

Conservative hedgers, on the other hand, do not lift the short hedge by buying back the futures positions when the price trend (based on the trend lines) turns back up. If prices move back above the level at which the short hedge was placed, they answer the margin calls and keep the hedge in place.

In markets in which major trends are prevalent, the trend line can be a powerful analytical tool. Figure 4.4 demonstrates, using a demonstrative chart pattern that exhibits several major trends. You should study the chart pattern in Figure 4.4 and think about precisely what is occurring. During the periods when prices are moving lower, a selective hedging program will have short hedges in place. When the buy signals are generated by moves up through the downtrend lines, the short hedges are offset by buying them back. Thus, the selective hedger is either short in the market and hedged or is out of the futures market and operating as a cash market speculator. *The selective short hedger is never long in futures.* For the producers of commodities or the holder of inventories, being long in the cash commodity and buying futures would mean they are speculators in both the cash and futures markets. The risk exposure is

[4]The long hedger, remember, is looking for protection against rising prices of an input or a raw material. For example, the firm engaged in buying and crushing soybeans will be interested in protection against higher prices of soybeans, its primary raw material. The long hedge will be demonstrated later in the chapter.

FIGURE 4.5
Demonstration of a Trend Line and Sell Signal on the December 1997 Corn Chart

tremendous if the markets turn lower. The selective short hedger should never, therefore, be long in the futures markets.

Figure 4.5 demonstrates application of a trend line for corn. The December 1997 corn chart (CBOT) was showing prices trending higher in the early months of 1997. Stocks in the U.S. and at the world level were relatively tight, with the estimated ending stocks for the 1996–97 crop year in the U.S. in the 1.2-billion-bushel area. In Chapter 3, the relationship between ending stocks as a percent of use and prices was analyzed and developed. That framework and the overall supply–demand balance were suggesting higher prices, and the market moved higher.

During February and March, the market started to react to changes in the supply–demand situation. Stocks at the world level were, in fact, being pulled lower by usage that exceeded production, but stocks were not moving toward dangerously low levels. That perspective, in combination with improving crop prospects from an expected large planted acreage in the U.S., started to have its influence. The direction of price trend changed during April as a close below the trend line signaled a major switch in the consensus of traders in the corn markets.

The sell signal that occurred during April was important to both the conservative and the selective hedger. As a conservative hedger, the producer's primary need was to decide how much of the projected crop to hedge. The need to get price protection established should be clear with the change in price direction. The typical reaction by conservative hedgers is to generate a very conservative estimate of production and hedge that much corn. Since the hedges are to be held until the crop is harvested, a cautious approach is used to ensure that the volume that is hedged will not exceed actual production.[5]

The selective hedger will hedge most or all of expected production. Since the hedges will be lifted later if a buy signal is generated, there will be less concern about restricting hedge volume. In this instance, many selective hedgers would have lifted the short hedges during July. The implicit downtrend line on the chart (connect the highs in mid-April and those in early June) developed over a long time period and would be seen as legitimate by most technical analysts. If the hedges *were* lifted, the producer as a selective hedger would have replaced the short hedges in early September when a new uptrend line was penetrated and a sell signal was generated.

Not all markets show major and sustained trends and demonstrate the easily identified sell and buy signals around trend lines that could be used to guide a selective hedging program. Other technical tools must be employed to guide both the conservative and the selective hedger in short-hedge and long-hedge programs when major price trends are not apparent on the charts.

The trend line is possibly the single most important tool that can be applied to a bar chart of futures market activity. In a market characterized by sustained price trends, the sell and buy signals generated by price moves through the trend lines can be very effective guides to a price-risk management program. A close below or above a major trend line will lead to a significant price move in 70 to 80 percent of the cases, so the signals are quite reliable.

Resistance Planes

When no major trends are present, the resistance plane becomes important. Since futures markets have difficulty in going to price levels that have not been reached before, the resistance plane at life-of-contract highs can prove especially difficult for the market to penetrate. Figure 4.6 demonstrates.

As the market rallies back toward the existing contract high, it will take a significant change in consensus of the underlying supply–demand balance to generate new contract highs. Technicians and chart analysts know this, and depending on their analysis of the supply-demand balance, they will be inclined to place sell orders just below or at the old contract price high. Astute market observers and analysts may not know exactly where the orders are placed, but they will know the sell orders are there. *It is an excellent opportunity for both short hedgers and speculators to enter the market on the short side and is the type of opportunity for which disciplined hedgers and speculators will watch—and wait.* Technically oriented traders holding long positions in the market will also tend to sell the approach to the contract highs to take profits from long hedges or speculative long positions. When profit-taking actions move the "longs" to the same side of the market as the potential "shorts," the market is highly likely to fail and turn lower. If it does not, it is clear that the consen-

[5]You should reflect on this point. If the short positions in futures exceed the level of production, then there is no gain in the cash market to offset losses in futures if the markets move higher. The loss in futures is then an out-of-pocket loss rather than an opportunity loss. Concern over this tends to make producers reluctant to hedge 100 percent of expected production.

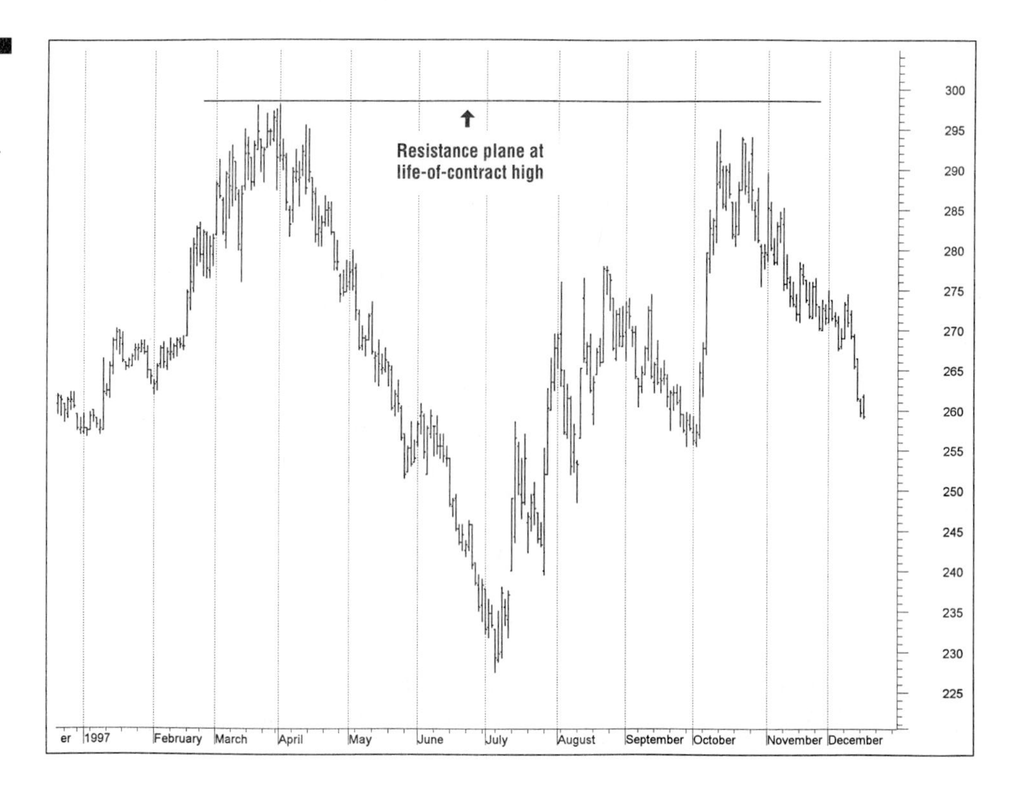

FIGURE 4.6
Illustration of a Resistance Plane at Life-of-Contract Highs

sus of traders is acknowledging some new and positive thrust in the market, and new contract highs will be reached.

In Figure 4.6, the resistance plane is drawn across the life-of-contract high. Other, less important, resistance planes can be drawn across price highs recorded below the contract highs. Figure 4.7 demonstrates where such intermediate levels of resistance occur.

Note that the futures contract demonstrated in Figure 4.7 has failed at prices well below contract highs. As soon as the upward momentum falters and two or three days of lower price highs are recorded, a resistance plane can be drawn across that recent price high. If the market drifts lower and prices periodically surge for a few days, there may be several layers of resistance recorded on the chart. *It is very important to recognize that the market will have difficulty moving up through those resistance planes.* After all, the market was unable to generate higher prices last week, for example, and failed. Why should it be able to go up through that level this week? If the supply–demand balance has not changed significantly, that recent price high could turn out to be formidable resistance.

The intermediate resistance planes become pricing objectives for both the short hedger and the speculator. There will be a cluster of sell orders at or just below those planes. Awareness of this and what is happening in the market is especially important to the producer who is looking for a chance to get price protection by placing short hedges after having missed an earlier opportunity at higher prices.

In terms of management, therefore, sell orders to place short hedges should be placed just below the first resistance plane that will be encountered if prices rally. If prices move up through that plane, additional sell orders can be placed just below the next resistance plane. Repeating this process will generate any added price pro-

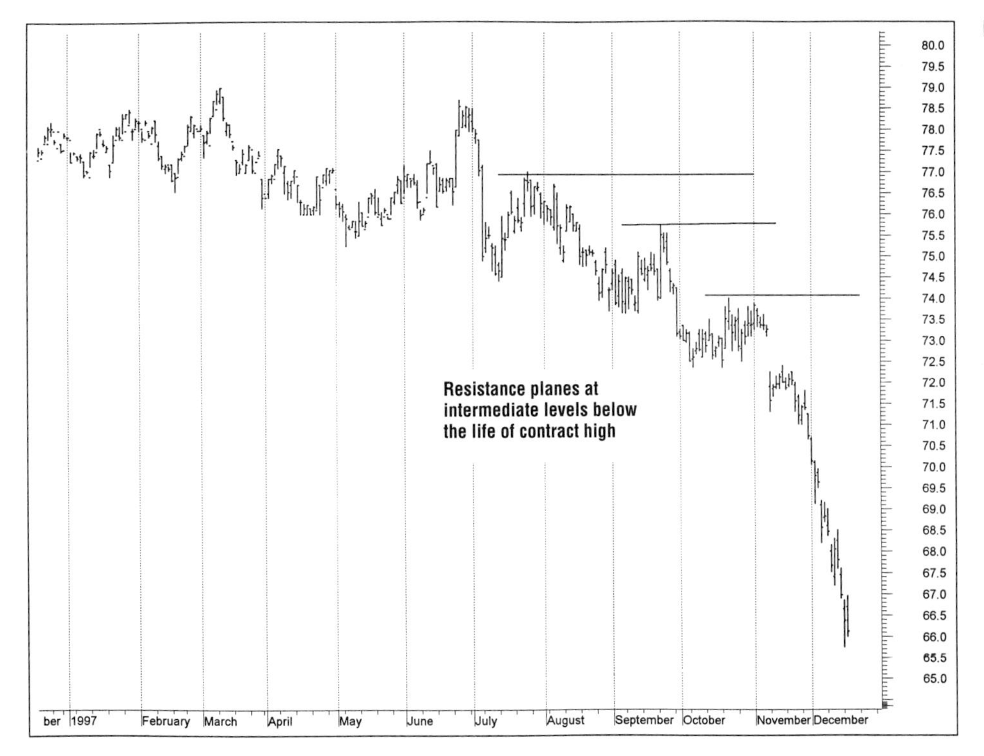

FIGURE 4.7
Illustration of Resistance Planes at Intermediate Levels Below the Life-of-Contract High

tection that is needed via short hedges on a scale-up basis as the market rallies. That is, new short positions are added on a rising price scale as the market moves higher. If the hedger fits the conservative hedger mold, this will mean answering margin calls on the initial positions, but there is nothing wrong with that approach. Actually, any producer or other decision maker who plans to price on a scale-up basis should want to see a margin call on the short hedges that were placed first. The margin call means higher prices are being offered, and that is certainly desirable.

Before proceeding, it is important to explain why the sell order should be placed *below* the resistance plane. We will come back to this point later in this chapter and in later chapters, but insight is needed here.

The potential short hedger, the potential short speculator, and the holder of long positions who is interested in taking profits want to see their sell orders filled. But there are lots of traders who want the same thing—sell orders to be filled. To increase the probability that the sell order will be reached and "filled," it is wise to drop *below* the resistance plane when placing the sell order. If there are lots of sell orders placed at and slightly below the specific price high that forms the resistance plane, the selling pressure could overwhelm the buying action—and the resistance plane might not be reached. *Dropping the sell order to prices below the plane helps ensure that the sell orders will be filled.* Suggestions on where the orders should be placed relative to the resistance plane will be included as the discussion develops for particular commodities.

The selective short hedger may opt to lift short hedges on a close above a resistance plane, viewing that penetration as tangible evidence that the underlying supply–demand balance has in fact changed and higher prices are likely. Buy orders are placed to offset the short positions in futures. When the resistance plane at life-of-contract highs is penetrated, that buy signal may be significant. It is very important that

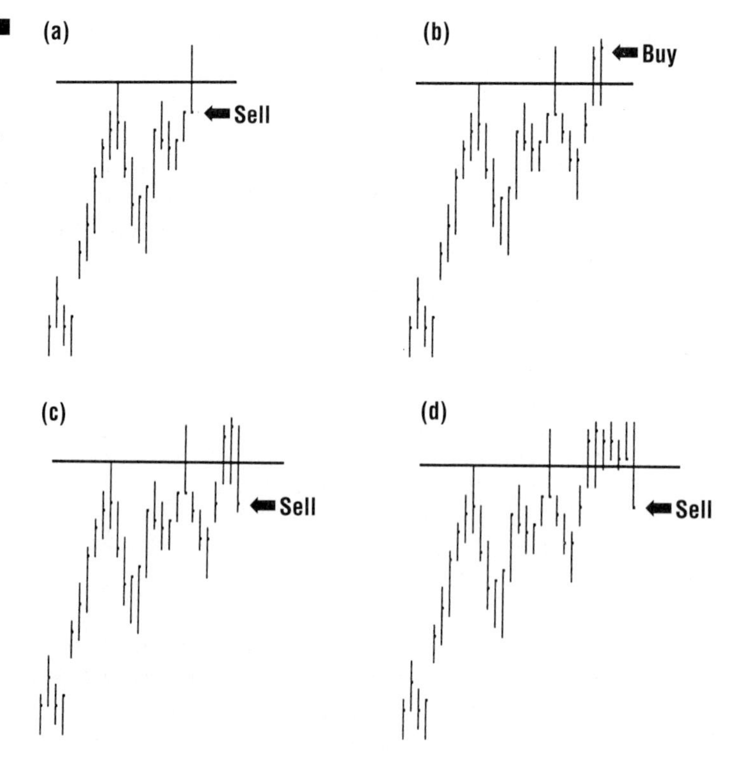

FIGURE 4.8
Demonstration of Possible Market Actions at the Life-of-Contract High Resistance Plane

the selective hedger have a plan to manage the action that occurs along the resistance plane at life-of-contract price highs, and this situation needs to be examined in detail.

Figure 4.8 demonstrates several possibilities. In panel (a), the resistance plane is penetrated, but the market does not close above the plane. The message is clear: The market has failed and prices will, in all probability, move lower. The tug-of-war on this particular day has been won by the analysts and traders who felt the supply–demand balance will no longer support higher prices. The short hedge placed on the approach to the resistance plane is likely to prove to be the correct action.

Panel (b) of Figure 4.8 demonstrates another very important possibility. The two consecutive closes above the plane suggest that the market has renewed strength, from improving demand and/or a contracting supply, and that higher prices are likely. The second of the two consecutive closes is a signal to the selective short hedger and the short speculator to lift the short positions.

This is an example of one of many moves in the markets that requires discipline. If the exchange on which the particular futures contracts are traded will accept a *buy-stop-close-only order,*[6] the hedge can be lifted or the long hedges replaced at the close on the second day. If the exchange will not accept the close-only order, the short

[6]A buy-stop-close-only order specifies a particular price level, and the buy action will be exercised if the particular futures contract *closes* above the specified price level. The order is filled just prior to the close. Among the major exchanges trading agricultural commodities, not all of the exchanges will accept and execute the buy stop-close-only order. Talk with your broker or contact the education and/or marketing departments at the exchanges to see whether the exchange accepts the order.

hedge can be lifted on the day after the second higher close is recorded by placing a buy order with the broker. Appendix 4A explains the buy-stop-close-only order and other types of orders and discusses the advantages and disadvantages of each. Each type of order will also be explained and demonstrated as this chapter develops.

In this instance, the buy-stop-close-only order is very effective. After the first close above the old price high is recorded, the close-only order can be placed early the next day, and the short hedges are lifted if a second consecutive higher close is recorded. To illustrate, assume the old price high on December cattle futures is $75.65. The market moves higher and closes at $75.90. A buy-stop-close- only order can then be placed as follows:

Buy 1 December live cattle at $75.65 stop-close only.

If a second close above the old high at $75.65 is recorded, the short hedges will be lifted near the close in terms of time and at or near the closing price.

At some exchanges, the close-only orders are not accepted. Brokers in the trading pits do not like the order because they have to make a decision in the last minutes of trade whether the market close will trigger the order and then try to get fills. The order puts pressure on the broker, and they would prefer not to make those final-minute decisions. In such circumstances, the producer's broker could be instructed to enter the market near the close of trade, if a second consecutive closing price at new higher prices looms imminent, and offset the short hedges in that way. Alternatively, as suggested earlier, the short hedges could be lifted the day *after* the second consecutive close above the old life-of-contract high price is recorded.

This type of action and related decisions are important. In a drought year, the new-crop December corn futures can blast through the old contract and trade up by $1 to $2 per bushel. In 1995 and 1996, a 1995 corn crop well below needs pushed corn prices up $3 a bushel as the marketplace rationed usage to keep us from "running out" of corn. More extreme possibilities are present in November soybeans if drought starts to threaten the soybean crop. *Such years provide extreme illustrations of what can happen to the conservative hedger who places a short hedge and intends to just answer the margin calls if the market trades higher.* In this type of market, many short hedgers will be forced out when the credit line for margin calls is not adequate to keep the hedge in place. If this forced, margin-related liquidation occurs near the price highs for the year, the result can be ruinous for the producer—especially if the market subsequently turns lower. Thus, the willingness to buy back the short hedges when the pattern in panel (b) develops can be very important.

Panels (c) and (d) illustrate added possibilities. In panel (c), the market fails on the day after the second close at new higher prices has generated a buy signal and is ready to turn lower again. This is unlikely, but can happen. In panel (d), several days are recorded before a decisive close back below the resistance plane across the old price high is recorded. In both cases, the market failures should be a signal for the selective hedger to consider replacing the short hedge. *The price move into new contract highs is not being sustained and the market action is suggesting that lower prices are likely.*

The critic might argue that all this is too much trading. But the objective is to be positioned correctly, to be either hedged or a cash market speculator, when a consensus emerges and the market moves in one direction or the other. The key to success in these instances is to have a marketing plan and know what you are going to do for the alternative scenarios the market can present—and to follow through with

FIGURE 4.9
Market Action around the Contract High on October 1997 Soybean Meal Futures

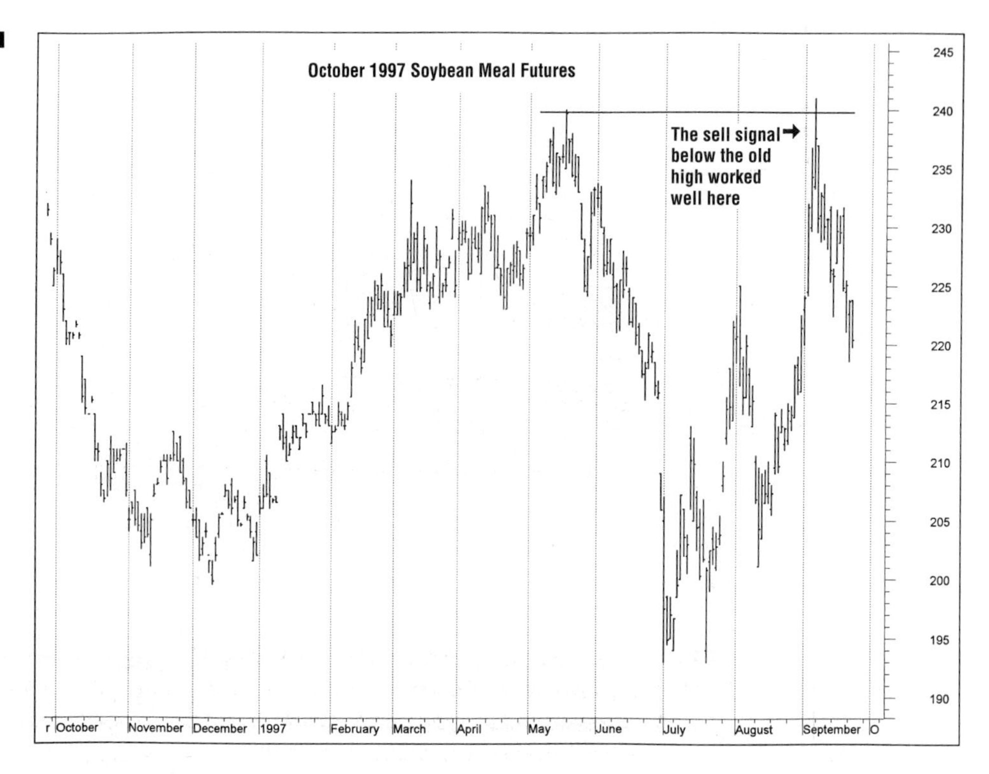

discipline.[7] An obvious alternative that will be preferred by some users is to adopt a conservative approach to hedging and get prepared to answer any margin calls.

The key point being made here is worth repeating for emphasis. *Unless some dramatic development in the supply–demand balance is emerging (a sustained drought, a government program to subsidize exports is announced, etc.), the odds favor a significant price decline after an approach to a life-of-contract high.* If the contract high is toward the upper end of the price range projected from fundamental analysis, the chances the life-of-contract high will "turn" prices significantly lower should be in the 70–80 percent range. Thus, the resistance plane at life-of-contract highs is an excellent place at which to place short hedges.

Figure 4.9 demonstrates the resistance plane at contract highs, and also shows how use of the markets can at times be a bit frustrating. The October 1997 soybean meal futures started a price surge in July and early August. The market had surged to $240 back in mid-May. That high was a new contract high, and the analyst for a company holding meal in inventory decided to sell a rally back toward that important resistance plane and placed an order to sell at $239, just below the high. Inventories had

[7]A marketing plan should involve the producer, the banker, and the broker. The objective of the plan should be worked out and provisions made for the actions that will be taken and who must approve those actions if, for example, the market starts to make new contract highs. There should also be a commitment by the bank to a credit line to answer margin calls. The role of the broker is one of understanding what is being done and executing the orders. Many banks and many brokerage firms have a standard plan or agreement that can be modified.

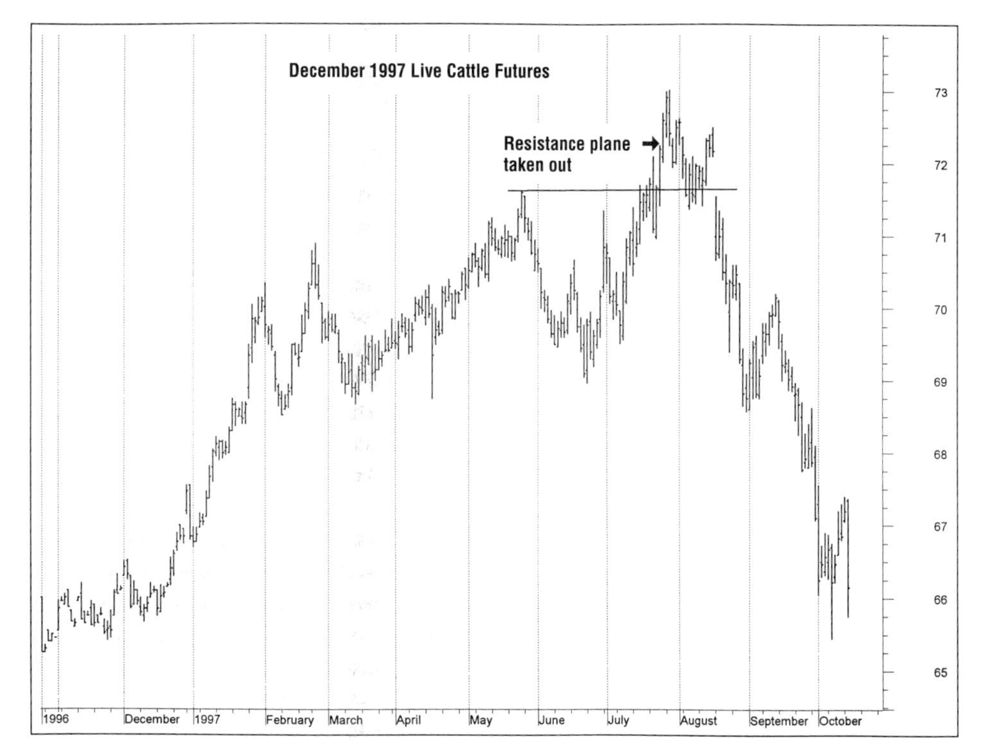

FIGURE 4.10
A Price Rally through Resistance on the December 1997 Live Cattle Futures

started to accumulate as the 1997 harvest for soybeans approached, and a top management strategy meeting on September 1 generated a decision to hedge the now accumulating inventories. The order to sell at $239 was filled on September 5, but the market pushed up to a new high of $241. Looking back, the decision looks good, but there were some nervous moments on September 5 as the responsible analyst watched the market and its close for the day.[8]

Many analysts will ignore closes essentially equal to the old high and wait to see if the markets can show a close significantly above that old price. In this case, those higher closes never developed, and the run to the new high at $241 was not sustained.

Figure 4.10 demonstrates a situation in which the old contract high did not stop a price advance. The December 1997 live cattle futures rallied in May to $71.62. The market then dipped to the $69.00 areas before starting a rally that eventually moved through the $71.62 high and up to $73.02 in late July. But those $73.00 prices were not sustained. Placements of cattle into feedlots during August ran above prior expectations and started to raise the possibility of supply levels above expectations that prevailed in early July. During the July rally, hedgers and speculators holding short positions bought back those positions, especially when a new life-of-contract high was recorded in late July. Note the rapid increase in late July as holders of short positions joined those wanting to be long in buying the market. Then, when the supply expectations changed, the prices plunged again.

[8]You might ask "why nervous?" and that is a pertinent question. After all, this is a short hedge on inventories and if the futures had moved up, the firm was "covered" by the increased cash market value. But our analyst might well be on an incentive system that reflects how well he or she did in managing exposure to price risk—and that makes you nervous!

In markets that do not exhibit sustained price trends, the resistance plane can become an important guide to pricing action. The resistance at life-of-contract highs can be especially formidable and a price rally toward the contract high may offer an especially attractive forward-pricing opportunity for the potential short hedger. It is important that the decision maker have a plan of action and follow through with discipline. This is especially important when prices approach the life-of-contract high resistance plane and the market actions turn volatile around those highs as a consensus for the correct price direction is being sought. The markets will challenge you on these days, but it is worth paying attention to these often tremendous opportunities.

Support Planes

The horizontal plane across the lows recorded by past trading days is the mirror image of the resistance plane and is labeled a *support plane.* There can be intermediate planes, but the plane across the life-of-contract low is the most important support plane. Figure 4.11 demonstrates.

On an approach to a support plane, selective hedgers will place buy orders. In the case of the short hedge, the selective hedger will consider lifting or offsetting the hedge. The long hedger, as a selective hedger, will consider placing the hedge. And it is near the support plane that the conservative long hedger will also look at placing hedges.

FIGURE 4.11
Illustration of a Support Plane at Life-of-Contract Low

If the market drops through an intermediate support plane, the conservative long hedger should consider placing more hedges on a scale-down basis and prepare to answer margin calls if necessary. Keep in mind that the conservative hedger will not lift or offset hedges, but there is still the flexibility of placing hedges in increments. On a dip to a support plane, the conservative long hedger might cover 20 percent of raw material needs, for example. If the market moves through the plane and approaches a second and lower support plane, an added 30 percent of needs might be covered via long hedges. Such an approach will usually involve margin calls, but it generates a scale-down approach that will provide a good weighted average cost of raw materials.

The selective long hedger will consider lifting the long hedge as the support plane is penetrated and the prospect for lower prices and therefore lower costs of raw materials emerges. The selective long hedger will now be interested in being a cash market speculator. On the other side of the issue, the producer or holder of inventories, acting as a selective short hedger, will consider replacing the short hedges that were lifted or offset on an approach to the support plane if the plane is penetrated and two consecutive closes below the plane are recorded.

This all sounds complicated, but it is not. You should keep in mind that the conservative hedger wants to place short hedges on a scale-up basis or place long hedges on a scale-down basis. If the markets continue to move, margin calls are answered and more futures positions are established. The support planes simply help identify the price levels at which the conservative long hedger should consider adding to futures positions.

For selective long hedgers, the support planes show the price levels around which they should consider long hedges. The support plane at life-of-contract lows will be especially important. *Management of positions around the life-of-contract low support plane is just as important as was management of positions at the life-of-contract high resistance plane.*

The mirror image of the scenarios presented in Figure 4.8 around contract highs are presented in Figure 4.12 for support planes at life-of-contract lows. To manage the selective hedge program effectively, special attention should be paid to the two consecutive closes in panel (b). Short hedges should be replaced and long hedges lifted on the second consecutive close in new, lower-price ground. The market is signaling that lower prices, often significantly lower prices, are imminent.

Panels (c) and (d) in Figure 4.12 require management much like that discussed for Figure 4.8, but in a mirror-image context. The close back above the plane, whether it occurs on the next day, panel (c) or several days later (panel [d]), is a clear signal that lower prices are not likely. The emerging consensus on the supply-demand balance blocks the need for lower prices, and a price advance is to be expected. On the closes back above the initial support plane, any short hedges that were placed (or replaced) should be lifted again and long hedges should be put back in position. You should reflect on what is being done in each panel of Figure 4.12. It is also useful at this point to start building a perception of how much discipline will be required to manage positions consistent with the actions suggested by Figure 4.12.

Figure 4.13 uses the February 1998 pork belly futures chart to demonstrate decisions around a support plane. Companies holding bellies in stock might consider buying back short hedges on the mid-August dip to the $68.72 late-June low. An approach to this $68.72 support plane would also be an excellent opportunity for firms that will need bellies for processing in early 1998 to place long hedges. Both firms would like to see prices "hold" at or above the $68.72 level, but the late-August action shows this

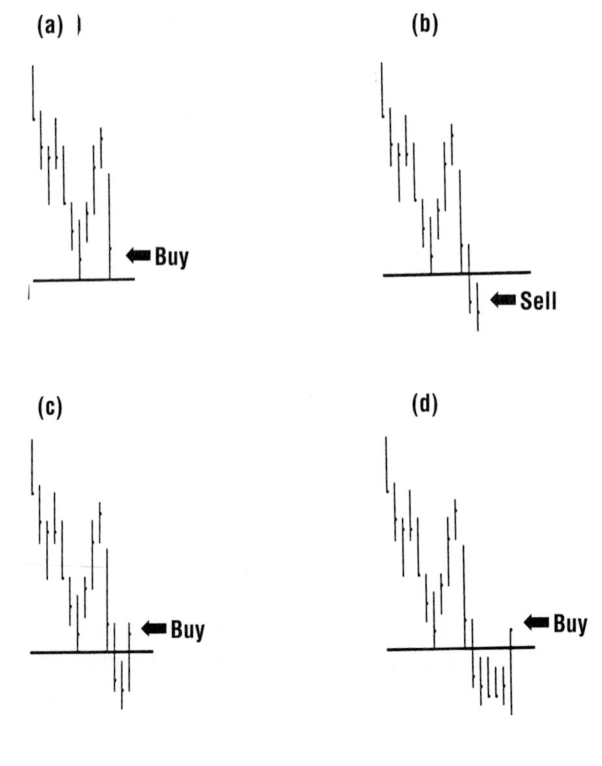

FIGURE 4.12
Demonstration of Possible Market Actions at the Life-of-Contract Low Support Plane

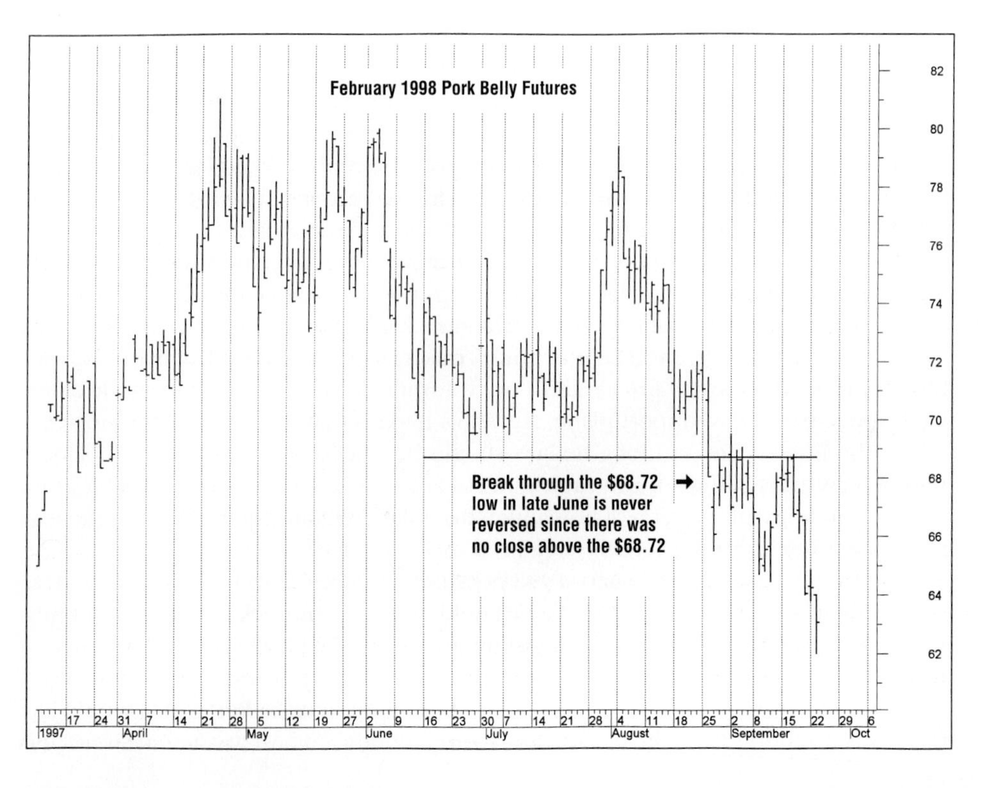

FIGURE 4.13
Demonstration of a Price Decline through Support on the February 1998 Pork Belly Futures

was not to be. Two consecutive closes below the $68.72 plane signaled lower prices, but the market gave up the price ground reluctantly. One close in early September was at $68.62, close enough to $68.72 to make the selective short hedger who had replaced short hedges to start to feel uncomfortable. But there was never a close above $68.72 and therefore no reason for the selective short hedger not to hold the replaced short hedges or for the selective long hedger to replace long hedges that were lifted on the break down through $68.72.

As suggested with the resistance planes, the support plane becomes very important when the market is not showing major and sustained trends. A moment's reflection will reveal that placing a long hedge on a dip toward the plane across a life-of-contract low has a much better chance of securing prices (and costs) near the lows for the year than does the trend-line approach. As an obvious corollary, lifting short hedges on a dip toward contract lows has an excellent chance of putting the selective short hedger in a position to benefit from being a cash market speculator in the uptrending market that often develops as the market tests the contract lows and then turns higher.

The reliability of the life-of-contract low support planes parallels those of the resistance planes. The chance that prices will move higher from contract lows is in the 70–80 percent range. Obviously, the reliability of the support plane is greater when it is near the bottom end of the price range coming out of the fundamental analysis, and the importance of fundamental analysis emerges again. The support plane will be much more likely to stop the price decreases if it is (1) low in the range of recent prices given the fundamental picture, or (2) low in a historical context. Serious market analysts who do their homework in the fundamentals and watch the charts recognize the huge opportunities around the resistance and support planes. Their actions in the markets, whether as hedgers or speculators, tend to create a self-fulfilling prophecy around those planes and make them reliable guides. Keep in mind a basic point: *The futures markets will have trouble reaching new price highs or lows unless the underlying fundamentals have changed in a major way.*

The support plane becomes very important as a guide to selective action by either the short hedger or the long hedger. As was the case with the resistance plane, disciplined action is important as prices challenge the support plane at life-of-contract lows. Proper management of the market plan in the often volatile price patterns around contract lows can result in the selective hedger being "positioned" correctly as a consensus develops and a new and major price move is begun.

Double Tops, Bottoms

When the market does as expected and turns lower near the resistance plane at contract highs, a *double top* is formed. When the market turns higher after approaching a support plane at contract lows, a *double bottom* is formed. Figure 4.14 demonstrates a double top, and Figure 4.15 shows a double bottom formed along a support plane at contract lows.

The reasoning for these two formations has already been presented. If the underlying supply–demand balance has not changed significantly in terms of the consensus of market participants, the market will have trouble going into new higher or new lower price ground. The expected failure along the resistance planes at contract highs

FIGURE 4.14
A Double Top on a Bar Chart at Contract Highs

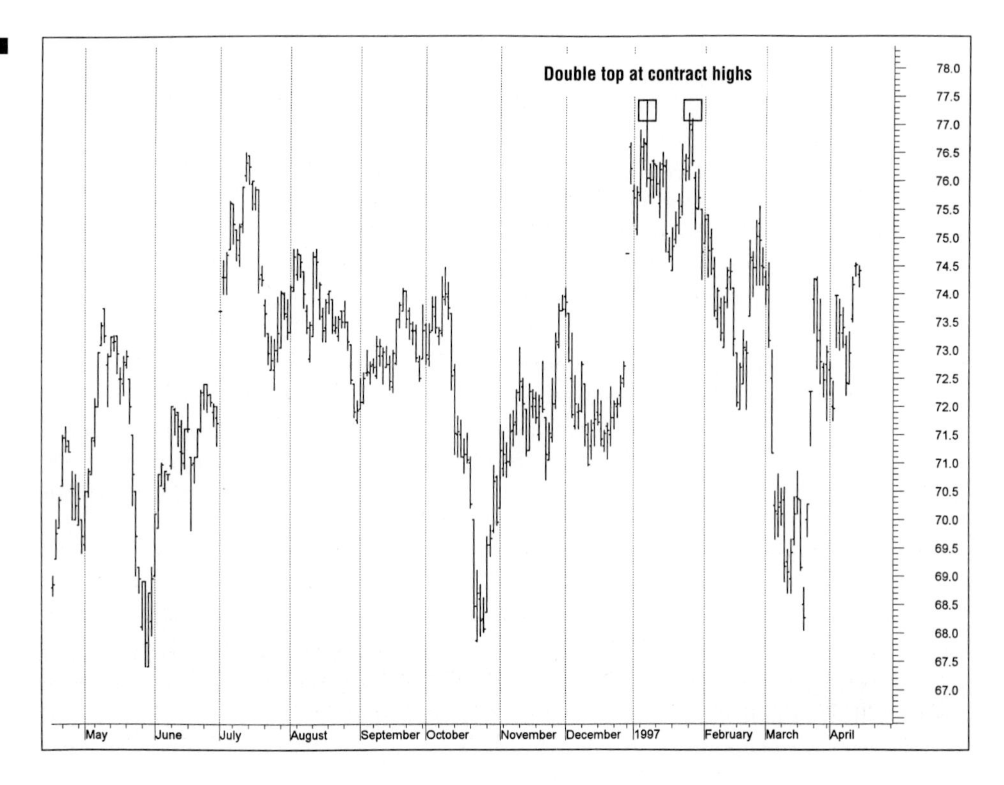

FIGURE 4.15
A Double Bottom on a Bar Chart at Contract Lows

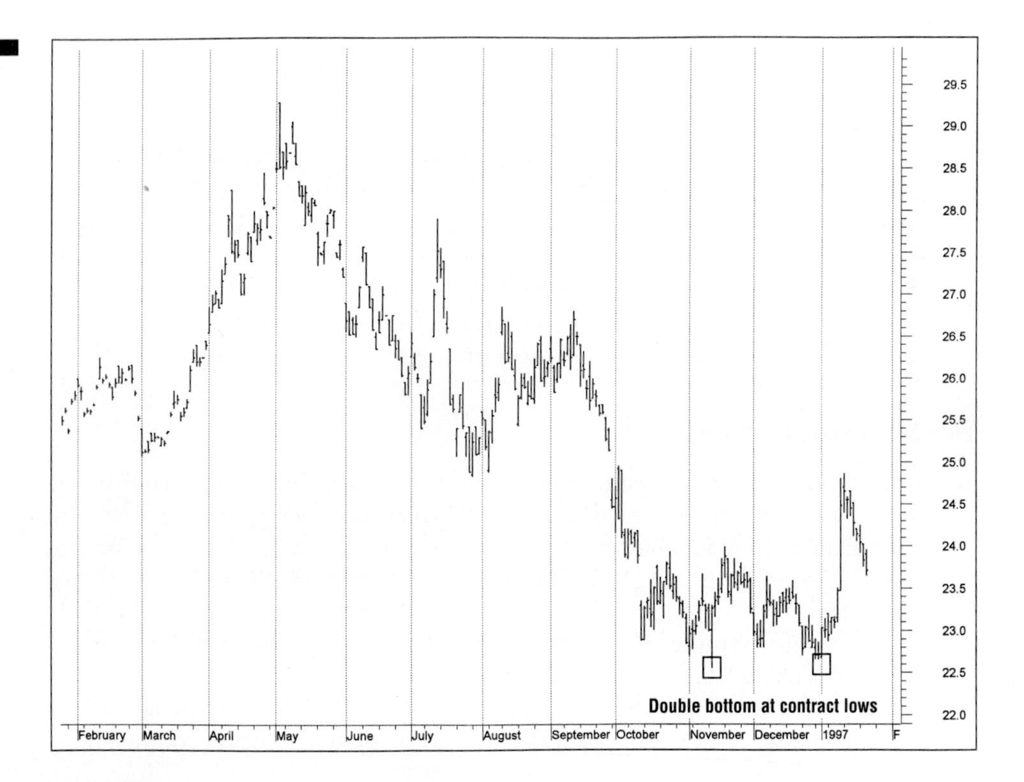

or along the support planes at contract lows creates the double tops and double bottoms, respectively.

The question of how much time is required usually emerges with these topping and bottoming formations. In some instances, the time span can be only a few days. On other occasions, there can be weeks or possibly even months between the price rallies that form a double top or the price dips that form a double bottom.

In terms of management, the sell or buy orders are placed just below the resistance plane or just above the support plane in anticipation of a double top or double bottom, respectively. When the market penetrates and closes outside the plane, the potential for a double top or double bottom is being negated, and the management strategies described for the resistance and support planes are appropriate. You should pause here and reflect again on how important it will be to have a plan—a written plan—that lays out exactly what will be done around the contract highs or contract lows.

As the potential for double tops and double bottoms starts to emerge on the charts, the question of where to place sell or buy orders again becomes very important. An approach to life-of-contract highs often presents a very attractive pricing opportunity to the short hedger, and there should be a strong desire to get a fill on any sell order that is placed. A *limit-price order,* an order that specifies a certain price level at which the potential seller is willing to go short, should be placed slightly *below* the old life-of-contract price high. As suggested earlier in the chapter, there will be a distribution of sell orders near the old price high, and the selling action associated with those orders may be enough to turn the market lower *before* the old price high is reached. *The probability of getting a fill will be greater if the market is not required to reach or exceed the old price high to fill the sell order.* This rule is generally applicable whenever the market is rallying toward a chart position that should be characterized by a cluster of sell orders. It is better to give up 2–3 cents per bushel in trying to forward-price wheat, for example, than to insist on the market matching the old high and never getting a fill on an order to sell in the presence of an excellent pricing opportunity. In the livestock futures, placing the sell order at a price \$.10–.15 per hundredweight below the contract high will help to ensure the order will be filled.

The February 1998 live cattle futures recorded a double top during July and August (Figure 4.16). The late July high at \$73.92 came at the end of a dramatic price climb. After about 14 days of drifting lower, the market surged higher to attempt a challenge of that old price high. A price of \$73.85 was reached in the next two days before the market turned sharply lower. *You can virtually see that distribution of sell orders waiting for the market to stage that last rally*, and selling pressure pushed the market quickly lower. The important point: a sell order at \$73.92 would not be filled. A sell order placed at \$73.80, just \$.12 below the \$73.92 high, would have been filled.

The double top or bottom is one of the most widely observed topping or bottoming chart patterns. These formations emerge along the life-of-contract high resistance planes or the life-of-contract low and support planes that were just discussed. To the disciplined trader, the double tops and double bottoms offer effective guidelines to the selective hedging program. Placement of orders will be important and can influence the chances of getting price protection established. In general, the orders should be placed such that the old price highs or price lows do not have to be reached or exceeded for the order to be filled.

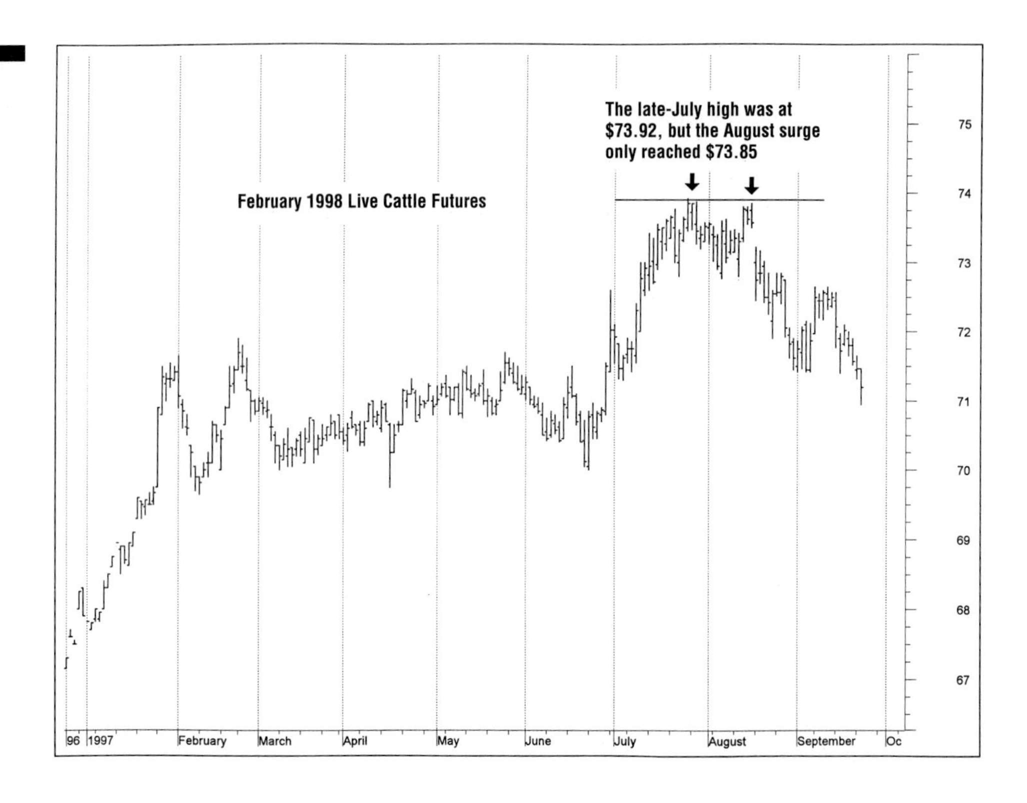

FIGURE 4.16 Double Top on the February 1998 Live Cattle Futures

Head-and-Shoulders Formations

One of the most common top or bottom formation is the *head-and-shoulders formation.* Figure 4.17 demonstrates a head-and-shoulders top. The first price rally forms a "shoulder" and then the "head" is formed as the market surges to new life-of-contract highs. The second "shoulder" is formed on a rally back toward the contract highs, but the rally fails—often at about the same level of price observed in the forming of the first shoulder.

The topping action is completed when the market closes below the neckline, a line formed by connecting the lows during the price dips on each side of the head of the formation. After the close below the neckline is observed, the minimum projected price move from the point of penetration of the neckline is the vertical distance from the top of the head to the neckline. This is the distance marked "A" on the chart, and this technique for projecting the magnitude of the price move is quite accurate and reliable. Years of observation suggest the price objective is reached or exceeded in 60–70 percent of the cases.

Traders employ different approaches to managing the placing of orders for this formation. One approach is to place a sell order, using a limit-price order, as the right shoulder starts to develop in anticipation of the completion of the entire head-and-shoulders formation. This tendency helps explain why the right shoulder often forms at roughly the same price levels as the left shoulder. A *buy-stop* order can then be placed above the top of the head to limit the risk for the selective hedger or the speculator who establishes short positions as the right shoulder develops.

The buy-stop order becomes a *market order* if reached by price thrusts from below. Since a market order is filled at the first available price, the short positions

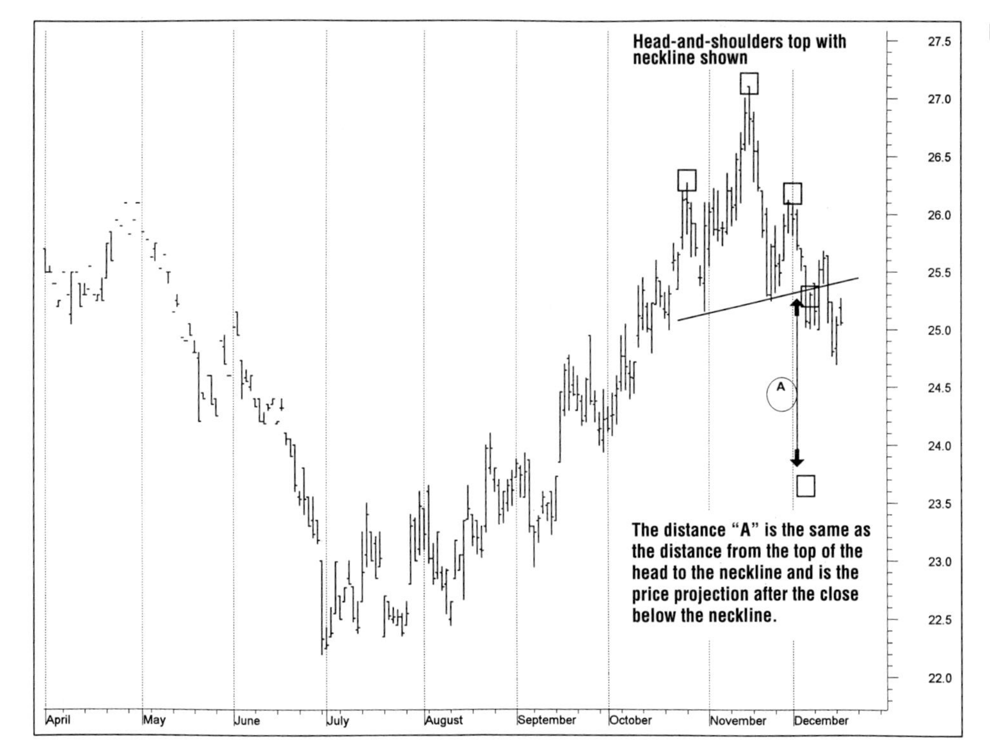

FIGURE 4.17
Illustration of a Head-and-Shoulders Top on a Bar Chart

would be offset and the exposure of the trader restricted accordingly. Since the top of the head is at the old life-of-contract high,[9] the astute trader may want to watch for two consecutive closes at new higher prices before lifting a short position, or use the buy-stop-close-only order discussed earlier. Remember, this latter order will offset the short positions only if the market *closes* in new higher price ground on the second day. You might want to review the discussion surrounding Figure 4.8 at this point.

Both selective hedgers and speculators like to sell in anticipation of a completed right "shoulder" in this formation. The life-of-contract high is typically close to where the sell order can be placed, and the risk exposure can be limited by using buy-stop orders or lifting the short positions if new contract highs do emerge. The speculator, in particular, should always look for significant profit opportunities that are characterized by the possibility of limiting the risk exposure if the trade turns out to be wrong. There is a significant opportunity associated with head-and-shoulders topping formations.

A second approach, one that is less likely to be a mistake but one that will establish short positions at lower prices, is to sell on a close below the neckline. Since this formation is so frequently seen by chart watchers, whether hedgers or speculators, there will be a widespread tendency to sell on a close below the neckline. Some deterioration in perceptions of the supply–demand balance (weaker demand,

[9]It *is* possible to see head-and-shoulders formations at intermediate price levels, of course. Technically, we might argue they are not then "tops" or "bottoms," but that argument is not productive. If the formations develop below a very old contract high, for example, it might be a very effective chart signal.

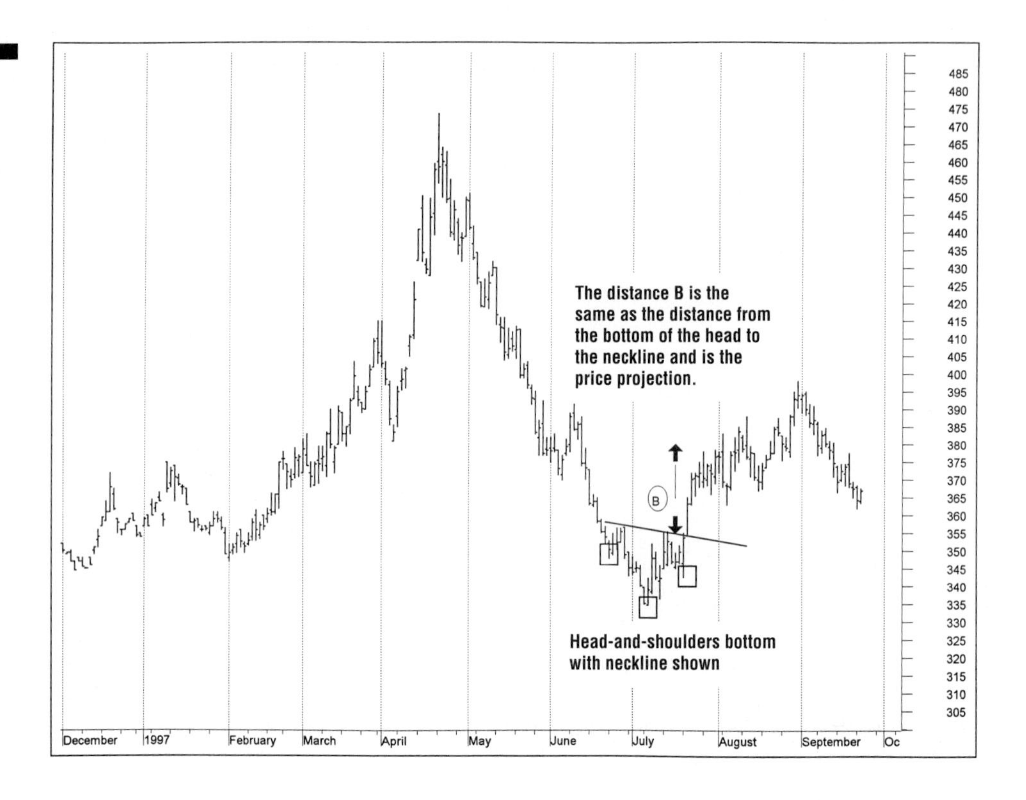

FIGURE 4.18
Illustration of a Head-and-Shoulders Bottom on a Bar Chart

increased supplies) has occurred to permit the topping formation to develop, and technicians read this message from the market by observing the pricing action on the chart. Prices will often drop quickly after the topping formation is completed. Keep in mind that traders who bought the market looking for higher prices will see the same sell signal, and they tend to offset their long positions by selling in concert with short hedgers and short speculators. The selling pressure can be intense for a few days, and sharply lower prices are likely.

Figure 4.18 demonstrates a head-and-shoulders bottom. The projection technique, after a close *above* the neckline (the distance "B" on the chart), is similar to that for the head-and-shoulders top. Management alternatives are similar. Buy orders can be placed in anticipation of the right shoulder being completed, or they can be placed to buy on the close above the neckline. A buy-stop-close-only order will work here on exchanges like the Chicago Mercantile Exchange, which accepts stop-close-only orders. The order can specify a price at or just above the neckline and it will be filled only if the market closes above the neckline, thus ensuring the head-and-shoulders bottom has been completed.

It usually requires at least 10 trading days to complete a recognizable head-and-shoulders formation, with 20–30 days being more common. On the other hand, the formation can span weeks or even months of trading activity. Figure 4.19 demonstrates a head-and-shoulders topping formation on December 1997 (Chicago) wheat that required about 15 days to be completed.

The price decline far exceeded the projection from the formation. A sell-stop order below the apparent neckline would have been filled near $4.30 in early May. The decline to $3.35 in early July brought a price move approaching $1.00 per bushel.

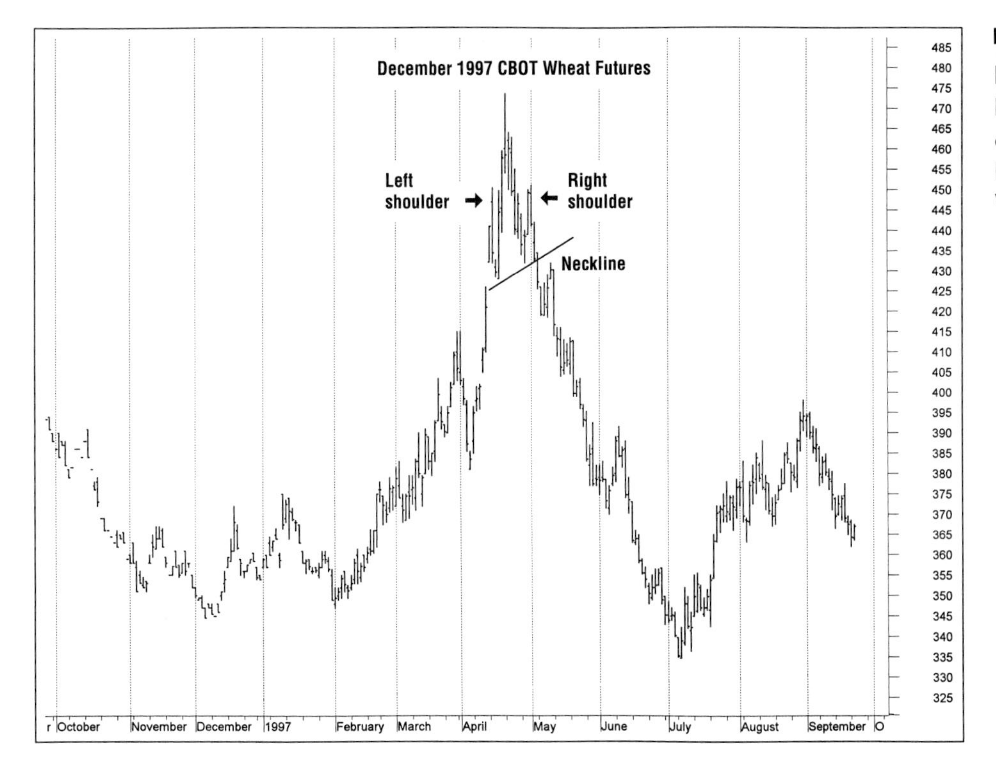

FIGURE 4.19
Illustration of a Head-and-Shoulders Top on December 1997 CBOT Wheat

The supply–demand balance was starting to show more nearly adequate supplies in the U.S. and at the world level, and the market needed to "discover" much lower prices. The brief rally in June would have pushed up through a steep downtrend line, and a more substantial buy signal occurred in mid-July.

There are two interesting features of this particular chart. First, it shows a circumstance in which short hedges or short positions placed by a sell order in anticipation of the right shoulder would have been filled at almost exactly the level reached on the left shoulder. Near $4.50, those levels were at least $.20 above the level a sell-stop order under the neckline would have realized. Note the right shoulder was exactly the same "height" as the one-day rally that formed the left shoulder.

The second interesting feature is the brief price recovery after the price dropped below the neckline. This is not unusual and is evidence of the tendency of the market to make an initial correction after a buy or sell order—a sell order here—is generated. A tendency for the market to pause and "correct" before going on a sustained price move can help the disciplined trader, whether a hedger or a speculator. If you miss the sell opportunity on the break through the neckline, the market will often give a second chance a few days later at prices almost as high.

Head-and-shoulder formations are widely observed, are reliable, and offer a price projection opportunity. Typically, the price move after the completion of the formation matches or exceeds the projected price move. Hedgers should always be alert for the emergence of the early parts of the formation and be prepared to take action.

FIGURE 4.20
Demonstration of a Key-Reversal Top on a Bar Chart

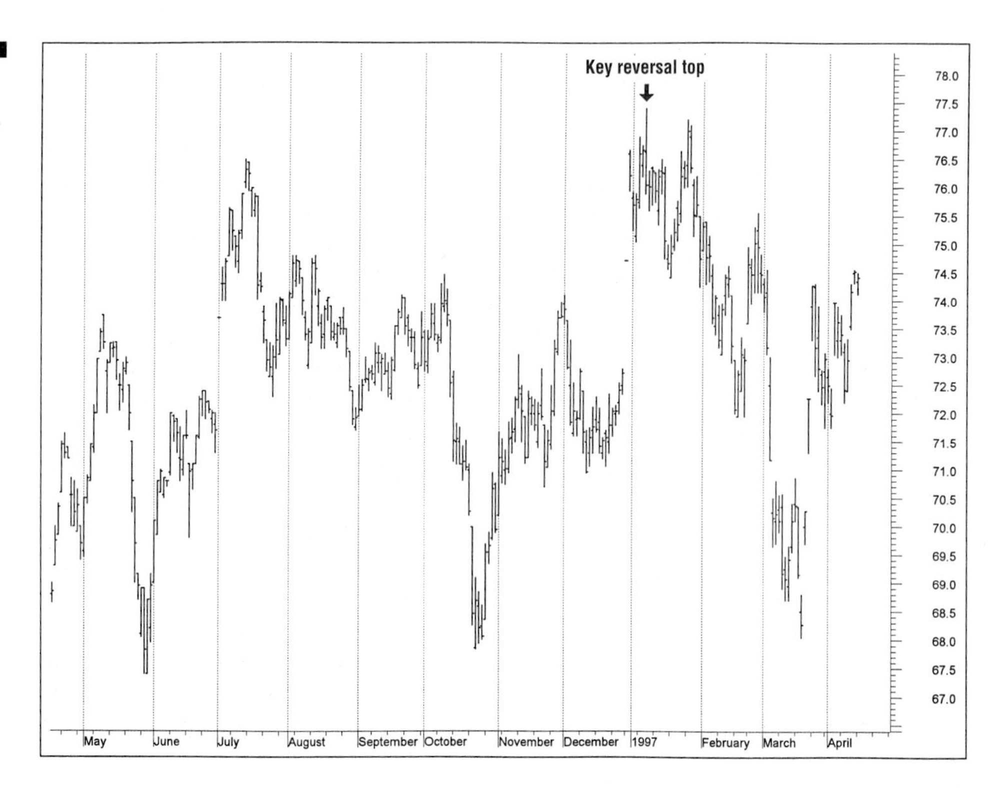

Key Reversals

A key-reversal top, for example, must be an *outside day,* must record a new life-of-contract high, and then show a lower close. The outside-day provision simply means that the trading range for the day exceeds the trading range of the previous day—a higher "high" and a lower "low" than those of the previous day. Figure 4.20 demonstrates.

The underlying psychology of the market becomes almost visible on the day a key-reversal top is recorded. The new high signifies an ability to record higher prices, but the lower close says the higher prices cannot be sustained with the emerging consensus of the supply–demand balance. The enlarged trading range required to record an outside day suggests the tug-of-war between bullish and bearish traders waged back and forth over a significant price range.

To understand the psychology of the market during the emergence of a key-reversal top, visualize the position of the bullish traders—speculators and long hedgers—who feel higher prices are likely. They buy the market and feel good when the market surges to a new life-of-contract high. But anxiety creeps in when the market falters and then closes lower for the day. On that lower close, sell-stop or sell-stop-close-only orders may offset the long position—and the bullish traders have turned sellers. If the long positions are not offset by stop orders either prior to or at the close of trade, the traders are likely to "cover" the next day by selling to offset the long positions that are now threatening to become losing trades. The psychology of the market switches—quickly.

In terms of management, several approaches are possible. Sell orders might have been placed, using a limit-price approach, under the resistance plane across an old high that would be filled on the surge to new highs. Alternatively, the short hedger or the speculator can wait until the lower close and the completed key reversal appears imminent and sell on the close or sell early the next trading day. In any event, fairly aggressive action is needed.

Key-reversal bottoms are also widely observed. The requirements here are the same, except a new life-of-contract low is required, and the close must be higher. Management procedures for the short hedger looking to lift a hedge, the long hedger interested in placing a long hedge, and the speculator are the mirror image of those described for the key-reversal top. Once again, relatively aggressive action is needed. A major price rally often follows a key-reversal bottom as new buyers enter the market and traders rush to offset short hedges or short speculative positions.

It is productive to introduce the notion of trading volume at this point. The topic will be covered later in the chapter, but most analysts would bring up trading volume when discussing key reversals. Trading volume is the number of contracts traded during the trading day. It is a measure of intensity. *High trading volume tends to confirm any signal being generated on the charts, and trading volume is watched closely on the days prior to and including a key-reversal day.*

Figure 4.21 shows a key-reversal top on October 1997 live cattle futures. Prices had surged during the midsummer months as the "grilling season" brought good movement of beef into consumption and the number of cattle ready to be sold out of feedlots recorded a seasonal decline. Note that the market, after hesitating for several days in mid-July, was able to move up through resistance at the May high just above \$69.50. A few days later, July 19 brought a surge to a new life-of-contract high at \$71.45 and a weak close of \$70.50, only slightly above the daily low of \$70.45. It was an outside day and showed all the requirements for a key-reversal top. The fact that

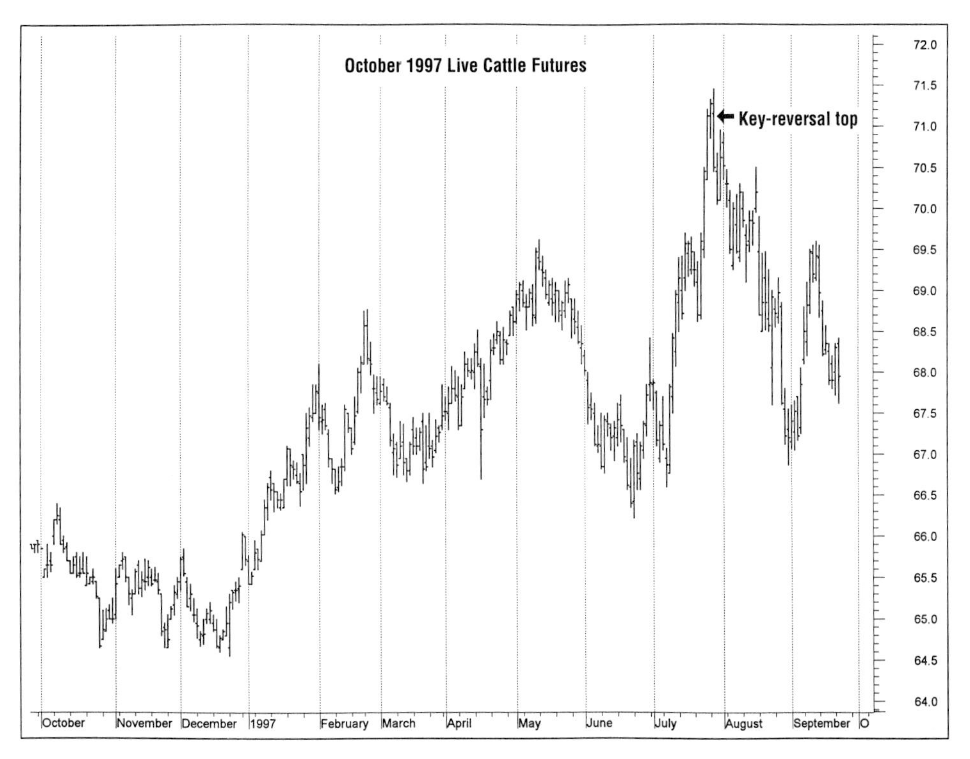

FIGURE 4.21
Key-Reversal Top on October 1997 Live Cattle Futures

the close was also below the *low* of the previous day will also please many analysts who are looking for a sell signal; some insist on this added confirmation. A check on the trading volume showed that 30,719 live cattle contracts were traded that day, a volume well above the levels for days both prior to and after this key-reversal day.

Obviously only one day is required for the key-reversal formation. Such a reversal pattern might be seen again within a few days or weeks. There is an intermediate key-reversal "bottom" on June 23 of the same cattle chart, only slightly more than a month prior to the topping action. Volume on June 23 was only 11,380 contracts, not a strong confirmation. The chances of a major price move after a key-reversal top or bottom is observed are in the 70–80 percent range, especially when the reversal is accompanied by high trading volume as was the case on July 29 with the top.

The key reversal is a reliable top or bottom formation, and is widely observed. As the key-reversal day runs its course, the psychological or behavioral dimension of the market is apparent. Price action in the days following a key-reversal top or bottom is often rapid and dramatic.

Hook Reversals

This formation is the same as the key reversal, but the new high or new low prices do not occur on an "outside day." Generally considered to be a less reliable indicator of an emerging reversal in price direction than is the key reversal, the hook reversal is nonetheless an important pattern. In general, *any* chart signal generated on an outside day (as with the key reversal) is viewed as a more reliable indicator than a signal emerging on a day that does not show a wider trading range than the previous day (as with the hook reversal). Management techniques are the same as for the key reversal. Figure 4.22 demonstrates a hook-reversal topping formation on the September 1997 soybean futures. On March 11, a new high was recorded on a thrust up to $8.03, but the market closed lower for the day. Trading volume was relatively high for the day, and increased to an unusually high level on March 12, the day after the sell signal was recorded. Prices drifted sideways to lower for several weeks before dipping below $6.20 near July 1 as the prospects for a large crop from a record U.S. soybean acreage began to emerge. Dry weather in key producing states (Illinois and Indiana) during July and August subsequently brought a price rally back above $8.00 just prior to contract expiration. (It is interesting to note that the surge in September went above the $8.03 high back on March 11 but never *closed* above that level.)

The reliability of the hook-reversal formation is in the 60–70 percent range, somewhat below that for key reversals. But the hook reversals are more widely observed and thus create more opportunities for observant hedgers and speculators.

Hook reversals are more common than key reversals, but are also generally seen to be less reliable. Management techniques are similar to those for the key-reversal top or bottom. These formations create an opportunity to sell near the highs and buy near the lows, and this is always appealing to many traders.

Island Reversals

The key to the island-reversal top or bottom is the *gaps* that appear on the chart. Later in the chapter, the chart gaps will be discussed in more detail. Here, it is sufficient to

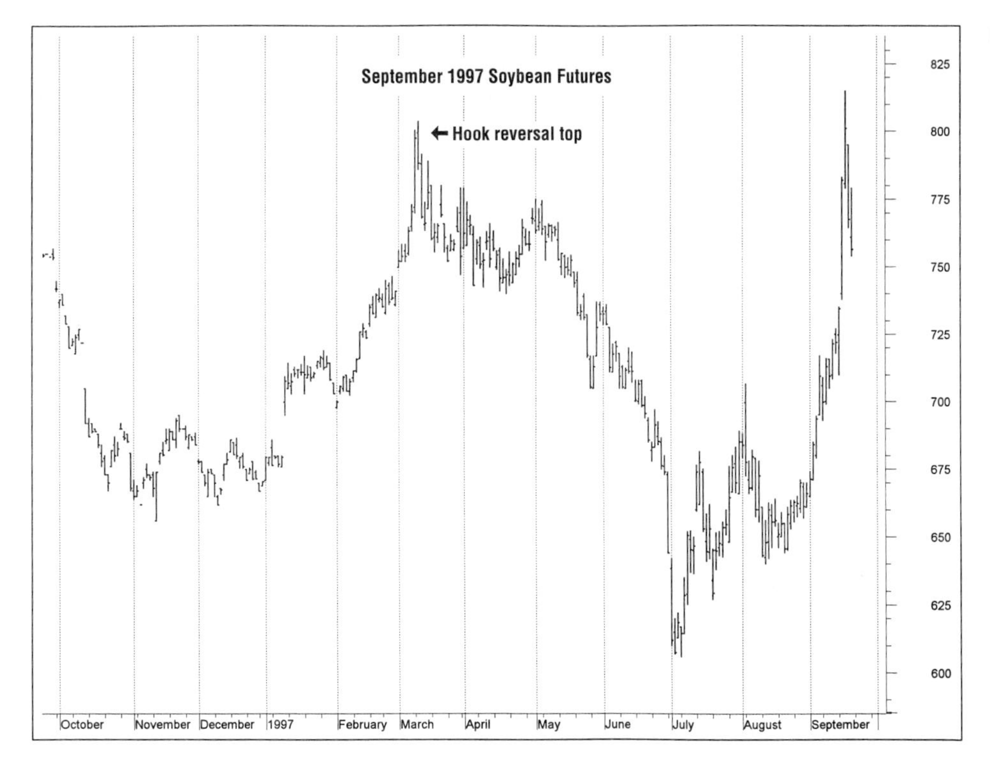

FIGURE 4.22
A Hook-Reversal Top on September 1997 Soybean Futures

note that a gap occurs in a rising market when the market is trying to go up so fast that it does not bother to trade over some price range, or in a declining market that is moving rapidly lower. Figure 4.23 demonstrates an island-reversal topping pattern.

Note the price gap as the market rallies, trades for a few days, and then records a second price gap as prices falter. The "island" of activity above the gaps can be viewed as a period in which the market is attempting to digest the available information and form a consensus on the needed price direction. In this instance, the market appears to realize with a rush that the higher prices cannot be sustained. Traders holding long positions start to offset, and hedgers and speculators looking to sell the market start to take action. The result is a significant topping pattern on the chart. In the case of an island-reversal bottom, of course, the island of activity is left behind at the temporary lower prices as the market starts a major rally.

Management of this formation is not easy. Often, it is not apparent that a top, for example, is being recorded until the second gap is left as the market drops. An effective approach is to move aggressively and sell after the island reversal is clearly present. This is especially true for the risk-averse producer or the producer who does not have the financial prowess to accept the risk of lower prices.

A second approach is to place a *limit-price sell order* just below the bottom of the second gap that leaves the island of activity. The market is likely to stage a corrective rally back up toward that chart gap before moving lower. A limit-price order at or just below the bottom of the gap is thus an alternative. An order placed below the bottom of the gap is more likely to be filled.

Once again, there is an opportunity to review the importance of where the order is placed. There *will* be a cluster of sell orders near the bottom of, or in, the chart gap as the market starts to stage any corrective rally. Technicians see this as an effective place to enter the market from the short side and will tend to place sell orders accordingly. To increase the chances of getting the limit-price sell order filled, the trader should look at placing the order just *below* the lower part of the gap. As a general rule of thumb, use at least 10 percent of the daily limit move in the commodity. That would mean 1 cent per bushel for corn (daily limit is 12 cents) and $.15 per hundredweight for cattle (daily limits are $1.50 per hundredweight), to illustrate. *Informed placement will increase significantly the probability of getting the sell order filled.*

When an island-reversal bottom emerges, the possible approaches are comparable. To lift short positions or to place long hedges, just buy the market aggressively. An alternative is a limit-price buy order just above the top of the gap that completed the formation. The market is likely to stage a corrective dip back down toward the gap before going higher. This notion of "corrections" toward chart gaps will come up again later in the chapter, and the market's tendency to correct is the key to placement of orders when island-reversal formations are involved.

Figure 4.24 demonstrates an island-reversal bottom on wheat futures. A small gap occurred in late June followed by an "island" of activity that lasted nine days before the market gapped up on July 14. Through July 21, the market corrected down into the July 14 gap, then gapped higher again on a price surge that reached to $4.04 on August 29. Again, note the chance to get a buy order filled that was placed near the top of the July 14 price gap. The market does tend to give you a second chance, but a buy order near the bottom of that price gap that completed the island-reversal bottom would *not* have been filled.

The reliability of completed island-reversal formations is in the 70–80 percent range. Often, it is recognition that a top or bottom has occurred that is important to

FIGURE 4.23
An Island-Reversal Top on a Bar Chart

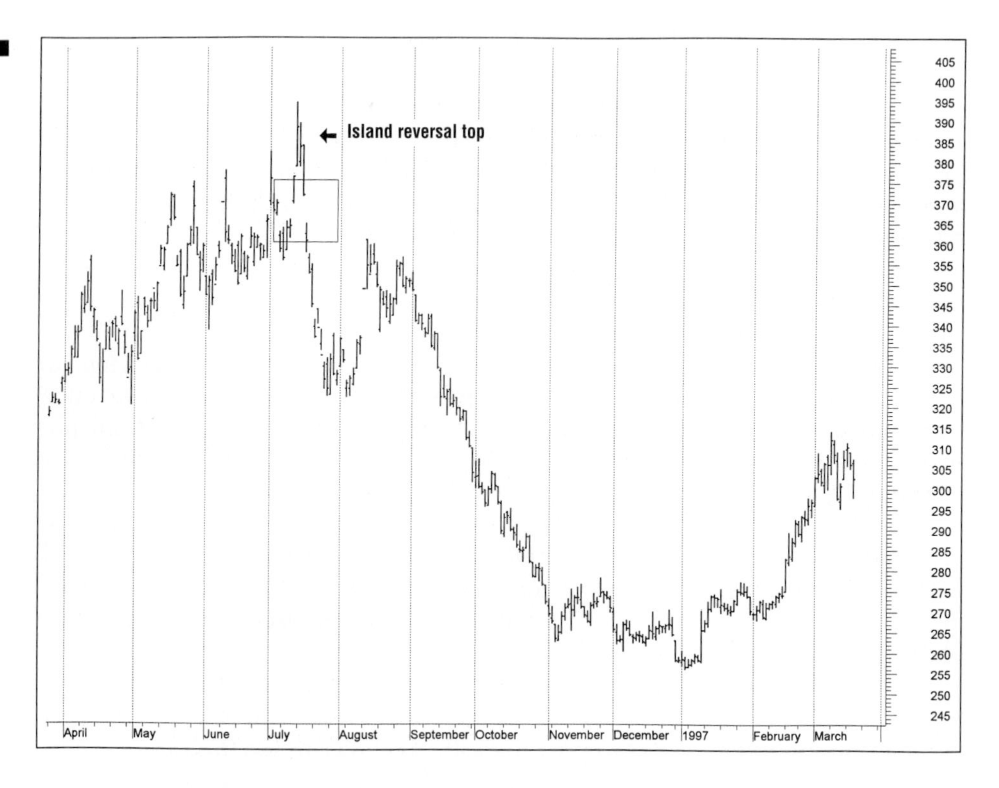

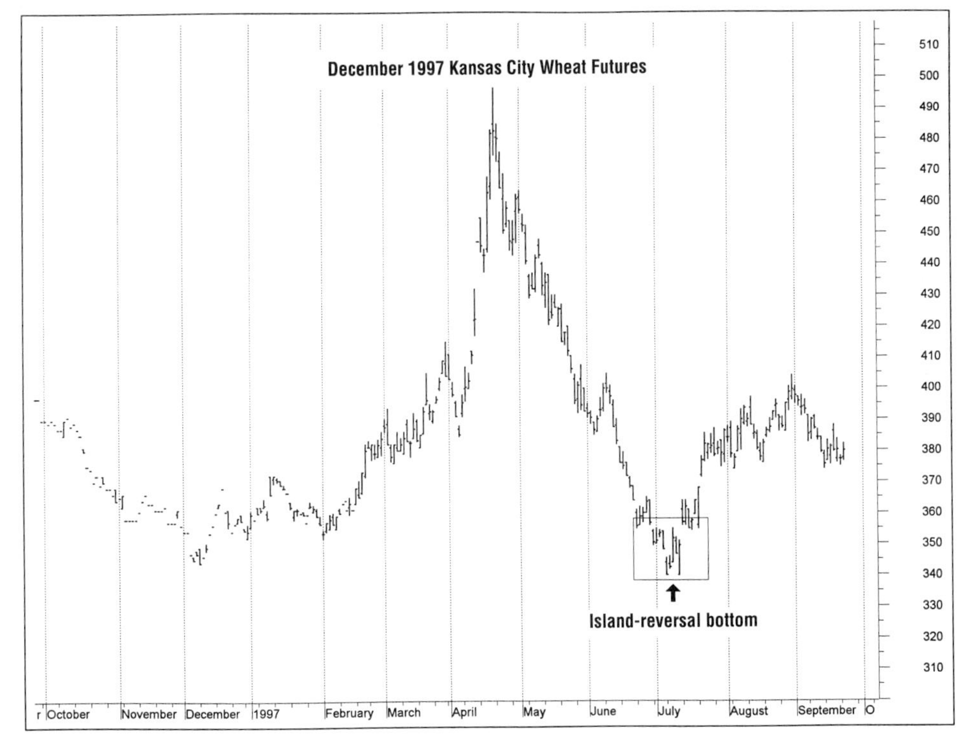

FIGURE 4.24 An Island-Reversal Bottom on December 1997 Kansas City Wheat Futures Chart

subsequent decisions, and the island-reversal patterns are particularly useful in that context.

> **The island-reversal pattern is distinctive in appearance. Chart gaps are involved, and the gaps become important. A sell order near the bottom of the gap that completes an island-reversal top, as prices start a corrective rally back up toward the chart gap, has a good chance of being filled at a favorable price level. A buy order near the top of the gap in an island-reversal bottom is a correct approach. The formations are very reliable when completed.**

Consolidation Patterns

Coverage to this point in the chapter has dealt with ways of spotting or predicting reversals in the direction of price trend. The trend line, when penetrated, signals a change in direction. The topping or bottoming formations, such as the key reversals, signal an end to a trend in prices.

Once a price move has started, the charts tend to register *consolidation patterns.* After starting a major move, the market often tends to pause and consolidate for a few days before resuming the increasingly apparent price trend. The consolidation phase takes on often recognizable shapes or patterns. The ability to recognize and interpret the consolidation patterns can be very important to the decision maker looking for a chance to add hedges, or the trader with positions in the market trying to decide whether the price trends that are apparent on the charts are approaching completion.

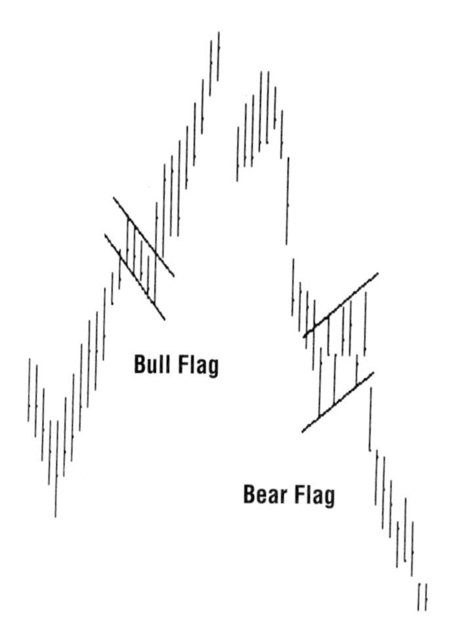

FIGURE 4.25
Demonstration of Bull and Bear Flags on a Bar Chart

In discussing the more important consolidation patterns, it is useful to review and emphasize the concepts of bullish and bearish in describing the markets. The term *bullish*, you will recall, suggests prices are moving higher or are expected to move higher. The term *bearish* suggests prices are expected to move lower.

Flag Formations Easily recognized and reliable formations, the *bull flag* occurs in the midst of a major uptrend, the *bear flag* in a major downtrend. Figure 4.25 demonstrates both.

The flag formations are "resting places" or periods of consolidation after the market has moved for several days. When a top occurs, for example, and a widely recognized sell signal such as a key reversal is observed, the trader using the charts to guide timing of actions (hedger or speculator) tends to establish short positions in the market. A few days later, the psychological dimension of the market starts to emerge. The recently established short positions are now profitable, and there is self-imposed pressure for the short trader to take profits by buying to offset the short positions. Buying to cover or offset the now-profitable short positions puts the trader who was a recent seller on the same side of the market as those traders who are buying because they think the market should move higher. A brief price rally develops. In a downward-moving market, the initial price plunge often takes the shape of a *flagpole,* and the period of consolidation develops a *flag* with an ascending shape. The flag portion of the bear flag is formed by ascending parallel lines sketched across the highs and lows of the consolidation pattern.

A close below the lower parallel of the bear flag signals a continuation of the downtrend in price. By projecting the length of the flagpole from the breakout point, a minimum projection of the extended price move down can be generated. Figure 4.26 demonstrates. This is a reliable projection, and the technical trader either places or holds existing short positions with a feeling of confidence that the downtrend will continue. In Figure 4.26, the length of the flagpole is labeled "A," and this is the magnitude of the expected price move when the market moves lower from the flag for-

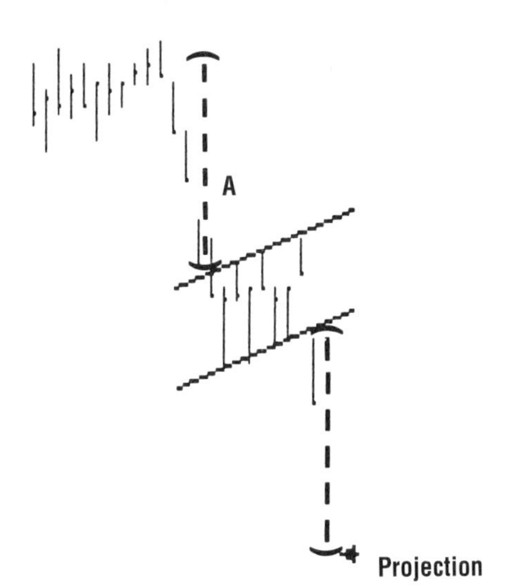

FIGURE 4.26
Price Projection from a Bear Flag as the Downtrend Is Continued

mation. The price range in "A" is extended downward from the point at which price moves below the flag.

A bear flag is demonstrated on the September 1997 corn futures (Figure 4.27). Note the characteristic shape of the consolidation pattern in late May and early June. The idea of a "flagpole" is not very clear on this chart unless you go back up to the action after the price decline through the support plane at the April lows near $2.85. The "flagpole" on the chart uses that level as a starting point. Note that prior to the late May–early June development there were several mini "bear flags" as the market paused every few days. Ultimately, as the prospects for a large 1997 crop filtered into the market, prices were pushed below $2.30 essentially matching the projection that could be gleaned from the chart. The July–August dry conditions that were mentioned earlier for Illinois and Indiana then prompted a price rally.

There is an interesting message here. If the producer knows nothing about charts and chart patterns, how is perspective formed? All the prices shown on the September corn futures are within a reasonable range that might emerge from an analysis of supply–demand fundamentals. Without the chart and the opportunity to observe the bear flag as it forms, will you as a producer feel comfortable holding short hedges? *The importance of technical analysis, very basic technical analysis, in the timing of pricing decisions is very apparent here.* The probability that a major price move will follow a break from a flag formation is in the 70–80 percent range, so the formations are quite reliable.

As the flag formations start to develop, it is important to monitor developments to determine whether it is indeed a consolidation pattern or some type of topping or bottoming formation. A price break down from a bear flag, such as the late May–early June pattern in Figure 4.27, could turn out to be a double bottom if prices hold across the low in the flag formation and start to turn higher.

The concept of trading volume as an aid to interpretation of the charts was introduced earlier in the chapter. If the price plunge through the bottom of a bear flag occurs on a high-volume day, the sell signal is authenticated. The June 16 price break from the bear flag on September corn was on a large but not huge volume—but above any volume since May 28. The fact that volume was smaller as the flag formed also tends to confirm it as a consolidation pattern rather than a bottom. Trad-

FIGURE 4.27
A Bear Flag on September 1997 Corn Futures

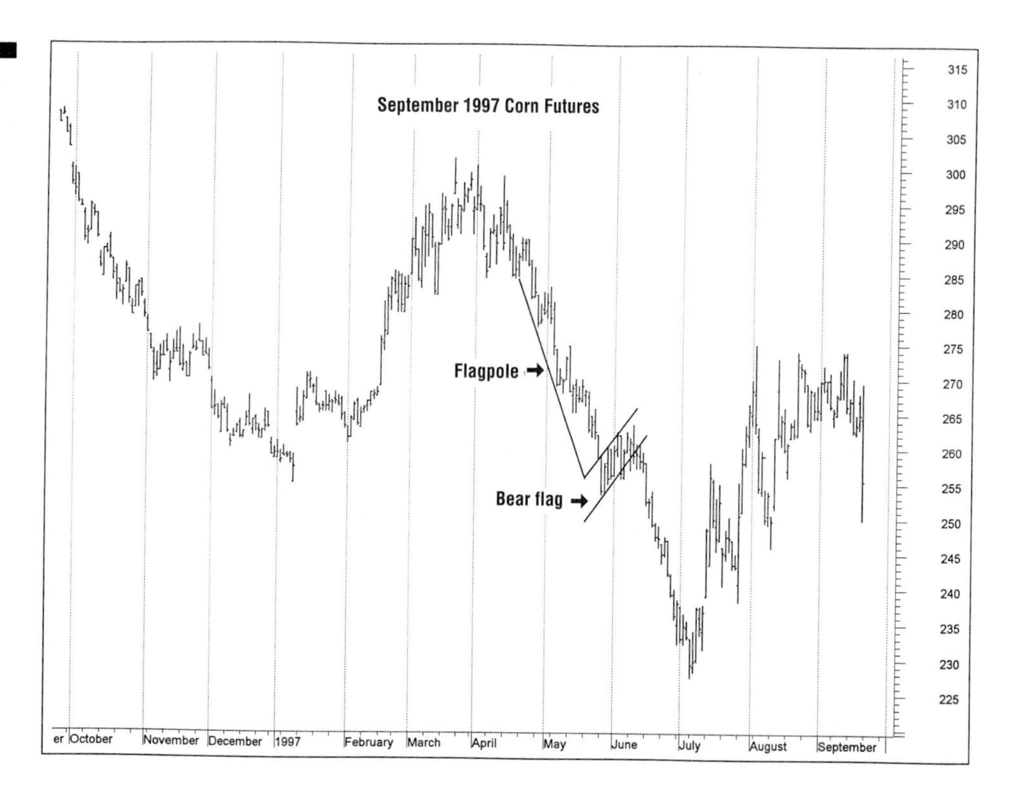

ing volume is in newspapers and is estimated within the day on most electronic services.

A second aid is to watch to see whether *open interest* is declining. Open interest is the number of positions in futures for a particular commodity that have not been offset or closed out. Assume a hedger enters the market and sells 10 corn futures contracts on a particular day. A new buyer, long hedger or speculator, comes into the market and buys the 10 contracts. Open interest would increase by 10 contracts if those new positions are in place at the close of trade at the end of the day, all other things equal.

The use of open interest will be treated in more detail later in the chapter, but declining open interest suggests that *short covering* is causing the observed bear flag. Such action argues in favor of a consolidation pattern rather than a bottom formation. When holders of short positions take profits and exit the market, open interest will decline. A decline in open interest tends to identify price increases as a *short-covering rally* and not a bottom in the market. Open interest *was* declining during the formation of the bear flag on the September corn chart.

If the prices are trending higher, a decline in open interest suggests it is a bull flag rather than a topping formation, especially if the potential consolidation pattern is significantly below life-of-contract highs. When holders of long positions take profits, open interest will decline. That action suggests a consolidation pattern and suggests that prices will continue higher rather than record some topping action.

Figure 4.28 shows a bull flag on October 1997 feeder cattle futures. The early-July pattern came before prices surged higher by about the same amount as the flagpole that stretches back down toward $78.70 would have predicted. The push up through the resistance plane across the highs near $78.70 in May and June gives a logical place to start the flagpole measurement.

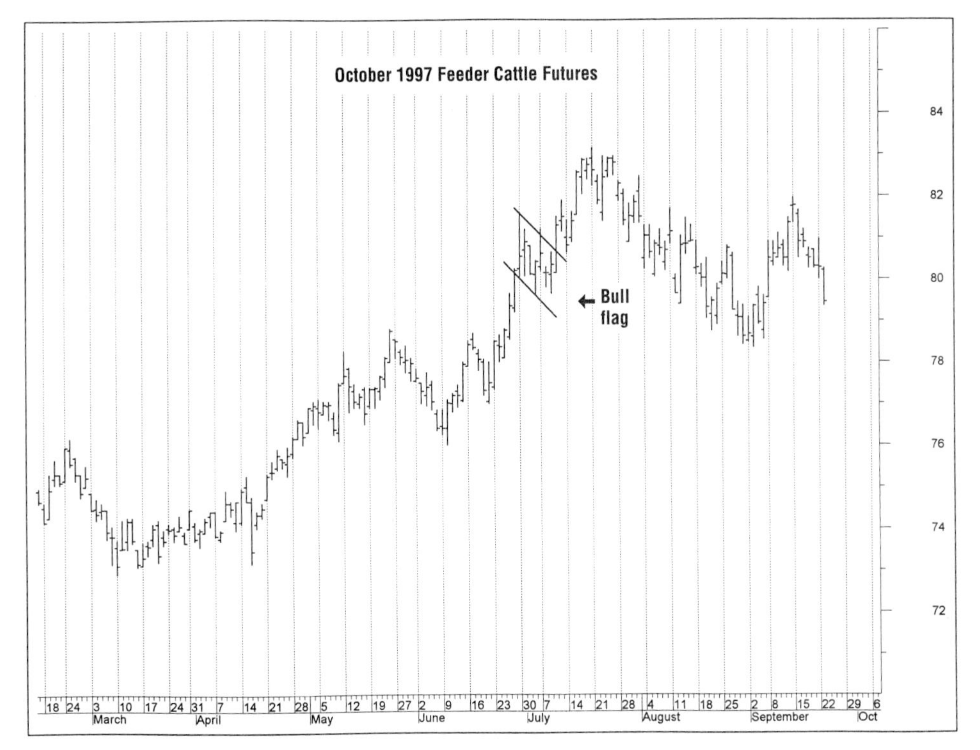

FIGURE 4.28
A Bull Flag on October 1997 Feeder Cattle Futures

Management of the flag formations is fairly apparent. A close below the line forming the bottom of a bear flag is a sell signal for hedger and speculator alike. It also confirms to the holder of short hedges that the position is correct and added price protection might be established prior to the anticipated further declines in price.

A close above the line forming the top of a bull flag is a buy signal for the long hedger and for the speculator inclined toward the long side. The selective short hedger who has short hedges in place would see the buy signal as an indication that the market is going significantly higher and consider lifting or offsetting the hedges. If long hedges are already established, holders of these positions keep them in place with renewed confidence that the market is going higher.

Flag formations are the most common of the consolidation patterns. The projected price moves are a minimum expectation and are generally quite reliable, in the 70–80 percent range. These consolidation patterns are important to traders trying to decide whether the prevailing trend has run its course and to producers looking to place late or catch-up hedges.

Triangles A second widely observed consolidation pattern is the *triangle.* Figure 4.29 demonstrates typical triangles in an upward-trending market, panel (a), and a downward-trending market, panel (b).

In general, the market is expected to move out of a triangle formation in the same direction that it was moving when the consolidation pattern started to develop. As with the flag formations, it is important to monitor open interest and other measures

FIGURE 4.29
Triangle Formations as Consolidation Patterns on a Bar Chart

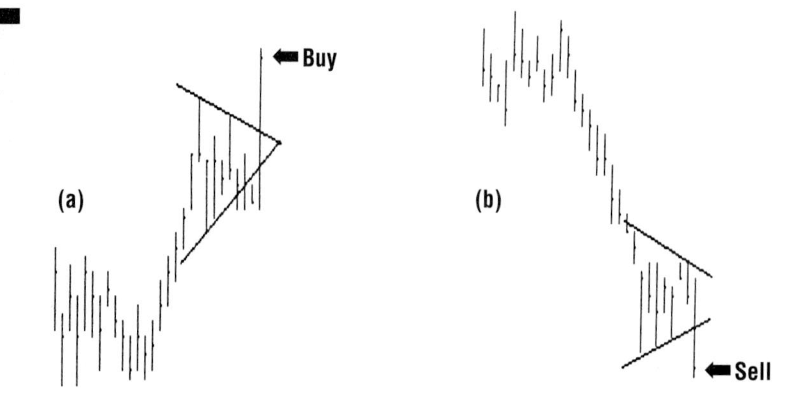

that help the trader to distinguish between the triangle as a consolidation pattern and the possibility of a top or bottom in the markets.

Figure 4.30 illustrates a possible means of projecting the price move from a triangle in an upward-trending market. The procedure involves a parallel line to the line forming the bottom side of the triangle and is more accurate if the breakout to higher prices occurs before prices move out into the apex of the triangle. In the example shown, the price move would be expected to reach point *X*, the point where the parallel (dashed) line and a vertical plane above the apex of the triangle intersect.

Management of the triangle formations parallels management of the flag formations. A breakout to the upside in an upward-trending market is a buy signal, and a message to keep long hedges in place, and so forth. If potential short hedgers interpreted the consolidation pattern incorrectly as a topping formation and sold the market, they would have to consider lifting the hedge if they were operating as selective hedgers. A break to the upside from an obvious triangle formation will, with 70–80 percent reliability, bring a major price advance. The reliability is the same for breaks below a bearish triangle.

Figure 4.31 shows at least two obvious triangles in a surging hog market. The October 1997 lean hog futures started a major price rally in anticipation of a bullish late-March *Hogs and Pigs* report. The postreport action was still to the upside, reaching a high of $77.00 on April 24. Triangles appear around $70 and again around $73.

The triangles are frequently seen as consolidation patterns. Prices typically move from the triangle formation in the same direction as the price direction when entering the consolidation pattern.

FIGURE 4.30
Projecting the Price Move from a Triangle Formation in an Upward-Trending Market

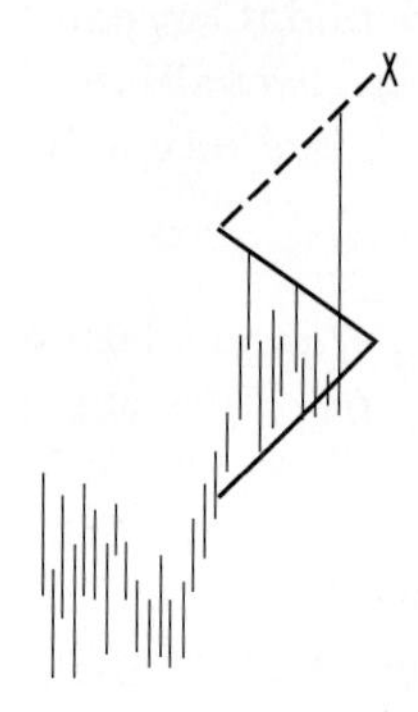

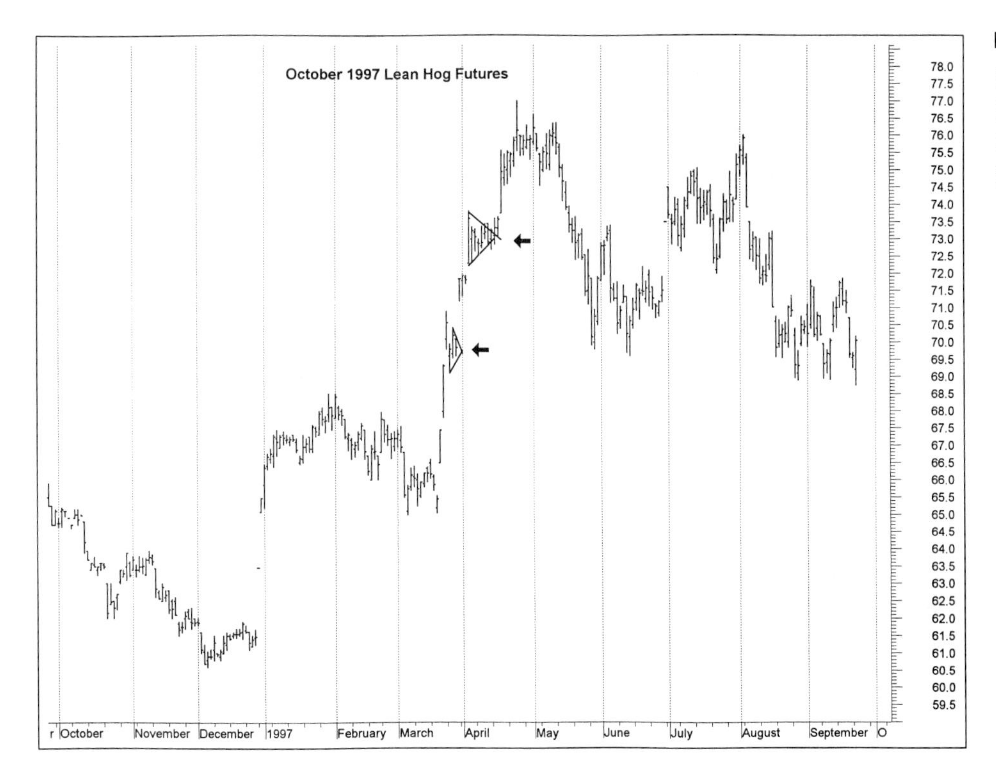

FIGURE 4.31
Bullish Triangles on October 1997 Lean Hog Futures

Pennants Best viewed as a small symmetrical triangle, many analysts differentiate the pennant from the triangle formation. Generally, the pennant is formed over fewer days, 10 days or less as a rule, than the triangle, which can take 10–20 trading days or more to complete. This suggests the "triangles" on the October hog contract in Figure 4.31 could have been called pennants or triangles.

There are no widely adopted means of projecting the price moves from the pennants shown in Figure 4.32. Some analysts use the staff of the pennant in the same manner as the flagpole on the flag formations. As with the triangle, the direction of price movement coming out of the pennant is expected to be the same as the direction of movement coming into the pennant. If this were not true, of course, *ex post*

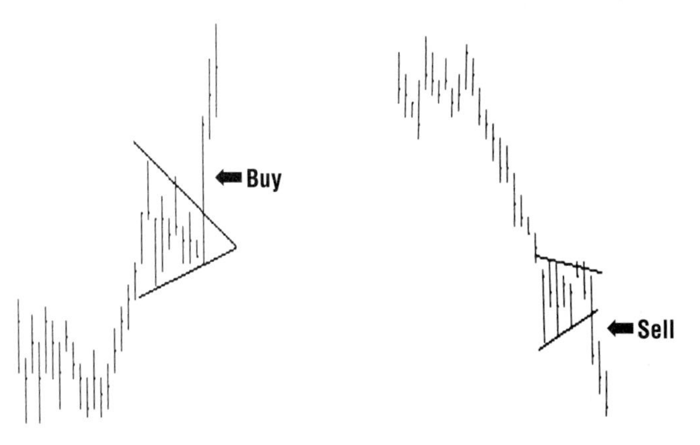

FIGURE 4.32
Demonstration of Pennants as Consolidation Patterns on a Bar Chart

observation would confirm that it was not a consolidation pattern at all, but either a top or bottom in the market.

Management of the pennants parallels the discussion of management of flags and triangles. This is a less reliable consolidation pattern than the flags or triangles, more nearly in the 60 percent area. Some analysts feel the direction of subsequent price moves is less predictable for the pennant relative to the flag formations. Nonetheless, a move out of a bull pennant is going to signal a significant price move up in a majority of the cases. No current charts are shown to illustrate triangles and pennants. Virtually any futures chart will show one or more patterns that appear as periods of consolidation and take the shape of either a triangle or a pennant.

Perhaps the least reliable of the consolidation patterns covered here, the pennants do suggest the direction of emerging price moves for many analysts. Significant price moves from the pennants are commonplace.

Congestion Areas Often, the consolidation occurs in what can only be described as congestion areas. Figure 4.33 demonstrates. The market pauses and trades for several days in a sideways pattern.

Here, we are dealing with periods of consolidation that fail to take on the distinguishing shape of a flag, triangle, and so on. Often a larger number of trading days is involved as the market pauses, takes stock of the prevailing and ever-changing base of information, and builds a consensus on the direction in which to move.

The top of the congestion area in an upward-trending market can be treated as a resistance plane. In a down market, the bottom constitutes a support plane. When the planes are "taken out" via a close above the resistance place or below a support plane, the correct management action parallels the earlier discussion on resistance and support planes. Once a move out of the congestion area is recorded, price often moves quickly up or down. *Prompt action by the trader is important.*

There is no universally accepted projection technique associated with the congestion areas, but many analysts will measure the vertical price range of the area and project that distance up or down. Some analysts prefer to use the horizontal width of the area as a projection device. Implicitly, this latter approach is saying that the longer the time period required for the market to digest information and decide which way to go, the bigger the price move will be.

Major price moves often evolve after price breaks out of a congestion area. In general, the longer the time period involved, the greater the price move is expected to be.

FIGURE 4.33
Congestion Areas as Possible Consolidation Patterns on a Bar Chart

Additional Chart Signals

Emphasis to this point has been on selected chart patterns and the complementary aids to their interpretation. There are additional chart developments, not patterns per se, that are very useful guides to the trader. Two examples are *chart gaps* and *corrections.*

Chart Gaps Often seen after the market has been "shocked" with new and surprising information, a gap appears on the chart when the market tries to move up or down quickly. The idea of a chart gap emerged earlier in the chapter in discussion of island-reversal tops or bottoms, but the chart gaps are important in and of themselves.

Several characteristics of market behavior in and around a chart gap become important. First, there is the previously observed tendency for the market to attempt to "fill" the gap. That is, the market usually attempts to come back and trade over the price ground that was skipped when the gap occurred. It is said that nature abhors a vacuum and that the futures market sees a price gap as a type of vacuum. Figure 4.34 demonstrates the December 1997 corn futures contract and clearly shows the tendency in the market to fill chart gaps by coming back and trading over that previously neglected price range. Note, in particular, the chart gap left in early July when the market gaps higher from the $2.37 level. A small down gap that appears a few days later *is* filled before the disciplined trader is rewarded, on the 11th day after the initial gap occurs, with a price correction down into the top of the chart gap. *Long hedgers waiting to take positions, potential long speculators, and short speculators and selective*

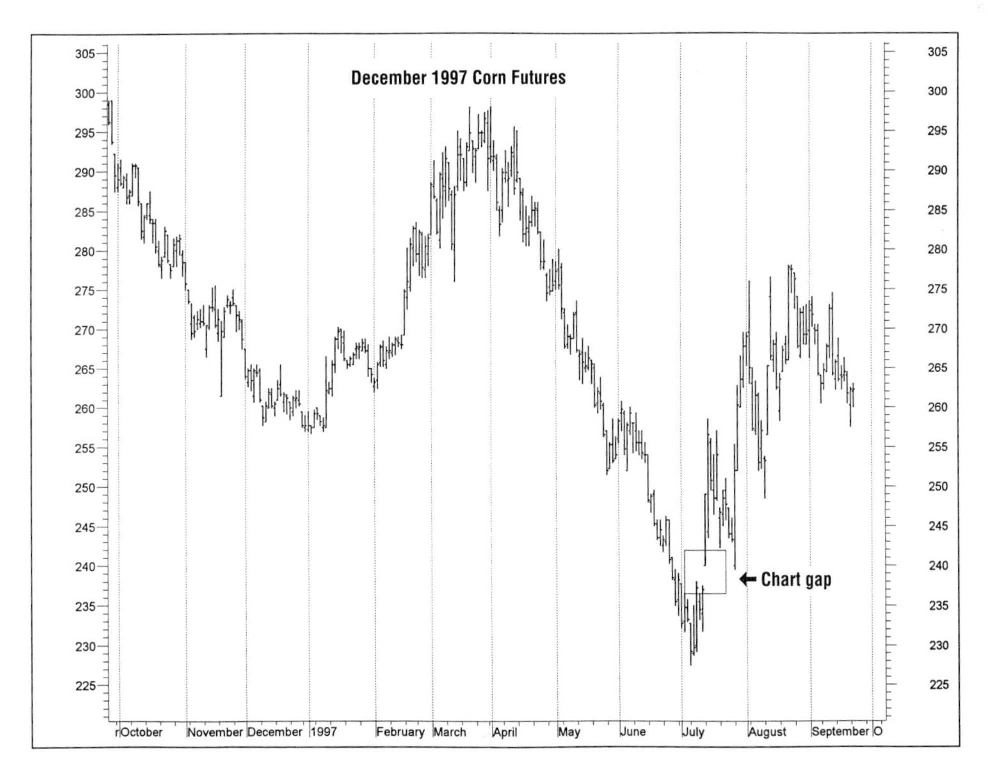

FIGURE 4.34
Market Action to Fill a Gap on the December 1997 Corn Futures Contract

short hedgers will all tend to have buy orders in the top of the gap. Not surprisingly, the market held there and traded sharply higher in subsequent days as the buying pressure overwhelmed the lingering selling pressure.

Remember, management of trades around the chart gap becomes important to the hedger and speculator alike. As suggested, the selective long hedger can use the dip back down toward the gap to add long positions or replace any long hedges that might have been removed as prices trended lower earlier in the year. Selective short hedgers can use the dip to the top of the gap to lift any short hedges before prices surge to higher levels.

When the trend in the market is down, both the short hedger and the bearish speculator will want to sell a price rally back up toward a gap area. The long hedger, caught in a downward-trending market, can use the rally toward the gaps to remove long hedge positions as favorably as possible, positions that now appear to be incorrect. The long hedger is then in a position to benefit from lower prices in the cash market, and those interested in short positions—hedgers and speculators—have seen a chance to get short before lower prices come.

The chart gaps provide yet another opportunity to hammer at the importance of management of orders. Consider, yet again, the position of a potential long hedger or long speculator who sees the market "gap" higher, as with the December 1997 corn in early July, and wants to get long hedges or speculative positions in place. As the market starts to correct back down toward the chart gap, there are sure to be buy orders in place, but it is impossible to know for sure just where the concentration of orders will be. But it is important to visualize what is happening and place orders accordingly. *The probability of getting a buy order filled is increased if it is placed at or just above the top of the gap.* An order to buy near the middle or bottom of the gap will likely require other more aggressive traders' buy orders at higher prices to be completed before the buy order you placed at the bottom of the gap will be reached. *Keep in mind that every trader, hedger and speculator, who uses charts sees the same chart formation.* Each will, or should, decide where the order should be placed to balance the price level at which they are willing to buy and the probability of getting the order filled. As an individual decision maker, you should try to envision the process that is going on and place orders that fit your market plan. This is especially true if the financial viability of the program could be at risk if the orders are not reached and filled and therefore no protection is established.

In a down market, of course, the market will try to come back up and fill the gap. In this instance, the short hedgers will be looking to add to their short positions and the bearish speculator will be inclined to sell the rally. There will be a cluster of sell orders near the bottom of the gap and in the gap awaiting the expected rally. *The chances of filling the sell order will be increased if it is placed at, or just below, the bottom of the gap.*

You should be aware that we could have selected futures charts to demonstrate *filling* of chart gaps. It is often the case that the gaps will be completely spanned and orders placed anywhere in the gap will be filled. The late April bear flag formation filled the small chart gap on that same December 1997 corn futures charts. But the circumstances demonstrated by the December 1997 corn futures in early July are not at all unusual, and they often present very important opportunities.

A second characteristic of market behavior around the gaps emerges when the gap is not filled within a few days (5–10 days) after it is created. When this occurs, the gap becomes a candidate to be a *breakaway gap,* and the technician watches for another gap to emerge as the market move gathers momentum. This second gap can

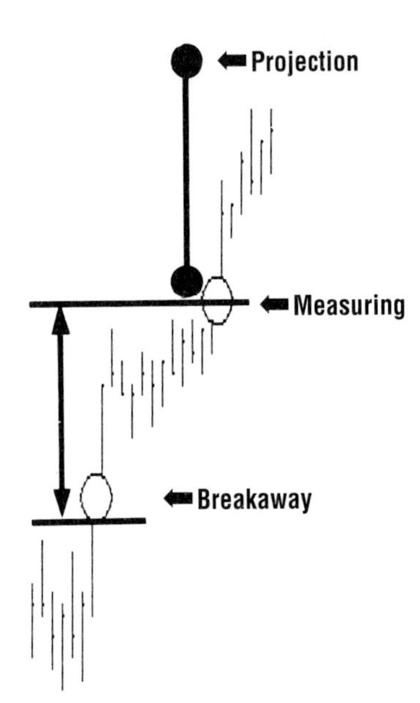

FIGURE 4.35
Demonstration of the Price Projection Using a Series of Chart Gaps

become a *measuring gap,* and we then have a means of projecting the market move. Figure 4.35 demonstrates using an upward-trending market. After the market moves quickly higher from some bottoming formation such as a head-and-shoulders, the second or measuring gap emerges when a broad consensus develops that the changing supply–demand fundamentals will support still higher prices. These dramatic moves are often based on both new buying, as speculators and long hedgers enter the market, and covering of short positions, as short speculators close out now losing trades and selective short hedgers buy to offset short hedges.

As suggested in Figure 4.35, the market is expected to move a distance measured by the start of the breakaway gap to the middle of the measuring gap. Often, after the projection is met and either topping or bottoming action begins (topping action in this instance), a third gap will emerge called an *exhaustion gap.* Many analysts use the presence of the exhaustion gap as a means of helping spot bottoming or topping action as the concerted action starts to disappear and uncertainty on future price direction starts to emerge. *The breakaway gap and the measuring gap are widely watched and widely used by technically oriented traders.*

You should note that the unfilled gap on the December corn chart back in Figure 4.34 is a candidate to become a breakaway gap. The second gap that appeared (around August 10) as the market moved higher becomes a measuring gap allowing us to project the magnitude of the market move. Extending up a distance from the bottom of the breakaway gap to the middle of the measuring gap, from $2.32 to $2.55, suggests prices will move up some $.23. Adding $.23 to $2.55 gives $2.78, essentially the level the market reached on August 21.

Placing buy orders near the top of gaps in ascending markets and sell orders near the bottom of the gaps in descending markets prepares the trader to manage his or her positions under the assumption the gap might become a breakaway gap. Often, what turns out to be a breakaway gap is partially filled before prices surge, and the orders to establish long positions or lift short positions will be filled in this instance. After the major price move starts, the traders are then ready to hold and manage their positions as they watch for a measuring gap. If the buy orders near the top of a gap in a surging market are not filled within a few days, it pays to be alert. Short positions that do not get offset by buying back would bring major losses to the speculator and major opportunity costs to the selective hedger.

On the corn chart, the top of the early July gap was \$2.40, with the gap occurring on July 14. Some two weeks later, on July 28, the market dipped down into the top of the gap, reaching as low as \$2.39½. Buy orders at the top of the gap were filled, and the market moved up to \$2.78.

> **Chart gaps on the bar chart are widely watched and widely used in managing a trading program. Observation of market actions over time will convince the chart watcher that many analysts employ the gaps in placing buy and sell orders, and those actions suggest a self-fulfilling prophecy dimension unless the fundamental picture changes significantly.**

Corrections The concept of *market corrections* has roots in many theories of market behavior, including the sophisticated Elliott Wave theory. The basic idea is that the futures market, as is the case with most types of markets, overreacts to an infusion of new or unexpected information. If the initial reaction carries too far and moves beyond what appears to be the proper market-clearing price, the market will then "correct" at least part of that overreaction.

Figure 4.36 demonstrates and records three possible levels of the expected corrections. Chart analysts tend to watch for corrections of 38, 50, or 62 percent. In Figure 4.36, the 38, 50, and 62 percent corrections of the price decline from *A* to *B* are calculated starting at *B*, the low price after the move down from the high at *A*. The corrections are then marked on the chart and the chart-oriented trader will start to monitor the corrections and prepare to sell the market. If the 38 percent level is exceeded, the analyst then watches for the 50 percent correction. If that is exceeded, the 62 percent correction is anticipated and actions to sell are made ready. If the correction of the price move from *A* to *B* exceeds 62 percent, there is then a tendency to conclude that the market is not just making a correction. Rather, the direction of

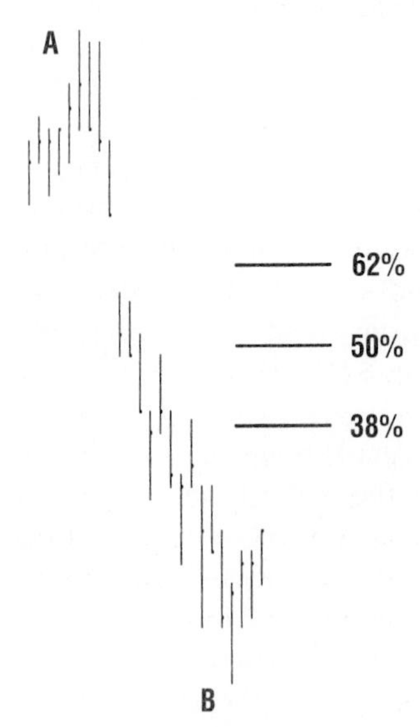

FIGURE 4.36 Illustration of the Expected Price "Corrections" After a Major Price Move

price trend is changing from down to up because the supply–demand balance has changed and short positions need to be lifted on some recognizable chart signal.

Understanding that the market will tend to correct and then managing that tendency can be very important. Consider the mental state of cattle feeders with a large inventory of cattle or producers holding soybeans in on-farm storage who see a major price break and realize they have no price protection on inventories that can be worth millions. As the prices plunge to lower and lower levels, panic tends to set in, and the tendency is to rush in and sell the market before it goes still lower.

If you are aware that the market will often correct to the upside, there is less self-imposed pressure to rush in and sell on a panicky basis. It is easier to be disciplined and wait if there is a reasonable degree of confidence that the market will make a correction. *Then, of course, it is important that you or any other trader needing short positions be disciplined enough to step up and establish short positions in an upward-moving market as the correction to the upside develops.* Knowing the market is likely to correct back to higher prices is the necessary condition for disciplined trading, and the 38, 50, and 62 percent corrections give possible pricing objectives. Being willing to establish short hedges when the now upward-moving market starts to falter around a 50 percent correction, for example, is the sign of a disciplined hedging program. This is very difficult for many producers!

Conversely, the long hedger or other trader looking to establish long positions or lift short hedges can expect to see a correction of a recent upward surge in prices. The upward-trending markets tend to be orderly in their ascent, exhibiting a five-wave pattern. Wave 1 launches the move from some bottoming pattern as the consensus on the supply–demand balance starts to change. Wave 2 is a correction of wave 1, and wave 3 then emerges as the often explosive move to higher prices. Wave 4 corrects wave 3, and wave 5 then must make a new high compared to the highs in wave 3. The corrective waves in a bull market, waves 2 and 4, often approximate one of the percentage corrections just discussed—the 38, 50, or 62 percent corrections.

Figure 4.37 demonstrates the application of this important tendency to correct. The wheat market staged a long downtrend as the June–July harvest approached in 1997. Remember, the market was coming off a 1995–96 period in which record high prices were recorded in the face of tight world stocks. The move was from $4.74 on April 21 to $3.35 on July 8, a price decline of $1.39. Adding 38 percent of the $1.39 back to $3.35 gives $3.35 + .53 and an initial objective of $3.88. On August 12, the market went to *exactly* $3.88 before trading lower for several days. The subsequent rally was an attempt, perhaps, to reach a 50 percent correction objective of $4.05, but the market failed after reaching $3.98 on August 29.

The wheat market, in September 1997, was apparently caught in a bear market. These markets tend to show a three-wave move, often labeled *A*, *B*, and *C*. *If* this is the correct reading of the chart, the move down to $3.35 was *A*, and the correction was *B*. The *C*-wave would need to trade down through the $3.35 level and record still lower prices. This might not be completed on the December chart, but it could occur on a later futures as the market moves into 1998 with a large expected wheat acreage after the 1996 farm bill legislation removed any acreage reduction requirement and as wheat acreage comes out of the long-term Conservation Reserve Program.

Bear markets, as suggested, are more likely to demonstrate a three-wave pattern, with wave 2 or *B* the corrective leg before prices move still lower. Every chart analyst would agree that the bear markets are quicker and more dramatic than the bull markets. A price increase that occurs across six months in a leisurely five-wave bull market can be wiped out by a bear move in six weeks or less.

FIGURE 4.37
Corrections on the December 1997 Wheat Futures Contract

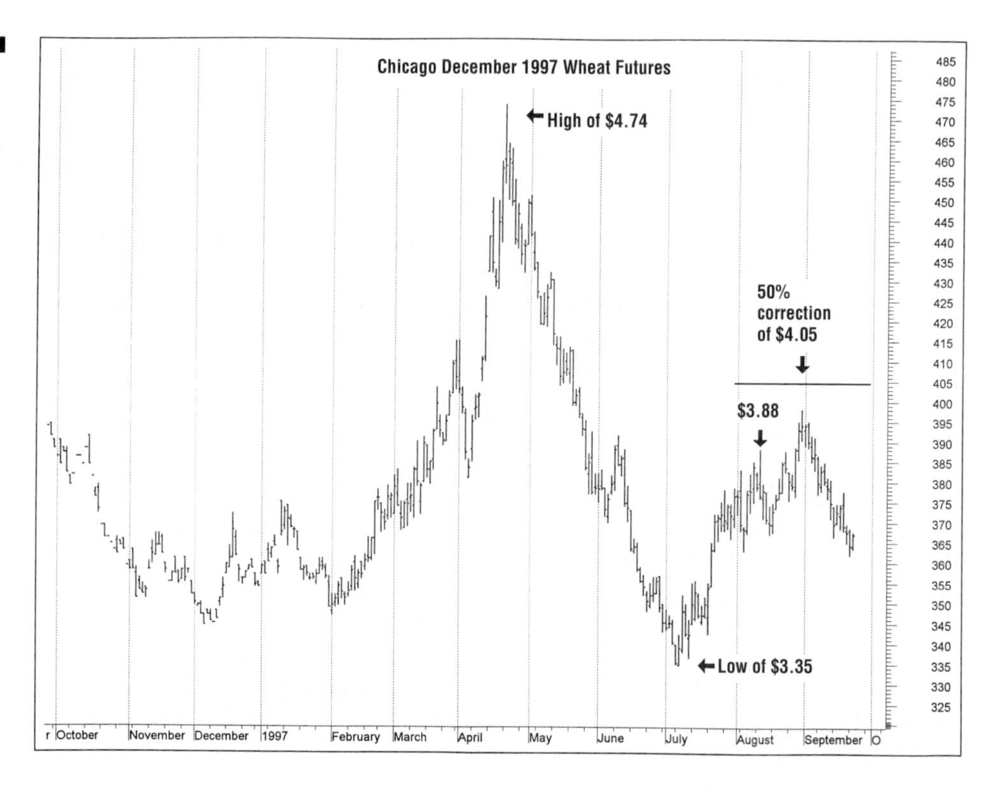

Exactly which correction to expect will depend on how weak or strong the underlying supply–demand balance is—and we see another specific example of how fundamental and technical analysis can be complementary. If the market is moving lower, for example, and all the new pieces of information on either the supply or the demand side are tending to be bearish, it might be prudent to start placing sell orders on signs of faltering prices near a 38 percent rally to the upside. If the rally does carry to higher levels, more short positions can be added at the 50 percent and the 62 percent levels. This approach would employ an element of the scale-up placing of short hedges that always has merit. But be careful: applying this to the December 1997 wheat chart, the market could never get all the way to the 50 percent correction. You might want to be more aggressive with short hedges when the attempt at a 50 percent correction falls short. This is a market with bearish supply–demand fundamentals. (And keep in mind the even dollar levels, such as $4, are *always* hard to penetrate.)

Conversely, if the supply–demand news is only mildly bearish, the producer might watch for something approaching a complete 62 percent correction before doing additional selling or starting a late, short-hedging selling program. Awareness of what is going on in terms of supply–demand fundamentals is important in making informed pricing decisions, and that awareness helps to instill more confidence in which corrective levels should be picked to start or resume a pricing program.

The futures markets tend to correct at least part of the major price moves before the trend resumes. Awareness of this tendency helps the trader to position orders correctly and helps bring the discipline needed to avoid the tendency to act or react on a panicky basis. The widely observed corrections in the futures market offer tangible evi-

dence of the complementarity in fundamental and technical dimensions of the markets.

COMPLEMENTS TO CHART PATTERNS

For virtually every chart pattern, every buy and sell signal, the volume and open-interest levels and/or changes in volume and open interest complement and reinforce the interpretation of the chart. Both concepts were introduced briefly earlier in the chapter, but more detailed coverage is needed. The user of the markets cannot afford to ignore what is happening in trading volume and in open interest.

Trading Volume

Trading volume is the total number of contracts traded for the day. *Volume is best viewed as an indicator or barometer of the level of intensity in the market.* In general, any sell or buy signal observed on the chart is going to be a more reliable signal if it occurs on a high-volume day. What is considered high volume is relative to the volume in recent trading days. Figure 4.38 demonstrates, with trading volume shown as a histogram at the bottom of the October 1997 live cattle contract used earlier to demonstrate a key-reversal top (see Figure 4.21). The chart is reproduced in Figure 4.38. Note the huge volume on July 29, the day of the reversal. The intensity in the market definitely confirmed this major top.

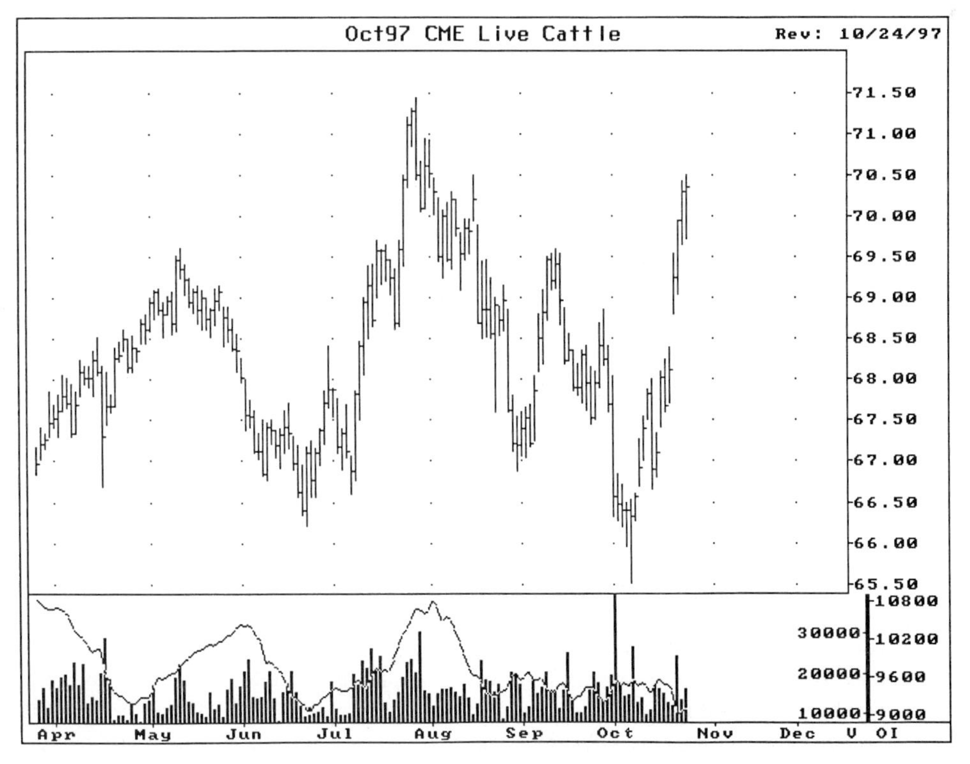

FIGURE 4.38
Demonstration of Trading Volume on the October 1997 Live Cattle Futures Contract

In using trading volume, several guides are in order.

1. Look at trading volume for all the months being traded for that particular commodity.
2. Interpret volume on a limit-up or limit-down day carefully.
3. Trading volume will often be unusually low prior to a holiday, prior to a three-day weekend, or on the day prior to a major report.

Looking at volume across all contracts is preferred because activity during the delivery period will distort the messages from the trading volume in the nearby futures contract. Volume in the nearby futures contract declines progressively as the market moves into and through the delivery period. The exchanges often require both hedgers and speculators to reduce large positions in the markets prior to or during the period. This type of required action in the nearby contract can distort the measure of intensity in the market as a whole if only the volume in the single nearby futures is being monitored.

On days that the market is "locked" limit up or limit down, there will usually be very little trading volume. If the market is limit up, for example, there will typically be few willing sellers. The volume on a limit-move day clearly should not be contrasted to volume in recent nonlimit days. It *is* important, however, to look at the volume on limit-move days compared to the normal trading volume. If trading volume during a limit-move day is 25 to 50 percent of the normal volume on a typical trading day, that tends to send a signal that some traders were willing to sell at limit up or buy at limit down. There is not likely to be much follow-through to the upside the next day if traders are willing to sell at limit-up prices. Conversely, if there is little or no volume, a significant follow-through to higher prices is likely the next day.[10]

It is best to ignore the light volume prior to long weekends, holidays, or prior to a major report. The short-term traders—scalpers, day traders, and so on—are not likely to be active on these days and the volume will usually be very light. Trading volume will often be light prior to major reports. All traders, and especially speculators, will be reluctant to hold positions in futures through reports that are prone to be surprises. An example is the quarterly *Hogs and Pigs* reports referred to earlier, reports that often bring surprisingly high or low numbers and prompt dramatic postreport price moves. Traders are often reluctant to establish positions just prior to such reports, and trading volume will tend to be unusually light as the market waits on the report.

Open Interest

As noted earlier, the number of contracts that are outstanding or that have not been offset at the end of the trading day is open interest. There is no binding relationship between trading volume and open interest. Large daily volumes can be traded by day traders and scalpers, or traders looking for small intraday price moves, but there is

[10]A related measure is the size of the *pool of unfilled orders* at the close on a limit-move day. If there is a large pool, follow-through to the upside the next day after a limit-up move is more likely. Most commodity brokers will have access to estimates of the size of the pool of unfilled orders on the limit-move days, and the estimates of the pools of unfilled orders are often presented on modern electronic market-news distribution systems. The information is thus available to every interested user of the markets.

zero change in open interest from the activity if all positions are closed out or offset before the close of trading for the day.

Several patterns in open interest can be important aids to the interpretation of the charts. Figure 4.39 demonstrates. Note the peak and then the decline in open interest (the line at bottom) a few days before the trend reversal in the market as the market records new price highs and then starts to move lower. Such a pattern often occurs, and reflects the adage that new money, which reflects new selling and new buying, is needed to sustain a price trend to the upside.

Note also the pattern in open interest on the November feeder cattle futures in Figure 4.40. The open interest starts to falter before the market makes new highs and records a possible double top above $84 in mid-July. The market had surged to a new high a few days earlier on July 16. *But open interest went down on July 16.* (Electronic systems allow you to pinpoint, in detail, what is happening on any particular day.) This faltering of open interest with prices still going up is a type of divergence and is a warning, often an early warning, that the market is ready to top.

Major trends, especially rallies, need new capital to continue, and increasing open interest indicates new capital is coming into the market. New traders are entering or existing traders are expanding their positions. In a bull market, new buyers are easy to find, but the market needs new sellers too. One group willing to sell in upward-trending markets is the short hedgers with price objectives at which they want to establish short positions. When that short hedging "capital" is exhausted, there may be no trader group left willing to sell the market—and open interest quits increasing. The market tops and turns lower.

A second use of developments in open interest comes during the short-covering price rally or the long-covering price correction to the downside. Figure 4.41

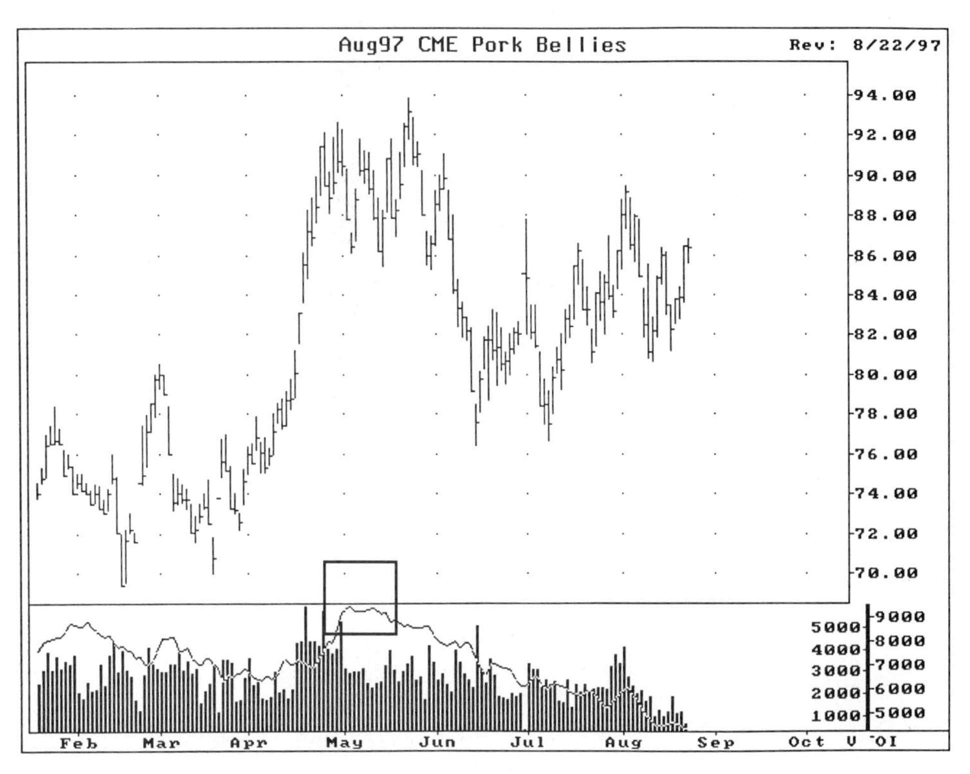

FIGURE 4.39
Demonstration of the Pattern in Open Interest to Anticipate and Confirm a Top in the Market

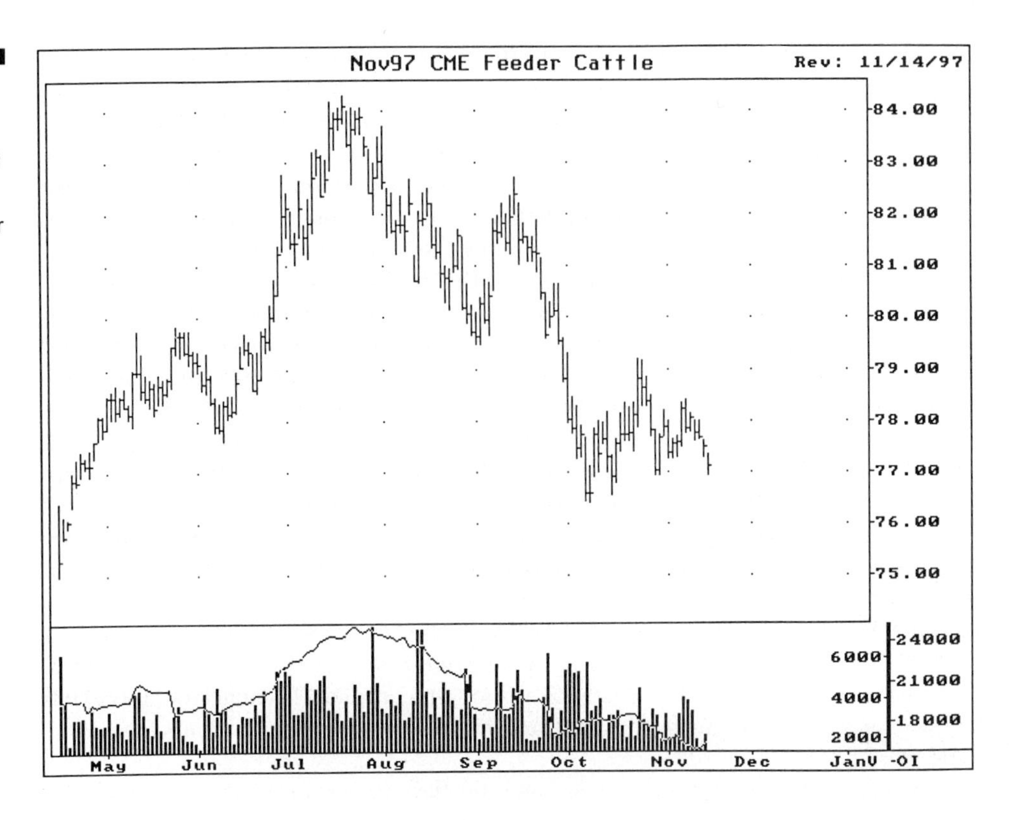

FIGURE 4.40
Demonstration of a Decline in Open Interest Prior to a Top, November 1997 Feeder Cattle Futures

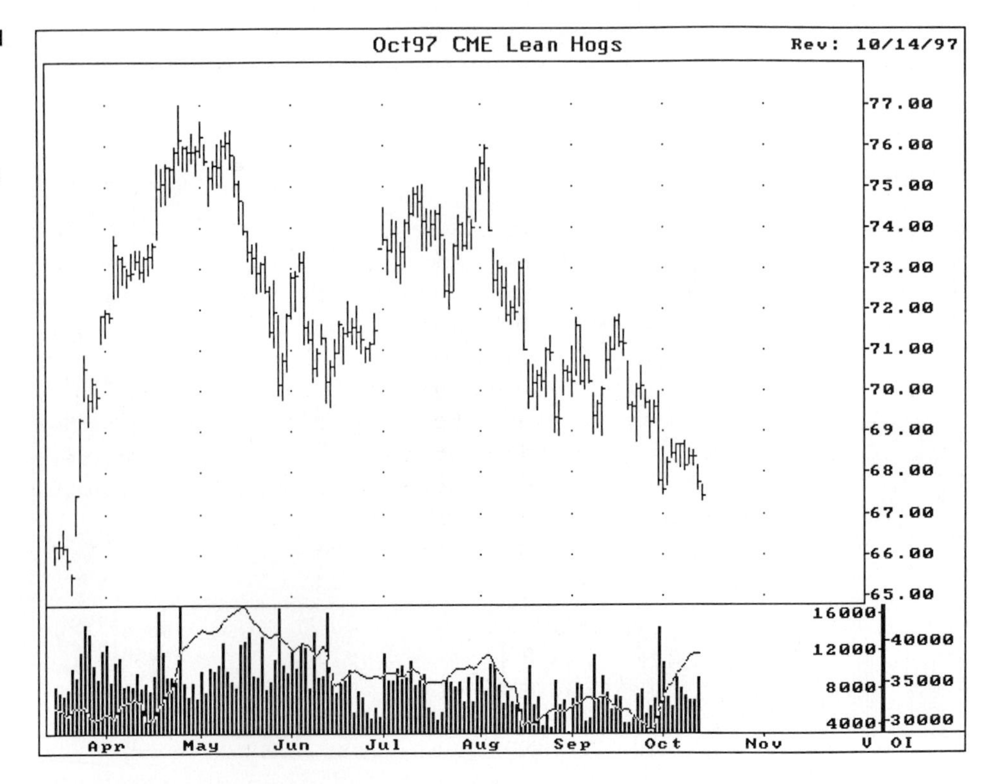

FIGURE 4.41
Demonstration of the Pattern in Open Interest During a Short-Covering Price Rally on the October 1997 Lean Hog Futures Chart

demonstrates one of the often-seen patterns using the October 1997 lean hog futures. In early August, the market gapped lower just below $74 and then spent seven days in a short covering rally as the market tried to correct back up toward the gap. Open interest plummeted as holders of now-profitable short positions bought them back. This pattern of events suggests that the early August developments are not a bottom in the making, and that lower prices are yet to come. Prices did move significantly lower.

The decline in open interest shown by the October hog futures is very important to chart analysts attempting to determine whether they are seeing a consolidation pattern in a major price move down or some type of bottoming action. This point was made earlier in the chapter. If it is but a short-covering or profit-taking price rally, it is likely that the market is just consolidating and will soon continue the trend toward lower prices. During such a correction, a correction that might take the chart form of a bear flag or some other consolidation pattern, open interest *must* decline as traders with short positions offset. *Short hedgers need to hold their short positions.* Potential long hedgers are given the signal to wait to establish long positions. *Lower prices are likely after the consolidation is completed, and long hedges can be set at lower price—and cost—levels.*

A dip in open interest prior to bottoming action in a downtrending market is less frequently observed and is generally a less reliable early warning of a price reversal, but open interest patterns should still be monitored. In the grains and oilseeds in particular, a decline in open interest as some type of possible bottoming action is being observed may be a tip that the large commercial firms have decided the price downtrend is over and are starting to offset their profitable short-hedge positions.

Open interest is an important complement to bar chart analysis. Changes in open interest can help the trader spot emerging tops or bottoms in the market and are invaluable aids in attempts to distinguish between consolidation patterns and tops or bottoms in the markets.

Relative Strength Index

Among the additional aids to the reading or interpretation of the bar chart, the most important subset are the tools that have been developed to measure the momentum in the market. A widely used measure of momentum is the relative strength index (RSI).[11] Table 4.1 records the procedure for calculating a 14-day RSI. The series employs the concepts of a *down* index and an *up* index. The calculations are based on the day-to-day changes in the market for a particular commodity. The modem electronic distribution services typically include the 14-day RSI and usually offer the flexibility of calculating RSI measures of different lengths.

The RSI measure can be calculated for each of the contracts for which a commodity is traded. Most commercial hard-copy chart services calculate the RSI for the nearby month and move to the next month when the delivery period for the nearby

[11]An excellent reference for more detailed coverage is the book by Wilder, *New Concepts in Technical Trading Systems,* listed in the references at the end of this chapter. Wilder is generally recognized as the leading authority in the development and use of the RSI and similarly conceived measures of momentum. For a broader and extensive treatment of the RSI and other technical tools, refer to the reference by Schwager.

TABLE 4.1
Procedure for Calculating a 14-Day Relative Strength Index

To calculate:

1. Record the last 14 day-to-day price changes based on closing prices.
2. Sum the negative and positive changes and divide each sum by 14 to create a "down average" and "up average," respectively.
3. Define Relative Strength Index as $(U)/(U + D)$ where U = up average and D = down average.
4. Employ RSI = $(U)/(U + D) \times 100$ to convert to percentages versus decimals.

futures contract commences. Some chart services show the RSI for the nearby and for the new-crop contract in the grains and oilseeds. With software packages designed to handle the calculations and updating, many traders watch not only the RSI in the nearby contract, but calculate the RSI for each futures contract in which they might be trading.

Figure 4.42 demonstrates use of the 14-day RSI with the October 1997 feeder cattle contract. The levels 70 and 30 are the widely employed thresholds to denote *overbought* and *oversold* markets, respectively. Conceptually, when the RSI is approaching the 70 boundary, the market is moving to a status in which all the traders who want to buy the market have done so. The market is running out of momentum, is becoming overbought, and a reversal in price trend to the downside is imminent. The trader looking for an opportunity to place short hedges, lift long hedges, or establish a speculative short position should be alert and ready to take action.

Moves to the 30 level by the RSI suggest the potential pool of sellers is being exhausted. The market is oversold. A reversal of a downtrend in price is likely, and the

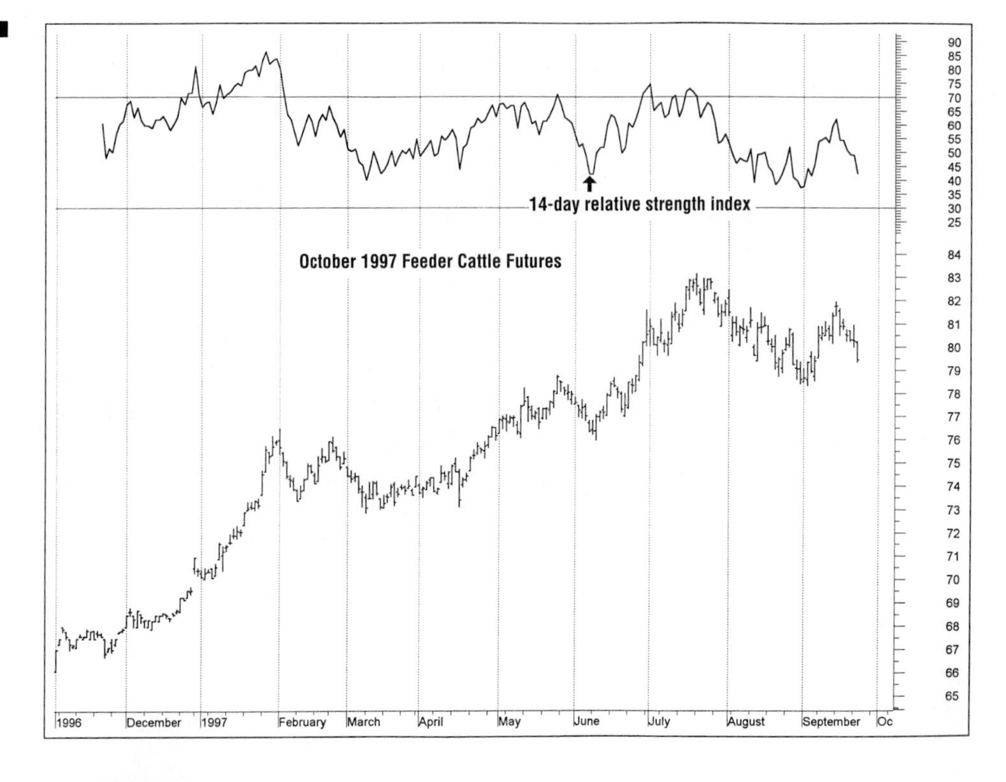

FIGURE 4.42
Use of the 14-Day RSI on the October 1997 Feeder Cattle Contract

analyst will start watching for bottoming action. Obviously, it is important to watch and see whether the rally is just a short-covering rally during a consolidation phase of a major downtrend. The patterns in open interest discussed earlier can be used in combination with the RSI to help ascertain whether the market actions are likely to be a bottom or just a consolidation phase.

The October feeder cattle chart, however, never records an RSI level of 30 and only occasionally records a level of 70. This suggests an obvious qualification of the measure. The 70 and 30 are general thresholds that may not hold for each commodity. More research is needed to optimize the thresholds by commodity, but not much work of this type is publicly available. Traders that have developed measures privately treat the information as proprietary. But observation will help. If you observed for a particular commodity that the futures tend to trade lower each time the 14-day RSI approaches 65, you might treat 65 as the "overbought" threshold (and watch to see if 35 works for an oversold status).

An alternative that many traders clearly prefer is to use a smaller number of days in the RSI calculation. A 9-day RSI is becoming widely used, and many software packages developed to analyze the markets allow you to set the length of the RSI and experiment. Figure 4.43 reports the same October 1997 feeder cattle chart with a 9-day RSI. Note it dips to the 30 threshold to indicate an oversold market on several occasions.

It is worth the effort. The RSI can provide significant help to traders looking to add discipline to their trading or price risk management program. Earlier, when discussing corrections, the possibility of panicky selling in a falling market (or panicky buying in a rising market) was introduced. Consider how much help might be accruing to the mental state of producers or other potential short hedgers when they see

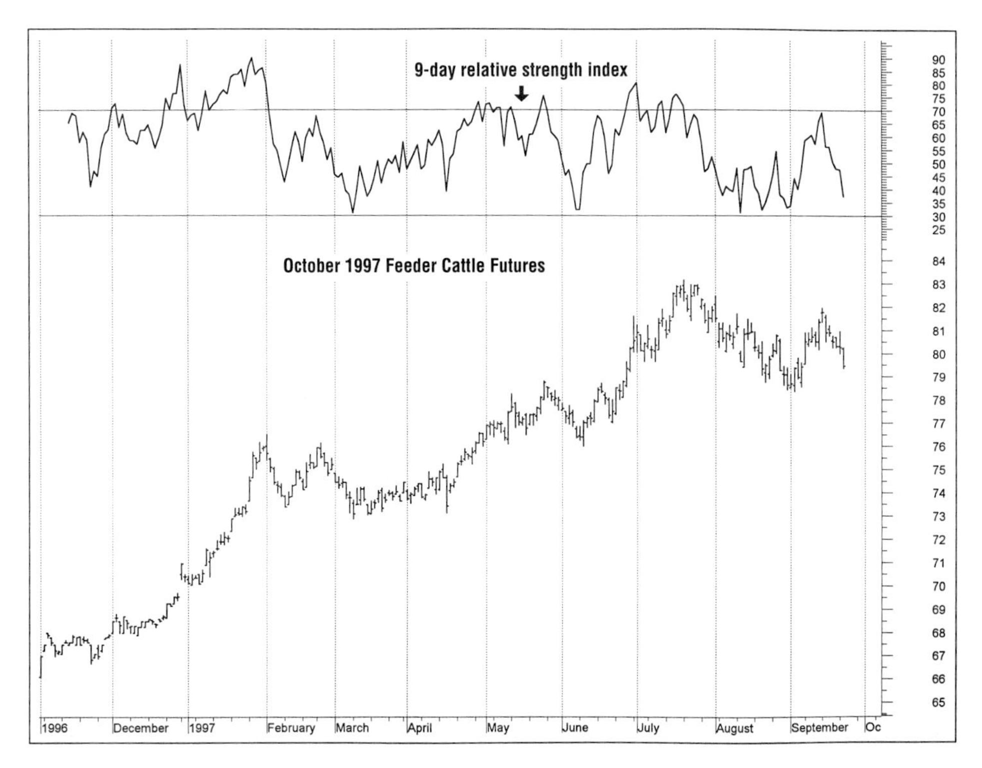

FIGURE 4.43
Use of the 9-Day RSI on the October 1997 Feeder Cattle Contract

the RSI below 30 and realize the market is oversold—and due for a rally. It is a lot easier to stand firm, be disciplined, and wait for the price rally to do some forward-pricing when a powerful tool like the RSI is supporting and reinforcing your analysis. A corrective price rally is likely when the RSI confirms that the market is oversold, and reasonable pricing objectives can also be established by calculating the 38, 50, and 62 percent corrections.

As with other chart patterns or technical tools, it is important to remember that the guidelines offered by the RSI become something of a self-fulfilling prophecy. If enough traders monitor the RSI and are therefore reluctant to step up and sell an already oversold market, the market is less likely to crash and go still lower. On the other side of the issue, if the fundamental supply–demand situation changes significantly and continues to change, the market will defy the RSI. We have seen corn prices climb steadily for 6–10 weeks as a drought intensified and the 14-day RSI remained above 70 during the entire time period. *No technical tool or chart pattern can defy a sustained and significant change in the underlying supply–demand situation.* We have, in the form of the RSI, another confirmation that the technical dimensions of the market cannot and will not dictate price direction that is contrary to an emerging and significant change in the supply–demand fundamentals. But those major and unpredictable shocks to the supply–demand setting occur infrequently and this means the RSI will be a major aid most of the time.

By watching what is happening to the price action on the chart and the RSI, a top or bottom in the markets can often be anticipated or predicted. *A divergence between the price action and the action of the RSI is an important signal that a price-trend reversal is coming, and this pattern is particularly useful to the disciplined trader.*

The same October 1997 feeder cattle chart demonstrates a diverging pattern between price and the RSI. In mid-July, prices surged higher and moved up through the resistance plane just below $82 at the late-June high. But the 14-day RSI, while above 70, did not make a new high in the mid-July action (Figure 4.42). *This divergence is powerful evidence that the market is at or close to a top, perhaps a major top.* The mid-July highs were, in late September, still the life-of-contract high.

Any pattern of divergence between price and the RSI that occurs above RSI values of 70 (in an uptrend) or below RSI values of 30 (in a downtrend) is extremely important. A pattern of divergence when the RSI values are between the 70 and 30 bounds should not be ignored, however. If the divergence occurs with RSI values in the mid-60s or the mid-30s, the behavior of the RSI can still be indicating that a top or bottom will develop.

The relative strength index is arguably the most important aid to the traders searching for help in bringing discipline to their trading programs. It is easier for the potential short hedger, for example, to avoid panicky selling of futures near the price lows when the RSI is suggesting the market will rally. The same assistance is present for the potential long hedger when an RSI above 70 suggests the market is overbought and that prices will not immediately move higher.

ANALYSIS OF THE 1996, 1997, 1998 CORN FUTURES

Technical analysis of the markets via the bar chart has been the focus of this chapter. Special attention has been paid to the buy and sell signals the chart patterns can generate. *There is a clear and intended implication that a disciplined trader who understands the basics in the art of chart analysis can be an effective selective hedger or speculator in the futures markets.*

Some traders are skeptical concerning the validity of technical analysis. At the more sophisticated level, the concerns are based on the notion that day-to-day changes in the markets are independent of each other. That independence, in turn, is based on the idea that the markets are efficient where the term *efficient* suggests that all available information is incorporated in the prices being discovered on any particular day. If the day-to-day changes are in fact independent over time, technical analysis of chart patterns will not work. The basic idea behind technical analysis is that what happened yesterday and on previous trading days will be a factor in determining what prices will do today and tomorrow.

The objective in this final section of the chapter, therefore, is to show you actual charts for corn for the 1996, 1997, and 1998 corn crops. Emphasis will be on the 1997 chart, but the discussion will be expanded to show the setting in which 1997 developments occurred and to discuss multiple-year hedging strategies when a bullish shock to the market occurs.

In late 1994, the USDA was considering whether to enforce a "set-aside" requirement for corn for the 1995 crop year. Ending stocks for the 1994–95 crop year had been boosted to estimates in the 1.4 to 1.5-billion-bushel range, a significant increase from the 686 million bushels in 1993–94. The flood-ravaged crop of 1993 was much smaller than usage, and stocks had been pulled down. Looking for the right balance between supply and demand to keep program costs in check, the USDA eventually announced a 7.5 percent set-aside requirement. This amounted to idling roughly 5 million acres.

The 1995 crop was planted late due to wet conditions in the late spring months, encountered less than ideal weather during the summer months, and then saw the yields from a late-developing crop reduced by an early frost. Early (May) 1995 USDA supply–demand reports were projecting a cash-price range of $2.30–2.70 for corn with ending stock estimates for the 1995–96 crop year at a relatively tight 998 million bushels. Farmers had reported plans to plant 75.3 million acres, less than many analysts had expected given the modest 7.5 percent set-aside requirement.

By December 1995, the 1995–96 ending stock estimates were down to 575 million and the cash-price range for the 1995–96 crop year was at $2.95–3.35. Stocks were *very* tight, and only a large 1996 crop could bring relief. Acreage planted in 1995 had skidded to 71.2 million acres because of wet conditions, but the May 1996 supply–demand report estimated 1996 acreage at 81.0 million acres, an expanded acreage that would be expected to produce a crop near 9.375 billion bushels and restore an ending stock level of 762 million for the 1996–97 crop year. (The May report put the estimate of ending stocks for 1995–96, the crop year ending August 31, 1996, at an incredibly small 317 million bushels—a historic low.) There were concerns we would "run out" of corn, and the market made sure that did not really happen by discovering—eventually—record high prices to ration usage.

This was a powerful "laboratory" within which to apply and test technical analysis. The markets were to test your discipline as well. It is revealing to look at the technical dimensions on the charts and see how well they worked.

Figure 4.44, the December 1996 corn chart, shows the price ramifications of that short 1995 crop. Cash prices started to move up in January and February and pushed

FIGURE 4.44
December 1996 Corn Futures

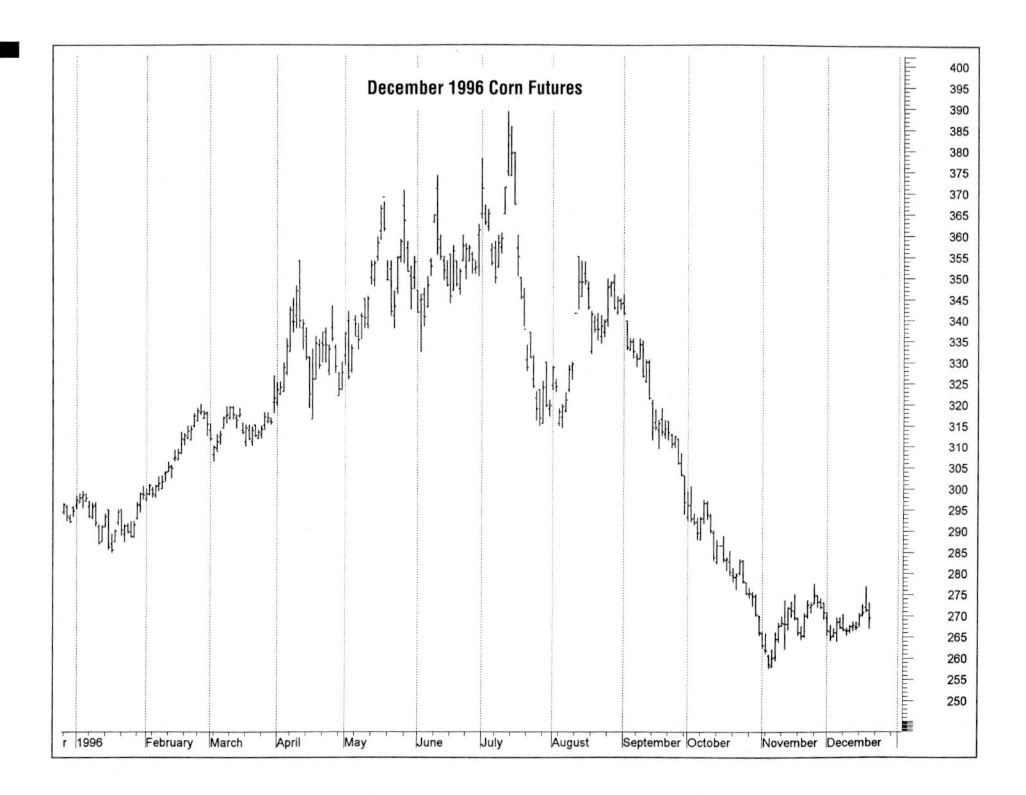

the new-crop December 1996 futures higher until mid-July. Cash prices were to eventually peak in August at levels above $6 in some high-deficit areas where concentrated poultry firms were located. The futures market was in a "wait and see" mode during the important pollination period of July, and then it waited to see what the yield-influencing weather patterns of August would do.

In this supply-driven bull market, the December 1996 futures started to make new highs in February and March. Long hedges placed on the second consecutive close at new highs during February would have established cost protection near $3.20. Much lower prices were available earlier during harvest of 1995. The moves on the preharvest futures such as the July were much more explosive, and long hedges would have been placed in those months to cover the intra-1996 needs. These long hedges would have been in place until a breakdown through the trend line on the chart during July would have generated a sell signal.

Short hedgers would have faced a more difficult management task. In April, May, and again in July, life-of-contract highs would have been "taken out" with two consecutive closes at new highs. Short hedges placed on rallies to those resistance planes should have been lifted (to be replaced on a rally to the new highs) or "margined" via margin calls. Note that during July, before the break of the uptrend line, there is the classic failure referred to earlier in the chapter. There were two consecutive closes above the high near $3.80 at the beginning of the month, but the close on the third day was not only very weak (near the low) but well below the old $3.80 high. *This failure of the market to respond to the buy signal denoted by the two consecutive closes at new higher prices is extremely bearish—and is a sell signal of major proportions.* After an August correction of some 50 percent of the July price plunge, the contract moved down to the $2.60–2.70 range in late harvest.

The insight and assistance in timing of pricing actions that can be gleaned from these charts could be repeated across the grains, the livestock futures, and the financial futures—across any of the commodities for which futures are traded. *It really does help to have a record of where the market has been, to be able to identify, even predict, changes in price direction, and to isolate reasonable and potentially attainable pricing objectives.* It is all logical, but it is not necessarily easy, and a caution is in order.

You would expect to see a relationship between the December 1996, 1997, and 1998 corn futures, and the relationship is definitely present. Remember, it is the production in calendar year 1996 that will be the corn stocks from which processing, exporting, feeding, and so on, is done until the harvest of 1997. If stocks are extremely tight, some corn can be imported from southern hemisphere producing countries where harvest is in March and April. Some of that was being done during the summer months of 1996 but not enough to change the supply–demand situation in the U.S. in any dramatic fashion. Until the corn crop is "made" in late summer of 1996 and stocks start to be replenished, the uncertainty in the marketplace will push price prospects for later years higher than normal.

The December 1997 corn chart in Figure 4.45 covers October 1, 1996, to late September 1997. Early-1996 developments looked very similar to those on the December 1996 chart. Summer 1996 highs moved above prior highs in February and March and price rallied through the summer months. The price boosting uncertainty for 1997 corn continued into September 1996, some two months after a price break on the December 1996 futures chart. The looming harvest was putting pressure on the 1996 corn prices—but the outlook for 1997 was still charged with uncertainty.

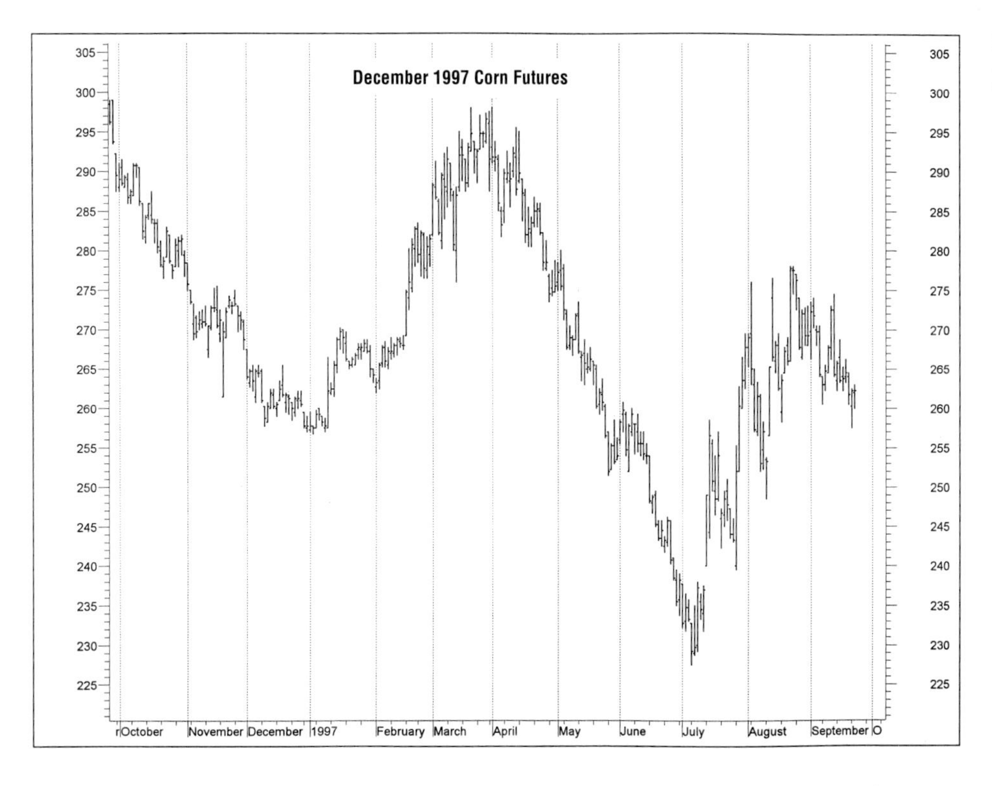

FIGURE 4.45
December 1997 Corn Futures

The supply-driven bull markets will always create pricing opportunities for later crop years. You will want to watch for these often significant opportunities before the inevitable supply response has time to drive the market lower again.

It *is* the case that the impact of the partial crop failure diminishes as you look further ahead. The July 1996 corn futures (not shown) was still "rationing" the short 1995 crop via discovering prices that went as high as $5.54, prices over $1.50 above the highs on the December 1996 corn futures. The December corn futures, of course, were discovering price *after* the 1996 harvest, damping the impact of the very tight preharvest's stocks. Note, too, that the 1997 chart showed price highs just above $3.10—well below the $3.90 levels on the 1996 chart and far below the record cash prices of the summer months of 1995. *But the impact was clearly still there: the best prices for the 1997 crop were offered in the late summer and early fall months of 1996.*

If you look still further ahead, into 1998, you would expect the impact of that short 1995 crop to be muted even further. It is, but the patterns remain. The December 1998 corn futures started trading in mid-August 1996 and recorded contract highs within the first few days of trading (Figure 4.46). This was the same month (September 1996) that showed the contract highs on the December 1997 chart. From that high until late September 1997, the 1997 and 1998 charts show similar patterns. The March-April price rally, the April-June price plunge, and the late summer rally are apparent on both charts. The September 1997 discovered prices for 1997 and 1998 corn were virtually identical.

FIGURE 4.46
December 1998 Corn Futures

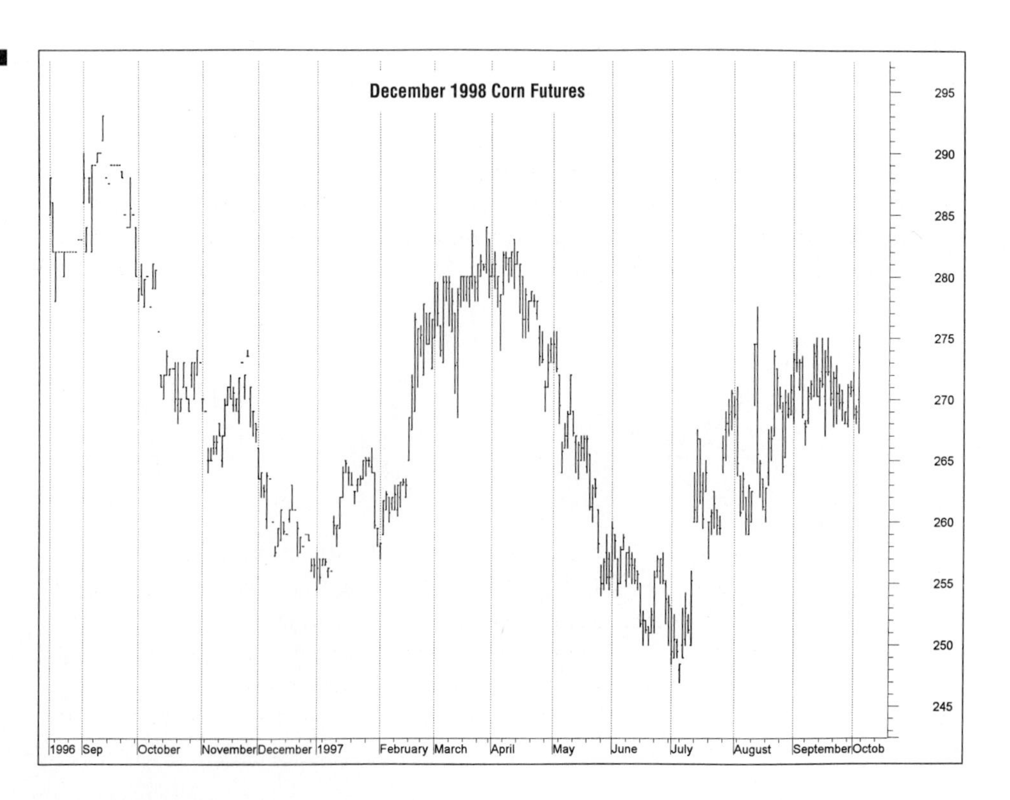

The three December corn futures charts, 1996, 1997, and 1998, clearly show a relationship. The possibility of multiple-year hedging emerges from that relationship. Techniques and how to pursue such strategies will be covered in later chapters. The intent here was to demonstrate how buy and sell signals on the closer charts in a time context, such as the 1996 corn, tend to prompt parallel moves on charts that are discovering prices for later years. Then, when the prices are pushed down by a supply response, it tends to happen on all of the charts.

The references at the end of the chapter by Kenyon and Beckman report on the effectiveness of this type of multiple-year strategy. They test a strategy that prices the crop for two years after the unusual market (1996 here) on the first signal back in 1996. Positions are placed in a late 1996 or 1997 futures and then "rolled" to the December 1998 contract at a later date. If you are interested in these multiple-year strategies, and you should be, the reports are worth requesting. Use the address or the e-mail address shown with the listing.

That's where the chart analyst has an edge. There is a long historical record that shows the markets *do* fall after sell signals at trend lines, that they *do* try to correct to the gaps, and there is a long historical record that shows the market *will* run into strong selling on an approach to the gap. Unless you have compelling fundamental analysis to suggest prices above $4 will in fact occur, evidence other than long-term forecasts of summer weather, the correct approach is to sell a close below an uptrend line or a corrective rally up toward a chart gap. But it is not easy to do, and we will return to this behavioral dimension often in later chapters as strategies are developed and explained.

LONG-TERM BAR CHARTS

The use of the bar chart will be demonstrated many times in later chapters. Before leaving this discussion, however, it is useful to discuss the long-term bar charts.

Most charts to this point are daily charts. The long-term corn chart was introduced briefly earlier. A weekly continuation or long-term chart can be constructed by taking the weekly high and low for the nearby futures and by using the close on Friday. When a particular futures contract matures and goes off the board during the week, the high and low from *both* the nearby contracts are used. To illustrate, assume the following high, low, and closing prices on Friday for the June and August live cattle futures during the week of June 18–22:

	June	August
High	$75.60	$74.80
Low	74.20	73.90
Close	75.30	74.50

On the weekly chart, the high for the June ($75.60), the low for the August ($73.90), and the close for the August ($74.50) would be used. June matures on the 20th and is no longer traded, so attention switches to the August.

A monthly chart can be constructed in a similar fashion. The high and low for the nearby (or the two closest futures if the nearby is maturing) futures are used with the close the last day of the month. The result is a continuation chart that records a historical pattern over a number of years.

Many of the chart patterns that appear on the daily charts are important on the long-term charts. This is especially true of trend lines and resistance and support planes. Many traders watch the lows and highs on the monthly charts for support and resistance, respectively. It is not at all unusual to see a bottom on a daily chart at or near the long-term support plane or to see a top at the long-term resistance.

For a longer-term perspective, then, the weekly and monthly charts can be very valuable. Longer-term hedgers would get guidance from the weekly charts, and the monthly chart can help in placement of hedges and on longer-term speculative trades.

SUMMARY

Technical analysis is the key to the correct timing of buy and sell decisions in the commodity futures markets. The technical dimensions of the market do not dominate the fundamental supply–demand dimensions, and no sustained technical pattern will develop that is contrary to the emerging and underlying supply–demand balance. But the discovered price can and will move, and trace out technical patterns, as the market seeks to discover the price that balances the forces of supply and demand. Within the limits to those price moves, *technical analysis can be an important guide to the timing of pricing actions.*

The bar chart is the most widely employed guide to the commodity markets. Trend lines, resistance and support planes, and various topping or bottoming formations can be employed in monitoring the direction of price trends or even predicting changes in the direction of the trend. Consolidation patterns assist the trader in making the correct trades or in timing of makeup pricing actions if an earlier signal and opportunity were missed.

The integration of *chart gaps* and the *expectation of predictable market corrections* assists the trader in bringing discipline to the trading program. It is easier to wait, be patient, and avoid panicky actions if there is reason to expect the market to generate reasonable opportunities a bit later.

Volume and *open interest* are important complements to interpretation of the bar chart patterns. The *relative strength index* (RSI) is another aid that helps decision makers bring discipline to their pricing programs and to avoid mistakes in a market that is reaching extremes in terms of price movement.

Once the decision maker determines the probable *price range and direction of price trend* using fundamental analysis, then *technical analysis* can be very effective in guiding the timing of actions taken as the fundamental picture is traced out during the year or other decision period.

The patterns *do* emerge on real-world charts, and the assistance the charts are presumed to offer *is* present. In the final analysis, *it will be the level of discipline in following through with a market plan that will determine how effective the technical analysis of the markets will be.*

KEY POINTS

- The *trend line* is perhaps the most reliable bar chart indicator. Closes below an uptrend line or above a downtrend line are *reliable indicators of a change in price direction* and generate buy and sell signals for the hedger and speculator.

- In markets in which major price trends are not present, *resistance planes* and *support planes* become valuable guides to pricing actions. The *planes at contract highs and at contract lows* are especially important, and trading actions around those planes require *discipline and a preset plan.*
- A number of *topping and bottoming formations* are widely recognized. Among these are *double tops and bottoms, head-and-shoulders* formations, *key reversals*, and *island reversals.* All indicate a change in the direction of the price trend, and some offer a means of projecting the magnitude of the expected price move.
- *Chart gaps* and *price corrections* provide a base upon which traders can develop objectives for pricing actions.
- Various *consolidation patterns* appear on the chart. They are helpful in deciding whether the market is starting a top or bottom or is just "resting" before a renewed price move.
- *Trading volume is a measure of intensity* and all buy and sell signals will be more valid when they occur on unusually high trading volume.
- *Open interest* can signal a coming change in the market. *In general, increasing open interest is needed to sustain major price trends.*
- *The relative strength index (RSI)* is a reliable measure of momentum in the futures markets. *Recognizing that the market is oversold or overbought and is ready for a correction helps bring the patience and discipline that is often needed.*
- Examination of current bar charts *shows the formations* and confirms the usefulness of the charts in *guiding the timing of marketing and pricing actions.*

USEFUL REFERENCES

Chicago Mercantile Exchange, *Trading Tactics,* CME, Chicago, 1986. A very basic and useful demonstration of chart analysis designed for the beginning user.

David Kenyon and Chuck Beckman, "Multiple Year Pricing Strategies for Soybeans," Virginia Cooperative Extension Publication 448-023, Virginia Tech, Blacksburg, VA, 1996. Order by calling Extension Distribution Center at (540) 231-6192 or e-mail to purcell@vt.edu.

David Kenyon and Chuck Beckman, "Multiple Year Pricing Strategies for Corn," Virginia Cooperative Extension Publication 448-024, Virginia Tech, Blacksburg, VA, 1996. Order by calling Extension Distribution Center at (540) 231-6192 or e-mail to purcell@vt.edu.

Jack D. Schwager, *Technical Analysis,* John Wiley & Sons, New York, 1996. The author demonstrates the way many of the technical dimensions of the market can be used in a disciplined trading program. Coverage includes oscillators, RSI, moving averages, and point and figure charting.

J. Welles Wilder, *New Concepts in Technical Trading Systems,* Hunter Publishing, Winston Salem, NC, 1978. This reference covers the development of the RSI and other measures of momentum.

APPENDIX 4A. TYPES OF ORDERS AND DEMONSTRATIONS OF USE

Many orders and many combinations of orders can be used in trading commodity futures. In this appendix, the more standard orders will be defined and illustrated. The development assumes that the trader is acting as a producer and is employing a selective hedging program, which means an objective of having short hedges in place when the price trend is down and being a cash market speculator with no futures positions in place when the price trend turns up.

Market Order: An order to take a position "at the market." The order would be written by the broker as

Sell 20 July wheat at the market.

The floor broker would sell 20,000 bushels of July wheat at the first available price offered by a buyer in the pits at the Chicago Board of Trade. Such an order would be used by the producer to hedge or forward-price 20,000 bushels of wheat when some preestablished pricing objective has been met.

Advantages of the order revolve around the assurance that the producer will get a "fill" and will be short in the market. The order would typically not be filled if the market is limit down and there are no willing buyers, but this is a rare occurrence.

The big disadvantage of the order is that the producer has no assurances as to the price levels at which the short positions will be established. If the order is placed with the broker with whom the producer is working when the July wheat is trading at $4.10, it is very possible that the actual short positions acquired will be lower than the $4.10 if the market conditions are volatile.

The market order should be used only when the trader feels that some action is imperative, and probably should not be used routinely in a selective hedging or other type of hedging program. The disadvantage of not having control over the entry-level price into the market is a serious shortcoming of the order.

Limit-Price Order: An order to take a position at a particular and specified price level or better. This order would be written as

Sell 20 July wheat at $4.10.

A sell or short position for 20,000 bushels of July wheat will be established only if the order can be filled at $4.10 or higher. While not always written that way, the order is interpreted in processing on the floor of the trading pits as "$4.10 or better." This is a widely used order in a scale-up pricing program such as

Sell 20 July wheat at $4.10,
Sell 20 July wheat at $4.20,
Sell 15 July wheat at $4.30.

The orders could be placed with the broker on a "good 'til canceled" basis (called GTC in the trade), and a total of 55,000 bushels of wheat would be hedged at prices from $4.10 to $4.30 if the market rallies to $4.30.

The primary advantage of the order is that the entry level into the market is controlled in terms of price level.

The disadvantage of the order is that it may not be filled and if the markets then trade lower, there is no protection in place. The $4.10 order will not be filled, of course, if the market never trades up to $4.10. There is some possibility that the trading range for a particular day could show $4.10 as the high for the day and the order may not be filled. It would have to trade at $4.10 long enough for all sell orders to be filled, and there is no guarantee of that.

The order is widely used in trading and hedging programs and is an effective and desirable order in that it controls the price levels at which the producer is willing to enter the market. The opportunity to place the orders such that hedging is accomplished on a scale-up basis is especially attractive.

Sell-Stop Order: An order that becomes a market order if touched from above. The order would be written as

Sell 20 July wheat at $3.95 stop.

In a selective hedging program, the producer might have reason to expect the market to move well above $4.00, but wants to be protected if that analysis is wrong and the market turns lower. By placing a sell-stop order at prices below where the market is currently trading, the protection against lower prices is in place. The order is widely used, for example, under an uptrend line with the thought that if the market trades down through the trend line, the price direction is reversing and short hedges will be desired.

The advantage of the order is that it allows the producer to "trail" an upward-trending market and be in a position to benefit from higher prices, but be protected if the market does turn unexpectedly lower. Like any order, the sell-stop can be placed on a "day" basis (which means it would have to be put in again the next day if desired) or on a GTC basis and moved up under an upward-trending market every day or every few days.

Among the disadvantages is the fact that the order becomes a market order when touched from above. It could be filled at levels below the price specified in a volatile market.

A second disadvantage is that the order offers no protection and can give unexpected fills to the beginning trader who does not fully understand the markets. For example, assume that the market shows a low of $3.97 on Thursday and an important and bearish report comes out after the market closes on Thursday afternoon. If the market opens sharply lower on Friday morning at $3.81, a sell-stop order at $3.95 that is placed on a GTC basis on Thursday is still in effect and would be filled at the first opportunity around $3.81.

The sell-stop order can be an effective order if it is fully understood. *It is appealing to be able to "trail" an upward-trending market and be in position to set short hedges if the market turns unexpectedly lower.*

Buy-Stop Order: An order which becomes a market order if touched from below. The order would be written as

Buy 20 July wheat at $4.19 stop.

Widely used by speculators who have short positions in the market and who want to limit their risk exposure if the market moves up against a short position, the order has a place in the marketing program for a selective hedger. For example, assume 20,000 bushels of July wheat were sold because the life-of-contract high was at $4.14 and the producer sold a rally toward that level with a limit-price sell order at $4.10. If the market trades as high as $4.19 and makes new contract highs, it is likely to trade sharply higher. The producer might wish to limit margin calls and be in a position to benefit from higher prices as a cash market speculator. A buy-stop order at $4.19 would offset or lift the short hedges if the market trades up to the $4.19 level.

The advantage of the order is that it allows placing a limit on the exposure from short hedge positions. For a selective hedger, this can be an important feature.

The disadvantages are the same as those for the sell-stop order. When the specified price level is touched from below, the order becomes a market order and might be filled at higher price levels. And if the market "gaps" higher due to a report and opens at $4.35, to continue the earlier example, the buy-stop order that is placed on a GTC basis at $4.19 will be filled near $4.35. The objective of limiting exposure from the short hedge positions is not met.

Buy-stop orders can be effective in selective hedging programs. They should not be placed so close to current trading levels, however, that the producer is constantly trading in and out of the market and is behaving more nearly like a speculator in futures.

Sell-Stop-Close Order: An order that will be filled only if the market closes below a specified price level. The order is written as

Sell 20 July wheat at $3.95 stop-close-only.

The order would be filled at the close of trade for the day if the close is below $3.95. The order is widely used by traders, including short hedgers, who wish to take a position in the market only if the market closes below some preselected level. The idea is that it is not where the market *trades* but where it *closes* that is important. For example, a regular sell-stop order that is being moved up under an uptrend line every few days will be filled if the market darts down through the trend line and touches the order from above even if the market then turns and closes sharply higher for the day. A short hedge placed under those conditions will not look attractive to the selective hedger.

The advantage of the order, therefore, is that it is filled based on the close of trading. As suggested, it is the close that is often viewed as the important signal for action in a selective hedging program.

The disadvantage is the same as any sell-stop order. It can be filled at a close that is well below the specified price level. On a day that the market closes limit down, the order will not usually be filled, of course. An important caution is in order here. The "sell-stop-close-only" order will not be accepted at all exchanges. The Chicago Board of Trade may not accept the order for the grain and oilseed contracts, even though the illustrations here are for wheat contracts. The Kansas City Board of Trade may take the order for its hard red winter wheat contracts. The Chicago Mercantile Exchange will accept the order for the livestock commodities. If the exchange will not accept the order, the only alternative is to have the broker enter a market order or a limit-price order near the close of the trading day and effectively accomplish the same thing. You should check with your broker to make sure that the order will be accepted by the particular exchange involved for various futures instruments.

Buy-Stop-Close-Only Order: An order that will be filled at the close of trade for the day if the close is above a specified price level. The order will be written as

Buy 20 July wheat at $4.19 stop-close-only.

Clearly, the order is superior to the buy-stop order if attention is being paid to the close rather than to the trading range for the day. A regular buy-stop at $4.19 would lift a short hedge that the producer would want to be in place if the market trades up to the $4.19 level and then turns and closes sharply lower for the day. The above order at $4.19 on a stop-close-only order would *not* lift the short hedge, of course, unless the market *closes* above $4.19. A review of a key-reversal top will quickly show the advantage of not lifting short hedges via a buy-stop order when the market then *closes* lower.

The advantage is that the focus of the order is on where the market closes. The disadvantage is that the fill could be well above the specified price level of $4.19 if the market does surge and close sharply higher on the day.

The same caution is in order: Not all exchanges will accept the stop-close-only orders, whether they are sell-stop or buy-stop orders.

Extensions and Constraints in Orders: Several constraints or conditions can be placed on the standard set of orders or used to enhance their effectiveness. The most widely used are discussed here.

1. *Market If Touched (MIT).* This condition converts a limit-price order, for example, to a market order. If the pricing objective of the producer is $79.50 in June live cattle, the MIT provision added to a sell order at $79.50 will help to ensure the order will be filled. The order becomes a market order if the $79.50 level is touched, and this protects against the possibility of the market showing a $79.50 high for the day but leaving some orders to sell at $79.50 unfilled. Fills will typically be at or near the specified $79.50 level because the order will be filled quickly and there is no time lag involved like there is in getting a regular market order into the trading pits. The order may not be accepted by the Chicago Board of Trade on the grains. Again, check with your broker.
2. *Good 'Til Canceled (GTC).* Discussed earlier, the GTC provision keeps the order in place until it is filled or canceled. It eliminates the need, for example, to keep placing the same order every day. A sell order on July wheat at $4.30 that has the GTC provision will be there and waiting on any rally that comes in the market up to the $4.30 level. Any order that does not have the GTC provision is canceled automatically at the end of each trading day if it has not been filled. Most brokerage firms cancel the GTC orders that have not been filled at the end of each month, however, and they would then have to be replaced. Check with your broker.
3. *One Cancels the Other (OTO).* Used when two orders are in place, the provision cancels one of the orders when one has been filled. For example, the producer might have in place at the same time a limit-price order such as "sell 20 July wheat at $4.10" and a stop order such as "sell 20 July wheat at $3.95 stop." The idea is to either sell on a rally to $4.10 and get hedges placed or sell at $3.95 via a sell-stop if the market turns lower rather than rallying to the $4.10 objective. If the $4.10 order is filled, the $3.95 stop order is automatically canceled and the

producer does not face the risk of selling more than had been planned and at two price levels.

4. *Time Constraints.* Virtually any order can be placed such that it is canceled if not filled by a particular time. A limit-price order to sell June live cattle at $79.50 could be placed so that it is canceled if not filled by 12:55 p.m. Central Time of a particular day. The producer might reason that if the market rallies near the close of the trading period, then still higher prices are likely the next day and they might not want the sell order filled in the last few minutes of trading. This is sometimes called a "fill or kill" provision in trade jargon.
5. *Price Constraints.* A constraint or limit can be placed on an order such as a "sell 10 June live cattle at $79.50 stop." If the producer does not want to see fills well below the $79.50 level as the order is converted to a market order in a downward-moving market, a "limit" of $79.40 can be placed on the fill for the order. This provision can be especially useful in a thinly traded market or a thinly traded distant futures contract where a significant increment of price move might be required to find someone who is willing to take the other side of the trade. The Chicago Board of Trade may not accept the use of "stop-limit" orders. Check with your broker.

The primary message here is that the trader must understand basic orders and be able to use them if the program is to be effective. Regardless of which orders are used, it is important to keep records on a day-to-day basis so that the producer is constantly aware of what active orders they have in the markets and make necessary adjustments over time. *It is always a good idea to discuss the order with the broker so both parties are sure that the order being placed by the broker is consistent with what the trader wants to do.*

CHAPTER 5

TECHNICAL ANALYSIS: ALTERNATIVES TO BAR CHARTS

Traders record market action in numerous ways and use numerous tools in managing a trading program. Bar charts are the most widely employed and widely watched alternative. Chapter 4 provided coverage of the bar chart and how it can be applied. But alternative tools are also widely used. The point-and-figure chart is a favorite approach of many traders and some traders use moving averages, a mathematical way to monitor the trends in the market. This chapter will explore selected alternatives to the bar chart and show how they can be used.

POINT-AND-FIGURE CHARTS

A sketch of a point-and-figure chart is shown in Figure 5.1. There is only a price scale. The time calendar used at the bottom of the bar chart is missing. Advocates of this charting technique will argue that only price action, not time, is important.

The important parameters on the point-and-figure chart are the *cell size* and the *reversal requirements.* The cell size is simply the value, in terms of price increments, of each of the cells or "boxes" shown on the chart. The reversal requirement refers to the magnitude of price changes required before a reversal in direction of the price trend will be allowed.

How effective this charting approach turns out to be in signaling correct buy and sell decisions for a particular commodity depends on the choices of cell size and reversal requirements.

Figure 5.2 shows a point-and-figure chart for a hypothetical commodity with a cell size of 4 cents and a three-cell reversal requirement. Each cell in the chart has a value of $.04, and the price scale is therefore set in multiples of $.04. The Xs indicate higher prices; the Os indicate lower prices.

The patterns shown in the figure suggest what is needed to record a price reversal. If prices have been moving up and then start to decline, the price move down must be sufficient to allow the chartist to drop a cell and then fill at least three cells. If these conditions are not met, no reversal has been recorded. A column of Xs, which

FIGURE 5.1
Demonstration of a Point-and-Figure Chart for Commodity Futures

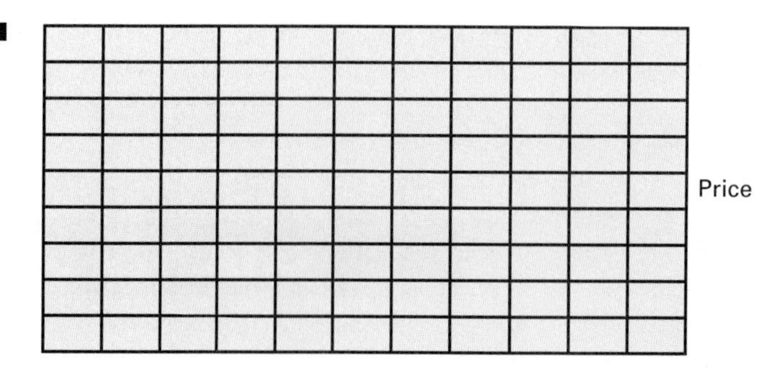

denotes higher prices, must therefore always extend one cell higher than a subsequent column of Os, which denotes lower prices. That is what is meant by "dropping a cell." Also, note that the minimum number of Os (or Xs) shown is always three. This is more precisely what is meant by a three-cell reversal requirement.

More detail on how this somewhat different chart is developed is needed, of course. To start a point-and-figure chart, look at the trading range for a particular day and continue on subsequent days to watch until the daily trading range is sufficient to fill three cells. When this trading-range requirement is met, fill the three boxes with Xs if the close is above the midpoint of the trading range, with Os if the close is below the midpoint.

Call the first day that three or more Xs, for example, can be plotted day 1. On day 2, given that Xs for higher prices are being plotted, look only at the high for the day. If the high allows one or more higher price cells to be filled, plot the added Xs and ignore the low for the day. The daily trading highs can keep the price direction positive and bring the plotting of Xs as higher price cells are filled. Continue the process until, for example, the high on day 5 does not allow the plotting of at least one higher price cell. On day 5, look at the low and do nothing if the low is not sufficiently low to allow dropping one cell and then plotting at least three cells with Os to the downside. If several days pass with no price low sufficient to record a reversal, keep watching and resume the plotting of Xs when the high of a daily trading range finally allows the uptrend to be continued by plotting a new higher cell.

Eventually, of course, higher prices are not available and the reversal requirement will be met and the trend will turn lower as Os are plotted. Then, watch the daily lows and plot Os until the day occurs in which a new lower cell is not filled and the reversal requirement is met. Turn the plot back up and start recording a new column of Xs, denoting that the price direction has switched from down to up.

The procedure looks complicated at first, but it is really quite simple and offers a quick way to track the market. Keep in mind that

There is no time on the chart;

The choices of cell size and reversal requirement will determine the amount of action on the chart. Clearly, if a large cell size and a large reversal requirement are selected, there will be many days during which nothing is plotted; and

Only the highs and lows of the daily trading range are employed. Once the charting process is underway, the daily close or settlement price is not used.

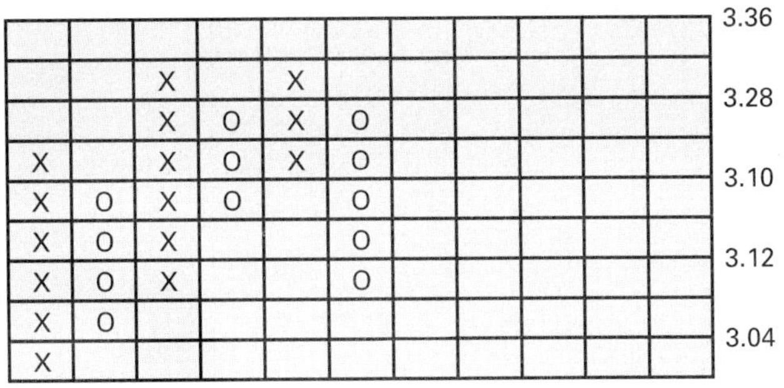

											3.36
		X		X							
		X	O	X	O						3.28
X		X	O	X	O						
X	O	X	O		O						3.10
X	O	X			O						
X	O	X			O						3.12
X	O										
X											3.04

FIGURE 5.2
A Point-and-Figure Chart with a Three-Cell Reversal Requirement and a $.04 Cell Value

In the context of Figure 5.2, the initial surge in price (left side of graph) never generates a daily high that fills the $3.24 to $3.28 cell. A daily high of at least $3.28 would be required. As market action was monitored, the $3.28 was not reached, but a daily low of $3.08 or less was eventually observed. A daily low of $3.08 or less allows one cell to be dropped and three lower cells to be filled, meeting the reversal requirement. A daily low of $3.04 or less was also eventually observed, but a reversal to the upside was recorded before a low of $3.00 was recorded (the $3.00 to $3.04 cell was never filled). The plotting then continues in a like manner with a column of Xs representing a rising market, a column of Os representing a declining market. A single column could cover several days, or even weeks, of trading.[1]

Examination of Figures 5.3, 5.4, and 5.5 demonstrates the importance of the selection of the cell-size parameters. If there is a widely accepted practice on reversal requirements, it is a three-cell requirement. An entire book by Cohen listed in the references at the end of the chapter deals with the three-point reversal requirement in market analysis. Both Figures 5.4 and 5.5 use a three-cell reversal requirement, but the cell sizes are significantly different. Figure 5.4 uses a $.04 cell size and Figure 5.5 uses a $.02 cell size. Figure 5.3, of course, shows the conventional bar chart and allows comparisons of the two approaches to charting. Most modern computer software packages designed to allow analysis of the markets will let you try various combinations of cell size and reversal requirements. For a particular commodity, you will quickly find the set that seems to capture the significant moves in the market. As a rule of thumb, if you are getting reversals daily, the cell size and/or the reversal requirement may be too small. If you are getting only one reversal or less per week, the cell size and/or reversal requirement are too large.

[1]It might help you to recognize that examination of the chart in Figure 5.2 would enable us to conclude that (1) the initial price rally never reached $3.28; (2) the subsequent and first decline never reached $3.00; (3) the second price rally never reached $3.36; (4) the second price dip never reached $3.12; (5) the third rally never reached $3.36; and (6) the last price dip, when plotting was stopped, had not reached $3.04. Thinking through these "conclusions" should help clarify how the chart was plotted. Also, go back through the fact that before the first price rally reached $3.28, a low of at least $3.08 was recorded and a reversal was allowed. Then, before a price low of $3.00 or less was recorded, a price high of at least $3.20 was observed, a reversal was recorded, and the second price rally on the chart was underway.

FIGURE 5.3
Bar Chart for December 1997 Corn

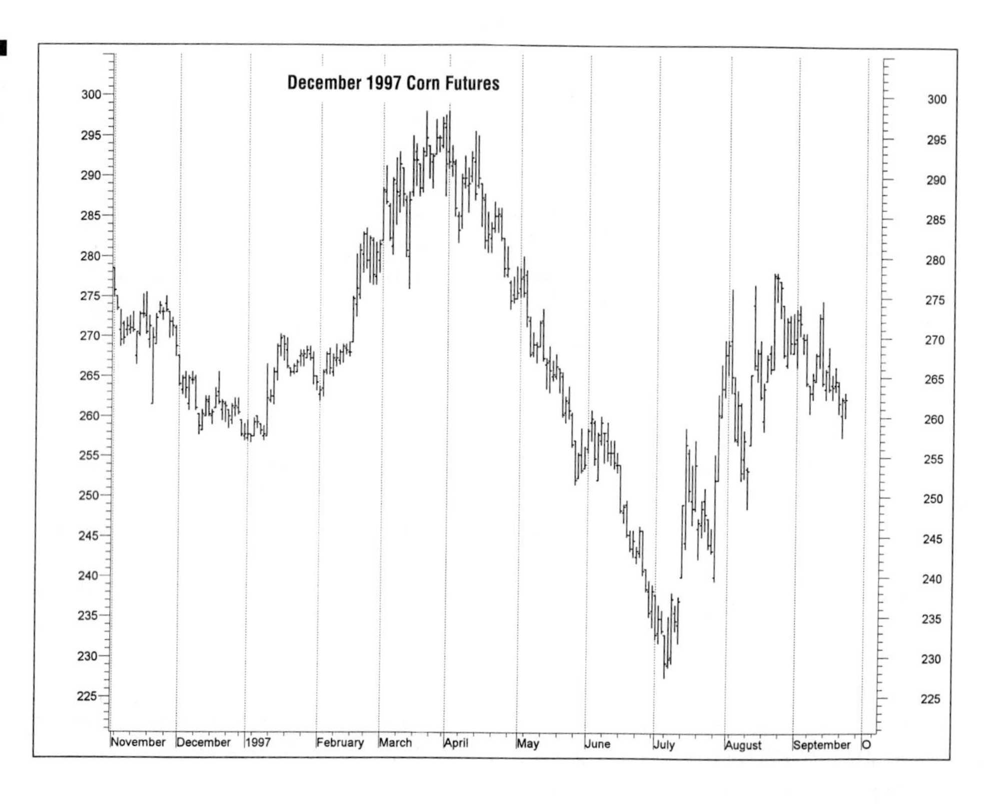

FIGURE 5.4
Point-and-Figure Chart for December 1997 Corn with a $.04 Cell Size and a Three-Cell Reversal Requirement

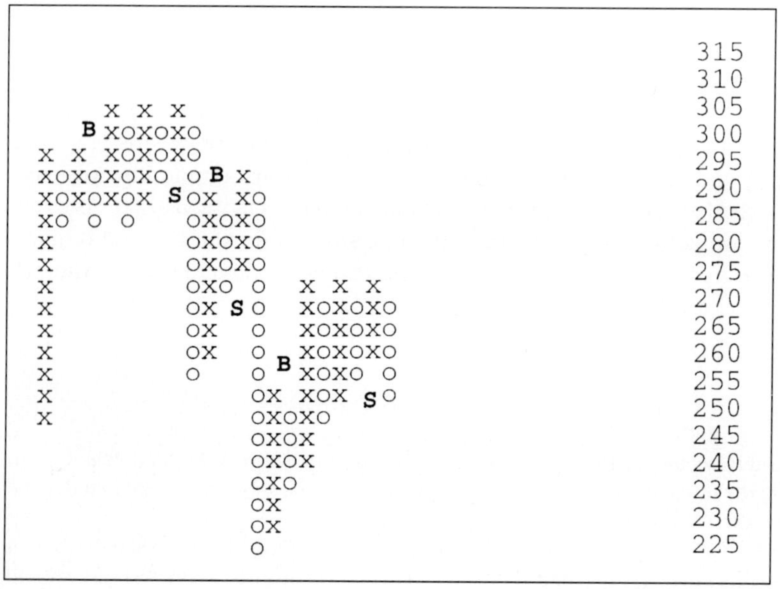

```
    X X X
   B XOXOXO
X X XOXOXO
XOXOXOXO O B X
XOXOXOX S OX XO
XO O O   OXOXO
X        OXOXO
X        OXOXO
X        OXO O  X X X
X        OX S O XOXOXO
X        OX   O XOXOXO
X        OX   O XOXOXO
X        O    O B XOXO O
X             OX XOX S O
X             OXOXO
              OXOX
              OXOX
              OXO
              OX
              OX
              O
```

```
315
310
305
300
295
290
285
280
275
270
265
260
255
250
245
240
235
230
225
```

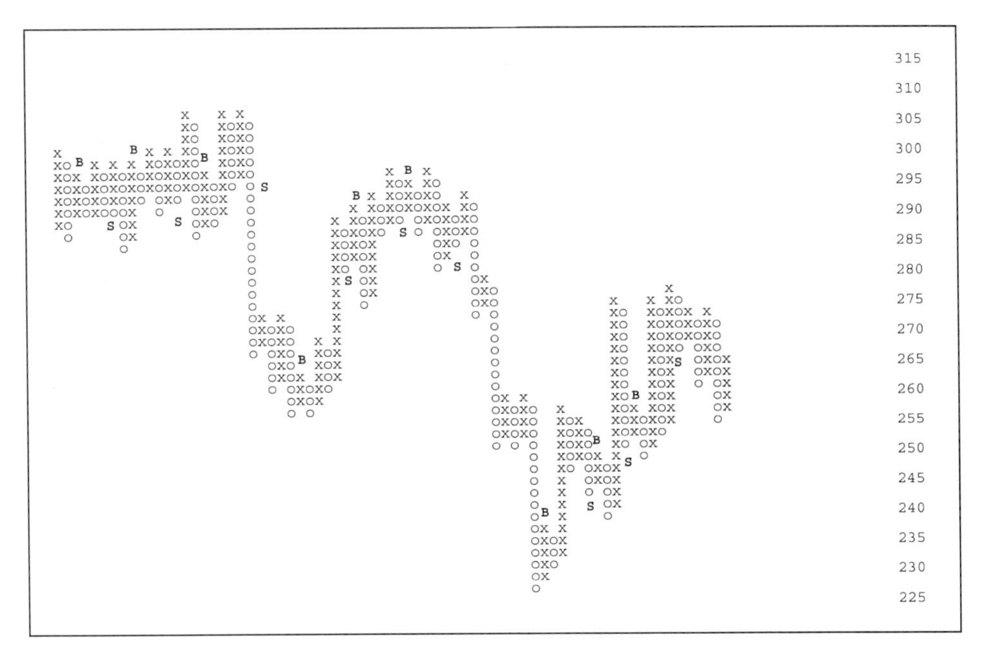

FIGURE 5.5
Point-and-Figure Chart for December 1997 Corn with a $.02 Cell Size and a Three-Cell Reversal Requirement

In the point-and-figure charting technique, a buy signal is generated when a market rally pushes a column of Xs above the previous column of Xs. A simple buy signal would therefore appear as follows:

```
O       X ← Buy
O X     X
O X O X
O X O X
O   O
```

Conversely, a sell signal occurs when the prices dip such that a column of Os moves below the minimum cell in the previous column of Os. A simple sell signal would thus be as follows:

```
O X   X
O X O X O
O X O X O
O X O   O
O       O ← Sell
```

Examination of Figure 5.4 indicates only three sell signals were generated during 1996 and into October of 1997. The first sell signal came in the fall months of 1996, during harvest, but the subsequent buy signal developed at a higher price. In a selective hedging program, this would have been a losing trade. The second and major sell signal occurred in April of 1997 when the topping action and price reversal shown on the bar chart (see Figure 5.3) developed. A third sell signal came in the early fall of 1997 when the prices dipped toward the $2.55 level.

Figure 5.5 shows a much more "active" chart. There were several sell signals during 1996 and a total of six during 1997. Even a cursory look suggests the $.02 cell size is the better choice if the charts are being used for a selective hedging program. The two completed round turns (sell and buy back) in Figure 5.4 would have essentially broken even (depending on exact prices realized) after paying commissions. Picking up the signals in Figure 5.5 during 1996 as the industry came off the record 1995 and 1996 prices, a sell signal at $2.92 was bought back near $2.66—a futures gain of $.26. The two subsequent "round turns" both lost money, about $.23 in total. The next and major sell signal came in late spring of 1997, with a sell signal near $2.83 and a buy near $2.40—a gain of $.43. The later August round turn lost about $.13, and the sell signal in early September would have the producer short as harvest approached.

In net, before commission costs of $.04 to $.06 per bushel, the program gained some $.33. Note this is close to what the conservative hedger or cash contractor would have realized if he had sold near $3.00 back in April of 1997 and still had short futures positions (or cash contracts) with the market near $2.65 to $2.70 in September. But there is a big difference: producers following the point-and-figure signals will buy back the hedge positions and be out of futures and in a cash speculative mode if a major and unexpected move to the upside occurs.

Figure 5.5 indicates the $.02 cell size would have been much more effective and the patterns also show the parallels between a point-and-figure chart with the correct parameters and the bar chart. The first sell signal on the point-and-figure chart came when the market declined through the trend line near $3.00 on the bar chart, a signal in October of 1996. The second and major sell signal came in April of 1997, corollary to the break down through the steep uptrend line on the bar chart. If you had used this point-and-figure chart instead of trend lines in a selective hedging program, the results would be roughly comparable.

Figure 5.5 can be used to show a feature that shows an advantage of the point-and-figure chart. The point-and-figure chart provides a means of projecting the price move after a major signal is generated. Once a topping or bottoming pattern is identified, the projection rule is

$$PPRO = CV \times RR \times NC$$

where

$PPRO$	= the price projection,
CV	= value of each cell,
RR	= reversal requirement in number of cells, and
NC	= number of columns in the top or bottom formation.

Some judgment will be required, but the pattern on the corn chart suggests topping action near $3.00 in the spring of 1997, and a topping pattern that includes 16 columns. Using that and subtracting the projected distance *from the top of the formation* generates a price estimate of

$$\$3.00 - (\$.02 \times 3 \times 16) = \$2.04$$

The projection is below the plotted low price of about $2.26 shown in Figure 5.5, but the projection does indicate a major price break is imminent. The bottom that dipped down toward $2.25 has four columns, projecting up to $2.49, and the market reached that level before recording another bottom on the dip to $2.38. That seven-column formation projects up to $2.80, and the next rally came very close to the $2.80 level.

The point-and-figure chart also uses a version of the trend line that is so popular with users of the bar chart. But on this chart a 45-degree line is very important. Figure 5.6 demonstrates the use of a 45-degree line in an ascending market in the summer of

1997. Note the pattern of higher lows that traces out a 45-degree line. Once the reversal toward higher prices near $2.25 is recorded, the 45-degree line can be placed on the chart and extended to the upside. When a cell is filled below that extended 45-degree line, this action will be treated as a sell signal, much like the close below the trend line on the bar chart.

As noted earlier, experience with the point-and-figure charting suggests that if little plotting is done in a trading week and reversals occur only once every three to four weeks on average, the cell size is too large, the reversal requirement is too tough, or both complications are present. Conversely, if extensive plotting of new cells is occurring on a daily basis, and reversals are being recorded two to four times per week, the cell size is probably too small and/or the reversal requirement is too easy to meet. If the parameters don't fit the market, then the projections are also likely to be wrong much of the time. As noted earlier, there are many point-and-figure advocates who use a three-cell reversal requirement and who then adjust the cell size for different commodities. If a three-cell reversal requirement is being used, experience suggests cell sizes of $.02, $.02, and $.03–$.04 for corn, wheat, and soybeans, respectively, and a $.15–$.20 cell size for cattle (live and feeder) and for hogs.

A moment's further reflection suggests there are many parallels between the point-and-figure chart and the bar chart. Some were identified in the preceding discussion. The sell signal generated by a price dip on the point-and-figure chart would be a break through a support plane on the bar chart of the same futures contract. The buy signal generated by a column of Xs extending above the highest cell filled with an X in earlier columns is the same action that generates a move up through a resistance plane on a bar chart. Which chart a particular individual might use will be a matter of preference. Many traders use both and experiment with the parameters on the point-and-figure chart to find an effective combination. *You should be reminded again, however, that the bar chart is still the more widely used charting technique.* Buy or sell signals generated on the bar chart are likely to be more widely monitored than buy or sell signals on a point-and-figure chart.

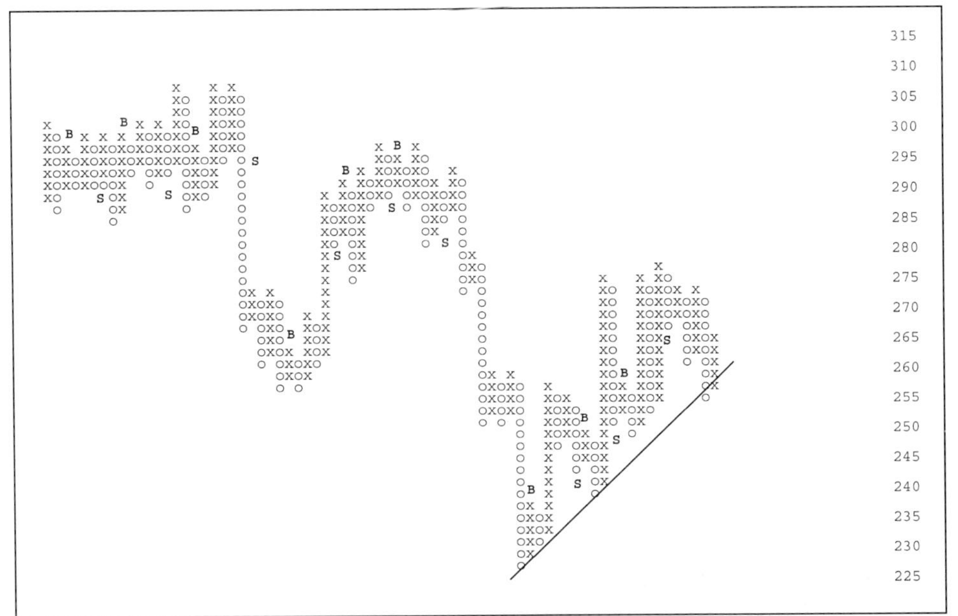

FIGURE 5.6
Use of the 45-Degree Line on the Point-and-Figure Chart

The point-and-figure chart is an alternative to the bar chart. There is no time shown on the chart. Analysts often like the clear and unambiguous buy and sell signals on the chart and a convenient price projection technique is available. The choice of cell size and reversal parameters will be important determinants of the effectiveness of the charting technique for a particular commodity, and you will need to look at the literature for guidance and experiment with different combinations on your own.

MOVING AVERAGES

Moving averages are a simple means of monitoring market trends. A three-day moving average can be calculated by adding the last three daily closing prices and dividing by 3. A 10-day moving average is simply the sum of the last 10 closing prices divided by 10. Each day, a new closing price can be added, the oldest close dropped, and new averages calculated. Modern charting packages, offered by advisory services and private distributors, have provisions that allow moving averages to be updated daily electronically. Many have the capability of plotting the moving averages on the computer screen, printing them either superimposed over the bar chart, or in a new chart "window" immediately above or below the bar chart plot.

The choice of what type of moving average system will be used will largely determine whether it is a complement to the bar chart or an alternative to chart analysis. *Moving averages in some form or type of application are used by a significant percentage of commodity traders.*

Many traders and market analysts use a single moving average, such as a 40-day or a 50-day moving average of closing prices, to help monitor the direction of trend in the market and to signal a trend reversal. This type of application can be complementary to bar chart analysis. When a topping action, for example, is seen on the bar chart and the long-term moving average stops increasing and turns down, the two developments tend to reinforce each other. The bar chart topping pattern is suggesting a reversal of price direction and the change in direction of the long-term moving average is suggesting the same thing. In some instances, the moving average turns down before a chart top is apparent and the analyst then begins to anticipate some type of topping action on the bar chart. In other cases, what looks like a top or bottom on the chart may never be confirmed by the moving average, and some analysts will then ignore the chart pattern as long as it is not confirmed by a change in the direction of movement in the moving average. *These types of uses of moving averages are largely a complement to bar chart analysis.*

Some traders use the price action relative to the long-term moving averages as a sell or buy signal. For example, a closing price that is below the 50-day moving average after a sustained period of higher prices is seen as a sell signal. A close above the average after a downtrend would be seen as a buy signal or, at a minimum, suggest that the trend in price is turning from down to up. Figure 5.7 shows a decisive close below a 50-day moving average on the 1996 corn chart during July. During August, this volatile and record-setting market showed closes back above the average, and some users would have lifted short positions. The short positions would have been restored later in September!

The more widely employed strategy using moving-average systems becomes an alternative to, or replacement of, bar chart analysis. In this type of usage, two moving

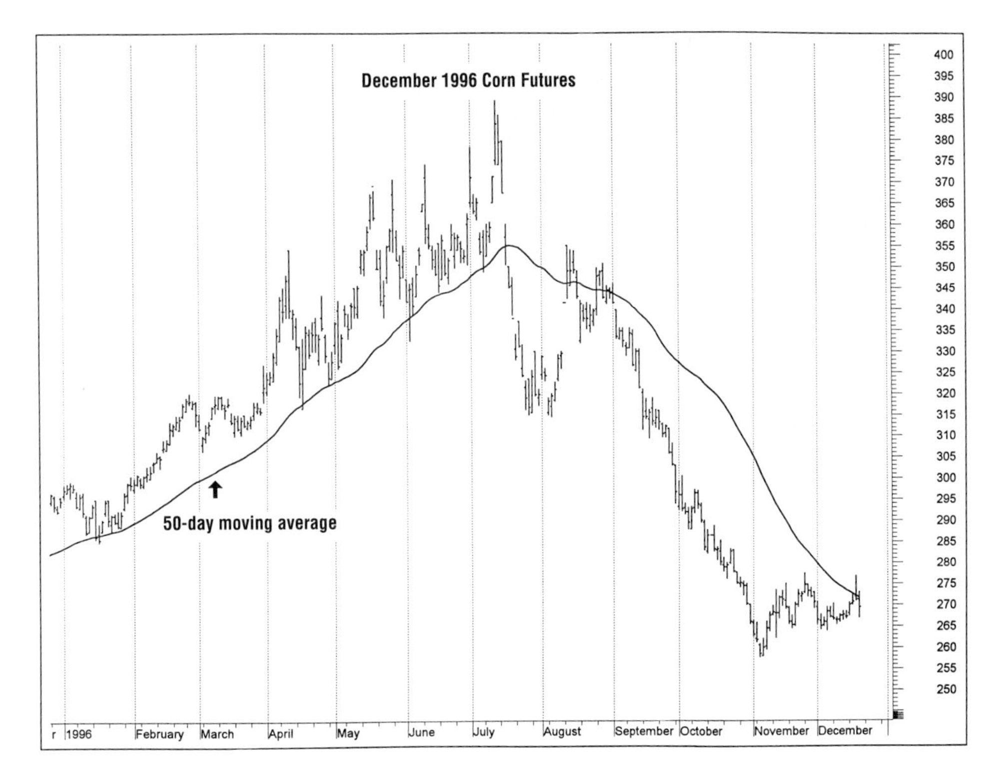

FIGURE 5.7
Use of the 50-Day Moving Average on the December 1996 Corn Futures Chart

averages are often employed and crossover action is used to generate buy and sell signals. Buy or sell signals are generated when one moving average crosses the second and a market bottom or market top is being confirmed.

Table 5.1 demonstrates with 3-day and 10-day moving averages. Conceptually, the idea is that in an upward- or downward-trending market, the shorter moving average tends to move faster and "leads" the longer average. When the market turns, the shorter average turns more quickly and crosses the longer and slower-moving average. It is this crossover action that generates the buy and sell signals. In the context of Table 5.1, a sell signal is generated the day the hog futures contract closes at $54.70 (the 12th entry in the "closing price" column) and the 3-day moving average drops below the 10-day moving average. Nine days later, with the closing price at $55.00, a buy signal is generated when the 3-day average moves back above the 10-day average.

Such a system of moving averages becomes a trend-following technique. *The market has to actually record a top or bottom and reverse direction before the moving averages will cross.* Trades will never occur at the exact price highs and lows for the year, and you must understand what the averages can and cannot do. The use of averages *can* impose a type of discipline to the trading program if you, whether hedger or speculator, cannot otherwise bring discipline to your program. The averages *cannot*, most informed observers would agree, outperform experienced bar chart analysts who handle their programs with discipline.

By observing the buy and sell signals in Table 5.1, you can, as a hedger, employ the moving averages in a selective hedging program. The sell and buy signals are clear and require no judgment. And, depending on the type of market behavior that develops, moving averages can be very effective as guides to selective hedging or to speculative programs.

TABLE 5.1
Demonstration of 3- and 10-Day Moving Averages for Lean Hog Futures: Calculations and Buy-Sell Signals

Closing Price	3-Day Moving Total	3-Day Moving Average	10-Day Moving Total	10-Day Moving Average Signal	
$54.10					
54.75					
54.90	163.75	54.58			
55.30	164.95	54.98			
55.65	165.85	55.28			
56.00	166.95	55.65			
56.10	167.75	55.92			
55.80	167.90	55.97			
56.05	167.95	55.98			
55.60	167.45	55.81	554.25	55.43	
55.10	166.75	55.58	555.25	55.53	
54.70	165.40	***55.13***	555.20	***55.52***	Sell
54.10	163.90	54.63	554.40	55.44	
54.20	163.00	54.33	553.30	55.33	
53.60	161.90	53.97	551.25	55.13	
53.10	160.90	53.63	548.35	54.84	
52.90	159.60	53.20	545.15	54.52	
53.20	159.20	53.07	542.55	54.26	
53.90	160.00	53.33	540.40	54.04	
54.40	161.50	53.83	539.20	53.92	
55.00	163.30	***54.43***	539.10	***53.91***	Buy
55.10	164.50	54.83	539.50	53.95	
54.90	165.00	55.00	540.30	54.03	
55.50	165.50	55.17	541.60	54.16	
55.90	166.30	55.43	543.90	54.39	

Management of orders to enter and exit the market is important. Market actions can be taken either the day the averages cross or the next trading day. If the 3-day moving average is moving lower and approaching the 10-day average, it is easy to calculate what closing price would prompt "crossover" action and a sell signal on any particular day. A *sell-stop-close-only order* placed at or just below the closing price needed to prompt the crossing action and generate the sell signal would place the short hedge or short speculative position the day the averages cross and the signal occurs. Alternatively, a *limit-price order* can be placed the next day. Long observation suggests there is very little difference in the effectiveness of the moving-average program due to which approach is used in placement of orders. On exchanges that do not accept stop-close-only orders, the order will need to be placed the next day, or the broker can be instructed to use limit-price or *market orders* near the close on the day the crossing action is imminent.

Moving averages work best in a market that presents major and sustained price trends. *When a major price move develops, the moving average system is extremely effective because (1) there will never be a major and sustained downward trend during which the short hedger is caught totally without protection, and (2) there will never be a major and sustained upward trend during which the selective short hedger is not out of the futures market and enjoying the benefits of being a cash market speculator.* Obviously, the same benefits are there for the long hedger. During major price moves up, the long hedger *will be protected against much of the increase in costs*. A corn user, for example, would never be caught unprotected during the long and sustained move to record high corn prices in 1995 and 1996.

Moving averages do not work well in choppy markets that show no sustained price trends. Frequent buy–sell signals may be generated, and the typically small losses on such trades may accumulate to very significant levels if the choppy patterns continue. Critics also point to the tendency for moving averages to see the consolidation patterns in upward-trending markets as tops, signaling short hedges prematurely that must subsequently be lifted at losses.

Figure 5.8 demonstrates the tendency for moving averages to signal a premature top, but it also shows the strength of the moving-average approach. Early in calendar year 1996, the 9- and 18-day moving averages generated sell and subsequent buy signals on December 1996 corn that amounted to very little in terms of losses or gains. Note the sell signals in January, March, late April, and early June when the 9-day average dipped below the 18-day average. A selective hedger following the moving-average signals would have placed short hedges on several occasions. The short hedges would have been typically lifted at a break-even or small loss less the costs of the round turn trade in futures. This is the type of action many critics point to in a choppy market where there are no sustained price moves or in a market that is trending higher (like the December 1996 corn) and working through periodic congestion areas.

In July, a sell signal was generated and the strengths of the moving-average approach started to emerge. The sell signal was well below the high in this volatile market, and the short position was lifted during August at essentially a break-even level. But a sell signal was generated near $3.40 in early September, and the short position was not lifted until mid-November near the $2.70 level. That round turn

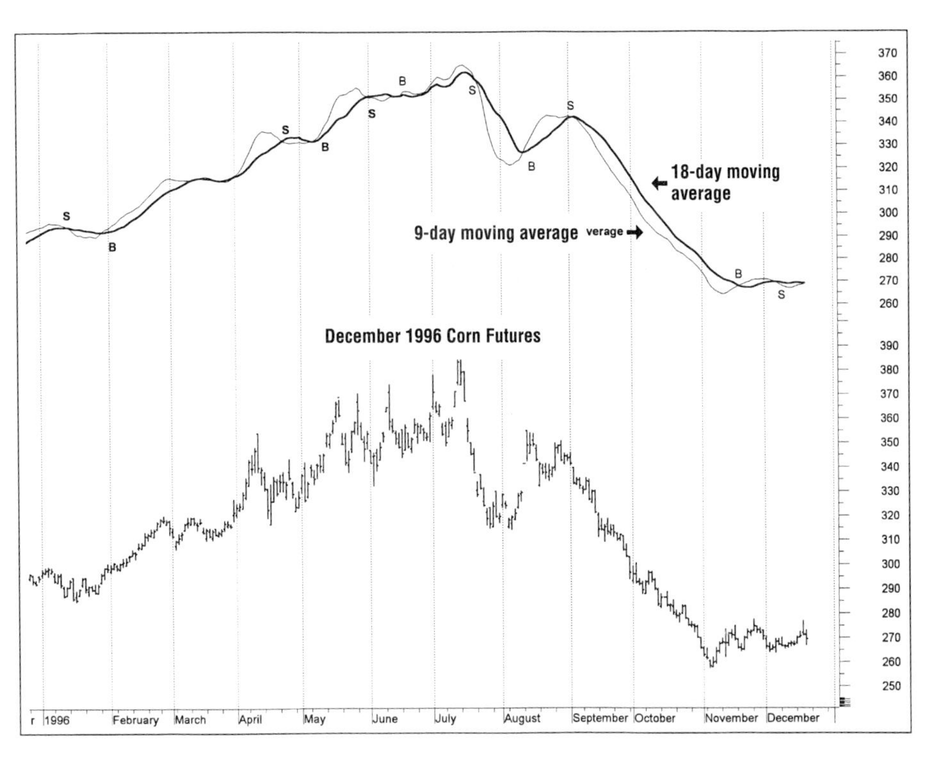

FIGURE 5.8
Bar Chart and Performance of 9- and 18-day Moving Averages for December 1996 Corn Futures

grossed \$.70, putting the net price well above \$3.00 depending on the cash-futures basis level employed.

An uptrend line on the bar chart would have generated a sell signal near \$3.50, but the August action would have bothered some chart readers. Do you lift the short hedges? A late-August buy signal would have been seen if you hook a trend line to the contract high and to the early to mid-August price rally. If short positions were lifted, do you have enough discipline to replace them later? Either a short-term uptrend line in August or the break through the early August lows would have generated sell signals. *But it takes discipline to replace those short positions*, and not all chart users would have followed through to that extent. The moving averages would put you short again in September and protect against that price plunge that was nearly \$1.00 per bushel.

In a market that is trending strongly higher, a selective short hedging program based on moving averages is likely to *lose* money in the futures account. *The strong point of the systems is that they allow the selective hedger to benefit from most of the increase in the markets by not having short hedges in place.* This involves managing exposure to price risk, and you will recognize there are times when you want to be a cash market speculator since you can then benefit, as a seller, from upward-trending cash prices. That is the idea in selective hedging.

Figures 5.9 and 5.10 show that the choice of moving averages is important. The averages track the December 1997 corn through 1996, a period in which a producer of corn might have been interested in extending price protection into a 1997 crop that had not been planted. Remember, prices were at record highs *during* 1996.

Figure 5.9 shows the 9- and 18-day moving averages. Look carefully at the choppy price action during March to August of 1996. The 9/18 combination is slow to gener-

FIGURE 5.9
Bar Chart and Performance of 9- and 18-Day Moving Averages for December 1997 Corn

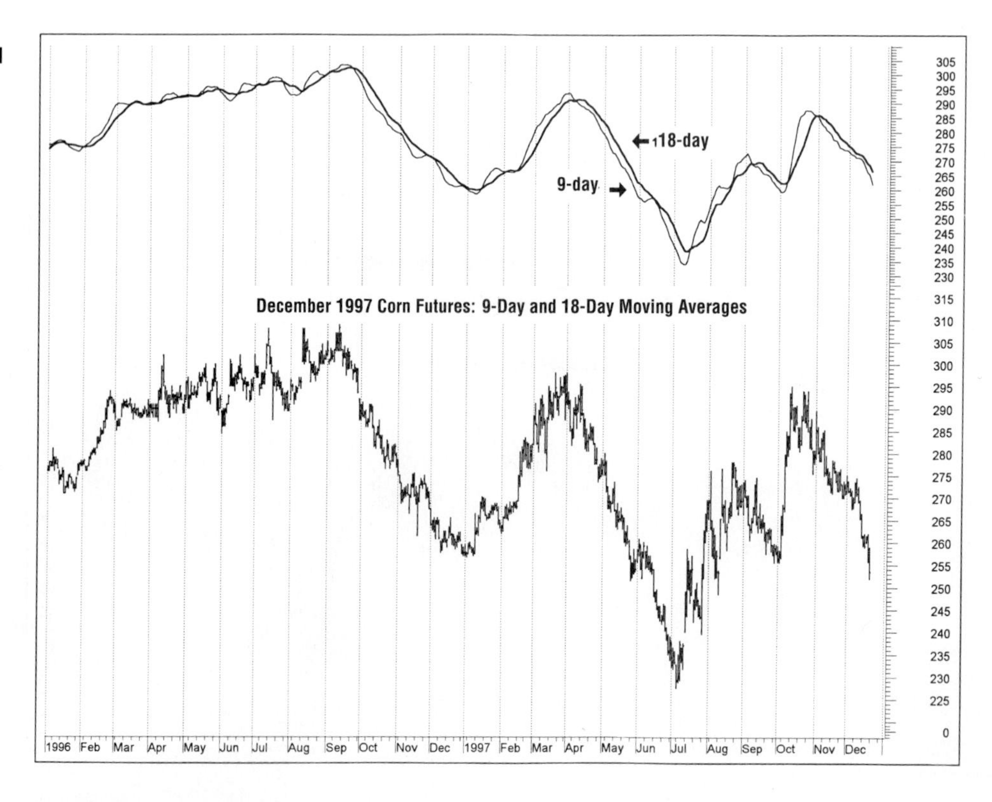

ate signals and will have trouble in this type of market. Note that the late May sell signal would have generated short positions that were bought back at a substantial loss in mid-June. The same thing happened again in late July and early August. Throughout this period, losses would have accumulated in the futures account.

The 4 and 9 combination in Figure 5.10 handled this choppy market much better. The mid-May sell signal and short positions would have been bought back at a break-even or better price in early June. The mid-July sell signal—coming several days earlier than the sell signal using the 9 and 18—would have been offset near August 1 at a profit.

Perhaps the most appealing feature of the moving-average approach to selective hedging is the tendency for such systems to reduce the variability in the net revenue flow to users without significant reductions in the mean level of revenue over time. In many instances, research has shown that the optimum set of moving averages can both reduce revenue variability *and* increase mean returns over time.

An example of the potential is shown in the work by Sronce and Franzmann at Oklahoma State. Employing both a third moving average to confirm buy–sell signals and a penetration rule—extensions discussed in Appendix 5A—the authors concluded that the moving-average strategies increased mean net returns *and* decreased the variability of net returns for hog producers. Miyat and McLemore, working with feeder cattle, reached essentially the same conclusion from research done in Tennessee. A number of other research efforts show the same results on other commodities.

In closing discussion on moving averages, three added thoughts are worthy of mention. First, we need to keep in mind that the effectiveness of moving averages is always tested on historical data. *There is no guarantee that the averages will be equally effective in the future.*

FIGURE 5.10
Bar Chart and Performance of 4- and 9-Day Moving Averages for December 1997 Corn

The second observation is related to the first. Over the past 20 years, analysis of the effectiveness of moving averages has been periodically updated by various researchers. *It is encouraging to note that the optimum set of moving averages in one period often performs equally as well or nearly as well in later periods.* That result suggests the averages that worked well across the past few years are likely to work well in the future. Conceptually, you would expect this to be the case unless the price action in the particular commodity changes significantly in terms of frequency and amplitude of price moves.

The third thought revolves around the need for you as a potential user to investigate moving averages, see if they fit your prevailing program and needs, and then make a commitment to the system. *If you tend to override the signals from the moving averages, then you should not use moving averages.* You have to "stay with" a moving-average strategy and give it a chance to work over time.

A number of modifications have been generated to improve the performance of the moving-average systems that employ crossover action to generate signals. One widely observed adjustment is the use of a third or leading moving average to confirm the sell–buy signals. Commodity Price Charts, for example, a widely distributed commercial charting service, shows a 4-9-18 combination. The 9-day and 18-day averages generate the signals, but they must be confirmed by the 4-day leading the 9-day up (down) the day the buy (sell) signal is generated by crossover action involving the 9- and 18-day averages.

Other analysts use optimized sets of moving averages, arguing quite logically that the market patterns vary from commodity to commodity and that no one set of moving averages will work for all commodities. If you wish to pursue this area still further, there are references at the end of the chapter. Appendix 5A extends the discussion concerning the correct set of moving averages and what modifications might help improve the performance of any selected set of averages. Neither the references nor the extended discussions are necessary conditions to moving to later chapters. They are offered solely to allow you, if you get intrigued with the potential of moving averages—and they *are* intriguing—to pursue the area with some focus and direction.

Moving averages can be complementary to the bar chart, but most traders see them as a replacement. In markets exhibiting sustained price trends, the averages can be very effective. In choppy markets with no major trends, moving averages will not work well. Users need to be aware of the strengths and weaknesses of these trend-following systems. Overall, the averages have the most potential for traders who otherwise cannot bring a disciplined approach to their trading activities, and that can be a powerful advantage.

SUMMARY

Among the alternatives to the bar chart, the *point-and-figure* chart is preferred by some traders because it generates clear and unambiguous buy–sell signals. It is easy to update and offers a price projection alternative whenever a top or bottom formation develops. The point-and-figure chart is essentially an alternative to the bar chart and offers few dimensions that could be seen as complementary to the bar chart. The choice of *cell size* and *reversal requirement* will be the key to the effectiveness of the point-and-figure charting technique.

The *moving averages* can be complementary to the bar chart, or they can be a replacement. A single moving average may be used to confirm trends or to signal changes in the trend that is visible on the bar chart. The more widely used application of moving averages involves two averages that are used to generate buy and sell signals via crossover action.

Which moving averages are employed will be important. Moving averages will seldom be effective in a choppy and sideways market, and they have a tendency to signal tops or bottoms when a consolidation pattern is developing. In selecting the length of the moving averages to be employed, the need is for a compromise between being responsive to market moves and avoiding the quick reactions that generate premature and incorrect signals. The *best averages will vary across commodities,* and no one set will be best for every commodity.

The primary advantage of moving averages is that they bring a kind of discipline to what might otherwise be undisciplined programs. In major trending markets where a user's emotions tend to get in the way, the moving averages are at their best. The *producer will never be totally unprotected during a major price decline, and the commodity user will never be totally unprotected during a major cost increase when moving averages are employed.* The averages can be a very effective guide to a selective short hedging or selective long hedging program.

KEY POINTS

- *The point-and-figure chart* is an alternative to the bar chart; it gives clear buy and sell signals and offers a price projection opportunity.
- How effective the point-and-figure chart will be will vary with the *cell size and reversal requirement* employed.
- A simple *moving average* is often used in monitoring *direction of price trend* in spotting changes in trend direction. As such, the use of moving averages is *complementary* to the bar chart.
- The most widely employed moving-average strategies use *two averages* that generate buy and sell signals via *crossover actions.* Moving averages used in this way are essentially an alternative to the bar chart.
- Moving averages are *not effective in choppy markets* that show no sustained price trends.
- *Moving averages* have a tendency to treat a consolidation pattern as a top or bottom, and this tendency *generates premature buy and sell signals.*
- The correct choice of moving averages will *vary with the commodity.*
- Moving averages have the potential of *bringing discipline* to otherwise undisciplined trading programs, and they can be *effective guides to selective hedging* programs. The moving averages will *provide protection against much of any sustained price move.*

USEFUL REFERENCES

W. A. Cohen. *How to Use the Three-Point Reversal Method of Point and Figure Stock Market Trading,* 5th ed., Chartcraft, Larchmont, New York, 1972. A detailed treatment of the three-cell reversal requirement approach to analysis of the markets.

C. D. Miyat and D. L. McLemore. *An Evaluation of Hedging Strategies for Backgrounding Feeder Cattle in Tennessee,* Bulletin 607, Tennessee Ag. Exp. Station, February 1982. The reference explores various strategies and shows performance of moving averages.

Jack D. Schwager. *Technical Analysis,* John Wiley & Sons, New York, 1996. The author presents a brief treatment of a single moving average in Chapters 3 and 8 and a more sophisticated coverage in combination with oscillators in Chapter 15.

Phillip W. Sronce and J. R. Franzmann, *Hedging Slaughter Hogs with Moving Averages,* Bulletin B-768, Oklahoma Ag. Exp. Station, July 1983. Alternative moving averages are used and both leading averages and penetration rules are investigated.

Chicago Mercantile Exchange, *Trading Tactics,* CME, Chicago, 1986. Demonstrates moving averages and point-and-figure charts.

APPENDIX 5A. MOVING AVERAGES: ADDED DISCUSSION, REFINEMENTS

As noted in the chapter discussion, the disadvantages of moving averages come from two primary sources. First, moving averages tend to signal a top too early. In an upward-trending market, the market will often show a consolidation pattern and then resume the upward trend. A bull flag, triangle, or a congestion area will often take several days to develop. In a system that uses two moving averages and relies on a crossover pattern for buy and sell signals, the moving averages may signal a top. The result is a placement of a short hedge in a selective hedging program. A few days later, if the upward trend resumes, a buy signal will be generated and the short hedge is lifted again—typically at a loss. The same thing happens when prices are trending lower. A bear flag might develop on the bar chart as a consolidation pattern emerges, but the moving averages might generate a buy signal and call it a bottom.

The second problem emerges during periods of choppy price action when there are no sustained price trends. Sell and buy signals may be generated as the congestion area or sideways trading pattern develops. Obviously, each of these trades could be a losing trade, and the user of moving averages can quickly become disenchanted.

The need is for an optimized set of moving averages that offers a compromise between the need for a system that (1) is sufficiently sensitive that it tracks the moves in the markets correctly, and (2) is not so sensitive that it gives frequent and premature signals. Based on research, experience, and observation, the following are crossover systems that appear to be effective for each of the commodities shown.

Commodity	Moving Averages
Soybeans	13–16
Corn	7–10
Wheat	4–9
Cattle	5–15
Hogs	7–10
Feeder cattle	4–8

There are no guarantees that these averages will work well in the future, but across historical data sets they typically perform better than a general set of averages such as 3–10 or 9–18 for the particular commodity listed. Since the frequency and amplitude of price undulations vary across commodities, it is logical to suspect that no single set of averages will work for all commodities. In general, the correct set of averages for each commodity will (1) generate fewer trades, and (2) avoid the mistakes of prematurely calling a top or bottom compared to a generic or general set of averages. You might want to start with the moving averages listed for each commodity and experiment with combinations near these in terms of length of averages.

Once a set of averages is selected for a particular commodity, several extensions or qualifications have a high probability of improving the performance of the averages selected. Most involve some type of confirmation for buy and sell signals that are generated.

A popular extension is to use a third moving average to confirm the signals generated by a base set of averages. For example, the widely watched 9–18 set of averages is often combined with a 4-day moving average to confirm the signals generated by the 9-day moving average crossing the 18-day average. Procedurally, any sell (buy) signal generated by the 9-day average moving below (above) the 18-day average must be confirmed by leading action by the 4-day average. That is, the 9-day dropping below the 18-day average is confirmed as a sell signal if the 4-day leads or is below the 9-day when the sell signal is generated.

The use of a confirming average is especially helpful in an upward- or downward-trending market when the 9- and 18-day averages would otherwise signal a top or bottom that would, *ex post*, turn out to be just a consolidation pattern. The short and responsive 4-day moving average would turn up quickly as the consolidation pattern on a bar chart is completed and the upward price trend is resumed. If the 4-day average is above the 9-day when the 9-day dips below the 18-day moving average, the decision maker ignores the sell signal. Conversely, if the 4-day is below the 9-day average when a buy signal is generated by the 9-day moving above the 18-day moving average, the buy signal is ignored.

Using the third and more responsive moving average as a confirming or leading indicator will eliminate some of the mistakes generated by the base set of moving averages. The number of trades, and the possibility of losing trades, is reduced for both the consolidation patterns in trending markets and the sideways patterns that are often present when there are no major trends present. The tendency to signal tops or bottoms mistakenly during consolidation patterns is reduced and the possibility of frequent and losing trades during periods of choppy and sideways price patterns is reduced.

You are encouraged to investigate and experiment with various combinations in addition to the 4- 9- and 18-day moving average combination. Our experience indicates that using a 3-day confirming average, for example, will improve the performance of the 5- and 15-day moving averages for live cattle futures.

A second extension is to use a penetration rule. The idea is to ignore sell and buy signals unless the shortest average moves through the longer average by at least a predetermined amount.

Research in the use of penetration rules is sketchy and is still evolving. It appears, however, that a penetration rule of $.0l–$.02 in corn and wheat, up to $.03 in soybeans, and $.10–$.15 per hundredweight in the livestock commodities will improve performance of the underlying moving averages. Sronce and Franzmann found $.14 per hundredweight to be best for hogs in the work mentioned in the chapter and in the reference list at the end of the chapter.

To elaborate, assume the 7- and 10-day moving averages are being used for corn. The decision maker might adopt a rule that a sell signal will be honored only if the 7-day moving average drops below the 10-day average by at least 1 cent per bushel. For a buy signal, the 7-day would have to be above the 10-day by at least 1 cent.

The logic of penetration rules is apparent. In an upward-trending market, the 7-day might dip below the 10-day by a fraction of a cent per bushel and be there for only 1 to 2 days if a consolidation pattern is developing. Using the penetration rule will eliminate some of the mistakes the moving averages make by calling such a consolidation pattern a market top. A possible disadvantage, of course, is that when a top or bottom does in fact develop, actions in taking a short or long position or in offsetting an existing hedge can be delayed for one or more days waiting for the penetration requirement to be met. The result can be the establishing of short positions at prices lower than the prices the day the initial sell signal is generated, or higher prices than those for the day the initial buy signal was generated.

A third alternative, one that our personal experience suggests merits consideration, is to use the relative strength index (RSI) in combination with the moving averages. It is very possible that a buy signal is generated during the volatile action that often characterizes topping action in the market, and the selective hedger would consider lifting short hedges in a seriously overbought market. The buy signal might come, for example, during a price surge that is accompanied by the divergent action by the RSI described in the chapter. If divergence between price action and the RSI is developing and/or if the RSI is above 70, the decision maker might reasonably ignore the buy signal generated by the moving averages and keep the short hedge or short position in place. If the RSI is below 30, indicating an oversold market, a sell signal from the moving averages would be ignored, of course.

An obvious positive dimension of using the RSI as a confirming indicator is the possible elimination of additional calculations. Many electronic market news services carry the 14-day RSI and update it after the close of trade each day. If there is easy access to the RSI, only the base set of moving averages need to be updated by the trader. Once familiarity with the averages is established, however, updating each day by hand or, preferably, by an electronic spreadsheet or other computerized routine is very simple and is not time consuming. Many of the more sophisticated electronic data services offer moving-average calculations and allow for some choice in the averages used.

Whatever the extension or refinement, it is important that the user adopt a system to the circumstances and *stick with the system.* Do not, for example, experiment with the size of the penetration rule during the decision period. At the end of the year or other decision period, more refinement might be considered—but not before.

Before leaving this subject area, it is worth emphasizing what moving averages can and cannot do for the decision maker. A system of moving averages generates clear and unambiguous buy–sell signals and thus eliminates the need to develop expertise in the art of chart reading. This may be especially appealing to selective hedgers who have trouble bringing discipline to their program. It can help them make decisions on when to place, offset, and replace hedges. And moving averages bring the huge benefit of ensuring that the hedge (whether short or long) will always be in place when major moves in the market occur that might bring financially ruinous pressure to the firm. Related, the hedges will never be in place when major and sustained moves develop that bring benefits to the firm from being in an unhedged or cash market speculative position.

When no sustained price trends are in place, however, the moving averages can tax the patience of even the most committed user. Frequent trades, incorrect buy–sell signals, and the losses that come with them can challenge both patience and the financial reserves of the firm for prolonged time periods. In the final analysis, you will have to decide whether the moving averages fit your management style and ability, financial position, and overall operation better than some alternative approach to price risk management. But don't take these systems lightly. Remember, they will *never* allow you to endure a major and sustained price break as a producer/seller or a major price rally as a user without protecting you for at least a substantial part of the price move. The record high grain prices in 1996 brought huge margin calls for short hedgers and essentially ruined some grain users, dairys, and poultry firms that were caught with no long hedges or other form of protection. Moving-average systems will *never* allow such extremes.

CHAPTER 6

PSYCHOLOGY OF THE MARKETS

INTRODUCTION

In earlier chapters, the term *discipline* was used on several occasions. Technical tools must be applied with discipline, the decision maker must be disciplined—the need for discipline comes up time and again.

The frequent references are justified. *Most experienced traders would agree that many approaches to the futures markets in risk management or investment programs can be successful if they are applied with discipline.* On the other hand, few, if any, approaches will be successful without discipline in use and application. In Chapter 5, moving averages were introduced as a possible alternative to bar charts for the user who has trouble bringing discipline to the analysis of the bar charts. Even then, the user has to be disciplined enough to follow the signals of the moving-average system and not start second guessing or ignoring the moving-average signals.

Understanding the psychology of the markets can help. The futures market is often labeled as a more psychological market than the cash markets, and it does tend to have a behavioral and psychological dimension. Understanding that dimension can help bring order to the interpretation of market patterns that otherwise sometimes appear to be chaotic. This chapter discusses the psychology of the markets, employs some of the technical dimensions developed in Chapters 4 and 5 by way of illustration, and helps you extend your understanding of the "why" of observed market moves.

THE HERD COMPLEX

An old adage advises you to "walk when other people run and to run when other people walk." The contributor of that piece of advice was clearly worried about just being a follower. The perspective is also the source of the "contrary opinion" or "contrarian" approach that is often mentioned by futures market traders. You see the tendency to follow the crowd frequently in the futures markets. Traders, especially infrequent and small traders, tend to get pulled into the markets after a price trend has been underway for some time. Often, it is the late entrants that provide the last surge of buying in the upward-trending markets or the last surge of selling in the downward-trending markets. *It is extremely important that the individual trader, whether a speculator or a hedger, understand what is happening in terms of the crowd-following tendencies.*

The pattern is almost always the same. A drought or some other supply-side shock launches a move to sharply higher prices in the grains and oilseeds. If the catalyst is an emerging drought, it takes several days for the awareness of what is happening to spread to the general public. An early surge of new buying hits a market that has already been boosted by the professional speculators and by commercial firms who need the corn in processing or for export. These full-time traders already know about the budding drought, and they act early. The next surge of buying, and the first surge by the followers, comes from the periodic traders who have active accounts and who are in a position to trade immediately. Open interest starts to increase, with much of the selling being done by hedgers willing to forward-price at the higher prices being offered.

Prices move higher, and a new surge of buying comes into the market. This time, the buying is often by traders who have rushed to open accounts or reactivate dormant accounts. The temptation to get involved is strong and the momentum starts to build. Several such surges of buying develop. Open interest may continue to increase, and daily trading volume tends to be large.

The final surge of buying tends to come from speculators who arrive very late to the scene and from holders of short hedge positions who are pressured into covering their positions. In the case of the speculator, often a small and inexperienced trader, the fear of missing out on a good thing is just too great. They calculate how much money they would have made if they had been in the market from an earlier point, and then tend to jump on the bandwagon at the last minute. The hedgers who get forced into covering their short positions are pressured by several factors.

As the markets move higher, the short hedger who priced at lower prices suffers a significant opportunity cost. Prices are higher, but the hedger is denied any benefits of those better prices by the short position in the futures market. Adding to the pressure they feel from the opportunity costs is the need to answer margin calls on the short positions as the market rallies. When no added credit line has been established for margin needs, or when any credit line that was established proves to be inadequate, the pressure can be intense. The result is often *margin liquidation* as holders of short positions are forced out of the market or are taken out by their brokerage firms when margin calls are not answered within prescribed time limits. The producer or other short hedger turns into a buyer along with the late rush of speculative interest. This "forced buying" contributes to the psychological dimensions of the market.

It is important to keep in mind that the producer comes to the markets to manage exposure to price risk. That need is an objective one, and should not be the source of a great deal of emotionalism. You should always keep an important point in mind. The position of a short hedger in a surging market is and always will be different from that of the short speculator. The hedger is covered in the cash market and is only suffering an opportunity cost. The hedged price or margin is secure subject to basis risk. The speculator is suffering an out-of-pocket loss and is not covered in the cash market. The emotions of the two traders should be very, very different, and we need to keep that in mind. You should try to put yourself in the place of the hedger and then the speculator in a surging corn, cattle, or cotton market. Do you feel your emotions would be different?

The futures market has a behavioral or psychological dimension. It is important that decision makers resist being followers of the crowd tendencies and bring discipline to their trading program. Adequate margin capital is essential if the short hedger is to avoid being forced to offset short hedges in a rising futures market. The emotions of the hedger should be different from the emotions of the speculator, and just how they should be different is worth a bit of thought at this point.

AIDS TO DISCIPLINE

Adopting the position of the short hedger in most instances for illustrative purposes, there are a number of aids that can help the decision maker bring much-needed discipline to the situation. The most widely used of such aids comes from the technical and not the fundamental analyses of the markets.

Open Interest

As noted in Chapter 4, price trends will not usually be sustained unless *open interest* is increasing. Any price rally that comes on declining open interest is a short-covering rally, and the last surge of buying in a strong bull market such as the one described earlier is no exception. When the open interest stops increasing or even starts to decline, the producer feeling pressure from margin calls and existing short hedge positions can start to relax. The market actions are suggesting a top is coming and, as a producer, you should resist the growing temptation to buy back the short hedges.

Figure 6.1 illustrates this situation using the July 1997 corn futures. The major rally during February was supported by a strong increase in open interest (line at bottom). A dip in open interest accompanied a brief profit-taking consolidation during March, and then the market surged to new contract highs near April 1. Open interest had already started to decrease. The new buying and selling that is always needed to sustain that last price rally had started to disappear. All the speculators or long hedgers who want to buy the market are in the market at a level that meets their objectives or needs, and there is no new money to finance still higher prices.

There is no binding relationship between open interest and trading volume, but the volume (histogram at bottom) also often tends to decline as the open interest drops off. The levels of intensity in the market are starting to diminish, and lower prices are likely. Figure 6.1 shows sharply lower trading volume in late March after open interest had started to decrease. There is a rule that says, "Do not buy a market rallying on quiet volume." In this instance, the smallest volume during the price rally in late March should have been, at a minimum, a warning to stay alert.

But let's get back to the use of open interest. *Producers who are feeling the pressure from margin calls and the self-imposed pressure from mounting opportunity costs have something observable that can help them resist the temptation to buy back the short hedges at high prices. When open interest starts to decline, the mar-*

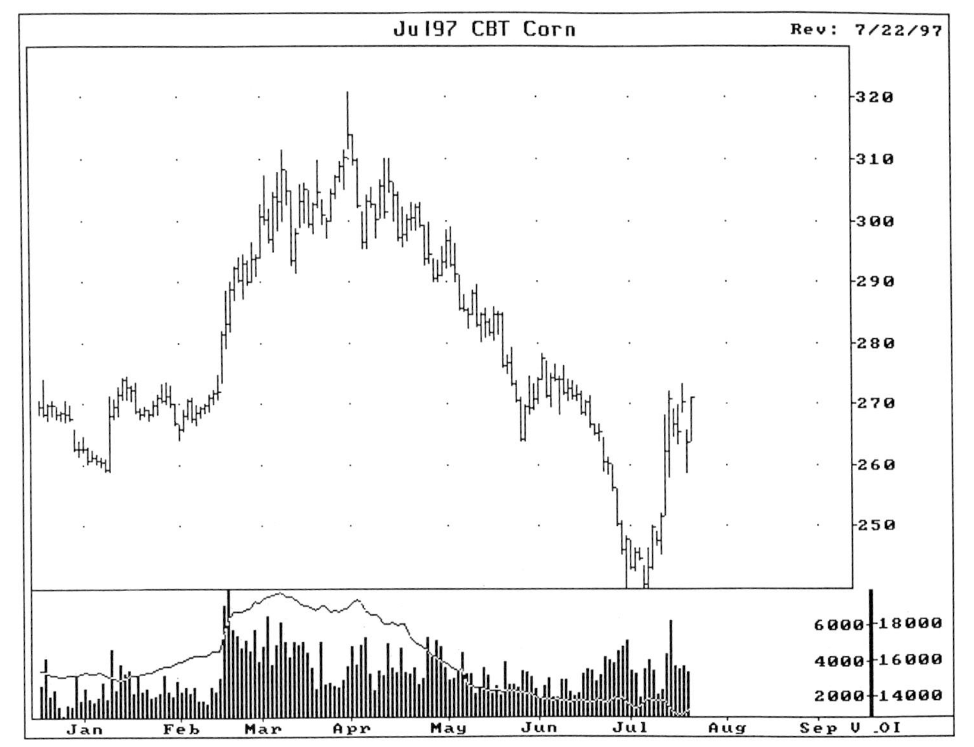

FIGURE 6.1
Open Interest Issues on the July 1997 Corn Futures

ket may be approaching a top. If you are an inexperienced or infrequent user of the markets, you, as a producer, should pause and attempt to decipher from the developments in open interest the behavioral dimensions of the market. When the experienced analysts and traders back away and are no longer willing to add to their long positions, open interest stops increasing. The message of the market then becomes relatively clear: prices are not likely to go significantly higher at this point.

In downward-moving markets, when the holder of long hedges starts feeling pressure, the message from open interest patterns may be less reliable. The tendency for price trends to come to an end when open interest starts to decline is still there, however. Declining open interest in a downward-trending market indicates holders of short positions are buying to take profits and do not appear inclined to push the market still lower. That action indicates their analysis of the fundamental supply–demand balance and of the technical picture is suggesting that still lower prices are not likely, at least not immediately.

The focus here is on open interest, but a matrix that also includes trading volume will help summarize this area. The matrix is

Price	Open Interest	Volume
S+	*I*	*I*
W+	*S* to *D*	*D*
S–	*I*	*I*
W–	*S* to *D*	*D*

where:

$S+$ = a strong positive price trend,

$W+$ = a weak positive price trend,

$S-$ = a strong negative price trend,

$W-$ = a weak negative price trend,

I = increasing,

S = steady, and

D = decreasing.

Developments in open interest can help you bring discipline to your program. A decline in open interest suggests a price uptrend is approaching an end, and this can help you as a short hedger avoid the temptation and the pressure to buy back short hedges. The same assistance is there for the long hedger in a downward-trending market.

Relative Strength Index

A widely used technical indicator in a family of indicators that attempts to measure the momentum in the market, the relative strength index (RSI) introduced in Chapter 4 is a powerful aid to the worried producer. If the 14-day RSI is approaching or has moved above the 70 level on the last surge to higher prices, the market is overbought. That means, remember, that most of the traders who want to buy this market have done so. The new buying and new selling needed to push prices still higher is simply not available. The electronic market news services you can now get allow you to try RSI measures of different lengths. The 14-day RSI is perhaps the most widely used, and it is used here for illustrative purposes.

Figure 6.2 demonstrates application of the RSI using the March 1998 feeder cattle futures chart. There are two apparent possibilities. First is the holder of inventories on pastures, or in backgrounding dry lots, who has short hedges in place and who is starting to feel the pressure of margin calls and the opportunity costs associated with the higher prices. Second is the user of feeder cattle, such as the feedlots, who has no cost protection in place and who is starting to worry about what still higher prices will do to the feeding margins.

The sharp price rally in late June and July was likely boosted initially by covering of short speculative and short hedge positions. As the market rallied, the 14-day RSI moved significantly higher. The RSI surged above 70 in late June, peaked, and then declined as prices softened. In mid-July, when prices made new highs, the RSI did not—a classic divergence. The market topped, traded down through an uptrend line that could be drawn hooking lows in mid-June and early July, and drifted sideways to lower.

The holder of short hedges can relax a bit when the RSI moves to 70 and higher. A correction to the downside will occur nearly 100 percent of the time, especially when the divergence shown here occurs. If the short hedges are to be lifted, they can be lifted at better prices as the price correction to the downside develops.

The potential long hedger can also relax. If cost protection is needed, there will be better chances to set that protection via long hedges at lower prices. The tendency

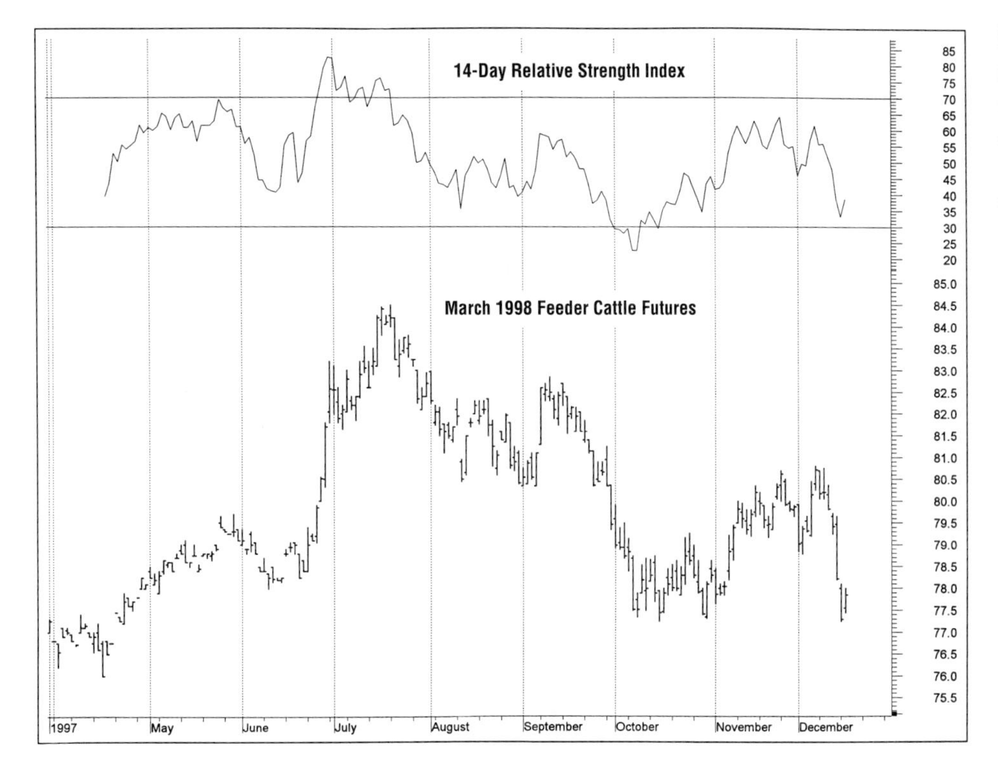

FIGURE 6.2
Relative Strength Index on the March 1998 Feeder Cattle Futures Chart

to panic and buy at what could turn out to be at or near a contract high is reduced by the RSI as an aid to a disciplined program.

The RSI is also effective in downward-trending markets. As the measure approaches 30 from higher levels or drops through 30, the market is oversold and is not likely to go significantly lower. Selective short hedgers looking for support for an inclination to lift short hedges and potential long hedgers looking for a place to buy the market are rewarded for watching the RSI. Obviously, the probability the markets will turn higher as the RSI drops to 30 or lower is higher if prices are approaching a life-of-contract low or are in the process of forming one of the bottoming formations discussed in Chapter 4.

Of all the aids to a disciplined trading program discussed in this chapter, RSI is arguably the most effective. While other measures of momentum are used by traders, the RSI is widely watched and brings an element of a self-fulfilling prophecy to the situation because it is so widely used. *To the anxious producer who is looking, often desperately, for some reason and for the discipline not to cover or buy back short hedges in a rising market, the RSI can definitely bring that much-needed element of discipline and confidence.* The technical reference by Schwager listed at the end of the chapter will be a good reference on other measures if you want to pursue this area in more detail.

The RSI is perhaps the most important aid in bringing discipline to the hedging program. It is highly effective. Recognizing the market is overbought or oversold helps the producer or user back off from the pressure to offset short or long hedges on a panic-stricken and emotional basis.

Positions of Traders

Released twice monthly, the report that shows the positions of various traders can be an added and useful aid to the producer. Table 6.1 illustrates the type of data that are provided. For convenience, data for only hogs and corn are shown. But as the table format implies, comparable data exist for all the commodity futures. For hogs, to illustrate, the data show that large speculators hold long positions in 5,860 contracts, up sharply (by 2,457 contracts) from the previous month. Continuing across the line, the large speculators have decreased their short positions (by 883 contracts). The large hedgers are moving to the long side of the market, and the small traders have reduced long positions sharply and are showing some movement to the short side of the market.

As a general rule, the small traders tend to be the followers and trade with little discipline. Conversely, the large traders—hedgers and speculators—tend to be more disciplined and are typically acting on better information and superior analysis of the markets.

It is often the small traders, for example, that show a significant shift to the long side of the market during the late stages of a major trend to higher price levels. Such action shows up in the positions held by the small traders as an increase in their total long positions, and this is often accompanied by a decrease in short positions. This pattern can be especially revealing if the large traders are showing a tendency either to get out of the market or to switch their positions to the short side of the market. In other words, the "long change" data for the small traders are positive and the "long change" data for the large traders are negative. If the "short change" data for the large traders are positive, then the large traders are abandoning the long side of the market and are moving to the short side. It is the small traders that are still trying to push the market higher.

The underlying premise is that the large traders are in fact better informed and/or more disciplined. But that assumption is not difficult to accept. Commercial firms in the market as hedgers, long or short, usually have an entire department charged with

TABLE 6.1
Format and Sample Content of the Monthly Commitments of Traders Report

	Noncommercial (Large Speculators)				Commercial (Large Hedgers)				Small Traders			
Commodity	Long	Change	Short	Change	Long	Change	Short	Change	Long	Change	Short	Change
Hogs	5,860	2,457	5,424	–883	2,464	81	2,579	–839	17,130	–3,396	17,451	864
.												
.												
.												
Corn	139,920	17,120	38,920	14,770	544,090	16,750	623,040	-50,090	450,975	29,340	466,025	98,530

Grains are shown in thousand bushels.

Livestock are shown in number of contracts.

The "change" entry is the change from the previous month.

the responsibility of analyzing the markets. Large speculators arguably have even stronger reasons than hedgers to be well informed and to be correct in their analysis of the direction and probable magnitude of price trends. After all, hedgers have a position in the cash product and are therefore covered by changes in cash prices when the futures market moves against their positions. If the hedger is wrong, the cost is typically an opportunity cost, and the hedged program can still be profitable.

Speculators have no protection on the cash side. Large speculators typically are trying to make a living from their trading programs, and those who are not good at their job do not survive. Therefore, when the large speculators start to shift their positions in the market, the message from these actions is worth noting.

Availability of the data limits application of the positions of traders. Bimonthly releases are too infrequent to help in day-to-day decisions. Most electronic news wires and advisory services give some daily information on who the large traders for the day are and what they are doing, but it is not always clear whether they are offsetting earlier established positions or establishing new positions. A reasonable strategy is to try to monitor what the large commercial firms and the large speculators are doing via their daily actions and watch those actions for an indication of what price change might be forthcoming. Those firms have their fingers on the pulse of the market if anyone does. Obviously, this suggests access to an electronic service. A service offered by an advisory firm that attempts to capture the essence of who is trading helps in the monitoring process.

In general, it is better to watch the positions of the large traders in the market. If the large traders are abandoning the long side of the market, a price uptrend is not likely to be continued. Conversely, a move away from the short side by the large traders suggests still lower prices are not likely. Monitoring what the large and small traders are doing can help bring discipline to the individual's pricing program.

Chart-Related Aids

The behavioral or psychological dimensions of the market are captured by the bar charts. Producers holding short hedges in an upward-trending market therefore have the entire array of chart-reading techniques at their disposal. A review of selected chart dimensions, and their application to the current illustration, helps to confirm that the producer can get help from the chart in deciphering the underlying market psychology. Chapter 4 covered the bar chart patterns, but did not cover the psychological dimension of the markets and how chart patterns can be used to help bring discipline to a trading program.

The futures markets tend to focus in on the eventual market-clearing price by going through price cycles or undulations. The markets characteristically react, overreact, and then correct for the overreaction.

Chapter 4 included discussions of the tendency for the market to correct at least part of a price surge before attempting to move to higher price levels. Producers holding short positions in an upward-trending market can reasonably anticipate that the market will soon pause, consolidate for at least a few days, drop back and retrace at least part of the price surge, and then move higher—if higher prices are in order given the emerging supply–demand picture. If a decision has been made to lift short hedges, the producer should try to be disciplined enough to wait for the correction to lower prices. The same charge is there for the potential long hedger—wait for the correction to lower prices before placing the long hedges.[1]

The use of expected corrections helps bring discipline to management of a market that is losing bullish momentum because the underlying fundamental picture is starting to change from improving weather, and so on. If the market makes significantly more than a 62 percent correction of the recent upward surge in price, the temptation to place long hedges or lift short hedges should be resisted. The market actions are suggesting that the direction of price trend is turning from up to down. Short hedge positions are the correct positions for the producer, and being a cash market speculator is the correct position for the selective long hedger.

As corrections are monitored, any signs of weakness that appear near a 50 percent correction should alert the hedger to the possible need to act. Examples of "weakness" would include declining open interest as the correction runs out of new buying and new selling, the RSI approaching the 70 or 30 levels indicating an overbought or an oversold status, or a sharp upward thrust in price on a particular day that completes a 50 percent correction to the upside and then a very weak close.

In a market that is surging to higher prices, chart gaps typically appear. Chart analysts expect the market to attempt to fill the gap by trading back into the gap area. Unless there has been an overwhelming change in the supply–demand balance, there is a feeling that there is no reason to fail to trade over a particular price range that might be represented by a chart gap. That tendency can be an aid to the producer caught holding short positions in what is starting to look like a bull market.

The June 1997 hog futures shown in Figure 6.3 demonstrates the chart gap and several related issues. The supply–demand balance did change in a major way in early January with the release of the quarterly *Hogs and Pigs* report. The report was a bullish shock to the market, and a huge chart gap resulted.

Given the shock to the base of information on supply and demand, the chart gap was not completely filled. In early March, buying support near the bottom of the gap started a new price thrust that turned into a major bull market in hogs.

There is an important message here for you with regard to the chart gaps. It will not work to sit and insist, in a mechanical way, on a full 62 percent correction of an upward surge in price before lifting short hedges or placing long hedges. *The fundamental information has changed significantly, and the chart gap can block anything approaching a full correction.* Here, aggressive action is needed to lift short hedges and to place long hedges. As discussed in Chapter 4, the buy order should be placed at or just above the top of the chart gap. That placement will greatly increase the chances of the order being filled, and the chart patterns in Figure 6.3 are an example of the need for effective placement of the buy order.

If margins will be a real problem at higher prices, the producer holding short hedges should be especially alert and aggressive as the price levels dip toward a gap area. Any signs that the market is starting to reverse the downward thrust above the gap, such as a dip toward the gap and then a strong close near the top of the daily trading range, should alert the producer to the possible need for a more aggressive pos-

[1]Remember, there are several ways to project how far the price surge will move before the correction starts. In Chapter 4, the use of gaps, consolidation patterns such as bull flags, and bottom formations such as the head and shoulders as projection devices was discussed. Point-and-figure charts can also be used to project the price move. References at the end of the chapter direct you to coverage of the sophisticated Elliott wave theory, which allows prediction of the length of each price wave in a major price move.

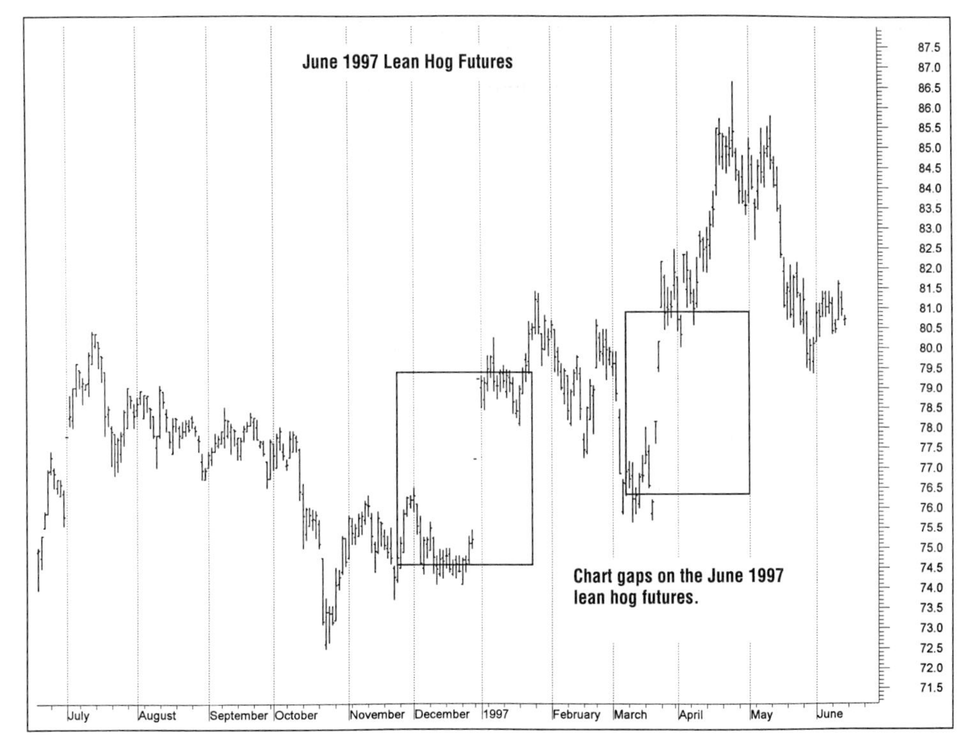

FIGURE 6.3 Demonstration of Chart Gaps on the June 1997 Lean Hog Futures Contract

ture. Limit-price buy orders might be raised to higher price levels or changed to market orders to ensure that the short positions will be covered and eliminate the margin call problems and opportunity costs that come with an expected new surge to higher prices.

Still another alternative that can help bring discipline to the program is the use of stop orders. *Generally, producers should be careful with the use of stop orders unless they are skilled chart analysts and are totally comfortable with a selective approach to hedging.* If the price surge is moving up toward life-of-contract highs, however, you should consider whether you are willing to answer margin calls if prices surge to new highs. If new highs do start to emerge, the stop-close-only orders can be effective on the commodities on which they are accepted. Get your banker involved in making these plans.

Analysts differ, but many argue that a short hedge position should be bought back on the second consecutive close above the previous life-of-contract high. A buy-stop-close-only order set at the old contract high and placed after the first close in new price ground will be filled at the close on the second consecutive close if new price highs do emerge. For example, assume that the prior life-of-contract high on December hog futures is $53.40. On Monday, as concerns about adequate supplies mount, the December contract trades up to $53.60 and closes at $53.55. On Tuesday, a buy-stop-close-only order could be placed at $53.40 or slightly higher, say $53.50. If the market closes at $53.65 on Tuesday, the short hedges would be lifted near $53.65 as floor brokers fill orders just prior to the closing bell. This order is widely used for futures traded at the CME. You should talk to your banker about the use of this order on grains or oilseeds at the CBOT or commodities such as cotton traded in New York.

Not all exchanges will accept this order, and you might have to let your broker take action late in the session or wait until the next day.

An alternative approach followed by some traders is to place the buy-stop-close-only order at $53.55. That would require the second consecutive close above the old high of $53.40 to be higher than the observed $53.55 close before the short hedges will be lifted. The latter approach is less likely to lift the hedges, of course, and the choice of order placement may well rest with where the market is in the projected price range coming from the fundamental analysis. If the $53.40 is very high relative to fundamental expectations, the approach of requiring the second close to be still higher and above $53.55 might be preferred.

Management of this type of price action, including what to do if the market later dips back below the old contract high, was discussed in Chapter 4. *The point here is that you as a producer should have a written plan in place as the contract high is approached, and follow through accordingly. With a previously conceived plan in place, worked out with your banker, you are much less likely to get caught up in the emotion of the market as new highs emerge.* It follows, of course, that you must have the discipline to follow through and execute the plan of action.

On the CBOT, as note earlier, the close-only stop may not be accepted. Facing an emerging drought, you must choose an alternative approach. Your broker may be willing to monitor the market and buy back the short position(s) as the second consecutive close at new higher prices looms imminent. If that approach is not possible, you can execute a buy order the first trading day after the two consecutive higher closes have been observed. Whatever the approach, it is important that a plan of action be in place and that it be followed with discipline.

There are numerous chart-related aids that can help the decision maker avoid getting caught up in the psychology of the market. Chart gaps, expected corrections, and so on, help the trader anticipate what could happen and plan a course of action that has an element of order and the much-needed trading discipline.

Alternatives to Bar Charts

Alternatives to the bar chart were discussed in Chapter 5. Point-and-figure charts and moving averages may be preferred techniques for decision makers who have trouble interpreting the bar chart and who would prefer buy–sell signals that do not require subjective judgment and interpretation.

The point-and-figure charting technique offers the advantage of clear buy and sell signals. You might wish to review pertinent sections of Chapter 5 at this point, and recall the parallels with the bar chart. The buy signal generated by the point-and-figure chart when prices move up through a previous high occurs when a resistance plane is "taken out" on the bar chart. The sell signal on the point-and-figure chart occurs when a support plane is "taken out" on the bar chart.

As resistance and support planes are being challenged, the user of the bar chart pays close attention to the closes around the plane. For example, the discussion in Chapter 4 and earlier in this chapter indicates that most analysts would want to see two conservative closes above a previous life-of-contract high before the market action is viewed as having generated a legitimate buy signal. The point-and-figure charting technique ignores the closing prices and the critic would argue the point-and-figure approach thus ignores valuable information. But there is another side to the issue. *It is the action around life-of-contract highs or other resistance or support planes that many users find frustrating about the bar chart approach.* On which day do you act? If prices close above the old high for two consecutive days, and then dip back below the old resistance plane, what do you do? In Chapter 4, the recommendation was to replace the short hedge or lift the just-established long hedge. But action around those planes is often volatile and the trading out, then back in, then out again, can be frustrating. The point-and-figure charting approach ignores all the action around the planes, and either records a buy signal—or nothing at all.

As suggested in Chapter 5, anyone interested in the point-and-figure approach should spend time watching and analyzing alternative cell sizes and reversal requirements. If the technique is constantly generating signals and moving the selective hedger in and out of the market, the objective of a disciplined hedging program is placed in question. If the cell size is too small and/or the reversal requirement is too easy to be met, there will be too much action for many potential users. The need here is for *discipline in the trading program,* and the point-and-figure chart can provide that discipline if it is configured to fit the particular market as discussed in Chapter 5.

Moving averages are a second and widely used alternative to bar charts. The primary advantage is the same as for the point-and-figure charts—the generation of clear and unambiguous buy–sell signals. The averages are based on simple arithmetic, and the sell signal that is generated when the shorter of two averages moves below the longer average eliminates the need for any subjective judgment.

The use of moving averages was discussed in Chapter 5, and the appendices to Chapter 5 presented possible refinements of moving-average systems. In the point-and-figure charts, the cell size and reversal requirements determine the frequency of buy–sell signals. With the moving averages, the length of the averages employed will determine the frequency of signals and trades.

In this discussion of the psychology of the markets, it is important that you appreciate what the averages can and cannot do and how they can bring discipline to the program. *One of the big advantages of moving-average systems is their ability to handle the major trends in the markets.* The major trends often give users of the bar chart trouble as they attempt to manage effective selective hedging programs.

Consider the position of the producer who correctly placed short hedges on a sell signal near the top of the market using a key reversal, trend line, or some other bar-chart approach. The market then moves lower for 10–15 trading days and starts to consolidate. But is it a consolidation pattern or is it a bottom? There is no clear pattern in the open interest, trading volume, or any of the aids to interpretation of the bar chart to give the producer a clear message. The RSI is in the 40–50 range and is giving no indication that the market is seriously oversold to suggest bottoming action is likely. In addition, you are now feeling the pressure of a profitable trade and the related desire to take profits. Your broker is advising taking the profits, waiting for a rally, and then replacing the short hedges. You yield to the pressures and to the well-meaning advice of the broker and buy back the short hedges.

A few days later, before any significant price rally has developed, the market starts to drop again. Now you face a very difficult decision: Should the hedges be replaced? *This decision is much more difficult in terms of the psychology of the moment than it appears on the surface.* After all, you have moved to a posture or a mind-set that says lifting the short hedges was the correct thing to do. Now, the market is showing patterns that suggest that mind-set must again be reversed. In a matter of a few days, perhaps just one or two days, you are having to adjust perceptions and price outlook from a bullish to a bearish posture—and to replace the hedge. *This type of adjustment is very difficult for many producers.* But if the hedges are not replaced, it is quite possible that a prolonged and financially devastating downtrend in prices can develop. It is not difficult at all to find an example of producers who "took profits" of $.05–$.10 per bushel from short hedges on corn and then remained out of the market and watched it drop $.50 to $1.00 per bushel. The same often-painful examples are present in soybeans, hogs, feeder cattle, T-bills—or any other futures commodity.

In strong price uptrends, there is a parallel and an equally difficult scenario to manage. The market rallies and reaches your initial price objectives. Short hedges are placed, and then the market starts to move higher. In this instance, the decision to lift short hedges at what would be small losses in the futures account is very difficult to make for many producers. It takes discipline to take the small loss in futures and open up the possibility of benefiting from a major price move up as a cash market speculator. There is always the temptation to "wait"—hoping for a 100 percent correction to the downside so that the hedges, if they are lifted, can be lifted at a break-even or profitable position. If the market surges, the tendency to wait is still there. Margin calls are answered, and it is often the pressure of extended margin calls that eventually prompts buying back the short positions. But by that time, the market may be approaching its final highs or be ready to top. The short positions are lifted at the wrong time and for all the wrong reasons. *There is little question that the most predominant illustration of a lack of discipline in selective hedging or speculative programs is the constant refusal to admit the market position is wrong and take the small losses while they are still small.*

The moving averages are not a panacea. They have obvious shortcomings, and those shortcomings were discussed in Chapter 5 in some detail. But the averages have obvious advantages if the decision maker has trouble handling effectively the scenarios discussed in this section.

In the downward-trending market, the moving averages can and do make mistakes in generating buy signals and dictating that short hedges be lifted in what turns out to be, *ex post*, just a consolidation phase of a major downtrend. But when the prices start to move lower, the moving averages will generate a sell signal that calls for the short hedges to be replaced. *You will never be completely unprotected during a major and prolonged price downtrend.*

During the strong uptrends, the moving averages perform a similar service. As a major price surge is resumed after a brief consolidation period, the averages will dictate that short hedges be lifted if they were replaced prematurely during the bull flag or other consolidation pattern. Those short hedges will, in most cases, be lifted at a small loss in the futures account. This probable result brings to the surface an important issue insofar as using moving averages are concerned.

As a potential user, you must understand the pros and cons of moving averages. If a decision is made to use moving averages to get away from more subjective chart interpretations, then you must follow through and honor the moving-average signals. If short hedges are indeed placed too soon in response to what turns out to be just a consolidation pattern in an upward-trending market, the short hedges will in fact often be lifted a few days later at a small loss. But the averages will require the selective short hedger to do what most have trouble doing—take a small loss in the futures position before it turns into a major opportunity cost as the markets move higher.

Many aspiring chart analysts simply cannot bring themselves to lift short hedges at a loss, and the result is often margin calls, major opportunity costs, and frustration as the supply–demand balance brightens and the markets trade persistently higher.

A different type of discipline is required in using moving averages, but discipline is still very important. If there is a tendency to overrule the moving-average signals, then you are back in the same subjective game that moving averages are supposed to eliminate. Frustration is the inevitable result, and if you *are* this type of decision maker, you should not use the moving-average approach. But if you like these approaches, it may be you have found your niche. Modern electronic news services either have or allow for software to allow you to look at moving averages on screen and test different combinations. The Wilder reference listed at the end of the chapter will also provide more detail.

Point-and-figure charts and moving averages are alternatives to bar charts that have the potential to impose discipline on the trading program. Both techniques generate clear and unambiguous buy–sell signals. But both techniques require a disciplined user who looks at the advantages and disadvantages, picks a system to use, and then is willing to follow through and honor the signals that are generated.

AN OVERALL PHILOSOPHY

Technical tools need to be employed in the presence of a perspective as to what goes on in the markets. There *is* emotion in the markets, the collective emotion of all the traders at a particular point in time. At all times, you must approach the market aware that the emotion is there and that the futures markets have a psychology of their own.

It is important that you not get caught up in the emotions of the market. Various approaches can be employed to guard against that possibility. *Hedgers who are interested in the long-term movements in the market, and speculators who are position traders looking for a major trend to follow, may adopt a posture of refusing to make a decision on whether to buy or sell and at what price during the trading session.* They argue, quite logically, that decisions on when and at what price trades will be made should be worked out in the more objective environment during the evening after the close of trade or early morning before the trading session opens.

Another guide employed by successful traders is the often-observed adage of "ride your winners and cut your losers." It is sound advice and is relevant to both the hedger and to the speculator who is a longer term or position trader.

Successful speculators must be willing to take losses while they are still small in order to survive and catch and "ride" the winners when they are right about price direction. It takes discipline, but a speculator should always place a protective stop and leave it in place as soon as a position is taken in the markets. If losses are constrained and the trader has the discipline to stay with the winning positions, trade can be successful and profitable if the prediction on the direction of price trends even approaches being 50 percent correct. A *sure sign of the lack of discipline in a speculative program is the moving of a stop as the market approaches the stop to keep from being taken out of the market.* The purpose of the stop is to offset the position, because it is apparently a losing position, and *the stop should not be moved to a dif-*

ferent price level. You cannot "outwait" the market, arguing implicitly that you are right and the world is wrong.

Hedgers often face the same issue. A short hedge position in a downward-trending market is a "winner" in that it is providing the needed protection as the cash market drops. As profit accumulates in the futures account, there is, as discussed earlier, a strong temptation to take profits. The idea, typically, is to capture the profits by buying back the short hedge and then replacing the short positions at higher prices when the market rallies.

But the price rallies may not come and, as suggested earlier, as a producer you often take a $.10 per-bushel profit from your corn futures positions and then sit on the sidelines and watch as the market drops an added $1.00 per bushel. It is very difficult for most hedgers to be disciplined enough to turn around and sell the market after having just convinced themselves they should buy back their short hedges because they think the market will rally. *Profitable short hedges should be lifted only on solid or reliable buy signals (such as a close above a major downtrend line) or on a dip to life-of-contract low support planes.* Even then, you should have a plan to get the short hedges back in place if the market does not rally as expected or proceeds to make new contract lows. If you are uncomfortable with exactly when lifting short hedges should be considered, you should go back and review the discussions on the management of chart patterns in Chapter 4. Also review material on alternatives (such as moving averages) in Chapter 5 and in this chapter. This is important!

To hedgers, the idea of using stops and eliminating or offsetting losing trades should take on a different interpretation than it has for the speculator. *The objective of the hedging plan is to gain effective protection against disadvantageous price moves.* If the market moves against the short or long hedge position, the hedger is covered in the cash market. There should be less urgency associated with offsetting existing futures positions that are starting to show a loss on the futures side of the hedge. In particular, this means producers should be careful placing stops too close to their entry position. All too often, the stop offsets the hedge position and leaves you exposed to major price risk as the market then moves in the direction that means the hedge was needed. A buy-stop that offsets short hedges on a modest price rally, for example, can leave you as a producer in the position of being a cash market speculator and totally exposed to price risk if the market then declines sharply. Theoretically, the short hedges can be replaced, but the point has been repeated many times that most short hedgers have trouble trying to muster up the courage to replace hedges that were lifted only a day or two earlier. The use of stops close to the entry position for the hedger is also likely to mean frequent trades for the hedger, and frequent trades are not the proper objective of the hedger.

In addition, as noted in earlier chapters, the tax status of the selective hedger with the Internal Revenue Service (IRS) is still not completely clear. You can be a selective hedger, but one criterion the IRS and/or the courts is likely to use to determine whether the trader is a hedger or a speculator is the frequency of trades. Very frequent trades can be viewed by the IRS as evidence of speculative activity.

It is important that decisions regarding trades in futures or options be made with as much objectivity and as little emotionalism as possible. A plan of action to cover every possibility in the markets should be worked out before the troublesome price pattern emerges. That plan might be more objective if it is prepared during a period when the features or option markets are not open.

SUMMARY

Futures markets exhibit a psychological or behavioral dimension. It is important that the trader, whether speculator or hedger, recognizes that this dimension is present and *brings discipline to his or her trading program.*

The first and most important requisite of a disciplined program is to *avoid the tendency to follow the crowd.* This is the basic thesis of the "contrary opinion" approach to markets. Emotions run highest, and the market activity often is the most volatile, when a major price move is approaching its completion. A pattern of consistently lifting short hedges or placing long hedges when a major price rally is approaching maturity is a sure sign of an undisciplined trader.

Several sources of assistance can help the trader avoid mistakes. The *relative strength index* can be used as an indicator of the momentum in the market and will often give early signals that a major price move is approaching its end. *Chart patterns* such as chart gaps and resistance planes, *the tendency for the markets to correct* at least part of a major move, and other technical indicators can help bring discipline to the trading program. *Point-and-figure charts and moving averages* offer alternatives to the bar chart that may be more appropriate for some decision makers because they generate more nearly objective and unambiguous buy–sell signals.

It is extremely important that you have an overall philosophy that is designed to avoid getting caught up in the emotions of the markets. A plan, preferably a written plan, that lays out what you will do prior to developments such as new contract highs or new contract lows is very important. That plan should be prepared in advance to help ensure the objectivity and lack of emotion that must characterize a disciplined trading program.

KEY POINTS

- The futures market has a *psychological* or *behavioral* dimension that suggests that certain patterns of activity are both predictable and likely to be repeated over time. To bring *discipline* to a trading program, the trader must *avoid the tendency to follow the crowd.*
- It is *primarily the technical dimensions of the market and technical analysis,* not fundamental analysis, that will help the trader adopt a disciplined approach to trade.
- Monitoring the *positions of traders* will help since the large traders often tend to behave differently from the small traders, and the large traders will generally be operating on the base of better information and from superior abilities as market analysts.
- The *relative strength index* is especially helpful in efforts to avoid buying (selling) the market that is approaching a top (bottom) and is ready to reverse the direction of price trend.
- A number of *bar chart patterns* such as resistance planes and chart gaps, and the tendency for markets to correct at least part of a major price move, are invaluable *aids to a disciplined trading program.*
- *Point-and-figure* charts and *moving averages* generate clear and unambiguous buy–sell signals and can be used as replacements for the somewhat subjective "reading" of bar charts.

- The trader should always have a *plan of action* before important price developments emerge to guide the actions that will be taken. It is important to keep in mind the *objective of the trading program* and to avoid frequent and undisciplined trades when the *proper objective* of the hedging program is to *effectively manage exposure to price risk.*

USEFUL REFERENCES

P. J. Kaufman, *Commodity Trading Systems and Methods,* John Wiley & Sons, New York, 1978. The author includes a chapter on "behavioral techniques." He discusses contrarian strategies and demonstrates the use of the Elliott wave theory and measurement of moves and corrections in futures markets.

Jack D. Schwager, *Technical Analysis*, John Wiley & Sons, New York, 1996. The author demonstrates the way many of the technical dimensions of the market can be used in a disciplined trading program. Coverage includes oscillators, RSI, moving averages, and point-and-figure charting.

CHAPTER 7

OPTIONS ON FUTURES CONTRACTS

INTRODUCTION

Trade in options on agricultural commodity futures began in the fall of 1984, and that development made a major contribution to the risk management tools at the disposal of the producer, holder, and user of agricultural commodities. The financial community has known about and traded options for some time. However, options on commodity futures contracts have had a more jaded and political history. Trading in options on U.S. agricultural commodities had been banned in the U.S. since 1936. In 1933, an attempt to manipulate the wheat futures market using options resulted in political pressure that brought on the 1936 ban. Much later, in the 1970s, trading of options tied to London commodity futures contracts became popular. Two scandals involving options led the Commodity Futures Trading Commission (CFTC) to suspend most trade in options in June 1978. Then, in September 1981, the CFTC approved a three-year pilot program for selected commodities. The program was deemed a success and option trade was expanded in 1984.

Options on futures contracts can remove two related and major barriers to the use of commodity futures in the forward-pricing of agricultural commodities. The first is the producer's constant fear that forward prices of future sales have been set too low or that forward prices (i.e., costs) of future purchases have been set too high. The opportunity cost of pricing lower or higher than would be possible at a later date has been discussed in earlier chapters. Producers often equate *bad outcomes*, in terms of opportunity costs, with *bad decisions*. Even if the forward price established is profitable, there is a tendency for producers to view the hedge set early at relatively low prices (or at relatively high costs) to be a bad decision. If the futures side of the hedge loses money, the tendency is to view the hedge as a mistake and to talk about losing money with the hedge.

A second and related barrier to direct use of the futures markets is the need to manage a margin account and answer margin calls as the market rallies against a short position in the futures. Neither producers nor their lenders have always understood the need for a special and additional credit line to answer margin calls. There are countless examples of producers being forced to offset short hedges due to the inability or lack of a willing creditor to provide the needed margin funds. Often, the market turns lower after the upward price move that forced the producer to offset the short hedges. A loss is incurred in the futures account and then the producer is without price protection as the market turns and trends lower.

A vivid example of the problems that can develop was the Chernobyl nuclear scare incident in the Soviet Union and Europe in the spring of 1986. Corn, wheat, and soybean prices soared, forcing many producers who had short hedges in place out of the market via margin liquidation. Prices then trended sharply lower for the rest of the year after the anxiety had subsided. This example is dated and extravagant, but this type of scenario is played out time and time again over the years. Frequently, over the course of a growing season or a feeding period, news about possible changing supply and demand conditions push commodity futures prices sharply higher only to be revealed later that the change was temporary or the news not as dramatic as was initially thought. Incidents such as Chernobyl are rare but the reactions by commodity futures prices to catastrophic events are not. Producers' and lenders' fears are not irrational, however. The summer of 1996 is a case in point. Grain and oilseed markets rallied strongly with continuing news about poor growing weather, strong demand by a recovering world and domestic economy, and downward revisions of grain stocks. Producers who forward-priced at very high prices early in the season then watched the market rally another 10 to 30 percent. Margin calls accompany these rallies, and they often prove difficult to manage and finance.

Purchasing options eliminates the need for a margin account and leaves open the potential of higher prices to the producer. They also leave open the potential of lower costs to the user who is interested in protection against rising raw materials costs. Options thus open up important new possibilities in marketing and price-risk management. *However, this opportunity comes at a cost.* To obtain an option, the producer or the commodity user must purchase it. The premiums are, in a realistic sense, the cost of the protection and flexibility that the options offer.

BASIC OPTIONS TRADING

This section will introduce you to the basics of options, options trading, and the use of options in forward-pricing. It is our experience that students and other beginners find options confusing. The main problem is the jargon. There is a lot of terminology specific to options. Table 7.1 lists and defines some terms. There is also a lot of flexibility in designing and executing options strategies. Cash and futures markets are relatively straightforward. Assets can be bought and/or sold in both. Things are not so simple in the options market. This section introduces you to the terminology and the basic components of options trade. Later sections in this chapter will elaborate on the different components and strategies.

An option gives the *right* to a position in the underlying futures contract but not an *obligation*. A purchaser of the option buys the right and the options seller sells the right. The idea is simple. Like an option on anything, the right granted by the option can be exercised, or the option can be abandoned. The cost of the option is incurred up front. It is the same whether the option is exercised or abandoned. It is important that you understand this distinguishing feature of options on futures before proceeding. We often treat options as contracts or as financial instruments. But what the con-

TABLE 7.1
Options on Futures Contracts: Terms and Definitions

Term	Definition
Option	A contract that gives the buyer the right but not the obligation to hold a futures contract at a specified price within a specified time period.
Call Option	The right to buy a futures contract at a specified price (strike price).
Put Option	The right to sell a futures contract at a specified price (strike price).
Strike Price	The price at which the option buyer can establish a futures position.
Premium	The cost an option buyer pays to an option seller for the option contract.
Write/Writer	Sell an option. Option seller.
Exercise	Covert the option contract to a futures contract.
Intrinsic Value	The value of an option if it is exercised immediately.
Time Value	The difference between the option price and intrinsic value.
In-the-Money	The market price for underlying futures contract is above the call strike price or below the put strike price. The option has intrinsic value.
Out-of-the-Money	The market price for underlying futures contract is below the call strike price or above the put strike price. The option has no intrinsic value.
At-the-Money	The market price for underlying futures contract is equal to the strike price.

tract does is implied in its name. An option is a *right* but not an *obligation* to take a position in the futures market.

A *put* option gives the buyer the right to a short position in the futures market. The seller, or *writer*, of a put option is assigned a long position if the option is *exercised*. A *call* option gives the buyer the right to a long position in the futures markets. The writer of a call option is assigned a short position if the option is exercised. Suppose a feeder cattle producer buys an $80.00 put option for November feeder cattle futures. The producer has purchased the right to a short position in November feeder cattle futures at $80.00/cwt. Intuitively, if the November feeder cattle futures trade below $80.00, the $80.00 put option will take on value because it could be used to acquire a short position in the November contract at $80.00. In practice, the feeder cattle producer will seldom actually exercise the option and take a short position in futures. Like the fact that delivery on futures contracts rarely occurs, options are rarely exercised. Rather, the market in that option will continue to trade so an offsetting position is taken. If the trader has bought a put option, selling a put option will net the options market position to zero. The same is true for call options. If a trader has bought a call option and then sells a call option at a later date, the net options market position is zero.

If the November feeder cattle futures trade down to $76.00, the right to a short position at $80.00 will be worth about $4. In this case the $80.00 put option could be sold for $4.00/cwt. and the producer can use these gains to offset the decrease in cash feeder cattle prices. However, if the November feeder cattle futures market increases to $83.50/cwt., then the right to a short position at $80.00 is worthless. The option is allowed to expire. The producer has bought price insurance. In a declining market the option covers some of the losses, and in an increasing market the option was price insurance that was not needed.

Cattle feedlot operators who need the feeder cattle might buy an $80.00 call option on the same November feeder cattle futures contract. They have the right to a long position at $80.00/cwt. If the November futures trade up to $83.50, the $80.00 call option takes on value, and the cattle feeders have the desired protection against higher feeder cattle costs. Conversely, if the November futures trade down to $76.00, then the right to a long position in the November futures at $80.00 will be worthless. The cattle feeders have bought cost insurance they do not need, and they are in a position to benefit from lower prices if feeder cattle can be bought below $80.00. However, they received the benefit of having protection against a sharp and unexpected increase in feeder cattle costs.

The returns to the two preceding options examples are shown in Figure 7.1 This figure is a returns diagram for the two options strategies. Returns diagrams are used frequently in discussions of options strategies. Futures prices are on the horizontal axis and returns to the options trader are on the vertical axis. We see that the $80.00 put has no value if the market price climbs above $80.00/cwt. The right to sell at $80.00 is worthless if the market is higher than $80.00. The $80.00 put has value if the market price falls below $80.00. The right to sell at $80.00 is worth $4.00 if the market price is $76.00. The put return line is kinked at $80.00. The $80.00 call has no value if the market price falls below $80.00/cwt. The right to buy at $80.00 is worthless if the market is lower than $80.00. The $80.00 call has value if the market price climbs above $80.00. The right to buy at $80.00 is worth $3.50 if the market price is $83.50. The call return line is also kinked at $80.

Because returns to options depend on the direction in which the market moves, they are easier to explain in a diagram than in discussion. Returns diagrams will therefore be used frequently in this chapter. As suggested above, the first exposure a person has to options is often confusing. The source of confusion with options stems from two things. First, the returns line is kinked. It isn't kinked with futures or cash positions because you either make money or you don't on a position. And second, you can buy both a right to a short position and a right to a long position. In futures and cash markets, buying is synonymous with a long position, but this is not the case with options. *You should review all returns diagrams and be comfortable with what each one is attempting to communicate before continuing*. Each diagram is as important as the related text.

The returns shown in Figure 7.1 are incomplete. To obtain the right to a short position, a put, or the right to a long position, a call, the trader must pay a *premium*. The premium is what the purchaser of an option pays for the right to enter into a futures position. Options are traded at various strike prices, in the process of trading puts and calls. The *strike price* is the futures contract price for which the option purchaser has a right. The strike price for the example in Figure 7.1 is $80. Exchanges set the strike-price intervals. At the CME, the exchange trades options at three strike prices above and below the current futures price in even-dollar units. The strike prices are thus in $1.00/cwt. intervals. The option premiums for each designated strike price, of a particular commodity futures contract and a given expiration month, are discovered in an open outcry auction on the exchange floor. The price discovery process is much like that for futures contracts themselves. Table 7.2 records the premiums for puts and calls at various strike prices for options on the October 1997, November 1997, and January 1998 feeder cattle futures. These prices were the closing premiums on October 2, 1997.

Look carefully at the premiums reported in Table 7.2. The premiums differ significantly depending on the underlying strike price and the amount of time between

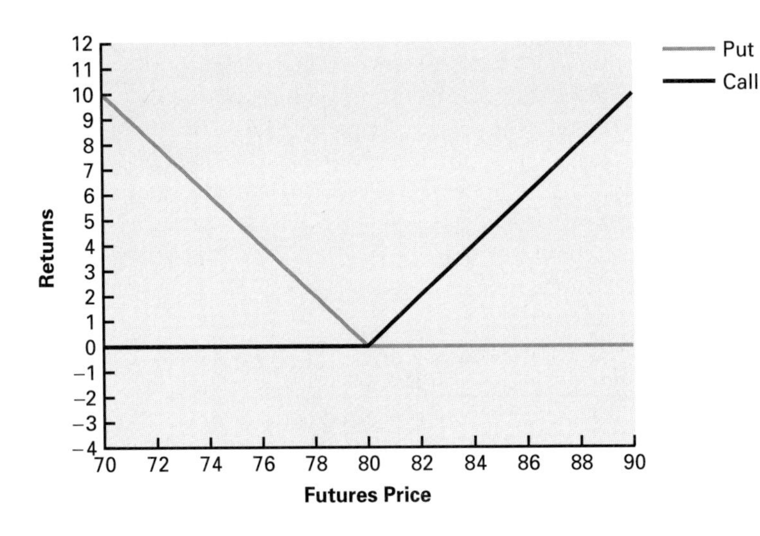

FIGURE 7.1
Returns Diagram from Purchase of an $80 Put and an $80 Call without Considering the Premium

the current date and the date the option expires. Options expire on the last day of the month prior to the delivery month for contracts with physical delivery. Options expire early in the delivery month for contracts that are cash settled. The feeder cattle contract is for 50,000 pounds or 500 hundredweight. Buying a $78.00 put would cost the trader $650 ($1.30/cwt. × 500 cwt.). Notice that the underlying futures contract, the November contract, closed at $77.85. The January contract closed at $79.27, so the $80.00 put on the January contract is more comparable. Buying an $80.00 put on the January contract would cost the trader $1,125 ($2.25/cwt. × 500 cwt.).

It is time to correct the returns diagram in Figure 7.1. Figure 7.2 incorporates the premium paid into the option returns diagram. In this example, we have constructed the returns to a put and a call both with an $80 strike price. To purchase either of these two options, assume we must pay a $2/cwt. premium. The $80 put has no value if the market price climbs above $80/cwt.. and the $80 put has value if the market price falls below $80/cwt. The right to sell at $80 is still worth $4 if the market price is $76 and the put return line is still kinked at $80.

TABLE 7.2
Premiums for Put and Call Options on the October 1997, November 1997, and January 1998 Feeder Cattle Futures Contracts Discovered on October 2, 1997

	October 1997 Close @ $77.40		November 1997 Close @ $77.85		January 1998 Close @ $79.27	
	Premiums		Premiums		Premiums	
Strike Price	Calls	Puts	Calls	Puts	Calls	Puts
75	—	—	—	—	—	—
76	1.65	0.25	2.37	0.55	—	0.80
77	—	0.57	—	—	—	—
78	0.45	1.05	1.15	1.30	—	1.40
79	0.20	1.80	—	—	—	—
80	0.10	2.70	0.45	2.57	1.52	2.25

FIGURE 7.2
Returns Diagram from Purchase of Put and Call Options with Premiums Included

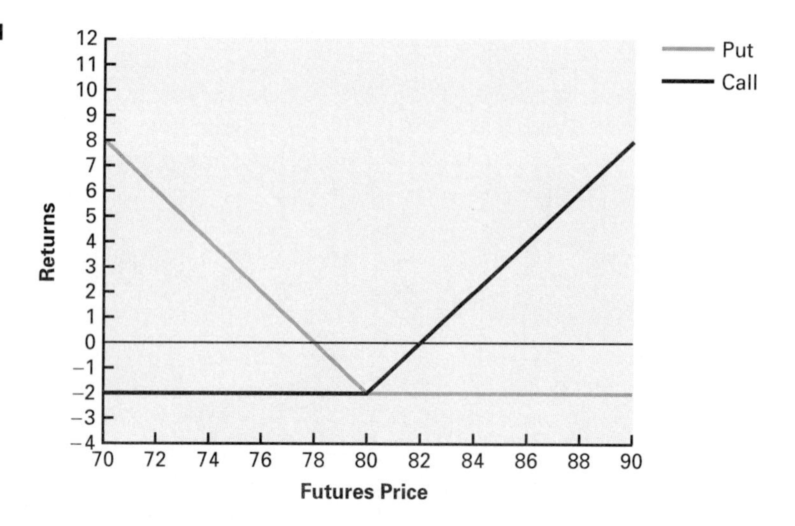

However, the return line is shifted down by the amount of the premium. The put strategy will give the trader a positive return if the market price falls below $80.00, but it will not be profitable until the market falls more than $2.00 below $80.00. The same idea holds for purchasing the call. The $80.00 call has no value if the market price falls below $80.00/cwt. and the $80.00 call has value if the market price climbs above $80.00. The right to buy at $80.00 is still worth $3.50 if the market price is $83.50 and the call return line is still kinked at $80.00. Again, the return line is shifted down the amount of the premium. The call strategy will give the trader a positive return if the market price climbs above $80.00, but it will not be profitable until the market climbs more than $2.00 above $80.00.

Much of the science to understanding options is in understanding the factors that determine the level of the premium. But before discussing these factors, more basic terms need to be introduced, some simple calculations need to be made, and some basic concepts need to be discussed. As mentioned earlier, the exchanges set strike prices in constant increments. For livestock and meat product futures, the CME uses $1.00/cwt. increments. The CBOT uses $.10/bu. for corn and wheat, and $.25/bu. for soybeans. If the December live hog futures contract, for example, is trading near $70.50/cwt., options will be listed with strike prices from $68.00 to $73.00 in $1.00 increments. Options for November soybeans that are trading near $6.50 per bushel would be offered in $.25 per bushel intervals above and below the $6.50 level.

Strike prices can be added to the trading set as needed. If the $80.00 strike price is the highest being traded for options on November feeder cattle futures contracts, the CME will add an $81.00 strike price if the futures start to trade above $78.00 and show potential for higher prices. There will always be strike prices that extend both above and below the trading level of the underlying futures contracts that are of interest to traders. In Table 7.2, strike prices from $75.00 to $80.00 were shown when the underlying October feeder cattle futures contract was at $77.40. Note that no trade occurred on that particular day in the $75.00 and $77.00 calls or the $75.00 puts.

An option strike price is *at-the-money* if the strike price is equal to the trading level of the underlying futures prices. An option is *in-the-money* if the strike price is at a level that would generate immediate return opportunities from the futures position if the purchaser exercised the option, took a position in the futures market, and then offset the futures position. A put option is in-the-money when the strike price is above the trading level of the underlying futures contract. For example, a $70.00 put option for the June live cattle futures contract is in-the-money if the underlying futures contract is trading at $67.50. The option, once purchased, could be exercised. The trader then holds a short position at $70.00 which they could then offset at the market price of $67.50. The returns from doing so would be $2.50, so the right to a short position has immediate value. Thus, the $70.00 put is in-the-money. Continuing with this example, a $65.00 put on the June contract would be *out-of-the-money*. If the underlying futures contract was trading at $67.50, there would be no immediate return possibilities if a short position was taken at $65.00, not when the futures would have to be bought back at $67.50. Intuitively, the premium on a $70.00 put will be much higher than the premium on the $65.00 put when the market is trading at $67.50.

Let's work another example referring to Table 7.2. The November contract closed at $77.85. Thus, the $78.00 put is in-the-money $.15 and the $80.00 put is in-the-money $2.15. The January contract closed at $79.27, so the $78.00 put on the January contract is well out-of-the-money and the $80.00 put is in-the-money by $.73.

Continuing with the live cattle example from earlier, the right to be short at $70.00 *has intrinsic* value when the underlying futures contract is $67.50. The right to be short at $65.00 has no intrinsic value. There will be a significant premium attached to the $65.00 put under these conditions only if we are dealing with a futures contract month that has a significant amount of time left before it expires. After all, if several months are left before expiration, the economic situation could change so that the futures contract could drop to $65.00 or lower. The $65.00 put has *time value* for that reason. This additional time value is also built into the $70.00 put premium as well. If the market drops to $65.00 or less, then a $70.00 put will have more intrinsic value. However, if the futures market stays around $70.00, the value of the $65.00 put option will decline to zero as the last trading day for the option approaches.

An option thus has *intrinsic value* and *time value*. The intrinsic value is determined by the strike price relative to the trading level of the underlying futures contract. A $70.00 put option has intrinsic value of $2.50 if the underlying futures contract is trading at $67.50. This is the case because the option could be exercised, the trader assigned a short position at $70.00, and then the position could be closed out by buying the futures position back at $67.50. A $65.00 put option would have zero intrinsic value if the underlying futures contract is at $67.50.

The time value is, of course, *a function of how much time is remaining before the option expires*. It is a reflection of the fact that the option is worth more the more time that is left to maturity. With more time, there is a greater chance that the futures market will move such that the option is in-the-money. *The time value component diminishes as the maturity data for the option approaches.*

You can see the decay in time value by looking at the premiums for feeder cattle options in Table 7.2. All of the feeder cattle futures prices are well above $76.00, so the premiums on all of the $76.00 puts reflect only time value, not intrinsic value. Notice that the premium for the January $76.00 put is $.80, the premium for the November $76.00 put is $.55, and the premium for the October $76.00 put is $.25. The October contract will expire in less than 30 days, the November in less than 60, and the January in less than 120. The premiums largely reflect the decay in time value due to the differences in how close the options are to expiration. However, the January futures contract is also $2.00 higher than the October or November contracts. If it were trading between $77.00 and $78.00, the premium on the January $76.00 put would be higher. But at $79.27, it has a smaller chance of falling below $76.00 and making the $76.00 put take on intrinsic value. A similar comparison can be made across the $78.00 puts. However, we need to remove the intrinsic value from the October and November $78.00 puts. The premium for the January $78.00 put is $1.40, the time value of the November $78.00 put is $1.15, and the time value of the October $78.00 put is $.45. Again, the time value decays for options closer to expiration. However, they are much larger for October and November $78.00 put options because the strike price is closer to being in-the-money. Here we have also observed something that will be discussed in later sections when we look at option premiums in more detail: in-the-money and out-of-the-money options are inexpensive relative to at-the-money options.

For a given strike price and time value, the *volatility* in the underlying futures contract becomes the most important determinant of the premium level. Highly volatile markets are risky markets, and the premium has to be large enough to entice traders to write or offer the option for sale. Someone has to be willing to offer or sell the $70 put on live cattle futures, to illustrate, before cattle feeders can buy a $70 put. The more volatile the market, the more premium that will have to be paid to get traders, usually speculators, to offer the $70 put options.

Volatility is particularly important and observable in options on grain futures contracts. If the summer growing weather is moderate and there is adequate rainfall for corn and soybean production, the premiums for options on the harvest corn and soybean contracts will reflect the normal time-value decay across the summer. However, if weather is unpredictable and it is questionable as to whether rainfall amounts were adequate, the time value in option premiums will increase with increased volatility in the futures market. Remember, this is an increase in the time value and not just the premium. We are netting out the level of the futures price relative to the strike price. The premium on puts will also be increasing because the futures prices will be increasing relative to the fixed strike price.

The producer who buys a put option to get protection against lower prices has to open an account with a commodity broker and pay the premium at the time the put option is bought. The commodity user who buys a call option to get protection against rising costs likewise has to pay the premium when the call option is acquired. *Thus, the buyer of a put or call option only has to pay the premium and is not subject to margin calls if the underlying futures contract move against the position.* Buyers of the $70 put in live cattle futures have a right to a $70 short position, for example, but if the futures go to $80, they can just let the $70 put option expire worthless. *There is no position in futures and no margin calls.* In a similar fashion, the buyer of an $86 call option in feeder cattle futures pays no margin calls when the market drops to $80 at the time of maturity for the option. The option can be allowed to expire worthless. Thus, the option premium is similar to an insurance premium. If the protection is not needed, the premium is forfeited and no action is required.

The rationale for this can be seen in Figure 7.2. The trader who purchases options has a known and limited loss. The maximum that you can lose from the option position is the premium. And the premium is paid up front. Thus, there is no need for margin requirements and margin calls.

While the buyer of options does not need to worry about margin calls, the seller, or writer, of options is exposed to an unknown amount of risk at the time the option is sold, and *must post margin requirements and answer margin calls*. For example, the trader who sells or writes a $70 put in live cattle futures is agreeing to guarantee some other trader, a put option buyer, a short position at $70. Selling a $70 put carries an obligation to take the other side, the long side of the $70 futures contract, so that

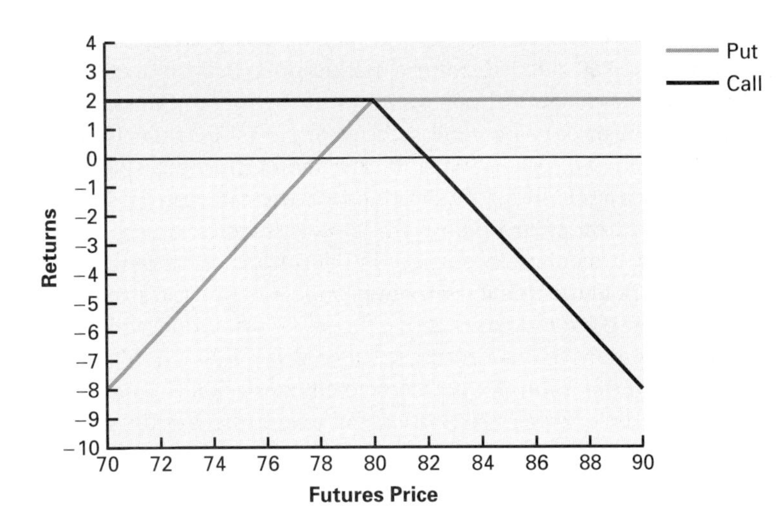

FIGURE 7.3
Returns Diagram from the Sale of Put and Call Options with Premiums Included

the buyer of the $70 put can be ensured a short position is available. Thus, selling a $70 put is comparable, in a sense, to being long in the live cattle futures market at $70.

If the market drops below $70, the buyer of the put can sell the option as the premium on the $70 put increases and registers the value of the right to a short position at $70. Alternatively, the buyer of the $70 put can exercise his or her option and ask for the short position at $70. A seller of the $70 put option, if the option is exercised, is then assigned a long position in the futures at $70.[1] It is apparent, then, that the seller of the $70 put will have to answer margin calls as the market drops below $70. In an account balance context, the sellers of $70 put options are treated essentially as if they had a long position in futures at $70. The margin requirements are often adjusted, but only if the put or call option is written or sold well out-of-the-money. *The key point here is that the buyer of options is not exposed to margin calls, but the seller of options is exposed to margin calls.*

We will return to the feeder cattle example used in Figure 7.2 and illustrate the returns diagram for options writers. In the example, we examined returns to a put and a call strategy with an $80 strike price. Let's look at the strategy from the viewpoint of the option writer. This is illustrated in Figure 7.3. The writer receives the $2 premium. The $80 put has no value if the market price climbs above $80/cwt. so the writer keeps the premium and has a $2 profit. The $80 put has value if the market price falls below $80. The right to sell at $80 is worth $4 if the market price is $76. The writer must pay this to get out of the obligation and has a $2 loss. The return line to the put writer is flat at the level of the premium when futures are above the strike price, the line is kinked at the strike price, and decreases one-to-one with changes in futures prices below the strike price. Selling the put will be profitable if the market price is above the strike price, will lose returns as the market price drops below the strike price, and will generate a loss when the market falls more than $2 below $80.

[1]If the buyer of a put does decide to exercise the option and take a short position in the underlying futures contract, the responsibility to take the opposite position will be assigned to the seller of the oldest put with the same strike price, determined by the date and time of trade.

The same idea holds for selling the call. The writer receives the $2.00 premium. The $80.00 call has no value if the market price falls below $80.00/cwt. so the writer keeps the premium and has a $2.00 profit before commissions. The $80.00 call has value if the market price climbs above $80.00. The right to buy at $80.00 is worth $3.50 if the market price is $83.50. The writer must pay this to get out of the obligation and has a $1.50 loss. The return line to the call writer is flat at the level of the premium when futures are below the strike price, is kinked at the strike price, and decreases one-to-one with changes in the futures price above the strike price. Selling the call will be profitable if the market price is below the strike price, will lose returns as the market price increases above the strike price, and will result in a loss when the market rises more than $2.00 above $80.00.

The buyer of a put option has two alternatives if the underlying futures contract price drops below the strike price for which the option purchased. First, the put option can be sold at the increased option premium that will accumulate as the futures market moves below the strike price. *The option does not have to be exercised and a position in the futures acquired for the price protection to work.* The option is simply sold as the end of the production or storage period approaches, and the net from the option buy–sell actions offset losses in the cash position.

The second alternative is to exercise the option and take a short position in the underlying futures contract. Occasional circumstances dictate that you should exercise the option and accept a position in the futures market.

One such set of circumstances can emerge for commodities that do not have a cash settlement delivery provision for futures contracts. The exchanges specify an expiration date for the options that is several weeks before the expiration date of the underlying futures to allow time for all the logistics that can be involved in the delivery process. For example, the June live cattle futures contract expires in the third week of June, but the options on the June contract expire in late May. To contrast, the options on the feeder cattle contract, which is cash settled, expire the same month that the futures contracts expire. If a cattle feeder has animals that are slower to finish than expected and therefore will not be sold until early June, he or she may well decide to exercise the option on a $70 put and ask for a short position at $70 in the June contract. This would be a good decision if fundamental and technical analysis suggest that the market is going lower in late May and early June. Protection is then extended into June when the cattle finish and are ready to be sold in the cash market.

A second but less likely reason for exercising the option occurs when the option is in-the-money and approaching its expiration date, but fails to show the premium appreciation expected given the decline in the underlying futures contract. By exercising the option and taking a short position in the futures, at the strike price, the full value of the decline in futures can be realized. The short position in futures can then be bought back. However, this action will not typically be needed. *Professional traders looking for arbitrage opportunities will ensure that the option value moves with the underlying futures as the maturity date of the option approaches.*

To illustrate, assume that the option premium for a $70.00 put option in live cattle has increased to $2.00/cwt. when the underlying futures are trading at $67.75/cwt. A knowledgeable trader could buy the $70.00 put for $2.00, exercise the option, take a short position at $70.00, and then immediately buy the contract back for $67.75. The gross return is $2.25/cwt. for an outlay of $2.00/cwt. which would net $.25/cwt. for each contract traded. The trader must also pay commissions but the $100/contract ($.25/cwt. × 400 cwt. / contract) of profit per contract traded would quickly be exploited by opportunistic floor traders. This process of arbitrage will push the put premium and the futures contract price up. Rational traders would do this until the profit does not cover commission costs.

One main reason for a trader not to choose to exercise an option is the associated increase in commission costs. Commissions on futures trades are assessed on the round-turn. For example, let's say a brokerage house charges retail customers commissions of $50 per contract. A trader who executes a trade will pay no commission until the trade is offset. But options are different. They may expire worthless. So our example brokerage house will charge something like $25 per contract when an option is bought and $25 per contract if the option is offset. Thus, if the option has intrinsic value at expiration, the trader's commission cost will be $50. If the options expire worthless, the commission is $25.[2] Now, instead of offsetting the option, suppose the trader exercises the option, takes a futures position, and then offsets the futures position. The commission costs will be $75, $25 from initiating the options position and $50 from offsetting the futures position. Generally, then, nonfloor traders do not worry about exercising options.

Options give important flexibility to users of futures markets. Concern over opportunity costs of entering futures hedges too soon at prices that turn out to be too low for a short hedge or too high for a long hedge is mitigated. Since the put option sets a price floor for the producer and the call option a cost ceiling for the user, the advantages of favorable prices or costs are still available to the buyer of options. In addition, the possibility of margins calls are eliminated. However, the purchase of an option requires payment of a premium, and the size of the premium is not inconsequential. The buyer of options can later sell them or allow them to expire worthless and no position in the futures market is required.

COMBINING CASH WITH FUTURES AND OPTIONS POSITIONS

This section examines the combination of cash, futures, and options positions. The results will typically be shown using a returns diagram. We will do this repeatedly through the remainder of the chapter, and it is useful to review the fundamentals of putting these diagrams together.

The purpose of the returns diagram is to communicate a lot of information in one place. The T-accounts in earlier chapters are useful for outlining and communicating the basic mechanics and ideas of a hedge. The mechanics are to sell in one market and buy in the other, and the concept is to offset losses in one market with gains in the other. However, we are now going to have to evaluate several potential strategies, not just compare a cash strategy to futures. We now have to add options to the mix and, potentially, options positions at different strike prices.

[2]Some brokerage houses charge the trader the second commission when the options expire worthless. It depends on the brokerage house.

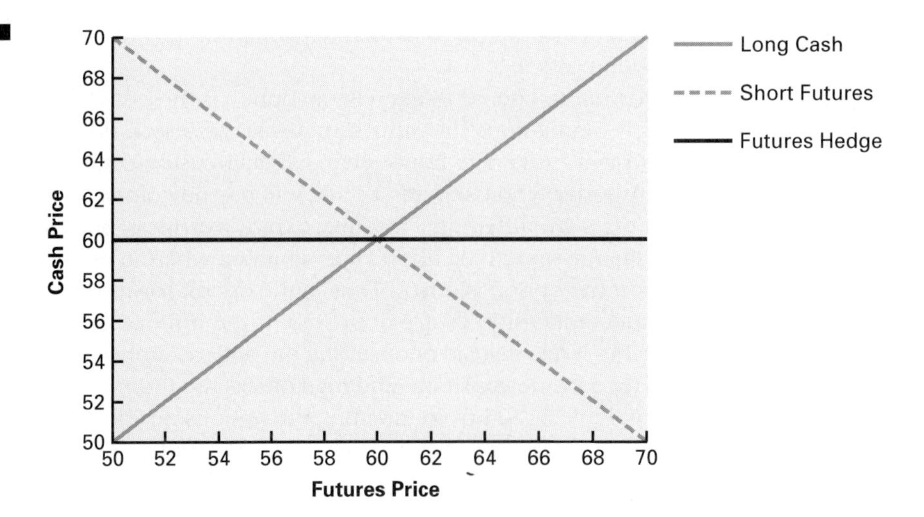

FIGURE 7.4
Long Cash, Short Futures, and the Hedge Forward Price

First, as in the earlier section, futures prices are denoted on the horizontal axis and cash prices on the vertical. You may notice that in other texts profits are listed on the vertical axis. This is the case here as well. The profits are profits from options or futures positions. They are profits after offsetting the futures or options position, *not profits that include costs from the production or storage of commodities*. Thus, the vertical axis can be viewed as cash prices or what we have been calling *net prices*. The upward-sloping 45-degree line on Figure 7.4 represents the value of a cash position. It also captures the relationship between the cash and futures market. As futures prices increase, cash prices also increases. The price levels denoted on the horizontal and vertical axes are important. The difference in the two prices reflects expected or average basis between cash and futures markets. If the basis for a commodity and location is zero, the two axes will match identical prices. The main information that the line in the figure communicates is the value of the cash position. As the futures market increases, the cash market also increases, and the value of the cash position increases.

A short futures position is also included in Figure 7.4. The short futures position is a downward-sloped 45-degree line. If the trader takes a short position in the futures market and the futures prices increase, the value of the futures position will decrease. Combining the cash and the futures position through a hedge results in an asset position that holds its value regardless of changes in market prices. *Losses in the cash market are offset by gains in futures and gains in the cash market are offset by losses in futures. The value of the hedge position is constant at the forward price chosen by the hedger.* The hedge is thus a horizontal line on Figure 7.4 at $60.00/cwt. To combine the returns from different positions, the lines on the returns diagram are summed vertically. For example, when the cash position has a low value, the futures position has a high value and vice versa. A weakness of the diagram is that it does not communicate basis risk. The net price will be different from the forward price by the amount of basis error. We ignore basis error in all of the return diagrams.

Next, let's combine a long put option position (you have bought a put) and a long cash position to Figure 7.4. The new diagram is Figure 7.5. Purchasing a put option is similar to taking a short position in futures. As the futures price falls below the strike price, the value of the option position increases. As the futures price rises above the strike price, the option has no value. However, the premium paid must be considered.

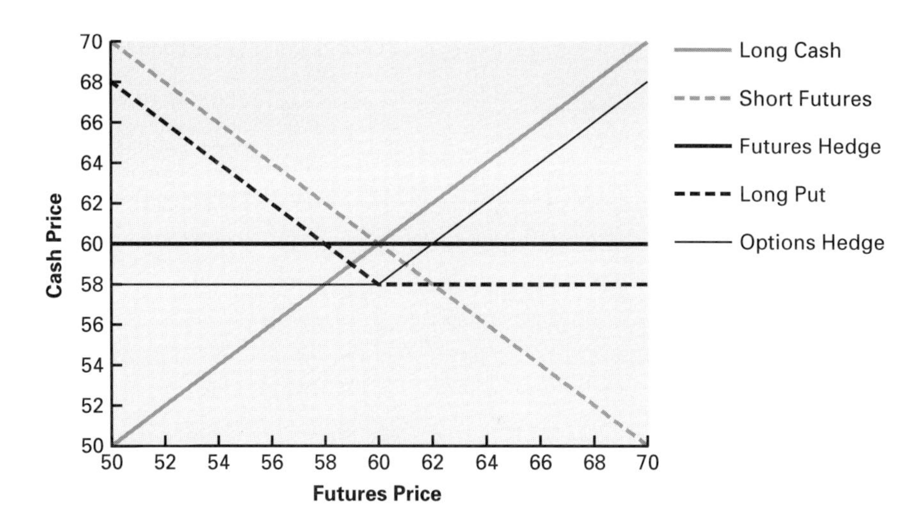

FIGURE 7.5
Long Cash, Long Put Option, and the Forward-Price Floor

Below the strike price, the combined cash and option position is similar to the hedge. However, it is below the hedge by the amount of the premium. Above the strike price, the combined cash and option position is similar to the cash market. The combined return is below the cash market by the amount of the premium. Purchasing a put option is similar to taking a short position in futures, but traders are limited in the amount of money they can lose in their brokerage accounts. They incur the maximum loss up front with the purchase of the option. The upside potential, it should be remembered, is greater than the premium cost.

Look closely at the diagram. It illustrates the strengths and weaknesses of cash positions, positions hedged with futures, and positions hedged with options. *In a falling market, the position hedged with futures has the best outcome. In a rising market, the cash position has the best outcome*. But this is the problem. *We cannot know the outcome before we make the price-risk management decision*. This is risk. So we do the best we can by knowing the production costs of our business, and incorporating fundamental and technical analysis into our decision making. *But one of the two positions is going to result in a poor outcome*.

Options change this result. Options blend a cash and a futures position. *The options position is second best in both cases. Options never result in the best outcome.* But if markets change significantly up or down, they will also not result in the worst outcome. And this is a big "if." Examine the diagram. Do options ever result in the worst outcome? Yes, if market prices do not change. The options position results in the poorest outcome when the final market price is close to the strike price.

As a final example, let's combine a short call option position and a long cash position in Figure 7.6. Selling a call is similar to taking a short position in futures. As the futures price rises above the strike price, the value of the option position decreases. As the futures price falls below the strike price, the option has no value. However, in this latter case the trader receives the premium.

Above the strike price, the combined cash and option position is similar to the hedge. However, *it is above the hedge by the amount of the premium from selling the call option*. Below the strike price, the combined cash and option position is similar to the cash market. However, *the combined return is above the cash market by the amount of the premium*. Selling a call option is similar to taking a short position in futures, but the traders are limited in the amount of money they can gain. They

FIGURE 7.6
Long Cash, Short Call, and the Net Price

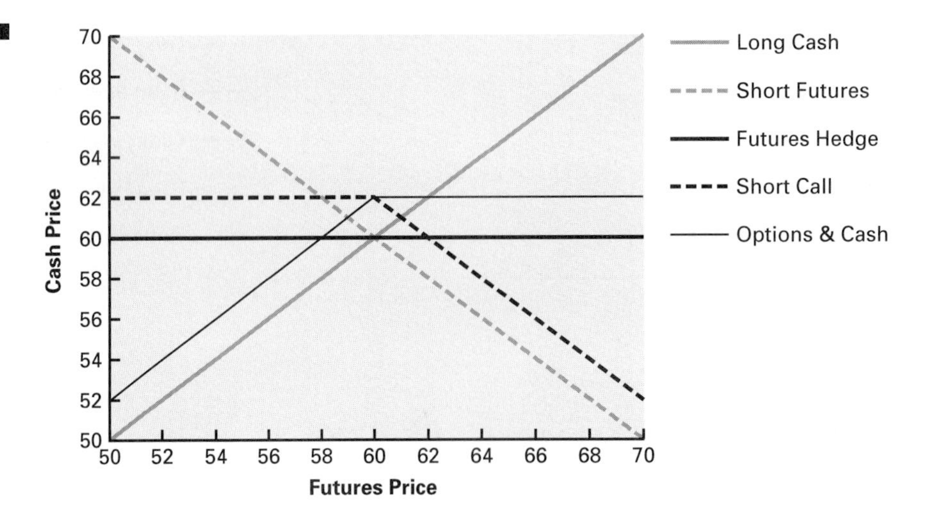

incur the maximum gain up front with the sale of the option. The downside potential is greater than the premium received. *This is not a hedge.*

Look closely at this diagram and compare it to the previous figure. Again, in a falling market, the position hedged with futures has the best outcome. In a rising market, the cash position has the best outcome. Options blend a cash and a futures position and are second best in both cases. But, now the options position has the best outcome if market prices do not change and the market price is close to the strike price. But, to achieve this good outcome, *the trader has no price floor. This strategy does not appear to be risk reducing.*

Return diagrams are important tools. We are vertically summing the lines that denote different asset positions. The picture then communicates the different returns to the trader under different market conditions—falling prices, rising prices, and scenarios in which price changes little. With a diagram, you can then start to think about picking the strategy that best fits your particular circumstances.

BASIC STRATEGIES FOR FORWARD-PRICING WITH OPTIONS

Using options to establish a forward price floor for producers and holders of commodities and a forward price ceiling for users of commodities is the objective of this section. More complex strategies are discussed in later sections. We will make use of returns diagrams to summarize the options strategies, but we now must be careful in numerically labeling the axes.

In simple terms, the buyer of a put option is looking for protection against falling prices and the flexibility of benefiting from higher prices. Assume a hog producer buys a $60.00 put option with a premium cost of $1.50/cwt. with the underlying October futures contract trading at $62.75/cwt. For ease in illustration, assume the expected basis is zero or that the producer sells hogs typically at a cash price that is very close to or the same as the futures price. Figure 7.7 shows the net price after

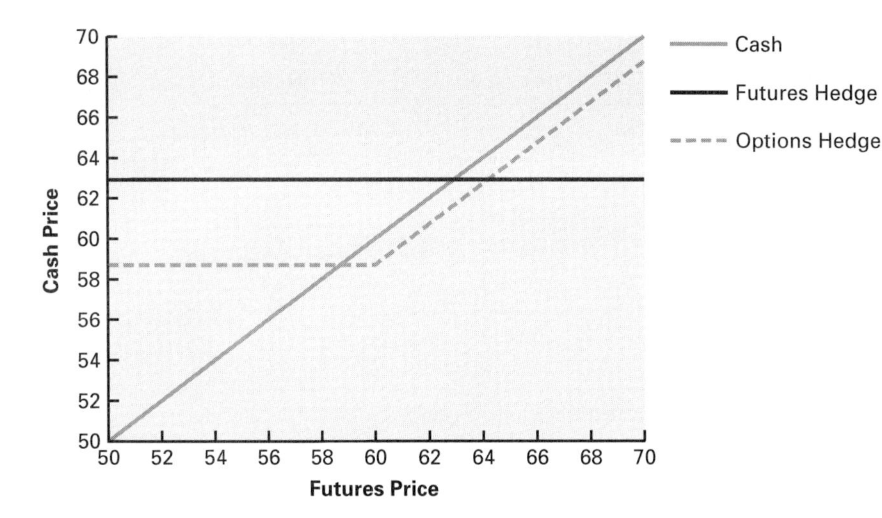

FIGURE 7.7
Net Price from Purchasing a $60.00 Put Option on Hog Futures Where Basis Is Zero and the Put Premium Is $1.50

accounting for the premium. The range of price chosen for the figure is from $50.00 to $70.00.

Note, in the figure, that the net price never goes below $58.50 as the market price, cash, and futures, move toward the $50 level. The price floor is at $58.50, or the $60.00 strike price plus basis of zero less the $1.50 option cost. The forward price floor (*FPF*) is calculated as follows:

$$FPF = SP + Basis - Premium$$

where

SP = strike price
Basis = expected cash-futures basis, and
Premium = option premium paid.

In this example, *FPF* = $60.00 + 0 – 1.50 = $58.50/cwt. If the expected cash-futures basis were not zero but –$1.00, then the forward price floor would have been $60.00 – 1.00 – 1.50 = $57.50.

The net price (*NP*) from the price floor strategy is the cash price received, less the premium paid, plus any premium or value that the option holds when the cash commodity is sold.

$$NP = Cash - Premium\ Paid + Premium\ Received$$

If prices fall, the $60.00 put takes on value. If the cash and futures market are trading at $54.00, the put will be worth $6.00. Thus, the net price is the cash price, $54.00, less the premium paid when the price protection was bought, $1.50, plus the current value of the option, $6.00, or $58.50. *The hog producer sells the hogs in the cash market and sells the $60.00 put option.*

If the cash and futures market rise above $60.00, the $60.00 put will be worthless at maturity. The net price received is the cash price less the premium paid. There is no premium received in this case. Suppose the markets rally to $68.00. The producer sells the hogs in the cash market for $68.00 and the $60.00 put expires worthless. No

TABLE 7.3
Net Prices from Selected Hog Marketing Strategies

Futures Price	Pure Cash	Hedge	Put Option Price Floor
$52.00	$52.00	$62.75	$58.50
54.00	54.00	62.75	58.50
56.00	56.00	62.75	58.50
58.00	58.00	62.75	58.50
60.00	60.00	62.75	58.50
62.00	62.00	62.75	60.50
64.00	64.00	62.75	62.50
66.00	66.00	62.75	64.50
68.00	68.00	62.75	66.50

Hedge and options were initiated when futures equals $62.75 and the $60.00 put premium equals $1.50. Expected basis equals $0.

price protection was needed and the option premium is much like the insurance premium on your health or your automobile. There are no positions in the futures market. However, the premium does reduce the net price. The net price received is the cash price, $68.00, less the option premium paid, $1.50, or $66.50.

Table 7.3 shows a simple comparison of strategies involving a hedge, a $60.00 put option, and a pure cash position. Figure 7.7 shows the hedge and the cash strategies along with the $60.00 put-option strategy. Above $60.00 the option strategy is superior to the hedge which forward-prices at $62.75. Below $60.00 the option strategy is superior to the cash position which falls with decreases in futures prices. However, the option strategy has the worst outcome if the market price is close to the strike price.

The same concepts and similar calculations are used to establish price ceilings for commodity users. The buyer of a call option is looking for protection against rising input prices and the flexibility of benefiting from purchasing at lower prices. Assume a cattle feeder buys a $2.80 call option with a premium of $.25/bu. on the underlying March futures contract trading which is at $2.80/bu. Expected basis is +$.20/bu. Figure 7.8 shows the net price (i.e., the cost) after accounting for the premium. The range of prices chosen for the figure is from $2.50 to $3.50. Notice the relationship between the futures price level and the cash price level. We have to be careful here to show the proper expected basis.

The price ceiling is at $3.25, or $2.80 plus +$.20 basis plus the $.25 option cost. The forward price ceiling (FPC) is calculated as follows:

$$FPC = SP + Basis + Premium$$

where

SP = strike price
$Basis$ = expected cash-futures basis, and
$Premium$ = option premium paid.

Note that the premium paid is added to get the price ceiling for a commodity to be purchased. The price floor is for a commodity to be sold. In this example,

$$FPC = \$2.80 + \$.20 + \$.25 = \$3.25/\text{bu.}$$

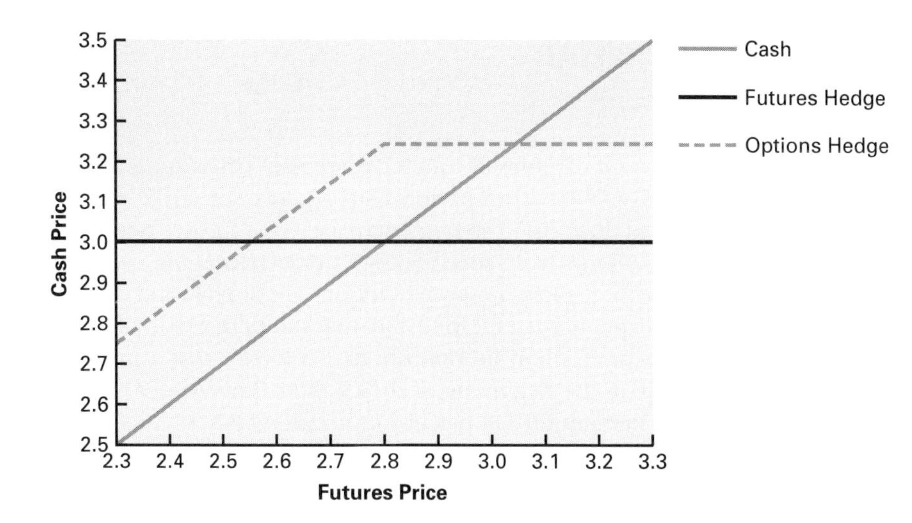

FIGURE 7.8
Net Price from Purchasing a $2.80 Call Option on Corn Futures Where Basis Is +$.20/bu. and the Call Premium Is $.25/bu.

The net price from the price ceiling strategy is the cash price paid, plus the premium paid, less any premium or value that the call option holds when the cash commodity is sold.

NP = Cash + Premium Paid – Premium Received

If prices rise, the $2.80 call takes on value. If the futures market is trading at $3.35, the call will be worth $.55. Thus, the net price or net cost is the cash price, $3.55, plus the premium paid when the price protection was bought, $.25, less the current value of the option, $.55, or $3.25. The cattle feeder buys the corn in the cash market and sells the $2.80 call option.

If the futures market falls below $2.80, the $2.80 call will be worthless at maturity. The net price paid is the cash price plus the premium paid. There is no premium received in this case. Suppose the futures market falls to $2.25. The cattle feeder buys corn in the cash market for $2.45 ($2.25 + .20) and the $2.80 call expires worthless. There are no positions in the futures market, but the premium does increase the net price. The net price paid is the cash price, $2.45, plus the option premium paid, $.25, or $2.70.

Later, we will examine more sophisticated options strategies that combine the use of both puts and calls in a single strategy. Go back through the examples if you are not ready to move ahead.

The forward price floor and the forward price ceiling are the most basic options strategies. To establish a price floor, the hedger purchases put options. The price floor is equal to the strike price of the put bought, plus basis, less the premium. A price ceiling is established by purchasing call options. The price ceiling is equal to the strike price of the call bought, plus basis, plus the premium. The net price for price floors or ceilings is the cash price received or paid net of the premium paid to purchase the options and any option premium received when offsetting the position.

COMPARING PRICE FLOORS, OR, WHAT OPTION TO CHOOSE?

The previous section went carefully through the steps of calculating the price floor, or price/cost ceiling, available via an option with a particular strike price, and then compared the returns to that option strategy with a cash or futures position. This section will examine price floors from several strike prices. Here, the usefulness of the returns diagram will become more apparent. Options are flexible and they offer many choices. The returns diagram is useful for evaluating the different choices.

Let's examine the price floor alternatives that a feeder cattle producer would have from the example in the beginning of the chapter. This feeder cattle producer has 700-pound calves coming off fall pasture in mid-November—it is early October now. The producer is worried about the fed-cattle market turning down and the corn market turning up. Both of these events will pressure feeder cattle prices lower. The futures and options markets are trading at prices reported earlier in Table 7.2. The expected basis for 700-pound feeder cattle is –\$4.00/cwt. in mid-November. The hedge forward price is as follows:

Hedge Forward Price = \$77.85 – 4 = \$73.85/cwt.

Remember that the forward price floor for put options is calculated as follows:

Forward Price Floor = *Strike Price* + *Basis* – *Premium Paid*

The forward price floors for \$76.00, \$78.00, and \$80.00 puts are therefore

FPF \$76.00 Put = \$76 – 4 – 0.55 = \$71.45/cwt.
FPF \$78.00 Put = \$78 – 4 – 1.30 = \$72.70/cwt.
FPF \$80.00 Put = \$80 – 4 – 2.57 = \$73.43/cwt.

The \$76.00 put offers less downside risk protection than the \$78.00 or \$80.00 puts, but we need to remember that because the strike prices and premiums are different, the upside potential is different as well. The \$76.00 put has no value if the market price is above \$76.00 at expiration. This means that the returns to this option strategy will follow the cash market more closely than the other two options if the market is above \$76.00. Observe the returns diagram in Figure 7.9. The price floor for each option is the strike price plus basis less the premium. The price floor is in effect when futures are less than the strike price, the option has value, and gains in the option value offset losses in the cash position value. Above the strike price the put option has no value at expiration, and the option strategy return equals the cash market price less the premium.

The returns diagram facilitates easy comparison of different strategies and strike prices. While it is easy to observe and compare the returns, it is not easy to choose among the strategies. *There is no clear best option choice based on any objective measure*. "Best" depends on the individual producer's willingness to accept risk. *The puts with different strike prices offer different degrees of downside risk protection and different upside profit potential, but there is definitely an inverse relationship between risk protection and profit potential.* The option with the most downside protection, the \$80.00 put, has the least upside potential. Likewise, the option with

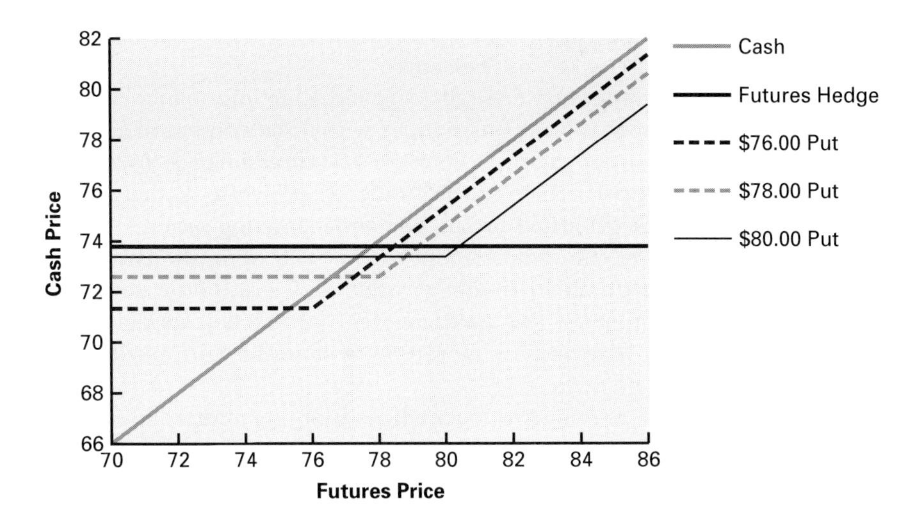

FIGURE 7.9
Feeder Cattle Price Floors with Different Put Option Strike Price Levels

the least downside protection, the $76.00 put, has the most upside potential. The $78.00 put offers an intermediate choice.

Options blend elements of cash positions and futures hedges and they offer flexibility. But there is no free lunch. Strategies with high price floors involve purchasing options that are well in-the-money. These strategies offer little upside potential and are relatively expensive. Strategies that offer good upside potential involve purchasing options that are well out-of-the-money. These options are relatively inexpensive but the price floor offers less protection. We also see the usefulness of the returns diagram. Five strategies were considered in an example, three put options, hedging with futures, and a pure cash position. The diagram shows the different returns under different final market prices, and the five strategies are easily compared.

WHEN TO USE OPTIONS

There is no question that options add attractive dimensions and flexibility to a producer's pricing program. Eliminating margin call problems and minimizing the opportunity costs of pricing too soon are attractive attributes. But it is important to recognize that a cost comes with this added flexibility. The buyer of the option pays the premium up front and that premium must be sufficient to attract a seller or writer to the market. *The option buyer is paying someone else to accept the exposure to market moves and margin calls that the option buyer is trying to avoid.* It follows, therefore, that options could be a relatively expensive way to price over time compared to other approaches.

The previous section explained rather clearly that there is no best option strike price to choose when establishing a forward price floor. The same thing holds for forward price ceilings. There is no best choice among the options using an objective measurement. This section discusses a framework for choosing among different options strategies. *There are logical economic factors to consider that will help you evaluate different options.* Just don't look for "the one right" answer. This section will also discuss when to choose hedging with options compared to hedging with futures contracts. This is a very common and practical question. There are basic procedures that can be used to help make this decision.

A "best" option strategy may emerge if you add other information to the question, or you place the different returns possibilities within the context of additional information. The most important factor to consider is your willingness to accept risk and your evaluation of the risk versus reward tradeoffs. A livestock and grain producer who is older, has well-established production and marketing systems, has little debt, has saved for retirement, and whose money is not tied up in the farming operation may rarely hedge through either futures or options. There is little need. Such a producer is used to accepting the risk associated with production agriculture and his or her financial situation, long-term or short-term, will not be hurt much if the current year is a bad outcome.

On the other hand, a young producer who is adopting new technology, has taken on large amounts of debt, and who has basically all of the family's resources tied up in the farming operation should seriously consider hedging. In fact, the bank that has loaned the producer most of the farm's capital may encourage or even require that the producer follow a hedging program. One or two bad price outcomes may create a financially insolvent operation. The producer may need to choose to live with many years of small positive returns and specifically avoid the large negative returns.

The first producer will want to participate mainly in the cash market. Think back to the returns diagrams presented earlier in this chapter, especially Figures 7.5, 7.6, and 7.7. This older producer can live with low prices, as long as there is also the possibility of high prices. The producer may also trade futures or options—or forward-price through cash contracts—but this will be done to exploit obvious profit opportunities. Futures and options trading, or cash contracting, is not something the producer will do every year or production cycle.

The second, younger producer will want to consider hedging aggressively. The younger producer will need to establish forward prices on much of the operation's production. This will be done mainly through the use of futures. Options price floors are always lower and this producer is interested in hedging much of the operation's production. The younger producer cannot wait for excellent pricing opportunities. Options may be purchased periodically when the price floor covers costs of production and an upside opportunity is present.

An entire spectrum of operations exists between the two example producers. And in that context, the rationale emerges for hedging a portion of production and marketing or hedging with more flexible pricing instruments, such as options. Producers with more intermediate financial conditions, debt, and willingness to accept risk will be the ones that find options attractive. Options offer downside price protection, but not as good as that of a hedge with futures contracts, while offering upside potential, but not as good as the cash market. Options are a tool that allows the producer to make forward-pricing decisions without guaranteeing that the decision will result in the best or worst outcome. If the price market moves away from the strike price, options will result in the second best outcome. Options are never the worst outcome, unless the market price trades close to the strike price level.

The discussion started out focused on risk and reward preferences. What is the producer's willingness to accept different levels of price flexibility? This influences whether he or she hedges at all, hedges aggressively, or chooses some intermediate ground. But the latter part of the discussion has taken on an element of price direction. *What is the likely future direction of changes in market prices?* The choice among cash, futures, and options for the producer with intermediate willingness to accept risk becomes clearer if you have an informed opinion about market direction. How do you become informed about likely future movements in price? The tools we used in previous chapters were fundamental and technical price analysis.

Whether the costs of an options-based program will exceed the costs of a futures-based program will depend on the characteristics of the underlying futures market. *In a choppy and sideways market, the options approach will tend to be more costly.* In a market with major and sustained price trends, in which significant margin calls can be involved, the futures-based program will typically be more costly on a per-unit basis in terms of actual cost outlays and/or opportunity costs.

Fundamental analysis can contribute to making informed choices among futures and options. The grain balance sheet is an important tool used to summarize supply and demand conditions in any grain market. Suppose an assessment of the corn balance sheet reveals that the grain stocks-to-use ratio is at its lowest level in 15 years. This was the situation in 1995 and 1996. At the same time, corn prices were trading at average levels. A small crop or strong demand could cause this market to move up significantly over the crop year. However, an excellent crop and weak demand could push prices down significantly below average. *This situation appears to favor the use of options over futures for hedging.*

Let's contrast this with the corn market situation that occurred in the mid-1980s. At that time, the stocks-to-use ratio was very high, approaching a ratio of 1.0 for some of the years. There is almost no upside potential in this type of market. Purchasing options would seem to be an expensive risk management choice. *The downside potential is far greater than the upside*, and it is actually much more likely that corn prices will stay at low levels. The best price-risk management strategy would then be to time forward sales on any rally and sell futures or use cash forward contracts.

Livestock producers can follow a similar decision-making process. The key is assessing the upside potential and the downside risk. A cattle feeder examines the most recent USDA *Cattle on Feed* report and finds that past placements into feedlots and the current inventory of cattle on feed were both above that of the previous year. It is likely, then, that future marketings will be large. But the feeder also reads that substitute meat prices are high, prices of pork and chicken, and that demand is strong. Further, feedlots are doing a good job of marketing cattle before they reach excessively heavy weights. While future cattle numbers will be up, the weight of those cattle are down compared to the previous year. How would you assess this market situation? The market appears to have upside potential *and* downside risk. *Thus, the cattle feeder would want to consider options.* Hedges with futures contracts would be a good choice if the market situation changes in that cattle feeders start holding finished animals to heavier weights. In this case, the market loses its upside potential.

The technical picture should also be included with fundamental analysis. Technical analysis often gives buy and sell signals before the information needed for fundamental analysis is released. Further, trend lines, support planes, and resistance planes often provide very basic and useful information about anticipated market directions. For example, it makes little sense to purchase an option price floor when the market is trading at a high level and is close to resistance at life-of-contract highs. There is little upside potential and the likely market move is down. This recommendation assumes there is no new fundamental information suggesting that new higher prices are warranted. It makes more sense to purchase a price floor when the market is trading at low levels yet is above support. Purchasing an option here may offer the producer peace of mind. The market is trending down and pressuring support. The producer missed the sell signals that occurred at higher prices before the downtrend began. The producer is also concerned about the market breaking through the support plane, perhaps a life-of-contract low support plane, and moving lower. It has broken minor support several times following the down-trend, and the price dropped quickly and significantly following each break.

Trends can also be worked into the decision of choosing between futures and options. Suppose the market price is trending up and then stalls. It has not broken the trend line but the producer is not convinced that the fundamentals warrant higher prices. The current market price (less the option premium) will cover production costs, so an option is purchased. The producer secures a price floor but if the technical up trend continues, the producer gains. Futures make more sense in a down-trending market. Upside potential requires a change in the technical picture. A good strategy is to hedge as the market price rallies to test the trend line. *One important thing to notice is that in all of the examples, there is always a blending of attitude toward risk and expected future price movements.* If the expected movements relative to the downside risk favor options, use options. Otherwise, use futures.

Advocates of selective hedging would argue that the producer does not have to endure substantial margin calls if the futures are used for hedging. When the market starts to rally, the short positions can be lifted. But a selective hedging strategy is not always easy to manage in practice. The producer has to be a good technical and fundamental analyst and must have the discipline to manage the selective hedging program effectively. Many producers will not be that effective in managing such programs. If that is the case, there are lots of reasons to argue in favor of an options-based program. If the market shows a sustained price rally, there are no difficult decisions to make. The put option expires worthless, and the producer benefits from any price rally that exceeds the option premium.

Pulling the points together, a general set of guidelines about the decision to use futures or options is as follows:

1. *Use options in markets likely to be characterized by large and sustained price moves.* If ending stocks are being reduced to a level that might force prices to rally significantly to ration usage, for example, options might be the preferred instrument in a price-risk management program. In livestock, the major price moves are often cyclical in nature. If prices are so low that some producers are liquidating breeding herds, any pricing for later in the year or the following year might be done with options in anticipation of a possible major price rally.
2. *Use options when there will be problems in arranging and financing a margin line.* The futures markets should *never* be used by decision makers who would have major problems financing margin calls.
3. Use options when the ability to manage a selective hedging program directly in the futures is questionable.

When do you use options instead of futures or cash marketing strategies? First, it depends on your willingness to accept risk and your evaluation of the risk and return. You must know your costs of production and marketing. Hedges with futures are the low-risk, and low-return, alternative. Risk and returns are both highest with cash strategies. Options are an intermediate choice, but hedging just a percent of expected marketings can also be an "intermediate" choice. Second, whether or not options are the "best" strategy also depends on

expected future price moves. Incorporate your futures-or-options choice with fundamental and technical analysis. Use options when the market offers upside potential or is expected to be volatile. Use futures when the downside looks most likely or there is little volatility. And remember, options are usually the second best outcome.

MORE ADVANCED OPTIONS TRADING

Three sections prior to this one, we introduced the returns diagram and illustrated the returns to purchasing puts and calls, and selling puts and calls. The options trading strategies were kept simple. We purchased either a put or a call, or we sold a put or a call. Later in the section, we combined these simple options strategies with a cash position. We will examine the returns to more advanced options trading strategies in this section and look at the returns from combining different options positions.

The point to this section is to illustrate basic options trading strategies. If you have any exposure to options or have read popular press articles on marketing strategies and outlook, you have no doubt seen a lot of fancy options strategies and terminology. These strategies include fences, windows, butterflies, straddles, strangles—the list is almost endless. People who write about options like to use jargon—words that don't really communicate. This section will present some of the most basic of these strategies and call them by their most common names. More complex strategies are usually variations of these. Some of the strategies are useful for hedging and the remainder are speculative. All strategies use puts and calls as building blocks. If you are going to trade options you need to be comfortable with their components. In addition, it is important to recognize the difference between strategies that are useful for hedging and those that are useful for speculation. If you understand the basic strategies, you will not fall into the trap that has caught producers who thought they were hedging but were actually speculating.

The first strategy involves purchasing a put and selling a call. The strike price is the same for both options ($60.00) and we assume the premium is the same at $2.00. Most of the time, this is a very reasonable assumption. Thus, the trader uses the premium received from the sale of the call to pay the premium on the put. At expiration, if the futures price is below the strike price, the call has no value and the put is in-the-money. If the futures price is above the strike price, the put has no value and the call is in-the-money. *This strategy is equivalent to taking a short position in the futures market and is often called a synthetic short hedge.* When combined with a long cash position, purchasing a put and selling a call creates a hedged position for the trader. However, without a cash position, the trader has a short position in the options market. If the futures price falls, the trader stands to make money, and if the futures price rises, the trader stands to lose. Figure 7.10 shows the results of this strategy for final futures prices from $50.00 to $70.00.

The trader can also create a synthetic long hedge in the options market. In this case, he or she purchases a call and sells a put. This is done at the same strike price ($60.00), and the put premium ($2.00) received is used to purchase the call. At expiration, if the futures price is above the strike price, the put has no value and the call is in-the-money. If the futures price is below the strike price, the call has no value and the put is in-the-money. This strategy is equivalent to a long position in the futures market. If the futures price rises, the trader will make money, and if the futures price falls, the

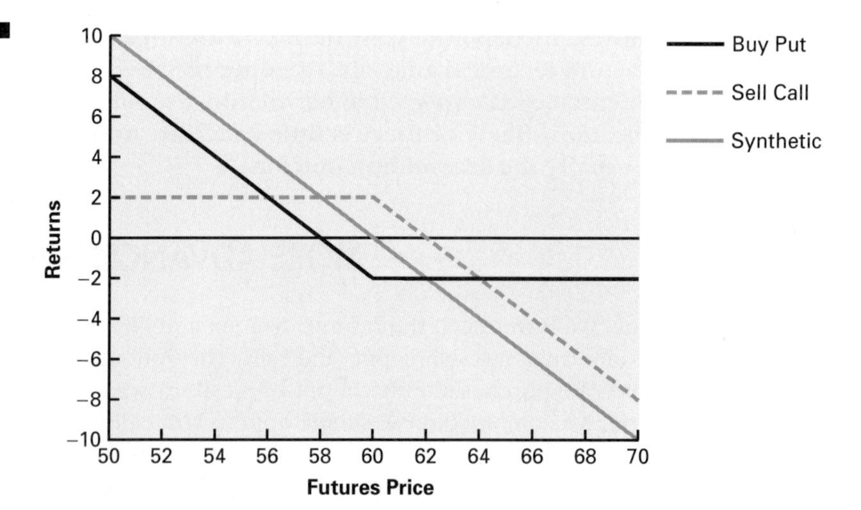

FIGURE 7.10
Synthetic Short Position Using Put and Call Options

trader will lose money. However, when combined with a short cash position, purchasing a call and selling a put creates a hedged position. Figure 7.11 shows the results.

The main point that synthetic hedges illustrate is that options can be used as building blocks. They are more basic than futures positions and can be used to create a position that looks like a futures position. However, they are more flexible than futures positions. The following strategies illustrate this.

How important is it that the trader uses options with the same strike price in a synthetic position? It is easy to answer the question by looking at the returns diagram. Assume the trader wants to construct a short position with options and that he or she purchases a put with a lower strike price than the call being sold. Assume both of the options are out-of-the-money. This assumption is not important but it makes the discussion easier. If the futures market is between the two strike prices, then both options are out-of-the-money and the trader does not have a short position. The trader only has a short position if the market falls below the put strike price ($56.00) or rises above the call strike price ($64.00). This is illustrated in Figure 7.12.

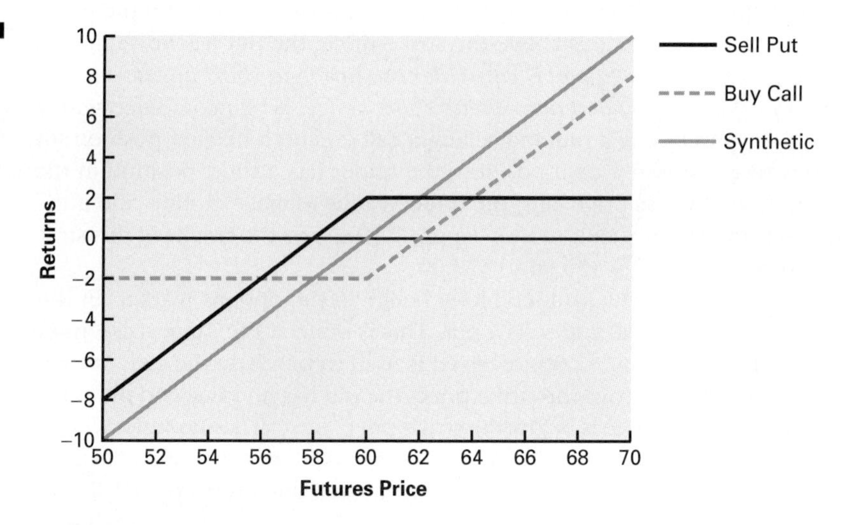

FIGURE 7.11
Synthetic Long Position Using Put and Call Options

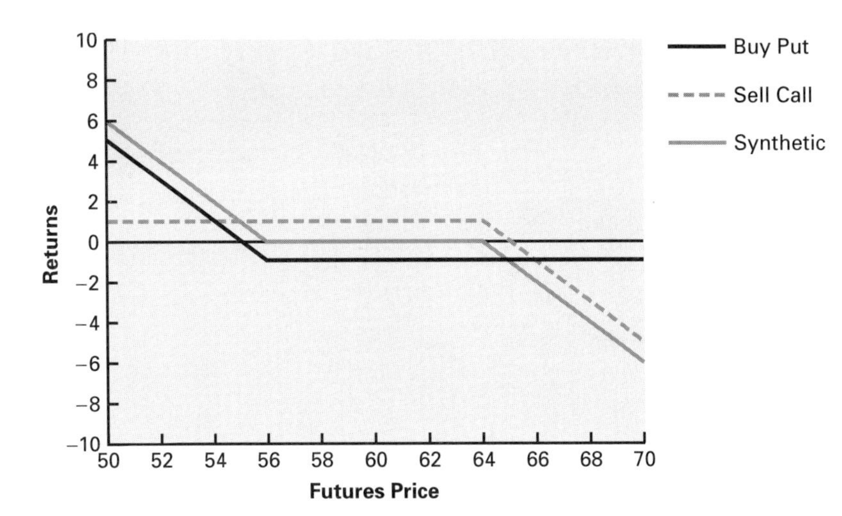

FIGURE 7.12 Synthetic Short Position with Different Strike Prices

The same thing occurs in the synthetic long position if the trader buys a call when the strike price is above the strike price of the put sold. The trader has a long position when the market is outside of either strike price and no position when the market is between the strike prices. This is illustrated in Figure 7.13.

What happens if the trader who is interested in a synthetic short position purchases a put with a strike price that is above the strike price of the call being sold? In this case, both of the options are in-the-money. In this case, the trader is essentially doubling the short position when the market is between the two strike prices. The trader has one short position if the market is below the call strike price and one position if the market is above the put strike price, *but he or she has two short positions if the market is between the two strike prices.* This case is confusing and would not be useful for risk management. None of the further discussion assumes that the trader uses options with strike prices that result in the position being doubled.

The next two types of strategies are useful primarily to speculators. The two strategies are called *spreads* and *straddles*. The details and mechanics of these two strategies are not very important for traders interested in risk management. However,

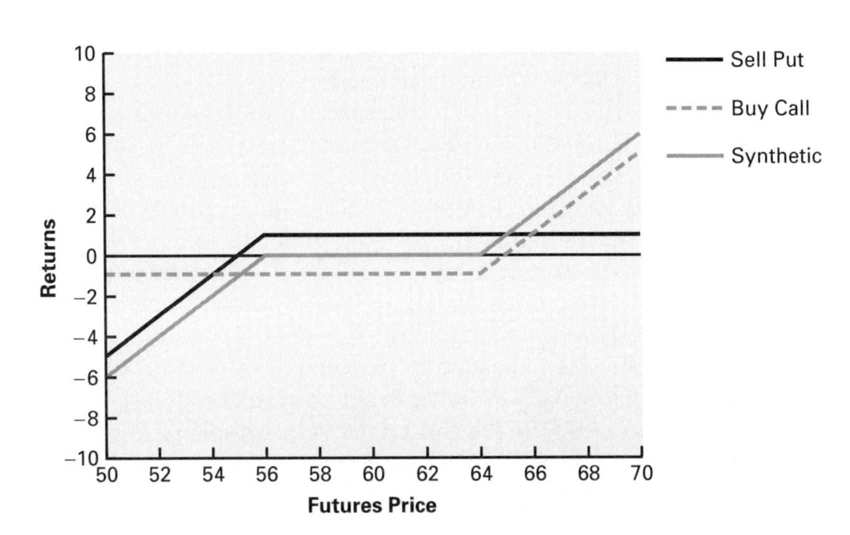

FIGURE 7.13 Synthetic Long Position with Different Strike Prices

FIGURE 7.14
Long Spread with Calls

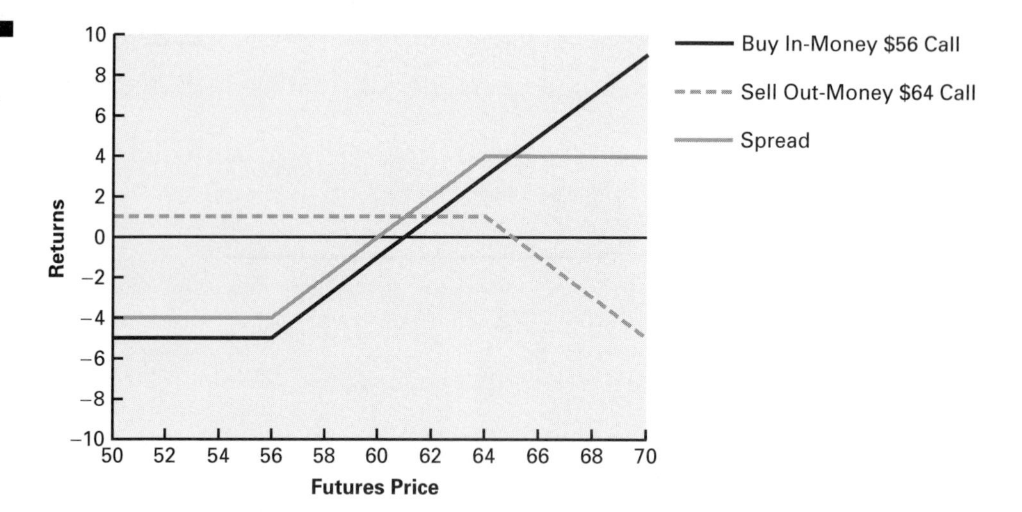

they are useful for illustrating how options positions can be pieced together to construct a larger strategy. They are similar in construction to the synthetic hedges and provide alternative examples. You should read the examples with this in mind.

The first strategy discussed is a spread. Spreads can be long or short. These are often called bull or bear spreads. In a long spread, the trader purchases an in-the-money call at a low strike price and sells an out-of-the-money call at a high strike price. The two options positions and the combined position are shown in Figure 7.14. At expiration, if the futures price is below the low strike price, the trader loses money because the premium paid for the in-the-money call is greater than the premium received for the out-of-the-money call. If the futures price is above the low strike price but below the high strike price, the trader will make money as the intrinsic value of the long call and the premium received from the short call are greater than the premium paid for the long call. If the futures price is above the high strike price, the trader will make money if the intrinsic value of the long call and the premium received from the short call are greater than the premium paid for the long call and the obligation to pay the intrinsic value of the short call. An important thing to see with a spread is that *the trader has a long position but the returns and losses are limited*. If the market moves up, the trader will make money and if the market falls, the trader will lose money. But losses are limited to the net of the two premiums. To achieve limited losses the traders have limited gains as well.

The trader can also construct a long spread through purchasing put options. The strategy involves purchasing an out-of-the-money put at a low strike price and selling an in-the-money put at a high strike price. The two options positions and the combined position are shown in Figure 7.15. If the futures price is above the high strike price, the trader makes money because the premium received for the in-the-money put is greater than the premium paid for the out-of-the-money put. If the futures price is below the high strike price but above the low strike price, the trader will make money if the premium received from the short put is greater than the premium paid for the long put and the obligation to pay the intrinsic value of the short put. If the futures price is below the low strike price, the trader will lose money because the intrinsic value received from the short put will be less than the premium paid for the long put and the obligation to pay the intrinsic value of the short put. Again, the important thing to see with this spread is that *the trader has a long position but the returns*

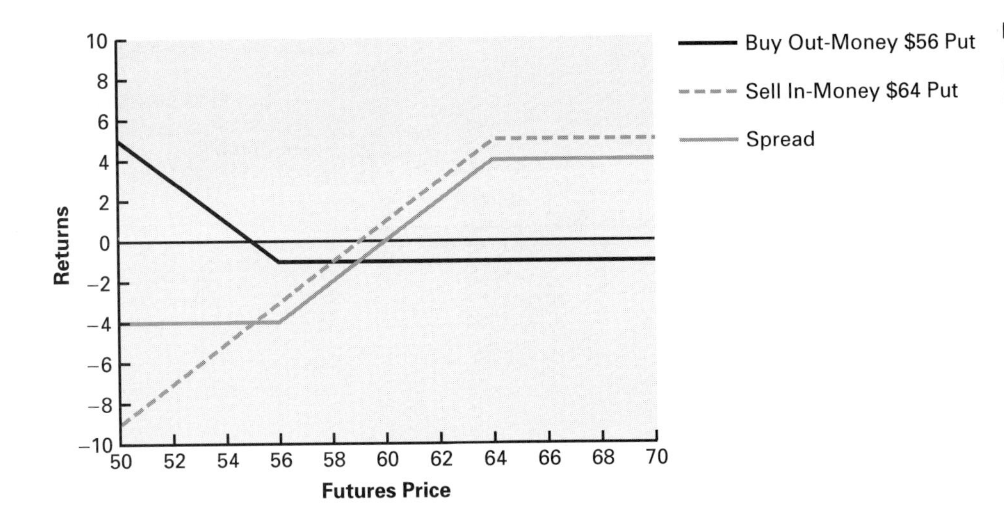

FIGURE 7.15
Long Spread with Puts

and losses are limited. If the market moves up, the trader will make money and if the market falls, the trader will lose money. But losses are limited to a net of the two premiums. Again, to achieve limited losses, the traders have limited gains as well.

Short spreads can also be constructed through trading two call options or trading two put options. Figures 7.16 and 7.17 illustrate the strategies. The strategies are very similar to the long spread. The difference is that the strategy reverses purchases and sales of in-the-money and out-of-the-money options. You should examine the return to each option and to the total strategy when the market price falls below the low strike price, is between the two strike prices, or is above the high strike price. You should examine both cases when either calls or puts are used. *The main point of the strategy is that a short position can be established and that this short position has limited loss potential.*

The second, and last, advanced trading strategy is the straddle and there can be long and short straddles. Spreads involve buying a put and selling a put, or buying a call and selling a call. Straddles involve buying a put and a call, or selling a put and a

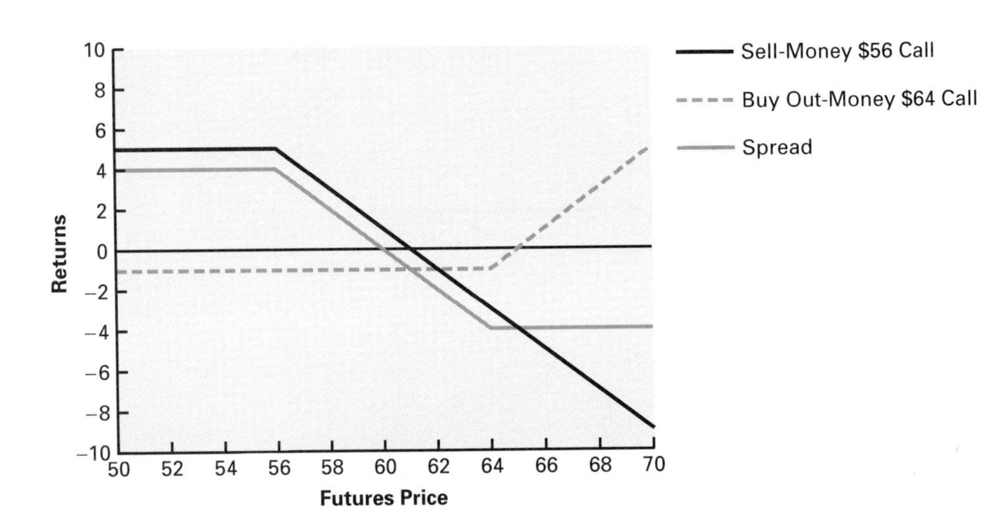

FIGURE 7.16
Short Spread with Calls

FIGURE 7.17
Short Spread with Puts

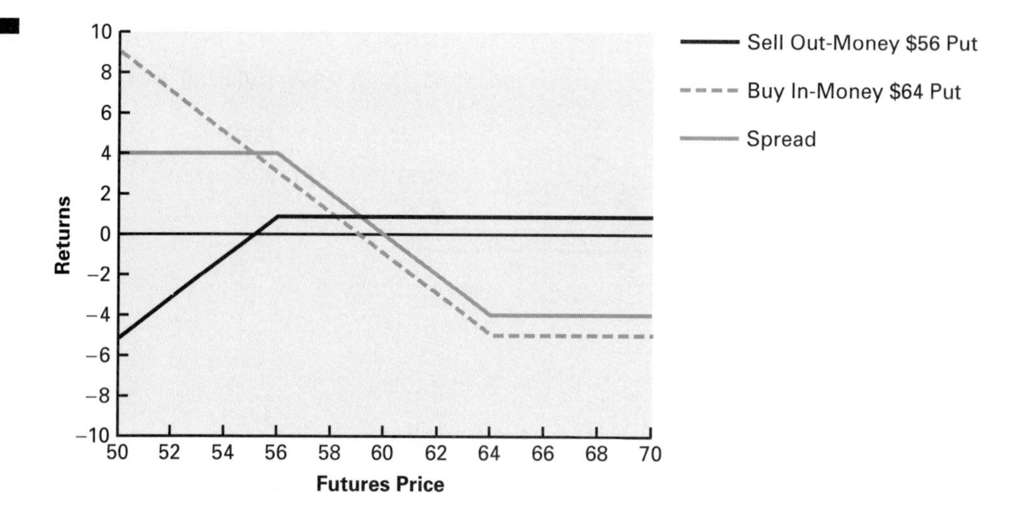

call. In a short straddle, the trader sells a put option and a call option. This strategy is illustrated in Figure 7.18. The put and call used both have the same $60.00 strike price. The trader obtains both premiums. If the market price at expiration is around the strike price, the trader makes money. If the market price falls or rises, the trader may lose. *The market price must be different from the strike price by the amount of the combined premium before the position loses.*

The long straddle, illustrated in Figure 7.19, is just the opposite of the short $60.00 straddle. The trader buys a put and a call. The put and call used both have the same strike price. The trader pays both premiums. If the market price at expiration is around the strike price, the trader loses money. If the market price falls or rises, the trader may make money. *The market price must be different from the strike price by the amount of the combined premium before the position makes money.*

The straddle strategy is attractive if the trader thinks the market will not move (short straddle) or will move sharply in either direction (long straddle). Further, the trader must perceive that these changes are likely relative to the level of the premium.

FIGURE 7.18
Short Straddle with the Same Put and Call Strike Price

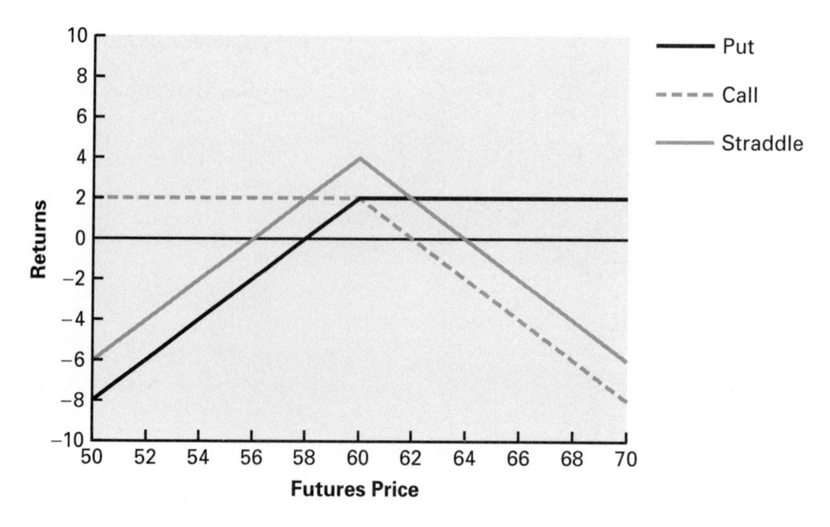

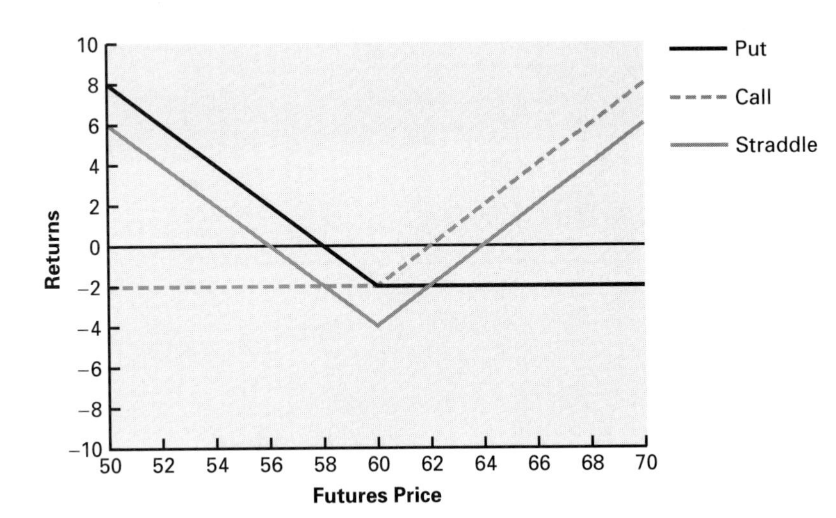

FIGURE 7.19
Long Straddle with the Same Put and Call Strike Price

In the short straddle, the trader perceives that both options are overpriced. The trader perceives that both options are underpriced in the long straddle.

How would the use of different strike prices change the straddle strategy? Again, we will place the call strike price above the put strike price to avoid doubling up the position. With the short straddle, the trader sells options which are both out-of-the-money. The trader receives a smaller total premium, but creates a larger range of futures prices where the strategy will be profitable. The same conditions hold for the long straddle. The trader pays a smaller total premium, but creates a larger range of futures prices where the strategy will not be profitable. Figures 7.20 and 7.21 illustrate these two straddles in which the strike prices are not equal.

The main point to learn from this section is the flexibility of options in putting together a position that will take advantage of different futures price movements, and consequently be exposed to different risks. Options can be used to construct positions similar to futures positions.

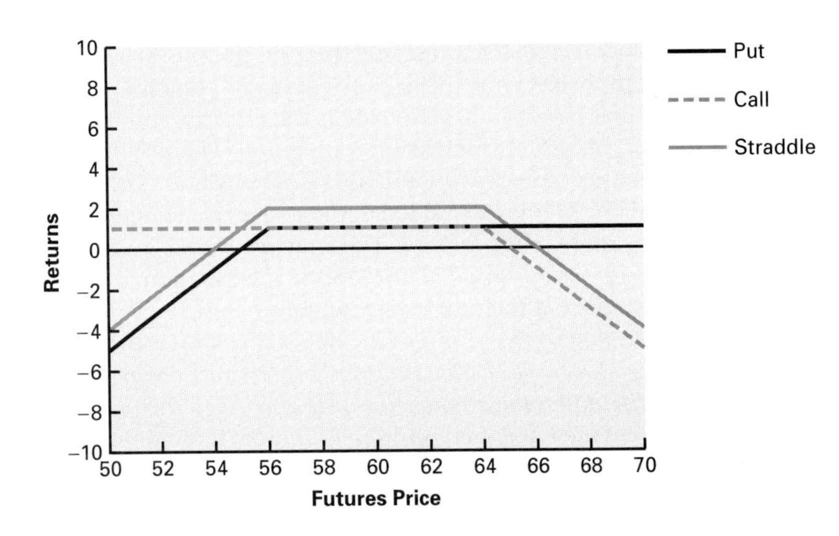

FIGURE 7.20
Short Straddle with Different Put and Call Strike Prices

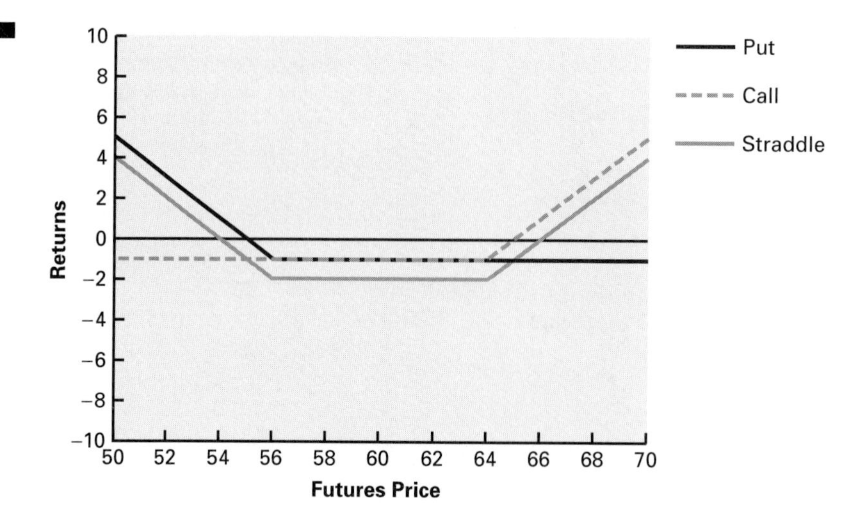

FIGURE 7.21
Long Straddle with Different Put and Call Strike Prices

We can take long and short positions with different options. But unlike futures positions, an options position is not limited to a long or short position.

ADVANCED STRATEGIES FOR FORWARD-PRICING WITH OPTIONS

This section will introduce you to combining more complex options trading strategies with a cash position. However, our perspective is that of forward-pricing and risk management. *The combined cash and options strategy should provide protection against downward movements in price to owners of the cash commodity and should provide protection against upward movements in price to users that must purchase the cash commodity in the future*. Strategies that do not do this should be classified as speculative. We will present common strategies that are of this speculative nature.

In order for an options position or strategy to provide protection against adverse price movements, that strategy cannot result in limited gains or unlimited losses. The term *unlimited loss* is perhaps too strong, but the point is that large losses are possible within the probable range of prices. The most common combined cash and options strategy that has this characteristic is the *covered call*. In this example, the trader has a long cash position and sells a call option. The sale of the call, which is a short position, is thus "covered" by the long cash position. The trader receives the call premium and this premium is added to the cash price to determine the net price. The strategy is captured in Figure 7.22. The strategy is price enhancing if the futures price is close to the strike price at the option's expiration. Further, the strategy enhances the cash price if the futures market falls below the strike price. The call is worthless at expiration so the return to the option position is the premium received. *However, as a producer, you have no price protection. The net price is above the cash price, but the downward potential of the strategy has no floor.* Further, the upward potential of the combined position is limited. If the futures price rises above the strike price, the call writer has an obligation to pay the difference. This obligation places a ceiling on the net price.

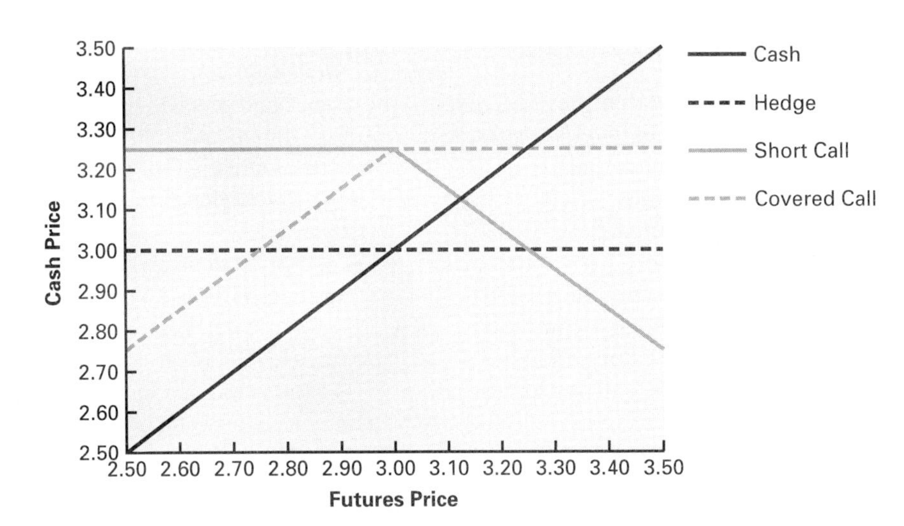

FIGURE 7.22
Long Cash, Long Call Option, and the Covered Call Net Price

Using a covered call strategy is not unreasonable. It is especially effective if the futures market is close to the strike price at expiration. Suppose the trader sells a $3.00 corn call for $.25 per bushel. Assume this option is at-the-money so the underlying contract is trading at $3.00. If the market does not change between the sale and expiration, the producer adds the premium to the cash price received for the corn. Many years, the harvest period corn futures contract goes off the board at the price discovered in late August. *You just need to recognize that it is difficult if not impossible to argue that this is a risk management strategy*. It is a *return-enhancing strategy* under some market conditions, but you are not protected from downside price risk. If the futures price falls to $2.50, you will receive $2.50 plus basis, plus the premium for the call. Likewise, if the futures price falls to $2.00, you will receive $2.00 plus basis plus the $.25 premium. *There is no floor other than the lowest price that can be discovered in the corn market.*

The limited upside potential of this strategy was part of the problem with the hedge-to-arrive (HTA) contract ordeal in the summer of 1996. The 1996 crop year saw corn prices rise to very high levels early in the spring, and in some areas of the country many producers entered into price-enhancing HTA contracts. After falling briefly off these record high prices, the markets then rallied higher on the news of poor growing weather, poorer than expected harvests, and strong demand. The call positions underlying these HTA contracts were losing money. The grain merchants that wrote the HTA contracts and offset them by selling call options received more and more margin calls. In many cases, they attempted to pass the margin calls on to the growers, and in any case, they required delivery of the grain to extract themselves from the strategy. Upon receipt of the grain, they could sell it at the high cash market price and pay the losses on the purchased calls. The remainder would be returned to the producer (less the grain merchant's margin) as per the contract. The producers had already received their premium from the calls.

Two problems emerged. First, *many grain merchants could not meet the margin calls*. In some cases, banks forced several grain merchants into bankruptcy after covering the margin calls. Second, *many producers could not, due to poor harvests, or would not, due to the low net price they would have received, deliver the grain required by the HTA contracts*. Producers complained in popular press articles that as the corn market approached $4 per bushel, they were receiving less than $3 for their corn under the HTA contract. All of the discussions of HTA contracts revealed that *a lot of producers do not understand the returns that come with a covered call strategy*. It is too bad that they had to learn the basic concepts the hard way.

In addition to the covered call, several options strategies discussed in the previous section have returns that are not useful from a risk management perspective. The short straddle is the most obvious choice. *The gains to this position are limited and the losses are unlimited.*

The long straddle is similar to a price floor and a price ceiling. For the producer or holder of a commodity, the purchase of a put provides downside price protection. Likewise, for the user of a commodity, the purchase of a call provides upside price protection. The second option in the position doubles up the returns to the cash position. For example, the call in a long straddle adds to the value of a long cash position. Similarly, the put in a long straddle adds to the value of a short cash position. The strategy may be useful from this doubling perspective, but it would be simpler to just purchase the type of protection desired.

Spreads, long or short, have little value for risk management. The gains to any spread are limited, as are the losses. It is good that the losses are limited, but we need the gains from the option position to offset the losses from a cash position. Thus, spreads are poor risk management tools.

The advanced options trading strategy that is the most useful, and the one we have only briefly discussed so far, is the synthetic futures position. The synthetic future position in which both options have the same strike price is not interesting. This position, long or short, can be obtained by selling one futures contract instead of two options, usually with lower commissions costs. The interesting strategy, and the strategy that cannot be obtained with a futures position, is the synthetic in which the strike prices are different. *This strategy is often called a fence or a window.*

The fence is a combination of a price floor and a covered call. You purchase an out-of-the-money put for downside price protection. You also sell an out-of-the-money call. *The call premium is used to offset a portion of the premium paid for the put.* While the call premium reduces the total cost of the floor price protection, it limits the upside potential of the combined cash and options position. A fence is contrasted with a price floor and a hedge in Figure 7.23. The feeder cattle futures contract price and premiums for options on feeder cattle futures contracts reported in Table 7.2 are used for the example.

The fence contains a price floor which is the put strike price, plus basis, minus the premium paid for the put, plus the premium received for the call. Assume the expected basis for 700-pound feeder cattle is –$4.00/cwt. in mid-November. The price floor is calculated as follows:

Fence Price Floor = *Put Strike Price* + *Basis* – *Put Premium* + *Call Premium*

The price floor for a fence with a $76.00 put and an $80.00 call is

Fence Price Floor = $76.00 – 4 – 0.55 + 0.45 = $71.90/cwt.

The fence price ceiling is the call strike price plus basis, minus the premium paid for the put, plus the premium received for the call. The price ceiling is calculated as follows

Fence Price Ceiling = *Call Strike Price* + *Basis* – *Put Premium* + *Call Premium*

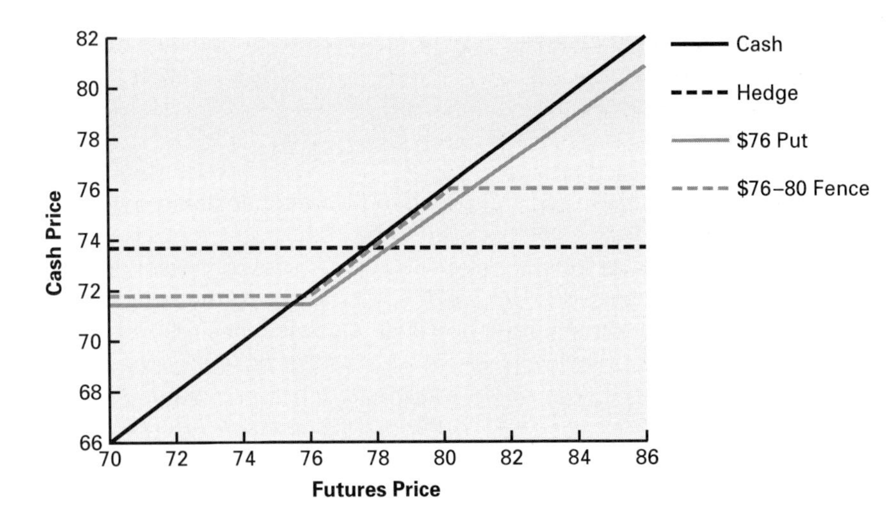

FIGURE 7.23
Fence, Price Floor, and Hedge Net Price Comparisons

The price ceiling for a fence with a $76.00 put and an $80.00 call is

Fence Price Ceiling = $80 – 4 – 0.55 + 0.45 = $75.90/cwt.

At expiration, the net price of the fence is the cash price, or the futures price plus basis, plus any premium that could be received by selling back the put, minus any premium paid buying back the call obligation.

Net Price = Futures Price + Basis + Net Put Premium + Net Call Premium

If the futures price falls below the put strike price, the put has intrinsic value and the call will be worthless. For example, if the futures falls to $72.00 then the $76.00 put will be worth $4.00 and the $80.00 call will be worthless.

Net Price = $72 – 4 + (-0.55 + 4) + (+0.45 – 0) = $71.90

Basis error will complicate this result. If the futures price rises above the call strike price, the call has intrinsic value that is an obligation and the put will be worthless. For example, if the futures rises to $86.00 then the $76.00 put will be worthless and the $80.00 call will be worth $6.00.

Net Price = $86 – 4 + (–0.55 + 0) + (+0.45 – 6) = $75.90

If the futures price is between the strike prices, both options will be worthless.

Net Price = $78 – 4 + (–0.55 – 0) + (+0.45 – 0) = $73.90.

We ignore basis error in these calculations. The producer is able to buy price protection below $71.90 for $.55, and $.45 of that can be deferred through selling a call. However, in addition to deferring $.45, the producer sacrifices price potential above $75.90.

These results are in contrast with the forward price offered by the futures hedge. The market is offering

Hedge Forward Price = \$77.85 – 4 = \$73.85/cwt.

This is the floor *and* the ceiling. Fences offer a lot of flexibility in terms of forward-pricing but there are trade-offs that need to be evaluated. A fence is similar to a price floor with a put option. The net price will not fall below a certain price. This allows the producer to manage risk. The fence is also similar to the hedge in that the producer gives up some of the upside potential. And the fence follows the cash market at more intermediate price levels. *There are no right or wrong choices with respect to choosing among the price floor, hedging with futures, and the fence. You need to assess perceived risks and opportunities, and make an informed choice among the alternatives.*

The concept of a fence is interesting because many agribusinesses use it in writing marketing contracts. The best example is with the hog industry. Over the late 1990s, almost all of the expansion in the hog industry has been through integrated operations. The most well-known of these are the firms that have built new production capacity in North Carolina and Oklahoma. These firms own the sow farrowing and meatpacking operations, and contract the feeding portion of the enterprise to other more geographically dispersed firms. The large integrated firm will then purchase the market hogs using a "fence" type pricing arrangement. The purchase price is tied to the futures price or some central cash market price. However, there is a price floor and a price ceiling. *These marketing contracts are essentially a combination of put and call options.*

Of the several good forward-pricing strategies that are more advanced than simple price floors or ceilings, the most useful is the fence. A fence involves the purchase of puts and sale of calls. The put establishes a price floor and the call sets a price ceiling. The premium received from the calls are used to reduce the cost of the puts. The key feature of a fence is that it provides downside price protection while allowing the hedger to capture benefits of price moves up to the call strike price.

OPTION PRICING WITH BLACK'S FORMULA

Mathematics is a powerful language, but it's not a language we use every day, and there are mathematical tools that most of us never use. However, a set of basic tools and formulas can be very useful for understanding why puts and calls are priced at the levels we see. For example, why is a \$78.00 put on November feeder cattle priced at \$1.30 per hundredweight in October? Why not \$1.00, or why not \$2.00? Is \$1.30 a good buy because it's too low, or should one sell it because it's too high?

A thorough understanding of the material covered in this section is not essential for making use of options in forward-pricing strategies. But the material is useful. The objective of this section is very similar to that of the chapters on fundamental and technical analysis. Fundamental and technical analysis are not essential for using futures to hedge. All you really need to understand is how to forward-price and basis. Rather, fundamental and technical analysis are important for understanding *why* we see the price levels and changes that we do. These analyses facilitate decision making. Likewise, this section is not absolutely essential for using options to hedge. It is helpful though, *in understanding the level and changes in options prices*. Technical analysis is relatively easy with its visual tools, and we used those tools mainly. It is harder with more mathematical tools, and much of technical analysis can be computerized. Fundamental analysis is relatively easy at some level. For example, it is straightforward to assess supply conditions from grain balance sheets and livestock reports. And, anyone with elementary training in economics can tell us the expected direction of price. If supply increases, then the price will fall. But the hard question is how much. This section asks the same question about options prices. Why do we see the premium levels and changes that we do? Answering the question "how much?" requires mathematics.

The simplest tool for calculating an option price is Black's Option Pricing Formula. The formula itself is written below. The user inputs a number of data variables and the formula returns a put or a call price. For example, we enter the strike price, the underlying futures contract price, a variable measuring volatility in the futures market, and the number of days before expiration, and the formula calculates an estimate of the premium. Before we consider the nuts and bolts of the formula, let's consider the concepts behind it. The mathematics look tough but they are actually summarizing a pretty simple idea.

Consider the following simple option example. You offer to sell to a buyer the right to receive $100 after observing the flip of a coin. The buyer must pay the premium for this right before the coin is flipped and if the coin lands heads-up, you will pay the buyer $100. If the coin lands tails-up, the buyer receives nothing. You keep the premium regardless of the outcome. For what value would you be willing to sell this option? First, we can bracket the answer. You would not offer the option for free, a zero premium. You will run out of money after a reasonably small number of coin flips. And you will not find anyone willing to pay $100 or more for such an option.

If you follow our thinking so far, you see that a fair price for the option is something close to $50. With a $50 premium, both sides can play the game without either side going broke in the long run. If you continue to follow our thinking, you would say, "I am not interested in being fair but in making money." So, you would want to sell the option at something above $50. However, you are in a room full of people trying to sell the same option to a large number of possible buyers. The pressure of competing for business brings your price down to something close to $50. *You would have to price the option at the expected value of its long-run payout.* There is a 50% chance of observing a head and having to pay $100, and a 50% chance of observing a tail and having to pay nothing. The expected value of the cost of obligation you are offering to pay in the option is thus $50. In a competitive market, the option price will roughly converge to this $50 premium.

The Black option pricing formula is basically the same thing. The real world of commodity prices is not as simple as the coin flip so the mathematics must be more complex to capture more of what is happening. In the real world, futures prices are the coin and there are more than two prices. What else is captured, and thus is important, in Black's Formula?

Four essential pieces of information go into Black's Formula.

1. The relationship of strike price to underlying futures contract price;
2. The amount of time before the option expires;

3. The volatility of the price of the underlying futures contract; and
4. The level of market interest rates.

We will consider each of these in turn after presenting the formula. We will discuss why each component is important and show through examples how each affects the option price.

Black's Option Pricing Formula for a call option premium is

$$C = e^{(-r \cdot t)} \cdot [FP \cdot cdfN(x_1) - SP \cdot cdfN(x_2)]$$

and Black's Formula for a put option premium is

$$P = e^{(-r \cdot t)} \cdot [FP \cdot cdfN(-x_1) - SP \cdot cdfN(-x_2)]$$

where

$$x_1 = \left[\ln(FP/SP) + (v^2 \cdot t)/2\right] / \left(v \cdot \sqrt{t}\right)$$

$$x_2 = \left[\ln(FP/SP) - (v^2 \cdot t)/2\right] / \left(v \cdot \sqrt{t}\right)$$

and

$cdfN(x_1)$ = standard normal cumulative density function evaluated at x_1
$cdfN(x_2)$ = standard normal cumulative density function evaluated at x_2
$e^{(\cdot)}$ = exponential function
$\ln(\cdot)$ = natural logarithm function

and

FP = price of underlying futures contract
SP = option strike price
v = volatility measure (%)
t = time to expiration (days/365)
r = risk-free interest rate (%).

We plug in values for FP, SP, v, t and r, and the formula returns an estimate of the premium.

Logarithmic and exponential functions are basic analysis tools and are programmed into all spreadsheets. Likewise, the cumulative density function is also programmed into commercial spreadsheet software. The normal distribution is characterized by two parameters, the mean and variance. The standard normal distribution has a mean of zero and a variance of one. The $cdfN(x)$ is the standard normal density function evaluated at the number x, which is

$$cdfN(x) = \int_{-\infty}^{x} \frac{1}{\sqrt{2\pi}} \exp\left[-\frac{s^2}{2\sigma^2}\right] ds.$$

If we set $x = 0$, then $cdfN(x) = 0.5$. Find the function in a spreadsheet you have access to and test this. Likewise, if we set $x = 1$, then $cdfN(x) = 0.8413$ and if $x = -1$, then $cdfN(x) = 0.3413$.

Let's work an example of Black's Formula and generate option prices. Assume we are interested in the premium of a soybean put. The strike price is $7.00 per bushel. and the underlying futures price is $6.90 per bushel. Volatility is 25 percent, there are 90 days until expiration, and the interest rate is 6.5 percent. We will use cents per bushel instead of dollars per bushel for the strike price and the futures price, so

$$FP = 690,\ SP = 700,\ v = 0.25,\ t = (90/365) = 0.2466,\ r = 0.065,$$

thus,

$$x_1 = \left[\ln(690/700) + (0.25^2 \cdot 0.2466)/2\right] / \left(0.25 \cdot \sqrt{0.2466}\right) = -0.0538$$

$$x_2 = \left[\ln(690/700) - (0.25^2 \cdot 0.2466)/2\right] / \left(0.25 \cdot \sqrt{0.2466}\right) = -0.1780$$

$$cdfN(-x_1) = cdfN(0.0538) = 0.5215$$
$$cdfN(-x_2) = cdfN(0.1780) = 0.5706$$

so that

$$P = -e^{(-0.065 \cdot 0.2466)} \cdot [690 \cdot 0.5215 - 700 \cdot 0.5706] = 38.998 = 39.$$

Black's formula estimates that the $7.00 put premium will be $.39 per bushel. Notice that the strike price and the futures price were expressed in cents per bushel so the premium will also be in cents per bushel. The premium is expressed in the same units as the strike and futures price used in the formula.

Let's work a second example for a call option. We are interested in the premium of a soybean call. The strike price is $7.00/bu. and the underlying futures price is $6.90 per bushel. Volatility is 25 percent, there are 90 days until expiration, and the interest rate is 6.5 percent. So, the inputs are the same

$$FP = 690,\ SP = 700,\ v = 0.25,\ t = (90/365) = 0.2466,\ r = 0.065,$$

and x_1 and x_2 are the same

$$x_1 = \left[\ln(690/700) + (0.25^2 \cdot 0.2466)/2\right] / \left(0.25 \cdot \sqrt{0.2466}\right) = -0.0538$$

$$x_2 = \left[\ln(690/700) - (0.25^2 \cdot 0.2466)/2\right] / \left(0.25 \cdot \sqrt{0.2466}\right) = -0.1780$$

but the cumulative densities are evaluated at different points

$$cdfN(x_1) = cdfN(-0.0538) = 0.4785$$
$$cdfN(x_2) = cdfN(-0.1780) = 0.4294$$

and the premium formula is

$$C = e^{(-0.065 \cdot 0.2466)} \cdot [690 \cdot 0.4785 - 700 \cdot 0.4294] = 29.115 = 29.1.$$

Black's formula estimates that the $7.00 call premium will be $.291 per bushel.

Black's Formula is reasonably accurate and is used fairly often by commercial option trading firms. There is a large body of new research that has been devoted to generalizing Black's Formula to more accurately capture the observed options premium levels seen in commodity markets. This body of research is growing. All of this research has led to more mathematically complex equations for the pricing formula. Many of these formulas are not expressions like the above equations—it is not always possible in these cases to plug in numbers for the variables and calculate the premium. The premium must be solved for via numerical techniques and computer simulation. Our point in mentioning this is that the Black Formula is the simplest to use. You are encouraged to work through the mathematics. All other formulas are much more work.

The main use of Black's Formula is to evaluate what price to bid or offer for different options, or to evaluate the premiums bid and offered in the options market. For example, our earlier calculations suggest that a $7 put on soybeans should have a premium of $.39, if the underlying futures price is $6.90, volatility is 25 percent, the time to expiration is 90 days, and the interest rate is 6.5 percent. If a trader observes that the market for this put is trading at $.35 then the option is underpriced, and the trader should purchase the option. Likewise, if a trader observes that the market is pricing this put at $.45, then the option is overpriced, and the trader should sell the option. This would be a speculative strategy. *It could also be used by producers or users of commodities to evaluate whether the options market is offering a good forward-pricing opportunity.*

A second use of Black's Formula would be to change various inputs and examine how these elements result in different premiums. These points have been discussed earlier in this chapter section, but only from an intuitive perspective. The formula will give us actual premiums and changes in premiums as we change the inputs. Thus, Black's Formula can be used from a learning perspective.

Refer back to the list of inputs that are used to calculate an option price. In addition to knowing whether we are pricing a put or a call, we need the strike price, the underlying futures price, time to expiration, volatility, and interest rate. The relationship between the strike price and futures price has been much discussed earlier in the chapter, so let's turn to the other factors.

Understanding the time decay of options premiums is important for successful use of options in forward-pricing strategies. Figure 7.24 shows the decay of the $7 soybean option with which we have been working. The put is $.10 in-the-money and the call is $.10 out-of-the-money. The lines represent premiums over the last year of the option and are calculated by holding the strike price at $7, futures at $6.90, and volatility and interest rate are held constant. The $7 put is $.70 at 365 days from expiration, $.35 at 70 days, and is $.14 at 5 days from expiration. The time decay is relatively smooth at first and then falls off more sharply as the option approaches expiration. Many beginning options traders learn the hard way that an option loses its value quickly as it approaches expiration.

The effects of increases in the volatility of the underlying futures contract are shown in Figure 7.25. Details about volatility and calculation of volatility are provided in the next section. It is an important topic in that volatility can have a major impact on the premium. So the volatility input that is used to estimate the premium is very important. The volatility that is used in Black's Formula is the standard deviation of the percent change in the price around its mean value. For example, in the soybean option example, a volatility measure of 25 percent was used. This is interpreted as follows. Two-thirds of the time, the underlying soybean futures price will change by 25 percent of its mean over the course of a year. If an average soybean price is $6.50,

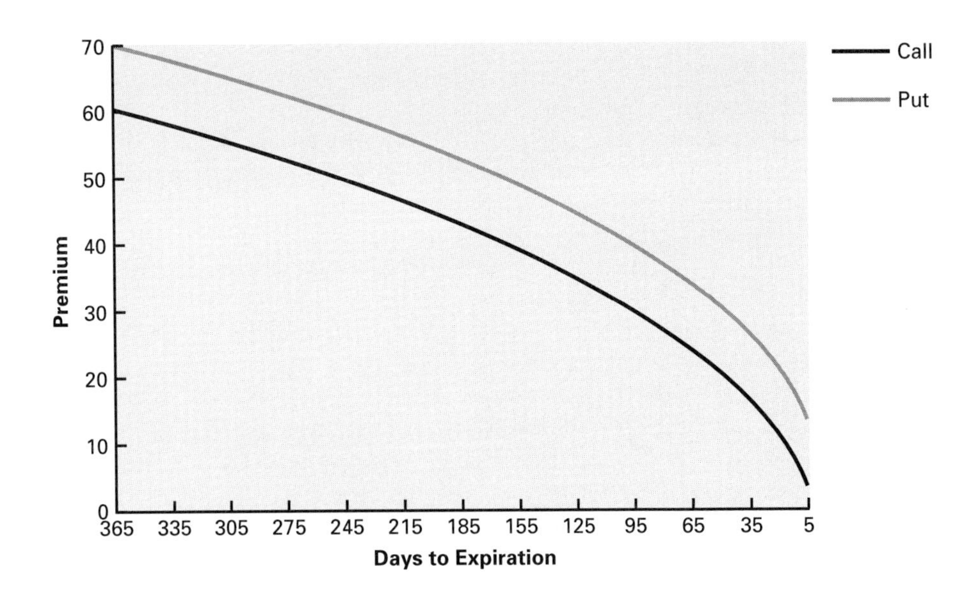

FIGURE 7.24 Time Decay of Options Prices

within a given year there is a two-thirds chance that the price will change by $1.625 or less. We see from Figure 7.25 that a doubling of volatility will double the premium, all else being constant.

Figure 7.26 illustrates the effects of changing interest rates on the premium. Interest rate has the smallest impact. The interest rate effect is an opportunity cost effect. The seller of an option receives the premium and this premium could be invested in an interest bearing account. The higher the interest rate, the more the seller makes from investing the premium. Sellers are able to offer options with lower premiums when interest rates are higher. We see in Figure 7.26 that as the interest rate increases from 7.5 percent to 10 percent, the put premium falls from just under $.39 to just over $.385. The impact is small, but the example is for an option that expires in 90 days. The impact is a little larger for options that are held longer.

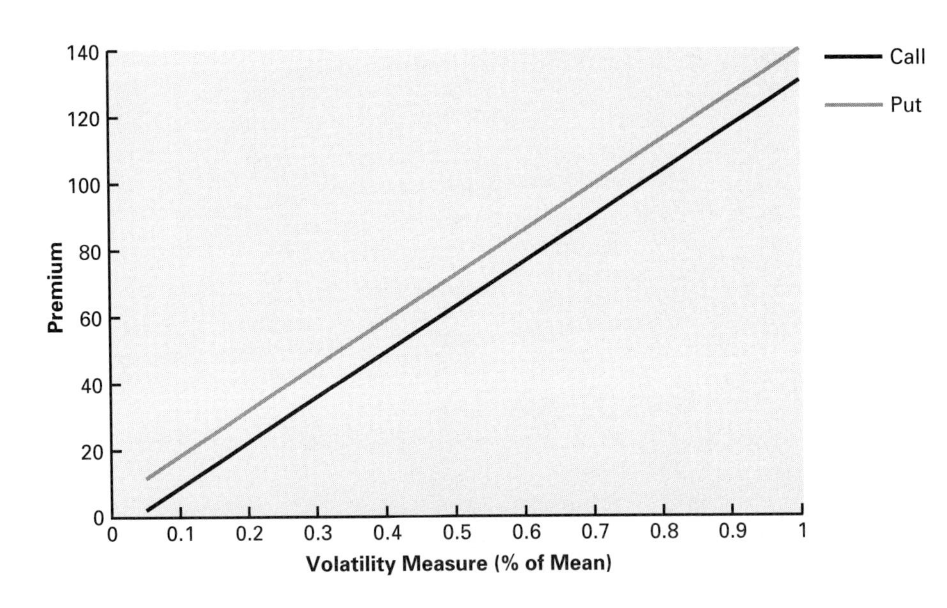

FIGURE 7.25 Effects of Futures Price Volatility on Options Prices

FIGURE 7.26
Effects of Changing Interest Rates on Options Prices

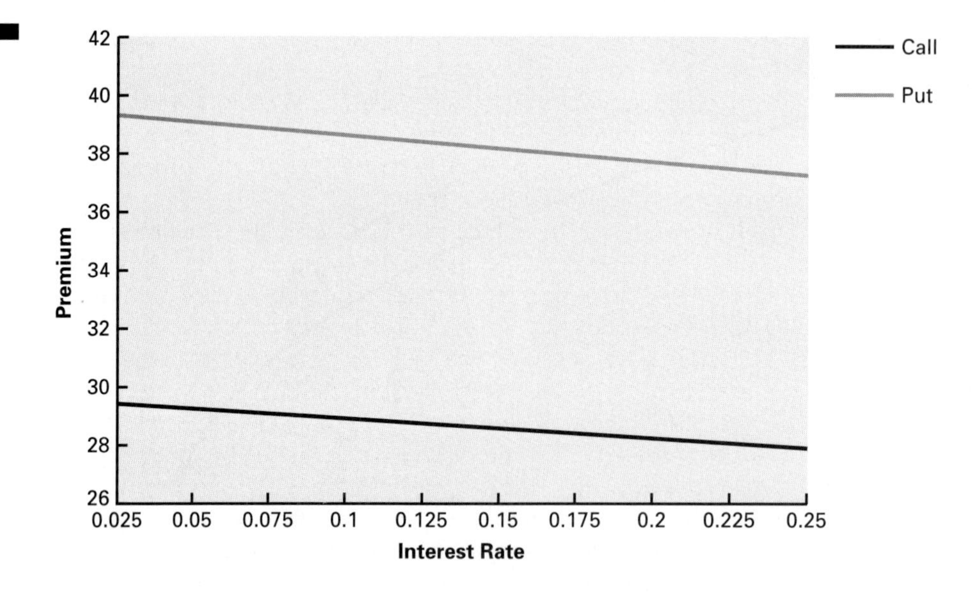

Figure 7.27 is one of the most important figures in the section. The figure is one of our familiar returns diagrams. It shows premiums for a \$7 soybean put with different levels of the underlying futures price. It also shows the premiums for this put when futures prices take on different levels at different times to expiration. For example, when the put is at-the-money, remembering we are examining the premium of a \$7 put, the put is worthless at expiration, the premium is 20¢ at 30 days from expiration, 28¢ at 60 days, 34¢ at 90 days, and 39¢ at 120 days. When the futures price is \$8, the put is out-of-the-money and worthless at expiration, the premium is \$.06 at 30 days from expiration, 3¢ at 60 days, 6.5¢ at 90 days, and 10¢ at 120 days. When the futures price is \$6.50, the option is in-the-money and the premium is \$.50 at expiration, 53.5¢ at 30 days from expiration, 59¢ at 60 days, 63¢ at 90 days, and 67¢ at 120 days. The figure summarizes a lot of the information about premiums, and how they

FIGURE 7.27
Options Returns Diagram with Different Days to Option Expiration

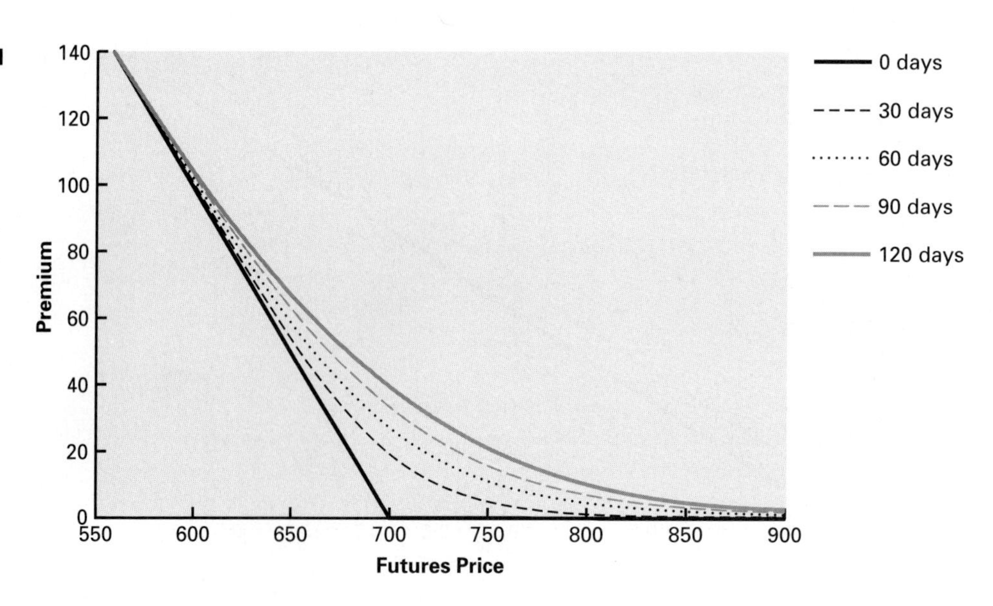

change as the futures price changes and how they change over time. You should spend some time on it.

Figure 7.28 combines the option premiums in Figure 7.27 with those of a cash position. We are vertically adding the premiums in Figure 7.27 to a 45-degree line. Figure 7.28 communicates the same information as Figure 7.27, only the returns are in terms of net cash prices and not option premiums.

Again, look closely at Figure 7.27, our familiar returns diagram. The line labeled 0 days to expiration is the option premium during the last trading day. This is the kinked return line that we have used so often, but we must remember that this is the return only on the last day of trading. Rarely will a trader in a hedging or speculative program carry an option up to the expiration date, so the option will have value when the trader liquidates the position. The familiar kinked line communicates the intrinsic value of the option. The vertical distant between this kinked line and the curved lines denotes the time value of the option. This time value is determined by the number of days before the option expires and the price difference between the strike price and the futures price of the underlying contract.

This remaining time value is important, and the figure communicates two things. First, when a producer lifts forward-price protection, the put option will almost always have some value. This is true even if the option is out-of-the-money. The option will only be worthless if it is out-of-the-money on the expiration day. Thus, net price calculations will almost always have some premium that is returned to the hedger.

Second, the difference between the strike price and the futures price of the underlying futures contract will change frequently and considerably over the life of the options contract. Figure 7.27 communicates how this will change the option premium. Think about a futures position first. The value of the position in futures changes one-to-one with changes in the contract price. A $.25 increase in the futures price will result in a $.25 loss on a short position and a $.25 gain on a long position. This point seems trivial, but compare it to the change in value of an options position at expiration. A $.25 change in the futures price will result in a $.25 change in the value of the option position only if the option is in-the-money. Otherwise, there is no change in the value. We are describing the slope of the lines in the returns diagram in Figure

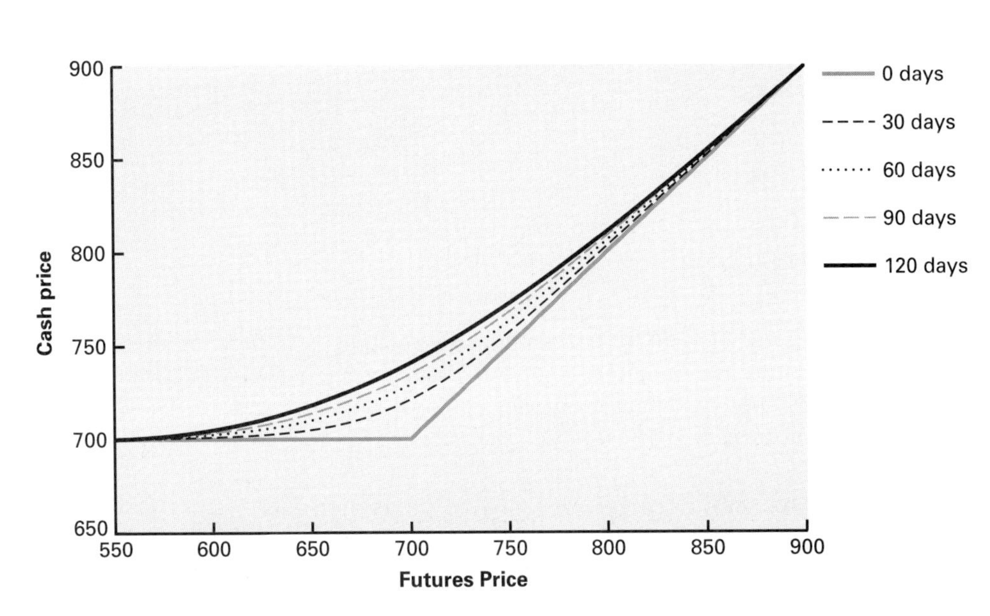

FIGURE 7.28 Returns Diagram for a Cash Position Which Includes an Option Position with Different Days to Options Expiration

7.27. The slope of the futures line is one (or minus one) and the slope of the option returns line at expiration is one (or minus one) or zero. However, we see from Figure 7.27 that if the option is not at the expiration date, the slope is usually not one, minus one, or zero. Rather, the slope is between the absolute value of one and zero, and *the option premium does not change one-to-one with changes in the underlying futures contract price*.

The concept of how options premiums change when the underlying futures price changes is called the *delta* by options trading professionals. Let's look closely at Figure 7.27 again. The option is a $7.00 put. Examine how the premium would change if the futures price, being well in-the-money, decreased from $6.00 to $5.75. Futures have decreased $.25 and the premium for this short position increases, depending on the time to expiration, by $.23 to $.24. The change in the premium could be the same if the futures price increased from $5.75 to $6.00. On the other hand, examine the change in the premium of an option well out-of-the-money, if the futures price increased from $8.75 to $9.00. Futures changes $.25 and the premium changed, again depending on the time to expiration, less than $.05. These are the extreme cases. The delta concept is more important for options that are closer to being at-the-money. If the futures price changes $.50 from $7.25 to $6.75, the option premium changes about $.25.

The practical implication of deltas being between one and zero is that option positions do not accumulate or lose value the way futures positions do. Options positions lose value more slowly and accumulate value more slowly. This can be an *advantage* or *disadvantage* to the hedger who uses options. The advantage will be discussed later in a strategy called "rolling up" price protection. The disadvantage is pretty straightforward. *If the producer has a long cash position and the cash and futures market prices are falling, the option position will not necessarily offset these losses one-for-one.* The only way to make sure that the option position offsets the cash position losses is to buy a put well in-the-money or buy a put that expires very close to when the cash position will be liquidated.

Black's Option Pricing Formula can be used to derive an equation for option deltas. Basically, the delta is the result of taking the derivative of the premium formula with respect to the futures price. The equations for the call delta (δ) and the put delta (δ) are as follows:

$$Call\ \delta = cdfN(x_1)$$

$$Put\ \delta = cdfN(-x_1)$$

whereas before

$$x_1 = \left[\ln(FP/SP) + (v^2 \cdot t)/2\right] / \left(v \cdot \sqrt{t}\right).$$

Using the $7 soybean put for an example, the volatility is 25 percent, there are 90 days until expiration, and the interest rate is 6.5 percent. We will choose three futures prices, one in-the-money, one at-the-money, and one out-of-the-money. The delta for this put when futures are at $6 is

$$x_1 = \left[\ln(600/700) + (0.25^2 \cdot 0.2466)/2\right] / (0.25 \cdot \sqrt{0.2466}) = -1.1797$$

$$\delta = cdfN(-x_1) = cdfN(1.1797) = 0.8809$$

or a 1¢ change in the futures price will result in a 0.88¢ change in the premium. When futures are at $8, the delta is

$$x_1 = \left[\ln(800/700) + (0.25^2 \cdot 0.2466)/2\right]/(0.25 \cdot \sqrt{0.2466}) = 1.1377$$

$$\delta = cdfN(-x_1) = cdfN - (1.1377) = 0.1276$$

or a 1¢ change in the futures price will result in a 0.13¢ change in the premium. Last, when futures are at $7, the delta is

$$x_1 = \left[\ln(700/700) + (0.25^2 \cdot 0.2466)/2\right]/(0.25 \cdot \sqrt{0.2466}) = 1.0621$$

$$\delta = cdfN(-x_1) = cdfN - (0.0621) = 0.4753.$$

A 1¢ change in the futures price results in a 0.48¢ change in the premium for this option when it is at-the-money.

We discussed the factors that influence option premiums in several of the previous sections of this chapter. This section provides more concrete information on how each factor and changes in each factor will impact price. Four essential pieces of information go into Black's Formula: (1) the relationship of strike price to underlying futures contract price, (2) the amount of time before the option expires, (3) the volatility of the price of the underlying futures contract, and (4) the level of market interest rates. With this information and Black's Formula, we can calculate the premium that should accompany any option. Further, we can change these pieces of information and examine the resulting changes in options premiums. Black's Formula provides a powerful and useful tool for pricing and understanding the pricing of options.

HISTORICAL VOLATILITY AND IMPLIED VOLATILITY

The volatility of futures contract prices is important in determining the proper price or premium for an option. The concept has a lot of intuitive appeal. *The more volatile a futures contract price is, the more likely the option will be to move into the money, and the higher the premium a writer or seller will want for that option.* Black's Option Pricing Formula clearly shows that as volatility increases or decreases, option premiums increase or decrease. Yet, we were not clear in the previous section on exactly how the volatility input used in Black's Formula was calculated. It will be covered in this section. The topic is reasonably difficult and subjective enough to warrant a separate section.

The treatment of volatility in options pricing is similar to the treatment of basis in hedging examples. In hedging examples, we used expected basis and actual basis. At the time the hedge is placed, the hedger forms a best guess at what the relationship will be between the cash and futures market on the date the hedge is lifted. It is usually calculated with historical data on cash and futures prices. However, the outcome of the hedge, whether or not the net price that was received equals the forward price chosen, depends on the actual basis at the time the hedge is lifted. A similar thing occurs with volatility.

We can use Black's Formula to calculate option premiums. We will need to put together an estimate of historical volatility. Data are gathered on past futures prices, and statistics are estimated that describe volatility. We use these statistics as best guesses about future volatility. Black's Formula will then tell us if current options premiums are too high, too low, or correct. Real world options prices may be quite different from that suggested by Black's Formula. This is because options premiums are discovered in a competitive open-outcry auction. Black's Formula is a model of the real world, and the model may not always describe how traders think. Also, different data can be used to calculate estimates of volatility. So instead of using historical volatility to calculate the premium, we could work backwards. We will use the actual premium and solve Black's Formula for the level of volatility implied by that premium. We can then compare the *implied volatility* to the *historical volatility* and see if the two measures are close in size or if the difference can be explained by supply and demand conditions.

The most common method of calculating historical volatility involves constructing a moving average of futures prices and calculating a standard deviation around that average. An example is presented in Table 7.4. A 21-observation sample of daily futures prices is used. A price ratio is constructed in which the price on the current day, day t, is expressed as a percent of the price on the previous day, day $t-1$. This is done for the whole sample and there are 20 ratios. The price ratio is then expressed in natural logarithm form. Look closely at the table. The natural logarithm function changes the price ratio to a series that shows the percent change from the previous

TABLE 7.4
Example Calculation for Historical Volatility in Corn Prices

Date	Price	Ratio	Percent Change (%Δ)	$(\%\Delta - \mu)^2$
9/1	$269.25			
9/2	272.25	1.0111	0.0048	0.000032
9/3	271.75	0.9982	–0.0008	0.000000
9/4	269.75	0.9926	–0.0032	0.000005
9/5	264.25	0.9796	–0.0089	0.000065
9/8	263.00	0.9953	–0.0021	0.000001
9/9	264.75	1.0067	0.0029	0.000014
9/10	267.75	1.0113	0.0049	0.000033
9/11	262.50	0.9804	–0.0086	0.000060
9/12	264.25	1.0067	0.0029	0.000014
9/15	365.75	1.0057	0.0025	0.000011
9/16	263.50	0.9915	–0.0037	0.000008
9/17	263.75	1.0009	0.0004	0.000002
9/18	264.50	1.0028	0.0012	0.000004
9/19	261.75	0.9896	–0.0045	0.000014
9/22	262.25	1.0019	0.0008	0.000003
9/23	262.25	1.0000	0.0000	0.000001
9/24	261.00	0.9952	–0.0021	0.000001
9/25	259.50	0.9943	–0.0025	0.000003
9/26	257.50	0.9923	–0.0034	0.000006
9/29	258.75	1.0049	0.0021	0.000009
Average/Sum			–0.0009	0.000287
Volatility				0.0606

day. This is a useful mathematical rule and trick. Next, we calculate the mean of the log ratio and use the mean to calculate the standard error of the 20-observation sample. The steps on this table are very easy to program into commercial spreadsheet software. The mean measures the average percent change in price between observations. For example, in Table 7.4 the average daily percent change in price is –0.09 percent. This is very close to zero so the market is not trending. Heuristically, the standard deviation measures the dispersion in the data. Sixty-eight percent of the percent price changes should be within the mean plus or minus one standard deviation. For example, 68 percent of the percent price changes should be within –0.0047 (–0.0009 – 0.0038) and 0.0029 (–0.0009 + 0.0038). The standard deviation of percent price changes is the measure of volatility in Black's Formula. However, the measure must be multiplied by 256, the number of business days in a year, to convert the daily measure to an annual measure. The formula for volatility is then as follows:

$$\sigma = \sqrt{\sum_{t=1}^{20} \frac{\left(\ln(P_t/P_{t-1}) - \mu\right)^2}{21 - 1} \times 256}.$$

The steps in Table 7.4 provide one measure for a 20-day period. A more complete data series on volatility would be constructed by dropping the last observation from the top of the table and adding a new observation to the bottom. This would be done for all prices in the life of the contract and would provide the trader with a history of volatility over the life of the contract. Likewise, the trader would need to calculate several series from different contracts. Traders need to develop this information or need to search out research that reports this information. It is essential and comparable to the hedger's need for basis information. With these estimates, the trader would then have a good database and knowledge about historical volatility. The trader is now prepared to use that information as inputs in Black's Formula.

A common alternative to the 20-observation moving average is a 30-observation sample. This illustrates the problem with historic volatility. There are a number of reasonable alternatives and there is no one right answer. Besides, if you are selling options, you would base your offers on expectations of future volatility, not just estimates of historic behavior. It *is* likely that the past will describe the future. For example, averages of past basis levels often do a very good job of predicting future basis levels. But practitioners should remember that this is what they are doing—using the past to make a best guess about the future.

Research and other reported information on futures price volatility are available but are not as voluminous as information on basis. But basis is different for every geographic region and season of the year, whereas volatility is measured on specific exchange traded futures contracts. Reasonable estimates of volatility are highly dependent on the commodity and season of the year. For example, soybean prices are much more volatile than corn prices. Wheat prices are more intermediate. Pork belly prices are very volatile, feeder cattle and hogs are next, and live cattle are the least volatile of the commonly traded livestock and meat contracts. Research on volatility tends to report average levels of volatility and then bounds. For example, corn volatility averages 20 percent, can fall as low as 10 percent, and can reach 30 percent to 40 percent in normal years. Drought years can increase volatility to 70 percent and 80 percent. The example time period used in Table 7.4 is not volatile, the measure is 6.1%.

Instead of calculating volatility measures using futures prices, some researchers use implied volatilities. Examine the inputs for Black's Formula in the previous section. All are well-known but one. The strike price is fixed for a given option, the futures price is reported, the time to expiration is known, and the proper interest rate level, while not very crucial to the problem, is easy to find. *The only thing that is tough is the volatility measure.* Instead of using historical futures price data, we could use the options price itself and solve Black's Formula backwards for volatility. This gives the level of volatility implied by the option price. We now have a large history of put and call premiums. This was not the case in the early to mid-1980s. This would be done for all premiums over the life of the option and would provide the trader with a history of implied volatility. Again, you would need to do this for options on many different futures contracts. With these estimates, you would then have a good database and knowledge from which to estimate future volatility and make options trades.

Calculating implied volatility is not as easy as calculating an option premium. Black's Formula cannot be inverted. We cannot solve for volatility as a function of the premium and the other factors. We have to use trial and error. Different volatility measures are used until the formula solves for a premium that is very close to that observed in the market. It is very easy to do this for a small number of observations on option premiums. Black's Formula usually needs to be evaluated four or five times. However, it is not a trivial exercise to do this for a large database. There are software programs which calculate implied volatility for large data sets and this data is also available from commercial data vendors.

Volatility to the options trader is as important as basis is to the hedger. And it is as hard to anticipate. We can use data to calculate historical volatility or use actual options prices to calculate implied volatility. Historical volatility is a moving-average calculation of the standard deviation of percent price changes from some previous number of days. It can then be used in Black's Formula to calculate options premiums. Implied volatility involves taking the option premium and solving Black's Formula backwards for the volatility measure. This is the volatility implied by the current price.

OTHER ADVANCED STRATEGIES FOR FORWARD-PRICING WITH OPTIONS

This last section in the chapter looks at combining options trading strategies with a cash position. The emphasis of this section is in making use of information gleaned from Black's Formula.

We learned about the delta factor in the section on Black's Formula. The delta measures the change in the option price given a change in the underlying futures contract price. Delta factors are somewhere between zero and one, but never reach zero or one except at the option expiration. An option that is well out-of-the-money will have a delta close to zero. Any change in the underlying futures contract price will result in almost no change in the option price. Because the option is so far out-of-the-money, there is little chance that the futures price will change enough that the option will move in-the-money. An option that is well in-the-money will have a delta close to one. Any change in the underlying futures contract price will result in essentially a one-to-one change in the option price. Any change in the futures price results in a change in the option's intrinsic value and this will be reflected in price. A delta less

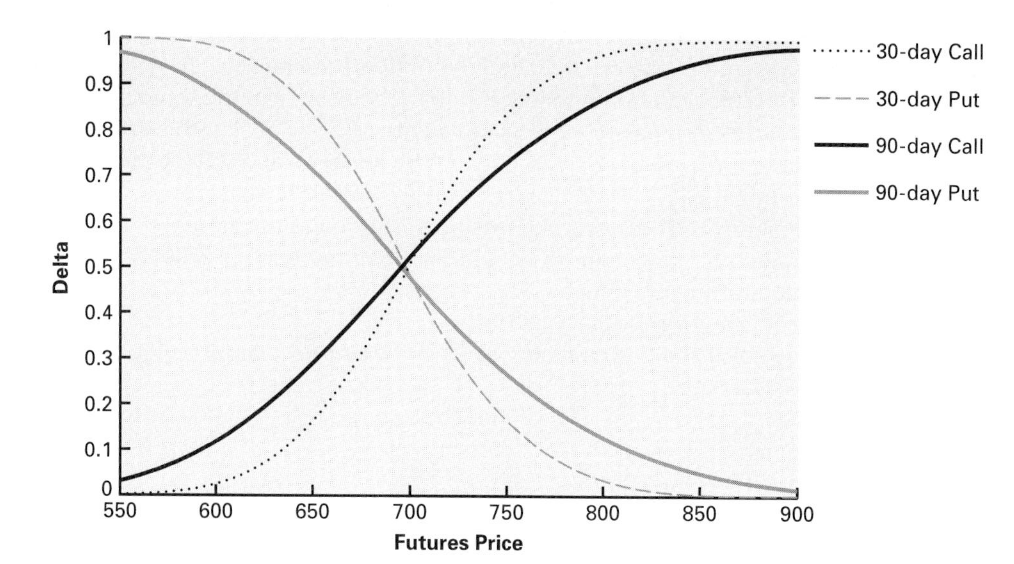

FIGURE 7.29
Deltas for Options That Are In-the-Money and Out-of-the-Money Different Amounts and Different Numbers of Days from Expiration

than one is a unique feature of options and is in contrast to that of futures. Figure 7.29 illustrates these relationships.

The "delta" for a futures contract is one. Any change in the futures contract price results in a direct change in the value of that position. The change in the value of an option position is less than one-to-one, however. As a hedger, you can use this to your advantage.

To illustrate, suppose a corn producer forward-prices his or her crop on the futures market. The December contract is trading at \$3.00, and assume the local basis is zero for ease in discussion. If the futures contract increases to \$3.20, the producer cannot lock in this new higher price. If the producer liquidates the hedge, the market must increase another \$.20 to \$3.40 before the producer can cover the \$.20 "loss," replace the hedge, and forward-price the crop at a net realized price of \$3.20. Even with fundamental and technical analysis, this is very difficult and it is certainly risky. To illustrate further, suppose the futures market increases to \$3.80. If the producer liquidates the hedge, the market must continue upward to \$4.60, enough to offset the \$.80 "loss," before the producer can replace the hedge with a net forward price of \$3.80. *Price protection from downside movements is also protection from upside movements when futures contracts are used*, and once the producer has pulled the pricing trigger, they have chosen a price. This is not the case with options.

Now, suppose as a corn producer you forward-price your crop with options. The December contract is again trading at \$3.00 and the local basis is zero. You buy a put option with a \$3.00 strike price for \$.40. The forward price floor is

$$FPF = \$3.00 + 0 - 0.40 = \$2.60.$$

If the futures contract increases to \$3.20, you or any other producer cannot lock in this new higher price exactly, but you can move the price floor up. The market has increased \$.20, but the options position will not have lost \$.20. The option started out with a delta close to one-half, and as the option moves out-of-the-money, the delta will decrease. The futures price changed \$.20 and the option premium will have fallen at least \$.09. This \$.09 change in the premium is due to the change in the futures price. However, time will have also passed so the put will lose time value. For this example, we will assume the put premium decreased \$.15. The hedger can sell back the \$3.00 put for \$.25 and purchase a \$3.20 put for \$.35.

The new price floor is the new strike price on the second put purchased. Refer to this as the new put, plus basis, plus the net premium of the first put which was offset, less the premium of the new put:

$$FPF = New\ Strike\ Price + Basis + (Net\ Premium\ Offset\ Put) - Premium\ New\ Put$$

or for the example

$$FPF = \$3.20 + 0 + (-0.40 + 0.25) - 0.35 = \$2.70.$$

The producer moved the price floor up \$.25 at a cost of \$.15, so the net increase in the price floor is \$.10. *This is not possible with futures. To move the price floor up \$.20, it costs the hedger using futures \$.20.* Let's continue with the example. Suppose the futures market increases further to \$3.80. The market price increases \$.60 but the put premium does not fall by that amount. In fact it cannot; the original premium was \$.35 and the maximum it can fall would be to zero. However, if the option has time value, it will not fall to zero. Assume the premium is \$.05. The producer can sell the put option back and again replace the hedge by purchasing another put option at a higher strike price. Suppose the hedger purchases a \$3.80 put for \$.20. The new forward-price floor is tied to the new strike price on the third put purchased. This is now the new put, plus basis, plus the net premium of the first and second puts both of which were offset, less the premium of the new put. For the example,

$$FPF = \$3.80 + 0 + (-0.40 + 0.25) + (-0.35 + 0.05) - 0.20 = \$3.15.$$

This options hedging strategy exploits the delta. The hedger is able to move the price floor because the loss to the options position is less than the gain in the futures market. The loss in the options market will equal the gain in the futures market only at expiration.

This strategy is called delta hedging, roll-up hedging, or rolling-up price protection. It requires the hedger to understand how delta and time decay impact the premium for options. These concepts are captured well by Black's Formula.

Black's Formula was also used to calculate the premiums in Figure 7.27 and the net cash prices in Figure 7.28. The formula can be used to anticipate changes in premiums and net prices from forward-pricing strategies.

The fact that options prices do not change one-for-one with changes in the underlying futures contract price can be used by the hedger to his or her advantage. Hedgers can replace lower price floors with higher price floors and the cost is not as great as the improvement in price protection. This is not possible in futures—the cost is equal to the improvement in price protection. This options strategy is called delta hedging and is revealed in Black's Formula.

DISTRIBUTION OF FUTURES PRICES AND RETURNS TO HEDGING

This final section of the chapter discusses a couple of points about probability distributions and their use. *It is optional reading and is not essential to later chapters.* Probability distributions are essential for options pricing. The example of the coin-flip option and the use of the standard normal cumulative density function should illustrate this. Black's Formula makes an assumption about the distribution of futures prices and futures price changes to derive the premium. It is important for you to recognize this. Generalizations of the assumption will change the formula and it is likely that future research will do this. In addition to their use in options pricing, probability distributions can be used to summarize the risk-reducing properties of hedges with futures contracts and options on futures contracts.

First, we will introduce and discuss the basics of distribution functions. Figure 7.30 illustrates the familiar bell-shaped curve of the normal distribution. The function we are graphing is the following:

$$pdfN(x) = \frac{1}{\sigma\sqrt{2\pi}} \exp\left[-\frac{(x-\mu)^2}{2\sigma^2}\right]$$

where we graph over different values of x. The normal distribution is summarized by the mean μ and standard deviation σ of the underlying random variable. For example, if price is the random variable, the distribution communicates that we will observe low values and high values of the price less frequently than intermediate values. The observations tend to be centered around the mean, which is another way of saying that the mean is the measure of central tendency. The standard deviation measures the dispersion in price around the mean. The larger the measure, the more variable price observations will be.

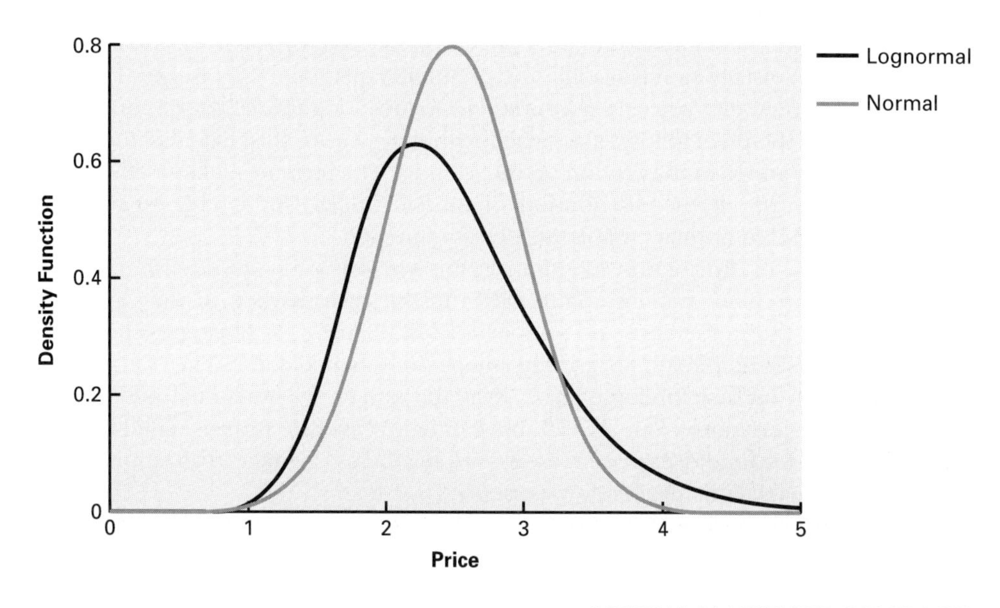

FIGURE 7.30
Normal and Lognormal Probability Density Functions

The density function has the appearance of a histogram. However, you should be careful in inferring probabilities from the function. They cannot simply be read off the vertical axis. The function must be examined over an interval. For example, the probability of observing a price less than the mean is

$$Prob(x \leq \mu) = \int_{-\infty}^{\mu} \frac{1}{\sigma\sqrt{2\pi}} \exp\left[\frac{(s - \mu)^2}{2\sigma^2}\right] ds.$$

This is the area under the curve to the left of the mean. Actually integrating this function is difficult. However, all commercial spreadsheet software have functions that will perform this task, given values of μ, σ, and the range for which you are interested.

While the bell-shaped normal distribution is used for many examples of random variables, this distribution does not describe futures prices satisfactorily. An alternative, which was used to develop Black's Formula, is the lognormal distribution. This distribution is also illustrated in Figure 7.30 to contrast it with the normal distribution. The function we are graphing is as follows:

$$pdfLN(x) = \frac{1}{\sigma\sqrt{2\pi}} \exp\left[-\frac{\left(\ln(x) - \mu\right)^2}{2\sigma^2}\right]$$

where

$x > 0.$

The lognormal distribution is slightly skewed. The mode, the most likely value, is to the left of the mean. Low values are more likely than high values, but we are more likely to see very high values than very low values. Commodity prices fit this distribution better than the normal distribution. Think of corn prices as an example. The average price is around \$2.50. A reasonable but very low price would be \$2.00. A reasonable but very high price would be something larger than \$3.00, possibly \$3.50. Commodity prices tend to follow skewed distributions.

Black's Formula assumes commodity prices follow a lognormal distribution. However, the normal distribution is used in the premium equations. Why is this? The reason is that if commodity prices are assumed to follow a lognormal distribution then the logarithm of the price follows a normal distribution. To see this, examine the equations defining x_1 and x_2 in the section on the formula. The formula makes a more realistic assumption about the distribution of futures prices, yet is able to use the well-known standard normal cumulative density function.

Our final use of distribution functions in this section will be to describe the distribution of a producer trading commodities in the cash market, in the cash and futures market, and in the cash and options market. Figure 7.31 illustrates the three sets of returns. For simplicity, we use the familiar bell-shaped curve. The distribution of cash returns is our base for comparison. Read the zero on the horizontal axis as average returns and deviations from zero as above or below average returns. The cash distribution has a mean and experiences above- and below-average realizations. The producer may make large profits or experience large losses.

The combined cash and futures returns are much less disperse. The producer forgoes the opportunity of large profits in return for removing the downside poten-

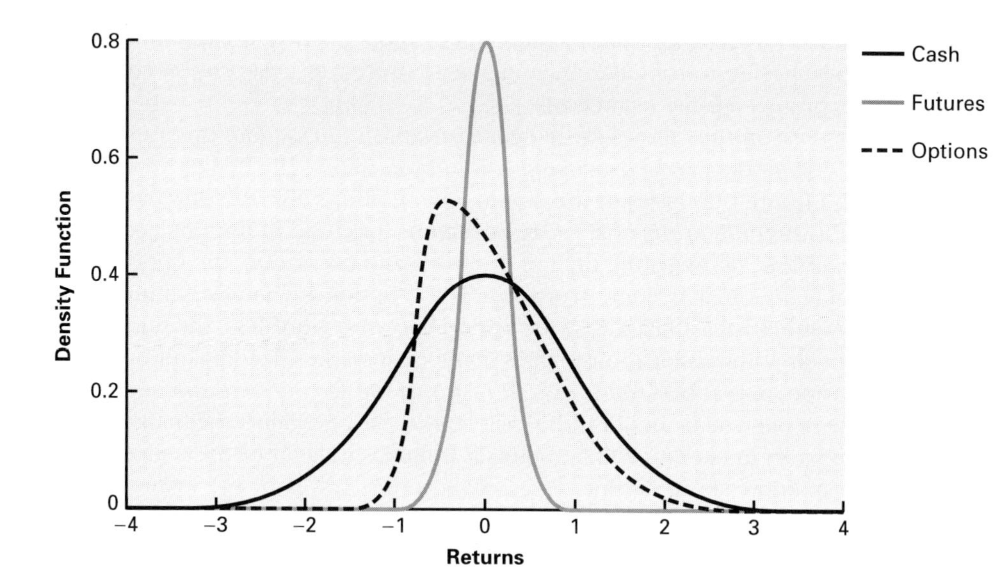

FIGURE 7.31
Cash, Futures Hedge, and Options Hedge Returns Distribution

tial. *The removal of this downside potential is risk management for the producer or holder of the commodity.* The hedger trades cash price risk for basis risk. By hedging, they remove cash price risk, but basis risk remains. If the remaining basis risk is as large as the underlying cash price risk, there is no point to hedging. The hedge is ineffective.

One thing that may be missing from the diagram is that the mean returns from the hedging distribution may be lower than returns from the cash distribution. Think back to the lognormal distribution. The most likely prices are lower than the mean because the mean is influenced by occasional run-ups in price. Thus, you need to mentally shift the hedge distribution slightly to the left. The amount of this shift depends on the commodity and location, and can only be answered by research.

The distribution of the combined cash and options strategy is itself a combination of the distribution from the cash strategy and the hedge strategy. When the producer purchases put options and realizes low prices, the producer is effectively hedging. The left side of the options distribution is similar to the futures distribution. This is the basis error and risk that we have been ignoring in the returns diagram. When high prices are realized, the producer is effectively in the cash market. Thus, the right side of the options distribution is similar to the cash distribution. However, the producer must pay premiums to purchase the options. This shifts the distribution to the left the amount of the average premium. The downside price protection is not as good as the hedge, but it is better than being unprotected in the cash market. The upside potential is not as great as the cash market, but it is better than no upside potential as is offered by the hedge. *Options are not a panacea; they offer alternatives not found in cash or futures strategies.*

SUMMARY

Options add flexibility to hedging and trading programs. The risk of incurring significant *opportunity costs* is reduced, and the often troublesome need to manage margins and *margin calls* can be eliminated. Producers and other potential users of the futures market who worry about these issues will be more likely to use the options in simple strategies that are restricted to buying puts to protect against price declines or buying calls to protect against rising costs.

Purchasing put options allows the hedger to establish a *price floor* on future marketings of the commodity, and *purchasing call options* allows the hedger to place a *price ceiling* on future purchases of the commodity. Options give the purchaser the *right* to but not the *obligation* of a futures position. Purchase of options costs the hedger the *premium*. Between the time of purchase and expiration, the option may accumulate value or it may become worthless. If the option accumulates value, this gain offsets losses in the cash market. If the option becomes worthless, the purchaser has lost only the premium and captures gains in the cash market. Trading options may be a good decision, but, at best, *options will result in the second best outcome*. The outcome of the hedge will be better if the cash market moves against the hedger, and the outcome of cash marketing will be better if it moves in favor of the hedger. But this is difficult to know ahead of time.

In addition to the traditional use of the options, more sophisticated option strategies can be used to allow the decision maker to *select price ranges* across which the *option strategy is superior to hedging with futures* and to select price ranges across which *exposure to cash price risk* would be preferred. Such strategies may involve managing a margin account, however, when options are sold by the decision maker. Further, the decision maker must exercise caution. Many sophisticated looking options strategies are weak price-risk management strategies.

The *fence strategy* is the classic, more advanced strategy. In this strategy, the hedger establishes a price floor through the purchase of put options and offsets part of the cost of the put by selling call options. The call options also place a ceiling on the strategy. Returns are "fenced" between the floor and ceiling. The strategy is widely used and is a legitimate price risk management approach.

The *returns diagram* should be used to examine the potential outcomes of different cash, futures, and options strategies. The diagram communicates quickly and clearly the usefulness of different strategies in protecting the decision maker from risk and offering return potential. *Hedgers should always recognize that risk and return are in conflict in risk management strategies. Strategies that offer high returns are also exposed to high risk. Strategies that offer low risk generally must accept low returns.* High-risk-low-return strategies also exist. But the point is that high-return-low-risk strategies *don't* exist.

Options trading should be integrated with solid fundamental and technical analysis. Hedging with options is attractive relative to futures if sustained market price moves are likely. Identifying when this is likely is the realm of fundamental and technical analysis.

Black's Formula is a useful tool for evaluating option prices. Premiums for puts and calls can be calculated with information on the *futures price relative to the strike price, volatility, time to expiration, and interest rate*. The tool can be used to assess if options prices are relatively high or relatively low, given historic volatility. This is similar to the evaluation of market prices with fundamental and technical analysis. The tool can give you buy or sell signals on options. The tool is also very useful for communicating how volatility and time to expiration affect option prices. For example, it can be used to show the decline in options premiums as the option gets closer and closer to expiration.

Volatility is one of the most important determinants of option price. Options traders need data on historic price volatility to understand option price movements. This is similar to the need for basis information. Historical volatility can be calculated or can often be purchased from commercial data providers.

The concept denoted as *delta* is important for options trading. Delta is the measure of how the option price changes following a change in the underlying futures contract price. The fact that deltas are between zero and one makes options trading more flexible than futures trading. Given a price move in the underlying futures, the trader will lose more in the futures market than in the options market. The hedger can use this to his or her advantage. The level of price protection can be increased for less than the cost of the change. This is not possible with futures.

Since options on agricultural commodities are still relatively new, strategies are still being developed and potential users continue to gain familiarity with this tool. Models are still being refined to calculate the proper premium for options given the underlying volatility in futures prices. For some of the more distant contracts, trade in options is very thin. Given the difficulties many producers and users of agricultural commodities have experienced in using futures, however, *it is reasonable to expect that use of the options will increase in the future*.

KEY POINTS

- The *put option* gives the user the *right to a short position* in the futures. The *call option* gives the user the *right to a long position* in the futures.
- Options are traded for various *strike prices* established by the exchanges on which futures are traded.
- The option *purchaser pays* a *premium* for the option. The option *writer*, or seller, *receives* the *premium* for the option.
- In a hedging program, strategies that buy options have the potential to *avoid exposure to opportunity costs*, and buying options to protect against price moves *eliminates margin calls* and the related need to manage a margin account.
- Option hedging strategies can be designed to combine a producer's cash position with many different combinations of options. Puts and calls can be both bought and sold. These strategies open up the possibility of *results superior to straight cash marketing and hedging with futures across different price ranges*. However, producers must be cautious in that they will be *exposed to price risk* across other and related price ranges. Sophisticated options strategies may be weak price-risk management strategies. These strategies will also often involve managing margins.
- The *returns diagram* is a good tool for communicating the possible results of different options strategies. The trader can examine the different profit or losses that will accompany the price changes between the time the options are traded and the time the position is closed.
- *Black's Formula* is a useful tool for assessing option prices. With information on the *futures price relative to the strike price, volatility, time to expiration, and interest rate*, the premium for puts and calls can be *calculated*.
- *Volatility* is an important determinant of option price. Traders of options *need data on volatility* to understand option price movements.

- Options will typically be the *preferred approach* in markets that have the potential for *sustained price moves* and in circumstances in which the *analytical ability and financial capacity of the potential user are limited.*
- Options on agricultural commodities are *relatively new*. Given the attractive features of options in eliminating *opportunity costs* associated with positions directly in futures and the possibility of eliminating the need for *margin calls*, the use of options is likely to increase in the future.

USEFUL REFERENCES

Fischer Black and Myron Scholes, "The Pricing of Options and Corporate Liabilities," *Journal of Political Economy*, Vol. 81, May/June 1973, pp. 637–659. The original work on option pricing.

Christopher A. Bobin, *Agricultural Options: Trading, Risk Management, and Hedging*, John Wiley and Sons, New York, 1990. An excellent introductory text that discusses the mechanics of trading and pricing options.

John C. Cox and Mark Rubinstein, *Options Markets*, Prentice-Hall, Englewood Cliffs, NJ, 1985. A reference for the reader who wishes to pursue the area, with coverage of the theory of premium determination and trading techniques.

David B. DuPont, "Buy, Sell: Do You Have an Option?" *Trading Tactics: A Livestock Futures Anthology*. Chicago Mercantile Exchange, Chicago, 1986. The chapter demonstrates a number of options strategies that can be used in livestock.

Rajna Gibson, *Option Valuation: Analyzing and Pricing Standardized Option Contracts*, MacGraw-Hill, New York, 1991. A reference that focuses on the theory and practice of premium determination.

John C. Hull, *Introduction to Futures and Options Markets*, 3rd ed., Prentice-Hall, Upper Saddle River, NJ, 1998. An excellent introduction to the mathematics of options pricing.

John C. Hull, *Options, Futures, and Other Derivative Securities*, 3rd ed., Prentice-Hall, Englewood Cliffs, NJ, 1997. An advanced treatment of the mathematics of options pricing.

David E. Kenyon, *Farmer's Guide to Trading Agricultural Commodity Options*, ERS, USDA, 1984. An excellent beginning-level treatment that stresses application.

CHAPTER 8

CURRENCY MARKETS AND EXCHANGE RATE FUTURES

INTRODUCTION

Exchange rate futures contracts were developed in the early 1970s, markets were opened in 1972, and in the U.S., these contracts are traded in the International Monetary Market (IMM) of the Chicago Mercantile Exchange (CME). Exchange rate futures were one of the CME's most successful early series of contracts. Now, many futures exchanges around the world trade exchange rate contracts in which foreign currencies are priced based on the local currency.

Exchange rate futures contracts exist alongside a very large and growing currency market. This currency market does not physically exist in one particular location. Rather, it exists within the communication system connecting the large banks of the world. With growing world trade, large flows of the different world currencies occur between these banks. More than $300 billion flows between the major banks and other financial institutions in Europe, the Far East, and North America. As a consequence, banks trade currencies in which their customers make deposits and require withdrawals. This market is referred to as the interbank currency market. International trade has been a significant portion of the world economy since the 1840s, and there has basically been a currency market for as long as there as been any form of electronic communication. The operation of this currency market improves with any progress in worldwide communications. The growth of this market in the 1970s accompanied improvements in satellite communications and was encouraged by the monetary instability of the time period.

Prior to the advent of exchange rate futures contracts, only large commercial banks transacted in the interbank currency market. The smallest transactions are in tens of millions of dollars, but transactions are more typically in hundreds of millions of dollars. These amounts are generally greater than the wealth of individuals or trading companies that have speculative interests. There is also no clearinghouse for this interbank market, no marking-to-market of currency contracts, and no margin system, so individual banks are very cautious in trading with individuals or other firms that are not other large commercial banks. Collateral is required, but arrangements are more flexible than in futures markets. This limited market access was targeted by the CME when exchange rate futures were introduced.

The interbank currency market is an interesting institution. One of the most interesting characteristics is that it is mainly a forward market. The banks trade currency primarily for future delivery. Currency to be delivered in 30-, 60-, 90-, and 120-day intervals are the most common. There are fewer transactions in currency for immediate delivery. Further, unlike futures contracts, the currency in these forward contracts is actually delivered. The banks trading the forward contracts desire to hold more or less of the transacted currency.

A volume of less than a million exchange rate futures contracts were traded each year between the introduction in 1972 and 1980. However, with the large trade deficits in the early 1980s, more than 15 million contracts per year were traded by 1984. The market volume peaked in 1992 with trade of close to 40 million contracts that year. Since the peak in 1992, this market has seen its volume reduced by half. This is not because of any reduced importance in international trade or currency flows. In part, the cause has been more stable world interest rates led by more stable U.S. interest rates and the resulting more stable exchange rates. Also, banks involved in international currency have reduced transaction costs and are offering more services to clients transacting in world currencies. Previously, firms were taking the do-it-yourself approach to exchange rate risk management. They were hedging and conducting other currency trades on the futures market. These firms generally had to take this approach because of the difficulty in transacting with large commercial banks. Banks are now capturing more of this market and are offering risk management services to smaller clients, but there are fewer small clients. Firms that are involved in international trade, and are successful, are becoming larger. In the face of these trends, volume declined.

Far and away the largest group of currencies traded at the IMM are those of western European countries. These include German marks, Swiss francs, and English pounds. The region is the largest trading partner with the U.S. In terms of individual countries as trading partners, Germany is generally the largest, followed by Japan. Thus, the Japanese yen has significant trading volume on the IMM, but it is generally smaller than the group of currencies from Western Europe. The final well-establish futures contract is for Canadian dollars. Trade in this contract has grown with the passage of the North American Free Trade Agreement (NAFTA). The IMM introduced a contract based on the Mexican peso in 1996, and if trends continue, trade in this contract will likely surpass even the active trade in the Canadian dollar.

CURRENCY AND EXCHANGE RATE BASICS

Every price in the currency market, forward or futures, is a relative price. To say that one U.S. dollar will purchase two German marks (DM), is the same as saying one German mark will purchase one-half a U.S. dollar. This is the exchange rate. The two currencies are expressed as a ratio. We may be interested in DM per U.S. dollar, or U.S. dollars per DM; it is a matter of choosing the denominator in the ratio. For example,

$$DM/\$ = 2 = 1/0.5 = 1/(\$/DM).$$

Exchange rate quotes are reported on many satellite information services that contain financial data. Exchange rates are also quoted in the *Wall Street Journal* and finan-

	Dollar	CdnDlr	D-Mark	Pound	Peso	Yen
U.S.	—	0.72738	0.56402	1.6159	0.12903	0.00822
Canada	1.3748	—	0.77541	2.2215	0.17739	0.01131
Germany	1.7730	1.2896	—	2.8650	0.22877	0.01458
U.K.	0.61885	0.45014	0.34904	—	0.07985	0.00509
Mexico	7.7500	5.6372	4.3711	12.523	—	0.06373
Japan	121.3	88.449	68.584	196.49	15.69	—

Source: *Dow Jones* as reported in the *Wall Street Journal*.

TABLE 8.1
Key Currency Cross Rates at the Close of Trading, 2 October 1997, in New York

cial newspapers in foreign countries. Exchange rate quotes are readily available, but the reader should be cautious in that often quotes are for forward markets—30 days ahead, for example. Spot markets are thinner. Thus, the reader should be careful to relate quotes with the proper time of delivery.

Table 8.1 contains exchange rate quotes from New York banks involved in the interbank market as sourced by *Dow Jones* and reported in the *Wall Street Journal* for October 2, 1997. These are spot market quotes. Deutsche marks were trading for 1.7730 per U.S. dollar because each mark was worth 0.56402 U.S. dollars. Likewise, Japanese yen were trading for 121.65 per U.S. dollar since a yen was worth 0.00822 dollars.

Table 8.2 contains prices for exchange rate futures contracts on the same day. The futures market exchange rates are slightly different from those of the forward market. The IMM operates in the U.S. and all transactions are in U.S. dollars. Thus, all exchange

	Open	High	Low	Settle	Change
JAPANESE YEN (CME) 12.5 million yen; $ per yen (.00)					
Dec	.8360	.8368	.8282	.8300	–.0061
Mr98	.8400	.8418	.8400	.8410	–.0062
June	.8585	.8585	.8530	.8522	–.0063
Est vol 16,019; vol Wed 18,522; open int 76,599, –981					
DEUTSCHE MARK (CME) 125,000 marks; $ per mark					
Dec	.5664	.5689	.5647	.5666	+.0001
Mr98	.5716	.5716	.5692	.5679	+.0001
June	—	—	—	.5726	+.0001
Sept	—	—	—	.5753	+.0001
Est vol 15,160; vol Wed 19,503; open int 60,998, –22					
BRITISH POUND (CME) 62,500 pds.; $ per pound					
Dec	1.6106	1.6140	1.6068	1.6108	+.0004
Mr98	—	—	—	1.6044	+.0004
Est vol 3,726; vol Wed 3,211; open int 28,044, +4					

Source: *Wall Street Journal*.

TABLE 8.2
Exchange Rate Quotes for IMM Futures Contracts on 2 October 1997

TABLE 8.3
IMM Exchange Rate Futures Contract Specifications

Currency	Expiration Months	Size	Price Quote
Deutsche mark	Mar, June, Sept, Dec	DM 125,000	$ / DM
Japanese yen	Mar, June, Sept, Dec	¥ 12.5 million	$ / ¥
Swiss franc	Mar, June, Sept, Dec	SF 125,000	$ / SF
British pound	Mar, June, Sept, Dec	£ 62,500	$ / £
Canadian dollar	Mar, June, Sept, Dec	C$ 100,000	$ / C$
Australian dollar	Mar, June, Sept, Dec	A$ 100,000	$ / A$
French franc	Mar, June, Sept, Dec	FF 500,000	$ / FF
Mexican peso	Mar, June, Sept, Dec	MP 500,000 New	$ / MP

rates are in terms of U.S. dollars per unit of foreign currency. All exchange rate futures contracts are for a fixed amount of the foreign currency. For example, the deutsche mark futures contract is for 125,000 marks and trades in terms of marks per dollar. The Japanese yen contract is for 12.5 million yen and trades in yen per dollar.

Table 8.3 provides a list of the exchange rate contracts trade on the IMM and the sizes of the different contracts.

EXCHANGE RATE TRADING EXAMPLES

Having discussed the size of the different exchange rate contracts and the units with which they are priced in the previous section, it is now relatively easy to work a few trading examples.

Suppose that in October a trader buys one December DM contract for $0.565/DM. In December, the contract is trading for $0.59/DM and the trader offsets the position by selling the contract. The net change in the value of the position is the value at the time of sale less the value at the time of purchase.

$$(\$.59/DM \times DM\ 125{,}000) - (\$.565/DM \times DM\ 125{,}000) =$$
$$\$73{,}750 - \$70{,}625 = +\$3{,}125.$$

The net change in the value of the position can also be calculated using the price change.

$$(\$.59/DM - \$.565/DM) \times DM\ 125{,}000 = \$.025/DM \times DM\ 125{,}000 = +\$3{,}125$$

The value of the dollar relative to the DM decreased or weakened. One DM could be exchanged for fewer dollars through the futures contract in October than in December. Thus, buying DM with dollars in October and then selling DM in return for dollars in December is profitable.

As a second example, suppose that in October a trader sells one December Japanese yen contract for $.0083/¥. In December, the contract is trading for $.0081/¥ and the trader offsets this position by buying the contract. The net change in the value of the position is

$$(\$.0083/¥ \times ¥\ 12{,}500{,}000) - (\$.0081/¥ \times ¥\ 12{,}500{,}000) =$$
$$\$103{,}750 - \$101{,}250 = +\$2{,}500.$$

The net change in the value of the position can also be calculated using the price change.

$$(\$.0083/¥ - \$.0081/¥) \times ¥\ 12{,}500{,}000 = \$.0002/¥ \times 12{,}500{,}000\ ¥ = +\$2{,}500$$

The value of the dollar relative to the yen increased or strengthened. One yen could be exchanged for more dollars through the futures contract in October than in December. Thus, selling yen in return for dollars in October and then buying yen with dollars in December is profitable.

If you have little experience with currencies or exchange rates you need to be careful working the trading or hedge examples in later sections. The common terminology used to explain currency and exchange rate movements is the reverse of the trading position that would seem to be implied by the movement. For example, if the dollar is strengthening, a short exchange rate futures position is profitable. Likewise, if the dollar is weakening a long exchange rate futures position is profitable. This is because the futures contracts are specified as U.S. dollars per unit of foreign currency. A strengthening U.S. dollar implies each dollar can buy more units of the foreign currency, or identically, each unit of foreign currency can buy fewer dollars. Therefore, the exchange rate specified as dollars per unit of foreign currency, like in the futures contract, will fall with the strengthening dollar. You are encouraged to rework the previous examples to obtain a solid understanding of exchange rate directions with respect to a strengthening or weakening U.S. dollar.

IMM futures contracts enable the trader to trade the exchange rate between foreign currencies and the U.S. dollar. To trade an exchange rate between two foreign currencies, the trader needs to trade a spread or to buy one contract and sell the other. For example, suppose that in October a trader buys one December DM contract for \$.565/DM and sells one December yen contract for \$.0083/¥. In December, the trader offsets the DM position at \$.59/DM and the yen position at \$.0081/¥. The gain on the DM position is +\$3,125 and the gain on the yen position is +\$2,500, so the net is a gain of +\$5,625. If, between October and December, the U.S. dollar could be exchanged for fewer DM and more yen, then the DM strengthened relative to the yen or the yen weakened relative to the DM as well. This relationship is known as a cross rate and is expressed as follows:

$$\frac{\$}{DM} \times \frac{1}{\frac{\$}{¥}} = \frac{\$}{DM} \times \frac{¥}{\$} = \frac{¥}{DM}.$$

In October this rate is

$$¥/DM = (1/0.0083)\ 0.565 = 68.072 \text{ or } DM/¥ = (1/0.565)\ 0.0083 = 0.01469$$

and it changed to

$$¥/DM = (1/0.0081)\ 0.59 = 72.840 \text{ or } DM/¥ = (1/0.590)\ 0.0081 = 0.01373$$

by December. Thus, between October and December, one DM can be exchanged for more yen. This implies a stronger DM and a weaker yen. The spread trade is equivalent to buying an exchange rate contract specified as ¥/DM or selling a contract specified as DM/¥.

EXCHANGE RATE HEDGING EXAMPLES

Knowledge of the trading examples in the previous section is a prerequisite for understanding a currency hedge. This section will present some example hedges, in which the futures exchange rate position is combined with a cash currency position. These examples can be confusing. You are probably used to thinking in terms of one currency—most likely U.S. or Canadian dollars. And you are used to asking the question, "Am I long or short the commodity?" But that commodity is priced in one currency. In the following examples, you need to ask the question: Is the trader long or short U.S. dollars? U.S. dollars are used because the exchange rate contract is priced in U.S. dollars. The trader is buying in one currency and selling in the other. They are simultaneously long the currency in which they are buying and short the currency in which they are selling. Thus, if they are buying in U.S. dollars and selling in foreign currency (i.e., exporting), they need to *buy* the IMM exchange rate contract to place a hedge. If they are selling in U.S. dollars and buying in foreign currency (i.e., importing), they need to *sell* the IMM exchange rate contract to hedge. Let's move on to examples for clarification.

The first example is that of an exporter. Suppose a grain trading firm signs a contract to deliver soybeans to a German processing operation in western Europe, and the company has grain purchasing facilities in the central U.S. The firm will be paid in DM per bushel upon delivery in March, will have to purchase the soybeans in dollars per bushel in late February and early March, and it is currently October. The contract written with the German processing firm establishes a price for the soybeans at DM 12.0 per bushel. Exchange rate futures for DM in March are trading at $.57/DM. The soybean price in U.S. dollars is $6.84 per bushel (DM 12.0 / bu. × $.57/DM). From this price, the firm must subtract transportation and other marketing costs and cover its profit margin. The firm will receive DM and pay in U.S. dollars. If the number of dollars that can be purchased with DM falls, the firm will lose money, or it will have its profit margin reduced. This example is illustrated in Table 8.4.

TABLE 8.4
Hedging an Exchange Rate in a Soybean Exporting Example

	Export Hedge for U.S. Firm	
Date	**Cash Market**	**Futures Market**
October	Contract to deliver 250,000 bushels soybeans and receive DM 3.0 million or DM 12.0 per bushel. Anticipate converting DM 3.0 million to $1.71 million or $6.84 per bushel which will cover cost of soybean purchase and delivery.	Sell 24 March futures contracts at $.57/DM covering DM 3.0 million.
March	Deliver 250,000 bushels soybeans and receive DM 3.0 million or DM 12.0 per bushel. Covert DM 3.0 million to $1.59 million or $6.36 per bushel.	Buy 24 March futures contracts at $.53/DM.
	$120 thousand loss.	$120 thousand gain.
Net soybean revenue $1.71 million and price $6.84 per bushel.		

We will hold the soybean price constant. This is realistic because the grain firm can hedge its future soybean purchases. Thus, we are assuming the net price paid for soybeans plus the marketing costs and profit margin is $6.84 per bushel. Now, suppose the exchange rate decreases from $.57/DM to $.53/DM between October and March. The firm is paid DM 12.0 per bushel as is called for in the contract. However, the contract soybean price in U.S. dollars has fallen to $6.36 per bushel after converting the DM to dollars. This will not cover the soybean purchase cost, marketing costs, and profit margin. The firm is long dollars and short DM in the cash position. The firm can manage the exchange rate risk through selling the exchange rate futures contract. Selling the exchange rate contract makes the firm short dollars and long DM. The cash and futures positions balance each other. Losses in the currency market are recovered in the futures market. Likewise, of course, windfall gains in the currency market would be lost in the futures.

The next example is from the perspective of an importer. Suppose a Japanese firm signs a contract to take delivery of corn from a U.S. grain trading firm. The Japanese firm will purchase the corn in dollars per bushel and will be paid in yen per bushel upon sale of the processed corn. The corn will be delivered, processed, and sold in May, and it is currently December. The contract written with the U.S. firm establishes a price for the corn at $4.00 per bushel. This price includes transportation and marketing costs. The corn will be delivered to Japan by the U.S. firm. Exchange rate futures for Japanese yen in June are trading at $.008/¥. The corn in Japanese yen is ¥500 per bushel ($4.00/bu. ÷ $.008/¥). This is the input cost for the Japanese firm. Suppose the firm has contracted with users of corn products (i.e., starches, sugars, and by-product) and that the net price per bushel to the firm after subtracting processing costs and a normal profit margin is ¥550 per bushel. The firm will receive Japanese yen when selling and pay in U.S. dollars. If the number of dollars that can be purchased with yen falls, the firm will lose money or will have a smaller profit margin. This is illustrated in Table 8.5.

The corn price is not an issue in this hedge because the Japanese firm has fixed both the purchase price and sale price through contracts. However, the firm *is* exposed to exchange rate risk. Suppose the exchange rate decreases from $.008/¥ to $.0075/¥ between December and May. The firm receives ¥500 per bushel to cover the

TABLE 8.5
Hedging an Exchange Rate in a Corn Importing Example

	Import Hedge for Japanese Firm	
Date	**Cash Market**	**Futures Market**
December	Contract to receive 500,000 bushels corn paying $2.0 million or $4.00 per bushel.	Sell 20 June futures contracts at $.008/¥ covering ¥250 million.
	Anticipate converting ¥250 million or ¥500 per bushel to $2.0 million to cover cost of corn purchase.	
May	Receive 500,000 bushels corn and pay $2.0 million or $4.00 per bushel.	Buy 20 June futures contracts at $.0075/¥.
	Convert ¥250 million to $1.875 million.	
	$125 thousand loss.	$125 thousand gain.

Net corn cost $2.0 million and price $4.00 per bushel.

cost of the corn, and they must pay \$4.00 per bushel as specified in the cash contract. However, the corn price received after converting the yen to U.S. dollars has fallen to \$3.75 per bushel. The firm cannot convert the yen received for the corn into enough U.S. dollars to cover the corn purchase price. The firm is operating at a net loss. The firm is long dollars and short yen in the cash position. The firm can manage the exchange rate risk through selling the exchange rate futures contract. Selling the exchange rate contract makes the firm short dollars and long yen. The cash and futures positions balance each other. Losses in the currency market are recovered in the futures market.

The similarity between the examples in Table 8.4 and 8.5 is that the trader receives foreign currency and spends U.S. dollars. Whether or not the firm is an importer or exporter is not the main issue. The issue is whether they are being paid or have to pay U.S. dollars. In the two examples, the firms are long dollars and short the foreign currency. Thus, they will need to sell the exchange futures contract where the contract is specified in U.S. dollars per unit of the foreign currency. If the firm is paid in U.S. dollars and pays out foreign currency, the opposite position in the futures market should be taken for a hedge. In this case, the trader will buy the exchange rate contract. We work this example next but will add a twist.

In the following example, suppose a pork processing firm in the U.S. imports hogs from western Canada, processes the animals into fresh pork, and then sells the pork to Japanese meat distributors. The firm is paid for the pork in Japanese yen and pays for the hogs in Canadian dollars (C\$). The U.S. firm contracts with a Japanese distributor to receive all of the pork processed through its facility during a two-week period in December. It is currently September. The firms agree to trade ¥425 million for that quantity of pork. The U.S. firm has also contracted with Canadian hog producers. These producers will provide hogs to their processing plant during this period. The hogs under contract will cost the processor C\$4.12 million. The Japanese yen exchange rate futures contract for December is trading at \$.0085/¥ and the Canadian dollar is trading at \$.70/C\$.

If the exchange rate between U.S. dollars and Japanese yen decreases, the U.S. dollar strengthens, or each yen buys fewer dollars, the firm will lose revenue on the pork. For example, suppose the exchange rate drops from \$.0085/¥ to \$.0080/¥. Revenue decreases from

$$¥425 \text{ million} \times 0.0085\ \$/¥ = \$3.6125 \text{ million}$$

to

$$¥425 \text{ million} \times 0.0080\ \$/¥ = \$3.4 \text{ million}.$$

If the exchange rate between U.S. dollars and Canadian dollars increases, the U.S. dollar weakens, or each U.S. dollar buys fewer Canadian dollars, and the cost of the hogs to the firm will increase. For example, suppose the exchange rate rises from \$.70/C\$ to \$.75/C\$. Costs increase from

$$C\$4.12 \text{ million} \times 0.70\ \$/C\$ = \$2.884 \text{ million}$$

to

$$C\$4.12 \text{ million} \times 0.75\ \$/C\$ = \$3.09 \text{ million}.$$

To hedge this position in two currencies, the U.S. firm would need to sell the Japanese yen exchange rate futures contract which expires in December and buy the similar Canadian dollar contract. The firm is hedging ¥425 million. The firm would trade 34 futures contracts (¥425 million / ¥12.5 million per futures contract = 34 contracts). The firm is hedging C$4.12 million. The firm would trade 41 futures contracts (C$4.12 million / C$100,000 per futures contract = 41.2 contracts). An example of where the exchange rate moves against the firm's cash position is shown in Table 8.6.

Thinking back to the basics of hedging, the hedger always avoids price risk and accepts basis risk. The same is true for exchange rate hedging. The general movements in relative currency values are captured in the exchange rate futures contract. Basis error for a currency hedge would be the difference between the exchange rate futures level and the actual exchange rate at which the processing firm must convert the Japanese yen to U.S. dollars and then U.S. dollars to Canadian dollars. This error is essentially zero in the futures contract expiration month. The currency market is very liquid, and currency is very inexpensive to deliver between various banks around the world. Currency is delivered or transacted between banks through electronic communication, and the cost is a very small percent of the amount of currency delivered. This is very different from physical commodities such as agricultural products, petroleum products, or minerals. Physical commodities have more illiquid markets and much higher market transaction costs; therefore, basis error and basis risk is often an issue. There are two things which are more important than basis error and basis risk for exchange rate contracts.

The first is the lumpiness of the currency contracts. Exchange rate futures involve large blocks of foreign currency. For example, in the previous Japanese yen and Canadian dollar hedge example, the processor would prefer to trade 41.2 Canadian dollar contracts. This is not possible. The larger the currency trade, of course, the less this is an issue. The difference between 41 and 41.2 contracts is not great, but the differ-

TABLE 8.6
Hedging an Exchange Rate in an Import and Export Example

	Hedge for U.S. Firm Importing Livestock from Canada and Exporting Meat to Japan	
Date	**Cash Market**	**Futures Market**
September	Contract to deliver ¥425 million in pork. Contract to purchase C$4.12 million in hogs.	Sell 34 December futures contracts at $.0085/¥ covering ¥425 million.
	Anticipate converting ¥425 million to $3.6125 million.	Buy 41 December futures contracts at $.70/C$ covering C$4.2 million.
	Anticipate converting $2.884 million to C$4.12 million.	
December	Deliver pork; receive ¥425 million. Purchase hogs; pay C$4.12 million.	Buy 34 December futures contracts at $.0080/¥.
	Convert ¥425 million to $3.4 million. Convert $3.09 million to C$4.12 million.	Sell 41 December futures contracts at $.75/C$.
	$212.5 thousand loss on ¥. $206 thousand loss on C$.	$212.5 thousand gain on ¥. $205 thousand gain on C$.
Net $3.6125 million received from pork and $2.885 million paid for hogs.		

ence can be important if the hedger desires to trade 1.5 contracts and has to choose between one or two contracts.

The second issue is that there is basis error if the hedge is not lifted at or close to the end of the expiration month of the futures contract. There are only four contracts traded for each currency per year. The contracts expire quarterly. This can be a problem if the trader desires to lift the hedge, for example, at the end of April and is trading the June contract. There can be a substantial difference between the exchange rate between two currencies across a two-to-three month period. But this difference is rather predictable. So we really should not call it basis risk, perhaps. This is the topic of the next section. It will answer the question: "What causes the difference between exchange rates for currency deliverable now versus three months from now or six months from now?"

Using exchange rate futures only looks a bit more complicated. Once you get the basic idea down, it all takes on intuitive sense. The example in which both the Japanese yen and the Canadian dollar were used is not really any more complex than the situation facing a cattle feeder who must manage exposure to risk in both selling price and input costs—and the basis is more predictable.

PRICING MODELS

This section presents two models used to explain the different exchange rates that will accompany forward or futures contracts that are deliverable at different times. For example, if it is currently October, can we explain why the exchange rate futures contract that expires in December is different from the contract that expires in March? Or, why is the 1-month ahead forward rate different from the 6-month ahead rate?

The two models used to summarize the difference in exchange rates are based on one underlying notion. The futures price for any financial instrument will equal the spot price plus the cost of carrying the instrument to the delivery date. In equation form,

$$S_t(1 + C_{t,T}) = F_{t,T}$$

where $F_{t,T}$ is the futures price of the financial instrument at time t of a contract that expires at time T, S_t is the spot price of the instrument in time t, and $C_{t,T}$ is the cost of carry for the financial instrument from t to T. As the futures contract approaches expiration, as t approaches T, then the carry cost shrinks to zero and the spot price and the futures price converge to

$$S_T = F_{T,T}.$$

This relationship also holds for financial instruments such as contracts on physical commodities. Futures prices for grains contracts reflect this carrying cost or cost of storage over the crop year. The difference between contracts on physical commodities and exchange rates is that at expiration, spot or cash price for the commodity will not necessarily equal the futures price because of the large costs of delivery. The cost-of-carry equation can be rewritten as follows if the cost of carry is a percentage:

$$S_t(1 + r)^{T-t} = F_{t,T}$$

where r is an annual interest rate and $T - t$ represents the length of time the instrument is held. For example, if the financial instrument is a bond that pays 8 percent per annum and costs the purchaser \$1,000, the futures price for the bond after it is held six months is

$$\$1000\ (1 + 0.08)^{0.5} = \$1039.23$$

and if the bond were held one and a half years, then

$$T - t = 1.5.$$

The first model used to explain the difference in exchange rates is called the Interest Rate Parity Model. The idea is that because the different world currencies are easily transferable, the people and banks that hold large reserves of currency will desire to hold the currency that offers them the greatest rate of return. The exchange rates will then adjust to reflect the supply and demand fundamentals for the respective currencies. The exchange rates will adjust so that, in equilibrium, the holders of currency are earning the same rate of return in all currencies. Thus, the term *interest rate parity*, means that equal interest rates emerge. Different world banks offer different interest rates for their home country currency. Exchange rates adjust so that investors are earning the same rate of interest in any currency. An example will illustrate this.

A currency trader living in the U.S. considers holding his liquid cash assets in both U.S. dollars and Canadian dollars. The trader lives in the U.S. and thus spends U.S. dollars but could easily exchange U.S. dollars for Canadian, place the currency in a Canadian bank, and exchange the Canadian dollars for U.S. when he needs cash. Further, there is no need to limit himself to the cash he has on hand; he could borrow. The trader considers the following: He could borrow the domestic currency (S_t^{dc}) at the domestic interest rate (r^{dc}), convert the domestic currency to the foreign currency at the spot exchange rate (S_t^{dc}/S_t^{fc}), invest the funds at the foreign interest rate (r^{fc}), and write a forward contract promising to deliver the foreign currency at the end of the investment period for the initial amount of currency plus the interest returned ($F_{t,T}^{fc}$). If enough traders do this, the spot and forward exchange rates adjust so that the amount borrowed, the left-hand side of the following equation, just equals the amount returned, the right-hand side of the equation, or

$$S_t^{dc}(1 + r_{dc})^{T-t} = \frac{S_t^{dc}}{S_t^{fc}}(1 + r_{fc})^{T-t} F_{t,T}^{fc}.$$

We need to rewrite this equation solving for the forward or futures price

$$F_{t,T}^{fc} = \frac{S_t^{dc}(1 + r_{dc})^{T-t}}{\frac{S_t^{dc}}{S_t^{fc}}(1 + r_{fc})^{T-t}} = S_t^{fc}\frac{(1 + r_{dc})^{T-t}}{(1 + r_{fc})^{T-t}} = F_{t,T}^{fc}$$

and we can see the forward rate is equal to the spot rate multiplied by a factor which is the ratio of the interest rates. This equation is a cost-of-carry equation between the spot and forward market where the cost of carry is

$$Cost\text{-}of\text{-}Carry = \frac{(1 + r_{dc})^{T-t}}{(1 + r_{fc})^{T-t}}.$$

It looks complicated, so let's work an example with numbers.

Suppose the U.S. interest rate is 8 percent, the Canadian rate is 5.1 percent, and the spot exchange rate between U.S. and Canadian dollars is 0.73. The forward rate deliverable in three months ($T - t = 90$ days / 360 days = 0.25) should then be 0.735. We see this from

$$\frac{(1 + 0.08)^{0.25}}{(1 + 0.051)^{0.25}} = 1.00683$$

and multiplying this factor by the spot rate gives

$$0.73 \times 1.00683 = 0.735.$$

We can also see this from the original cost-of-carry equation

$$(1 + 0.08)^{0.25} = \frac{1}{0.73}(1 + 0.051)^{0.25}\ 0.735$$

If the U.S. interest rate increases to 9 percent from 8 percent then the forward exchange rate will increase relative to the spot rate.

$$0.7367 = 0.73 \times 1.00915$$

The model also does more than link the current spot rate to a forward rate. The same relationship holds between two forward rates

$$F^{fc}_{t,T_2} = F^{fc}_{t,T_1}\frac{(1 + r_{dc})^{T_2-T_1}}{(1 - r_{fc})^{T_2-T_1}}$$

where the forward or futures contract that is deliverable in T_1 expires before the contract deliverable in T_2.

The second model used to explain exchange rates over time is a slight variation on the Interest Rate Parity Model. The second model is called the Purchasing Power Parity Model and it uses the idea that nominal interest rates across countries are not as important as real interest rates. Real interest rates are a ratio of the nominal rate and the expected inflation for the currency in question

$$(1 + r_n) = (1 + r_r)(1 + E(i))$$

where r_n is the nominal rate, r_r is the real rate, and $E(i)$ is the expected inflation rate. In the Purchasing Power Parity Model, the interest rates used in the cost-of-carry equation to figure out the ratio of the spot to the forward exchange rate are the real interest rates

$$(1 + r_r) = \frac{(1 + r_r)}{(1 + E(i))}$$

of the two respective currencies. This model will give identical answers to the Interest Rate Parity Model if the inflation rate is the same for both currencies. If the inflation rates are different, it would be important to use the real interest rates.

Now, how can the hedger use these two models? The models are useful for predicting basis for hedges that are not lifted in the exchange rate futures contract expiration month. For example, suppose a trader is interested in hedging a Canadian dollar transaction that will take place in October. The December Canadian dollar contract is trading at $.73/C$, the U.S. interest rate is 8 percent, the Canadian rate is 5.1 percent, and inflation rates in the two currencies are the same. The hedge will be lifted three months prior to contract expiration, so the cost of carry is

$$\frac{(1 + 0.08)^{0.25}}{(1 + 0.051)^{0.25}} = 1.00683$$

and the forward rate for October based on December futures is $0.725 = 0.73 \div 1.00683$. The basis is –$.005/C$. The number may appear small, but the larger the block of currency in the transaction, the more important this basis difference will become.

The two pricing models are useful for explaining the differences between exchange rates in which the underlying contracts expire at different times. This is useful for predicting basis in nonexpiration months. The models help explain the premiums and discounts across contracts, or the price of one contract relative to some other contract. In many respects, these relationships are stable and generate an ability to handle any basis risk that is not present for physical agricultural commodities. All this makes life more manageable for the multinational firm—and for you as a beneficiary of world trade.

EXCHANGE RATE FUNDAMENTALS

The previous section discussed why different currency contracts have different exchange rates, but all of the discussion is about exchange rate for any contract of a particular currency relative to one other contract for that currency. The pricing models explain premiums and discounts but do not explain the overall exchange rate level that is seen. This question has to be dealt with in the realm of fundamental analysis and is the topic of this section.

Currency fundamentals are more substantial and complex than fundamentals for physical commodities. Understanding exchange rate levels and changes requires knowledge of concepts in international trade and macroeconomics. However, currency is much like any other commodity in that its price is determined by the interaction and balance of supply and demand. If the supply of a currency increases, the price will fall. If the demand for a currency increases, the price will rise. And we must remember that the price of a currency is a relative price; it is the value of U.S. dollars relative to deutsche marks, pounds sterling, or yen.

The most important factor determining exchange rates between world currencies and those of particular countries is the sum of all commercial and financial transactions between one country's people and government and the rest of the world. The sum of these transactions is the *balance of payments* for that country. These transactions include imports and exports of goods and services, investments in foreign countries and investments by foreign people in countries or foreign governments, and dispersion or receipt of foreign aid. The balance of payments includes all cross-border transactions. The *balance of trade* is a subset of the balance of payments and refers to imports and exports of goods and services. Items in the balance of trade are more short-term consumption items. These items are readily measurable and include agricultural commodities, computers, automobiles, and tourism spending. Balance of trade also includes general transfers of currency by people, businesses, and governments. The sum of all imports of goods and services adjusted by, or net of, exports of goods and services in the balance of trade is known as the *current account balance*. In addition to the short-term current account transactions, the balance of trade includes long-term transactions in land, building, equipment, equities, and debt instruments. The net of imports and exports of long-term transactions is known as the *capital account balance*.

Exchange rates settle to equilibrium levels given world trade flows in short-term goods and services, flows of long-term capital investments, and reciprocal flows of currency. Likewise, changes in world trade flows will result in exchange rate changes. Generally, decreasing balance of payments will weaken the currency of the country whose imports are increasing relative to exports. Increasing imports relative to exports increases the amount of currency held by the rest of the world. If that currency can only be spent in an importing country, people and businesses in other countries will have to exchange it for their own currency and the rate at which that occurs will decline. For example, if Australia increases imports from Japan relative to exports to Japan, then Japanese businesses will hold more Australian dollars which they will need to convert to yen. The more Australian dollars they hold in aggregate, the fewer yen they will be able to exchange for each dollar. *Much of the variation in exchange rates can be explained by this excess supply and excess demand argument that comes from examining the balance of payments.* This is especially true for the smaller world currencies.

However, this relationship does not necessarily hold for the major currencies of the world—the deutsche mark, Swiss franc, British pound, and Japanese yen, and especially the U.S. dollar. The majority of world trade is conducted using these currencies. Let's revisit the example from the previous paragraph to illustrate. Suppose U.S. imports from Japan increase relative to exports to Japan. Again, Japanese businesses will hold more U.S. dollars. However, this does not necessarily translate into a weaker dollar. These Japanese businesses, if they are international firms, can use the U.S. dollars in conducting business with other firms if the other firms are willing to accept dollars. Thus, the excess supply and demand argument that comes from examining the balance of payments is supplemented by a *transactions demand* for individual currencies.

What factors cause changes in the balance of payments? There are four general factors what we can study to anticipate changes in the balance of payments and exchange rates. Or at least we should understand these factors so that we may understand currency market reactions. The first factor is the *economic conditions* of the country of interest. We are referring to the private sector. Is the economy in the private sector strong? Is domestic demand strong relative to domestic supply? If so, people and businesses will consume more imports and more of domestic production that previously was exported. This will decrease the current account balance, and the increased flow of currency out of the country can weaken that currency. The counterpart to this argument is that a strong economy will attract foreign investment and this will increase the capital account balance. However, the current account is in general bigger than the capital account.

The second factor is *government debt* or the willingness of the government to accumulate debt. The economic argument here is exactly the same as that discussed in the previous paragraph, but the source of demand is in government expenditures. Because the determinants of government spending are different from the determinants of private spending, this factor is worth separate mentioning. If a government is willing to increase debt, or spend more relative to taxes collected, it will consume more domestic production and may consume imports or cause the private sector to consume more imports. This will decrease the current account balance and the increased flow of currency out of the country can weaken that currency. This is precisely the major cause of U.S. dollar devaluation in the mid-to late 1980s.

The third factor is real interest rate or *interest rate relative to inflation*. A country and currency with high nominal interest rates relative to inflation will attract foreign investment and capital. The flow of currency into the country will increase the capital account balance, strengthen the demand for the currency, and strengthen the value of the currency. However, over time the inflow of currency willing to purchase debt instruments will result in declining interest rates.

The fourth factor that impacts balance of payments is the *price level and productivity* of a nation. A country with relatively low prices and high productivity will be in a position to export relatively more than it imports, and thus will incur a balance-of-payments surplus, which will likely lead to a strong currency. In the long run, a strong currency will limit the ability of that country to export, but in the real world these imbalance may persist for some time.

There are two remaining factors that have strong impacts on exchange rates that do not necessarily work directly through the balance of payments. The first of these are actions by the government or central bank, and the second are market expectations. *Government and central bank actions* can have a large impact on exchange rates. Actions can have a direct impact on the import and export balance and balance of payments through trade policies. Different tariffs, quotas, and treaties will change trade flows. Central banks also often buy and sell large quantities of their own and foreign currencies. This is usually done to maintain target exchange rates and can be done to support or depress certain currencies. The long-term success of central banks in maintaining target exchange rate levels is questionable. Market adjustments in exchange rates are in response to trade flows and in the underlying financial condition of the economy and currency in question. However, in the short term, central bank intervention is very effective in controlling exchange rates.

It should be no surprise that the value of our dollar is determined by fundamental supply and demand forces. After all, the exchange rate and the interest rate that become important factors in all this are "prices" of a type. The interest rate is a direct price of money, and the exchange rate is a "price" relative to other currencies. The basic supply–demand framework will help you in understanding all of this.

SUMMARY

The market for currency is one of the largest in the world and it is not located in any particular place. The market exists within the communication system of large world banks during their business hours. The market follows the daylight around the world. Further, much of the volume of trade in this market is for the future delivery of currency. This market exists alongside the different world exchanges on which currency futures contracts are traded.

Exchange rate futures contracts traded on the Chicago Mercantile Exchange allow hedgers and speculators to trade different world currencies in relation to the U.S. dollar. The German mark and Japanese yen have the largest volume, followed by the currencies of other western European countries, British pounds, Swiss francs, and French francs, currencies from other North American countries, Canadian dollars and Mexican pesos, and Australian dollars. Currency futures contracts are subject to a large volume of trade worldwide. Futures contracts are almost always traded in the various futures exchanges around the world.

Exchange rates are a relative price. For example, an exchange rate is the number of U.S. dollars that will purchase one unit of a foreign currency. Caution must be used in speculating and hedging with exchange rate futures. The relationships at issue are not common to domestic commerce. For example, a strengthening of the U.S. dollar implies that one U.S. dollar will purchase more units of foreign currency. This implies that the exchange rate, measured as U.S. dollars per unit of foreign currency, will decrease. The converse is true for a weakening dollar.

Exchange rate futures contracts can be used to forward-price purchases and sales in foreign currency, and sales and purchases in U.S. dollars. The important question to ask is: "Is the firm long or short the U.S. dollar relative to foreign currency?" If the firm spends dollars and receives foreign currency, it is long dollars and short foreign currency. If the number of dollars a unit of foreign currency will buy decreases, then the firm loses money. In this case, to hedge the transaction, the firm needs to sell the appropriate CME exchange rate contract. If the firm receives dollars and spends foreign currency, it is short dollars and long foreign currency. If the number of dollars a unit of foreign currency will buy increases, then the firm loses money. In this case, to hedge the transaction, the firm needs to buy the CME exchange rate contract.

There is little if any basis risk in the exchange rate futures market. The currency markets are very well traded and delivery costs are low for those specializing in currency trade. However, only four exchange rate futures contracts are traded for any particular currency in the course of a year. There will be differences between the cash and futures exchange rates for nonexpiration months. These differences can be predicted by the Interest Rate Parity Model and the Purchasing Power Parity Model. If the differences can be predicted, the basis difference is not risk.

The Interest Rate Parity Model suggests that exchange rates adjust between countries so that the interest rates move to the same level. The person holding cash is paid the same interest rate no matter the currency that is held. Any premium between the current exchange rate and some future exchange rate is due to higher interest rates in the domestic country than the foreign country. The Purchasing Power Parity Model is an improvement on the Interest Rate Parity Model. The improvement is that interest rates are measured in real and not nominal terms. The difference between the two models is not important if inflation is stable and at similar rates between countries.

Understanding exchange rate fundamentals requires knowledge of interna-

tional trade and macroeconomic concepts. The main determinant of exchange rate is the balance of payments within a country. Balance-of-payments deficits will weaken a currency and the supply of that currency increases on world markets. Balance of payments surpluses will strengthen a currency, and the supply of that currency decreases on world markets. The balance of trade is the main component of the balance of payments and is the net of goods and services imported versus exported.

Economic events that can lead to a stronger currency include weakening domestic consumption, increasing foreign investment, decreases in government spending, high interest rates relative to inflation, and improving productivity. All of these events can lead to increases in the balance of payments. *Actions by central banks can also have large impacts on exchange rates.* Banks buy and sell very large amounts of their own currency, pushing the price of that currency, the exchange rate, up and down in response to the changing supply.

The desire by the world economy to hold and trade in a particular currency can also have a strong impact on the exchange rate of that currency. The case in point is the U.S. dollar. Much world trade involves the exchange of U.S. dollars. Changes in this transaction demand may push the value of the U.S. dollar up or down.

Recently, the volume of exchange rate futures contract trade on the CME has declined substantially. This is likely due to a period of relative stability in currency exchange rates associated with a period of low inflation and stability in monetary policy. However, the importance of world trade continues to grow. *Events such as the devaluation of the Mexican peso, and the more recent devaluation of Asian currencies, highlight the continued need for these exchange rate futures.*

KEY POINTS

- *Exchange rate futures* contracts allow hedgers and speculators to trade *different world currencies in relation to the U.S. dollar*. The German mark and Japanese yen have the largest volume, followed by the currencies of other western European countries, and currencies from Canada and Mexico.
- *Futures contracts for exchange rates* are traded in a number of different futures exchanges *around the world*.
- Exchange rates are a *relative price*. For example, an exchange rate is the *number of U.S. dollars that will purchase one unit of foreign currency*.
- *Caution must be used* in speculating and hedging with exchange rate futures. The relationships at issue *are not common to domestic commerce*.
- A *strengthening U.S. dollar* implies that the *exchange rate will decrease*. A *weakening U.S. dollar* implies exchange rate futures contracts will increase.
- The important question for hedging is: Is the firm *long or short the U.S. dollar relative* to foreign currency? If the firm pays out dollars and takes in foreign currency, it is long dollars and short foreign currency. The firm needs to *sell the CME exchange rate contract* to hedge the transaction. If the firm takes in dollars and pays out foreign currency, it is short dollars and long foreign currency. The firm needs to *buy the CME exchange rate contract* to hedge.
- There is *little if any basis risk* in the exchange rate futures market. However, there will be differences between the cash and futures exchange rates in nonexpiration

months. *These differences can be predicted* by the Interest Rate Parity and Purchasing Power Parity Models.

- *Exchange rate fundamentals* involve the supply of and demand for different currencies. If the *supply of a currency increases* on the world market, its price will fall. If the *demand increases*, its price will rise. The supply of currency is determined by the *balance of payments for a country*.
- The *strength of the domestic economy*, *level of government spending*, *inflation relative to interest rates*, and *productivity* all impact exchange rates. However, *central bank actions* often have the biggest impact on exchange rates.

USEFUL REFERENCES

Steven C. Blank, Colin A. Carter, and Brian H. Schmiesing, *Futures and Options Markets: Trading in Financials and Commodities*, Prentice-Hall, Englewood Cliffs, NJ, 1991. Good intermediate text with coverage of financial futures. Good coverage of interest rate futures.

Robert T. Daigler, *Managing Risk with Financial Futures: Pricing, Hedging and Arbitrage*, Probus Publishing Company, Chicago, 1993. A solid and comprehensive text with many examples. The author is presenting many research results. Very useful for interest rate, stock index, and exchange rate examples.

Robert E. Fink and Robert B. Feduniak, *Futures Trading: Concepts and Examples.* New York Institute of Finance, New York, 1988. A good basic introduction to financial futures.

John C. Hull, *Introduction to Futures and Options Markets*, 3rd ed., Prentice-Hall, Upper Saddle River, NJ, 1998. An introduction to the mathematics of financial futures.

Robert W. Kolb, *The Financial Futures Primer*, Blackwell Publishers, Malden, MA,1997. An excellent introduction to financial futures. Many trading examples are presented. Some of the important subtleties of financial futures are also discussed.

Paul Krugman, *The Age of Diminished Expectations*, 3rd ed., The MIT Press, Cambridge, MA, 1994. An excellent book that discusses current topics of issue for the U.S. and world economies. The text is very useful for understanding the reactions of exchange rate, stock index, and interest rate markets.

Raymond M. Leuthold, Joan C. Junkus, and Jean E. Cordier, *The Theory and Practice of Futures Markets*, Lexington Books, Lexington, MA, 1989. A more advanced treatment of futures and hedging. Extensive focus on interest rate, stock index, and exchange rate futures.

Charles W. Smithson and Clifford W. Smith, Jr., with D. Sykes Wilford, *Managing Financial Risk: A Guide to Derivative Products, Financial Engineering, and Value Maximization*, Irwin Professional Publishing, Chicago, 1995. A fairly simple treatment of more advanced financial management topics. The focus is mainly on options and derivative pricing with interest rate, stock index, and exchange rate examples.

CHAPTER 9

INTEREST RATE FUTURES

INTRODUCTION

Without a doubt, interest rate futures contracts, which are contracts for debt instruments, are the most successful type of futures contracts. A quick review of the futures quote page in the *Wall Street Journal* will show this. The volume of trade in futures contracts on debt instruments is almost larger than that for all other contracts combined. The 1970s was the decade of physical commodities, currencies emerged with prominence in the early 1980s, stock and other index contracts have made strong gains in volume in the late 1980s and 1990s, but debt instruments were the contracts of the 1980s and 1990s. The volume of trade in these futures instruments continues to grow.

Trade in interest rate futures contracts began during October 1975 when the Chicago Board of Trade (CBOT) introduced the Government National Mortgage Association (GNMA or "Ginnie Mae") contracts. Since this time, exchanges have introduced a large number of interest-rate-based contracts. Many of the contracts have been failures, attracting very little volume and trading activity. However, several contracts have been huge successes.

The growth in volume of trade in interest rate futures contracts was due to the increased risk created by the interest rate volatility beginning in the late 1970s, which then carried through to the 1980s, and resurfaced to an extent in the mid-1990s. Prior to the 1970s, inflation and interest rates were stable and predictable. The late 1970s witnessed double-digit inflation rates. Increases in interest rates were implemented by the U.S. Federal Reserve in the early to mid-1980s to slow the economy and reduce inflation. The late 1980s and 1990s witnessed periodic adjustments to interest rates to both slow and stimulate the economy as inflation was brought down to the current relatively low levels. Revenue and cost uncertainty for lenders and borrowers following these substantial changes in interest rates have made hedging essential. The large movements in interest rates have also attracted speculative capital which has helped this market grow.

The 1980s and 1990s also saw an industrialization and integration process in the markets for capital and debt. Markets for debt are now always national and usually global. This process has been facilitated by standardization practices in capital markets and improved communication systems. Individual debts and debt instruments are standardized, pooled, and traded. Borrowers receive needed capital, lenders receive an income stream for a defined time period with known default risk, and financial institutions serve large roles as intermediaries—coordinating the interaction of those wishing to borrow and those wishing to loan capital. Financial institutions standardize and pool mortgages, bonds and commercial paper, and sell the resulting instruments to investors and businesses seeking interest returns on funds available to loan. Financial institutions purchase government and corporate bonds, bills, notes and other debt instruments and sell these to customers desiring income streams.

Lenders of capital face the risk of having the value of the debt instruments that they hold decrease. If a firm that operates a mutual fund purchases a debt instrument, for example a government bond, that yields 5.5 percent per annum and after two years the interest rate increases to 6 percent, there is an opportunity cost of lost income. Likewise, borrowers of capital are at risk of having the value of debt instruments they are under obligation to repay increase. If a food processing business issues a debt instrument, such as commercial paper, that yields 6.75 percent per annum after 180 days and the interest rate decreases to 6.5 percent before funds are returned, then there is an opportunity cost of paying higher than market rates. Further, the financial intermediaries that coordinate the flow of capital and debt instruments between lenders and borrowers are subject to substantial risk. This risk creates demand for vehicles to hedge the many types of debt instruments. This risk also attracts speculators.

All of the significant trade in U.S. interest rate futures occurs at the International Monetary Market (IMM) of the Chicago Mercantile Exchange (CME) and the Chicago Board of Trade (CBOT). The two exchanges have essentially specialized in the trade of different products. The IMM trades debt instruments with very short maturities. Eurodollar deposits and Treasury Bills are the most popular contracts. By contrast, the CBOT trades debt instruments with long-term maturities. These instruments include treasury bonds and treasury notes. The CBOT also offers a contract on 30-day Federal Funds, which are a short-term debt instrument, and a longer-term Municipal Bond Index. However, the volume of trade in these products is small.

Eurodollars are debt instruments for 90-day dollar deposits held in a European bank. The interest paid on this debt instrument is the London Interbank Offered Rate (LIBOR). Treasury bills are short-term U.S. government securities with typical maturities of 90 days and one year. The 90-day T-bill is a very active market. T-bonds and T-notes are U.S. government securities that make biannual interest payments to the holder. T-bond futures contracts are for bonds that mature in no less than 15 years. Separate futures contracts are offered on T-notes which mature in 10, 5, and 2 years. The 10-year T-note is the most active contract.

The remainder of this chapter will focus on the most active futures contracts. The short-term contracts considered are Eurodollars and treasury bills. The long-term contracts discussed are treasury bonds and treasury notes. Eurodollars and T-bills are priced in the futures market in a very similar fashion. The concepts that apply to one, apply to the other. The same is true for T-bonds and T-notes. The contracts are priced in a very similar fashion, and the general concepts that apply to one apply to the other. The main differences between the contracts are in their usefulness in hedging debt of various lengths of maturity. Our discussion will focus on this aspect of the futures contracts and trading. Table 9.1 presents the specifications of these interest rate contracts.

Much of the complexity in the different interest rate contracts emerges in the

Instrument	Exchange	Par Value	Maturity	Quote	Basis Point Value
Eurodollars	CME	$1 million	90 days	100-yield	$25
Treasury bills	CME	$1 million	13 weeks	100-yield	$25
Treasury notes	CBOT	$200,000	2 years	points & 32nds	$62.50
Treasury notes	CBOT	$100,000	5 years	points & 32nds	$31.25
Treasury notes	CBOT	$100,000	10 years	points & 32nds	$31.25
Treasury bonds	CBOT	$100,000	15 years	points & 32nds	$31.25

TABLE 9.1
Debt Instrument Futures Contract Specifications

delivery process. Significant delivery of the actual debt instruments takes place and the contracts are designed so that a multitude of different instruments within each category can be delivered. For example, the variety of treasury bonds that can be delivered on the CBOT T-bond contract is quite large. And much of the complexity in the contract emerges in determining this delivery process and valuing the variety of instruments that can be delivered on a contract that has a fixed specification. This chapter will not address this issue. Like other futures markets, delivery on interest rate futures contracts is the realm of professional traders who concentrate on providing this service. The focus of this chapter is on the mechanics of trading each contract and on use of the contracts in a hedging program.

INTEREST RATE AND DEBT PRICING BASICS

This section discusses what the different debt instruments are, how these debt instruments are priced, and how interest rate futures contracts for each instrument are priced. The instruments and pricing procedures are rather common to the various markets for debt instruments. But the procedures are rather different from contracts for physical commodities, exchange rates, and indexes.

We will begin with the contracts on debt instruments with short-term maturities. The two contracts of interest are for Eurodollars and treasury bills. Debt instruments with short-term maturities usually call for the payment of a fixed amount of cash at maturity. This is the *par value* of the instrument. For example, a T-bill may pay $1 million to the holder after 90 days. The amount of money that the holder must pay for the debt instrument, the *purchase price* or *value*, varies with the interest rate. The more the purchaser pays, the lower the interest rate of the instrument. Again, we are talking about actual T-bills and not the T-bill futures contract.

The interest rate for T-bills is quoted in terms of a *discount yield*. The discount yield is the interest rate associated with the T-bill. T-bills have a fixed par value at maturity, and the price is determined by the par value less the discount implied by the discount yield. The price is determined as follows:

$$\text{Price} = \text{Par Value at Maturity} - \text{Discount Yield} \times \text{Par Value} \times \frac{\text{Days-to-Maturity}}{\text{Days per Year}}$$

$$= \text{Par Value} - \text{Discount} \times \frac{\text{Days-to-Maturity}}{\text{Days per Year}}.$$

The discount yield equals the par value less the price, or the discount, as a percent of the par value converted to an annual interest rate

$$Discount\ Yield = \frac{Par\ Value - Price}{Par\ Value} \times \frac{Days\text{-}per\text{-}Year}{Days\text{-}to\text{-}Maturity}$$
$$= \frac{Discount}{Par\ Value} \times \frac{Days\text{-}per\text{-}Year}{Days\text{-}to\text{-}Maturity}.$$

For a T-bill with a $1 million par value, 90 days to maturity, and a discount yield of 6 percent, the price is

$$Price = \$1{,}000{,}000 - 0.06 \times \$1{,}000{,}000 \times \frac{90}{360}$$
$$= \$1{,}000{,}000 - \$15{,}000 = \$985{,}000.$$

Or if the discount yield can be recovered for the same T-bill with the price of $985,000,

$$Discount\ Yield = \frac{\$1{,}000{,}000 - \$985{,}000}{\$1{,}000{,}000} \times \frac{360}{90}$$
$$= \frac{\$15{,}000}{\$1{,}000{,}000} \times \frac{360}{90} = 0.06 = 6\%.$$

Like T-bills, Eurodollar deposits also have a fixed par value at maturity, a discount yield, a discount, and a price. However, the interest rate is quoted in terms of an *add-on yield*. The add-on yield is the ratio of the discount and the price of the instrument multiplied by a second ratio which converts the first to an annual interest rate. The add-on yield is

$$Add\text{-}On\ Yield = \frac{Discount}{Par\ Value - Discount} \times \frac{Days\text{-}per\text{-}Year}{Days\text{-}to\text{-}Maturity}$$
$$= \frac{Discount}{Price} \times \frac{Days\text{-}per\text{-}Year}{Days\text{-}to\text{-}Maturity}.$$

So, for Eurodollar deposits with a $1 million face value, 90 days from maturity, and a discount yield of 6 percent, the add-on yield is

$$Add\text{-}on\ Yield = \frac{\$15{,}000}{\$1{,}000{,}000 - \$15{,}000} \times \frac{360}{90}$$
$$= \frac{\$15{,}000}{\$985{,}000} \times \frac{360}{90} = 0.0609 = 6.09\%$$

Notice, the T-bill and Eurodollar deposits both have the same discount yield, but the reported interest rate is higher for the Eurodollar than for the T-bill because of the reporting convention.

Futures price quotes for the T-bill and Eurodollar contracts do not report the yields directly but use the IMM Index. The index is 100 minus the interest rate associated with the instrument or

IMM Index = *Futures Price* = 100.00 – *Interest Rate*.

IMM Index for the T-bill futures contract uses the discount yield

IMM Index = *Futures Price* = 100.00 – *Discount Yield*

so that if the yield is 6 percent, the futures price is

IMM Index = 100.00 – 6.0 = 93.00.

The price of the futures contract is quoted as *93.00 points* or 9,300 *basis points*. The interest rate is in percentage terms, 6 percent, and the price of the instrument is 93.00 percent of par. Basis points are 1/100th of 1 percent.

IMM Index for the Eurodollar futures contract uses the add-on yield

IMM Index = *Futures Price* = 100.00 – *Add-On Yield*

so when the yield is 6.09 percent, the futures price is

IMM Index = 100.00 – 6.09 = 92.91.

The price of the futures contract is 92.91 points or 9,291 basis points. The price of the instrument is 92.91 percent of par. In these two examples, the reported interest rate differs by nine basis points.

Table 9.2 presents futures price quotes for these two contracts reported in the *Wall Street Journal* for October 2, 1997. Notice the extent of the maturities traded for Eurodollars and notice the volume and open interest of both contracts.

An interesting feature of this pricing method is that the IMM Index decreases with increases in the interest rate. For example, the price quoted for nearby T-bills in Table 9.2 is 95.03 points. Suppose the next day the price decreases to 94.90 points. This implies the interest rate has risen from 4.97 percent to 5.10 percent. The implication is that beginners need to be careful working through trading examples. You are not trading the interest rate directly. Rather, the IMM Index is pricing the contract as a percent of the par value of the contract size. This is roughly the same convention with which the underlying debt instrument is traded.

Long-term debt instruments are very different from their short-term counterparts, are priced rather differently, and the pricing is more complex. Again, we are referring to the debt instrument and not the futures contract. Short-term debt instruments are almost always *pure discount bonds*. The investor pays a price for the instrument that is some discount to the par value and is returned the par value at maturity. For example, a firm may buy a T-bill with a par value of $10,000 after 90 days for $9,875. After 90 days the T-bill is redeemed for $10,000 and the investor has earned 5 percent per annum. Long-term debt instruments are rarely pure discount bonds. Investors usually receive period interest payments termed *coupons*. For example, a mutual fund may purchase a treasury bond with a $100,000 par value, 15-year maturity, and an 8 percent per annum biannual interest payment. The bond entitles the firm to receive

TABLE 9.2 Futures Prices for Eurodollar and T-Bill Futures, October 2, 1997

	Open	High	Low	Settle	Change	Yield Settle	Yield Change	Open Interest
EURODOLLAR (CME) $1 million; pts of 100%								
Oct	94.25	94.25	94.24	94.25	—	5.75	—	24,890
Nov	94.22	94.22	94.22	94.22	+.01	5.78	–.01	12,262
Dec	94.19	94.21	94.19	94.20	+.01	5.80	–.01	589,353
Mr98	94.15	94.18	94.14	94.16	+.02	5.84	–.02	424,920
June	94.07	94.11	94.06	94.09	+.03	5.91	–.03	315,650
Sept	93.99	94.04	93.98	94.01	+.03	5.99	–.03	242,998
Dec	93.88	93.92	93.86	93.90	+.03	6.10	–.03	223,182
Mr99	93.87	93.91	93.85	93.88	+.03	6.12	–.03	150,575
June	93.82	93.86	93.81	93.84	+.03	6.16	–.03	119,948
Sept	93.79	93.83	93.78	93.81	+.03	6.19	–.03	98,821
Dec	93.72	93.76	93.71	93.74	+.03	6.26	–.03	86,455
Mr00	93.72	93.76	93.72	93.74	+.03	6.26	–.03	71,511
June	93.69	93.73	93.68	93.71	+.03	6.29	–.03	59,705
Sept	93.66	93.70	93.65	93.68	+.03	6.32	–.03	50,295
Dec	93.59	93.64	93.59	93.62	+.03	6.38	–.03	40,627
Mr01	93.59	93.64	93.59	93.62	+.03	6.38	–.03	35,800
June	93.56	93.61	93.56	93.59	+.03	6.41	–.03	31,675
Sept	93.53	93.57	93.53	93.56	+.03	6.44	–.03	29,616
Dec	93.46	93.52	93.46	93.50	+.04	6.50	–.04	17,451
Mr02	93.46	93.52	93.46	93.50	+.04	6.50	–.04	16,673
June	93.43	93.49	93.43	93.47	+.04	6.53	–.04	13,091
Sept	93.40	93.46	93.40	93.44	+.04	6.56	–.04	11,286
Dec	93.34	93.40	93.34	93.37	+.03	6.63	–.03	7,678
Mr03	93.34	93.40	93.34	93.37	+.03	6.63	–.03	6,432
June	93.30	93.36	93.30	93.33	+.03	6.67	–.03	4,853
Sept	93.32	93.33	93.31	93.30	+.03	6.70	–.03	4,290
Dec	93.22	93.26	93.22	93.24	+.03	6.76	–.03	5,206
Mr04	93.22	93.26	93.22	93.24	+.03	6.76	–.03	4,436
June	93.19	93.23	93.19	93.21	+.03	6.79	–.03	6,251
Sept	93.16	93.20	93.16	93.18	+.03	6.82	–.03	3,955
Dec	93.10	93.14	93.10	93.12	+.03	6.88	–.03	4,360
Mr05	93.10	93.14	93.10	93.12	+.03	6.88	–.03	2,005
June	93.05	93.10	93.05	93.08	+.03	6.92	–.03	2,522
Sept	93.03	93.07	93.03	93.05	+.03	6.95	–.03	2,068
Dec	92.97	92.99	92.97	92.99	+.03	7.01	–.03	1,730
Mr06	92.97	92.99	92.97	92.99	+.03	7.01	–.03	2,799
June	92.93	92.95	92.93	92.95	+.03	7.05	–.03	1,819
Sept	92.89	92.91	92.89	92.91	+.03	7.09	–.03	1,750
Dec	—	—	—	92.85	+.03	7.15	–.03	1,485
Mr07	—	—	—	92.85	+.03	7.15	–.03	1,271
Est vol 355,105; vol Wed 449,623; open int 2,733,875, +23,-111								
TREASURY BILLS (CME) $1 million; pts. of 100%								
Dec	95.03	95.04	95.03	95.03	+.02	4.97	–.02	5,031
Est vol 1,187; vol Wed 688; open int 8,645, +270								

Source: Wall Street Journal.

coupon payments worth \$4,000 twice per year (\$8,000 = \$100,000 × 8% coupon) for 15 years, and then the bond is redeemed for the par value.

How much should the firm pay for this bond? The price of the bond is determined with the following formula which calculates the present value of a stream of interest payments with a lump sum end payment. The bond price is

$$Price = \left[\sum_{t=1}^{2n} \frac{I_t}{(1 + \frac{1}{2}i)^t}\right] + \frac{Par\ Value}{(1 + \frac{1}{2}i)^{2n}}$$

where I_t is the biannual coupon payment, i is the annual market interest rate, and the bond matures in n years. The summation is the present value of the interest payments. Notice the biannual payments. The second part of the right-hand side is the discounted par value. It is important to notice that there are two interest rates in the valuation. The bond pays a coupon which is the interest rate of the bond. However, the price of the bond, or what the bond is worth, depends on the market interest rate. The following examples will make this clear.

Suppose a bond has the par value of \$100,000, matures in 15 years, pays an 8 percent per annum coupon, and the current market interest rate is 8 percent. The price of the bond is

$$Price = \left[\sum_{t=1}^{2\times15} \frac{4000_t}{(1 + \frac{1}{2}0.08)^t}\right] + \frac{\$100{,}000}{(1 + \frac{1}{2}0.08)^{2\times5}}$$

$$= \$69{,}168 + \$30{,}832 = \$100{,}000.$$

The price of the bond is equal to the par value because the interest paid by the bond is equal to the market interest rate. This market interest rate is the opportunity cost of holding the bond.

Now, suppose that the market interest rate is 6.5 percent. The price of the bond with an 8 percent per annum coupon is

$$Price = \left[\sum_{t=1}^{2\times15} \frac{4000_t}{(1 + \frac{1}{2}0.065)^t}\right] + \frac{\$100{,}000}{(1 + \frac{1}{2}0.065)^{2\times5}} = \$114{,}236.$$

The price of the bond is at a premium to the par value because the bond returns 8 percent per annum but the market rate is 6.5 percent. The convention used in the bond market is that the price of this bond would be quoted as *114.24 percent of par*.

Suppose that the market interest rate is 9.25 percent. The price of the bond with an 8 percent per annum coupon is

$$Price = \left[\sum_{t=1}^{2\times15} \frac{4000_t}{(1 + \frac{1}{2}0.0925)^t}\right] + \frac{\$100{,}000}{(1 + \frac{1}{2}0.0925)^{2\times5}} = \$89{,}967.$$

The price of the bond is at a discount to the par because the bond returns 8 percent per annum but the market rate is 9.25 percent. The convention used in the bond market is that the price would be quoted as *89.97 percent of par*.

A further reporting convention used in markets for long-term government securities is that instead of using decimals in the percent of par quotations, the market uses *points and 32nds of par*. For example, a treasury bond may be priced as 114-08. This implies

114-08 points and 32nds = 114 and 08/32s points = 114.25 percent of par.

Likewise, if the T-bond is priced as 89-31, then

89-31 points and 32nds = 89 and 31/32s points = 89.97 percent of par.

It is not possible to take a quotation in points and 32nds of par, where par is for an 8 percent coupon bond, and use a simple formula to convert this to an interest rate quotation. Instead, you would have to solve the formula for the market interest rate as a function of the price, par value, and using the 8 percent coupon payments. This is not possible.

But, we can do one of two things to calculate the yield, or interest rate, implied by a bond price quotation. First, we can use trial-and-error with the bond price formula. We plug different interest rates into the formula and see which one gives us a price closest to the quotation. This is very easy to do with a spreadsheet. We can program the bond price formula into a spreadsheet and rapidly examine prices for different interest rates. Second, we can use the following formula which approximates the yield:

$$Yield = \frac{Coupon\ per\ Annum + \dfrac{(Par - Price)}{Years}}{\dfrac{(Par + Price)}{2}}$$

Working with the numbers of the first example, we see that the approximation gives

$$Yield = \frac{\$8,000 + \dfrac{(\$100,000 - \$114,236)}{15}}{\dfrac{(\$1000,000 + \$114,236)}{2}} = 0.0658 = 6.6\%$$

which is just slightly different from the true yield of 6.5 percent.

The long-term debt instruments traded at the CBOT are reported in points and 32nds of par. The contracts are specified for T-bonds and T-notes which pay 8 percent per annum biannual coupons. Deliverable T-bonds should mature in not less than 15 years. Contracts are offered on T-notes that, at the expiration of the contract, have maturities of 10, 5, and 2 years. Table 9.3 presents price quotes for T-bond and T-note futures contracts reported in the *Wall Street Journal* for October 2, 1997. Again, notice the volume of trade and open interest.

Like, the short-term contracts, pricing the futures contracts in points and 32nds of par results in the price of the contract decreasing with increases in the interest rate. Again, if you are not familiar with bond pricing, you need to be careful working through trading examples. You are not trading the interest rate directly. Rather, you are trading the contract as a percent of the par value. This is what is done in the underlying T-bond and T-note markets.

TABLE 9.3 Futures Prices for T-Bonds and T-Notes, October 2, 1997

	Open	High	Low	Settle	Change	Lifetime High	Lifetime Low	Open Interest
TREASURY BONDS (CBOT) $100,000; pts. 32nds of 100%								
Dec	116-10	116-28	116-06	116-20	+9	118-08	100-08	644,599
Mr98	116-03	117-24	115-21	116-10	+9	117-24	104-21	38,468
June	115-31	116-06	115-31	115-31	+10	116-14	104-03	4,238
Sept	—	—	—	115-21	+9	115-00	103-22	1,957
Dec	—	—	—	115-12	+9	114-18	103-13	4,669
Est vol 420,000; vol Wed 618,284; open int 693,986, +36,496								
TREASURY NOTES (CBOT) $100,000; pts. 32nds of 100%								
Dec	110-20	110-30	110-18	110-25	+5	110-30	104-10	377,787
Mr98	110-09	110-18	110-09	110-14	+5	110-18	105-24	14,729
Est vol 82,183; vol Wed 110,086; open int 392,519, +9,683								
5-YR TREASURY NOTES (CBOT) $100,000; pts. 32nds of 100%								
Dec	107-24	107-30	107-22	107-26	+2.5	107-30	104-005	234,928
Est vol 36,615; vol Wed 55,187; open int 235,001, +6,195								
2-YR TREASURY NOTES (CBOT) $200,000; pts. 32nds of 100%								
Dec	103-21	103-227	103-205	103-22	+1.0	103-227	102-265	43,212
Est vol 2,000; vol Wed 3,108; open int 43,212, +777								

Source: Wall Street Journal.

The volatile interest rates of the high-inflation years in the late 1970s and early 1980s gave huge impetus to interest rate futures. All of a sudden, business firms faced very significant risks they had not faced before. When risk of this magnitude emerges, there is great demand for a way to manage exposure to that risk. This is the role of futures trade.

INTEREST RATE FUTURES TRADING EXAMPLES

Bonding pricing is not all that difficult, but it is rather different from that of other assets. However, trading futures contracts on debt instruments is relatively straightforward once you understand the basics of bond pricing. We will work a few trading examples. These trading examples could be speculative or could be part of a hedge.

Suppose that in October a trader sells one December T-bill contract priced at 95.00 points. In December, the contract is trading at 94.50 points. The trader offsets the position by buying the contract. The net change in the value of the position is the price of the instrument at the time of sale less the price of the instrument at the time of purchase, which is

$(95.00 \text{ points} \times \$1 \text{ million}) - (94.50 \text{ points} \times \$1 \text{ million}) \times (90 \text{ days} / 360 \text{ days}) =$
$(95.00\% \times \$1 \text{ million}) - (94.50\% \times \$1 \text{ million}) \times (90 \text{ days} / 360 \text{ days}) =$
$(\$950{,}000 - \$945{,}000) \times (90 \text{ days} / 360 \text{ days}) =$
$(\$5{,}000) \times (90 \text{ days} / 360 \text{ days}) = \$1{,}250.$

The net change in the value of the position can also be calculated using the price change.

$(95.00 - 94.50) \text{ points} \times \$1 \text{ million} \times (90 \text{ days} / 360 \text{ days}) =$
$50 \text{ basis points} \times \$1 \text{ million} \times (90 \text{ days} / 360 \text{ days}) =$
$0.50\% \times \$1 \text{ million} \times (90 \text{ days} / 360 \text{ days}) =$
$(\$5{,}000) \times (90 \text{ days} / 360 \text{ days}) = \$1{,}250$

The discount yield increased from 5 percent to 5.5 percent between October and December, and the trader made money by selling the T-bill contract. The procedure for calculating the gain or losses from trading the Eurodollar contract are exactly the same as for the T-bill.

Rather than performing the preceding calculations, it is easier to keep track of the change in the number of basis points associated with a position—the sell price minus the buy price—and then multiply this change by the value of a one-basis-point change in the contract. A one-basis-point or a 0.01 percent change in a contract with a $1 million par value would be $100, but we need to remember that the debt instrument matures in 90 days.

$\$1 \text{ million} \times (90 \text{ days} / 360 \text{ days}) \times 1 \text{ basis point} =$
$\$1 \text{ million} \times (90/360) \times 0.0001 = \$25$

Thus, the value of a one-point change in the T-bill or Eurodollar contract is $25.

As a second example, suppose that in October a trader sells eight March Eurodollar contracts at 94.14 points. The next day, the contract is trading at 94.16 points, and the trader offsets this position by buying eight contracts. The net change in the value of the position is

$(94.14 - 94.16) \text{ points} \times (\$1 \text{ million} \times (90 \text{ days} / 360 \text{ days})) \text{ per contract} \times$
$8 \text{ contracts} =$
$-2 \text{ basis points} \times \$25 \text{ per point per contract} \times 8 \text{ contracts} =$
$-\$50 \text{ per contract} \times 8 \text{ contracts} = -\$400.$

The yield decreased from 5.86 percent to 5.84 percent between the two days, and the trader lost money by selling the Eurodollar contracts.

Since T-bonds and T-notes are quoted in points and 32nds, one basis point is 1/32nd of a point. The change in the value of a contract following a one-basis-point change in the quote is

$\textit{Par Value} \times 1/32 \text{ of } 1 \text{ point} = \textit{Par Value} \times 0.03125\%.$

The par value of a T-bond, T-note (10-year), and T-note (5-year) is $100,000, so a one-basis-point change is worth $31.25 per contract. The par value of a two-year T-note contract is $200,000, so the value of a one-basis-point change is double. It is easier to work with T-bond and T-note prices if they are converted to decimals. In decimals, a one-full-point change is worth \$1,000 (32 32nds per full point × \$31.25), and in this case a one-basis point change is worth \$10. Be careful keeping track of the units.

Suppose that in October a trader sells one December T-bond contract at 116-16, which is a yield of 6.285 percent. In December, the trader offsets the contract at 115-08, a yield of 6.403 percent. The net change in the value of the position is

(116-16 – 115-08) points and 32nds × \$31.25 per 32nd point per contract =
(1-08) points and 32nds × \$31.25 per 32nd point per contract =
(32+8) 32nds × \$31.25 per 32nd point per contract =
(40) 32nds × \$31.25 per 32nd point per contract = \$1,250

or converting the 32nds to decimals,

(116.5 – 115.25) points × \$1,000 per point per contract =
(1.25) points × \$1,000 per point per contract = \$1,250.

The yield increased from 6.285 percent to 6.403 percent between October and December, and the trader made money by selling the T-bond contract.

The important thing to remember with interest rate futures contracts is that you are trading the discount to par. If the price that represents the discount to par increases, the discount is decreasing, and the interest rate is decreasing. This point is worth repeating, especially since much of this section has dealt with the details of debt instrument pricing. Do not lose the larger perspective because of these details. The price you are trading for the debt instrument underlying the futures contract is some measure of the discounted market value of the instrument. As interest rates rise, the discounted market value and the futures price will fall. Likewise, as interest rates fall, the discounted market value and the futures prices will rise.

INTEREST RATE HEDGING EXAMPLES

This section presents several example interest rate hedges. As with all hedging, the trader has a position in the cash market or will have a position in the cash market that will change in value if the price of the underlying asset for a debt instrument changes. In this case, the assets are loaned or borrowed money and capital. The trader has purchased or will purchase a debt instrument, or is or will be under obligation to repay a debt instrument.

A person or business who has borrowed or will loan is long spot market interest rates. This trader has sold or will purchase a debt instrument. If interest rates increase, the trader benefits. For example, if a business borrows money at a fixed interest rate and the market interest rate increases, the firm benefits. Likewise, if a trader promises to deliver a T-Bill in the future and interest rates increase, the trader will have an unanticipated decrease in the cost of meeting that obligation. A person or business who has loaned or will borrow is short spot market interest rates. This trader has purchased or will sell a debt instrument. If interest rates decrease, the trader benefits. For example, if a business loans money at a fixed interest rate and the market interest rate decreases, the firm benefits. Likewise, the trader holds a T-Bond and interest rates decrease, the trader can sell the T-Bond for more than the purchase price. An important component of interest rate hedging involves recognizing whether the spot position is long or short—does the trader stand to gain or lose from in increase or decrease in interest rates? This must be determined first.

A second component involves matching the cash instrument with an appropriate futures contracts. What is the maturity of the debt instrument? Long-term, intermediate-term, and short-term interest rates can be rather different and can respond to different market events. Traders may be interested in buying or selling debt instruments with maturities from the very short-term, such as 30 days, to the very long-term, 25 years, for example. And as we have seen, there are futures contracts for short-term debt, T-bill and Eurodollar futures contracts, intermediate term debt, two-year and five-year T-notes futures contracts, and long-term debt, T-bond futures contracts.

It is also important to recognize that the maturity of the debt instrument and the length of the hedge are two very different time components of the hedging problem. For example, suppose a trader knows that in three months she will have $1 million with which to purchase 90-day T-bills. The trader will purchase 90-day T-bills in three months and that rate may or may not be the same as the rate for 180-day T-bills today. The market consensus of the rate that the trader will receive is the current price of T-bill futures that expire in three months. The maturity of the debt instrument and the length of the hedge may be the same. For example, suppose a business with a one-year loan with a floating interest is interested in hedging or locking in a fixed rate. The firm can take an opposite position in the futures market of a contract that the loan rate follows. Then, the length of the hedge is equal to the length of the loan. But the T-bill example earlier shows that the two time periods are different things. The point we are making is that there are many time periods and time horizons to consider with debt instruments. You need to keep the maturity of the debt instrument and the length of the hedge separate in your mind.

Let's work a couple of basic hedges. The first hedge is for a portfolio manager who will purchase T-bills in the future. It is March 15 and the manager learns that he will have $985,000 on June 15 with which to purchase 90-day T-bills. The June contract is trading for 93.75 points, or a discount yield of 6.25 percent. The future price implies the manager can lock in the price of one T-bill with a par value of $1 million for $984,375. The manager takes a long position and buys one contract. The complete hedge is summarized in Table 9.4.

Suppose that on June 15, the June T-bill contract is trading at 95.25 points and the yield on T-bills is 4.75 percent. The yield implied by the futures price is also 4.75 percent. The manager sells one June T-bill contract to offset the long position. The net on the futures trade is +150 basis points and the return is $3,750. The manager pays $988,125 for a 90-day T-bill with a par value of $1 million and uses the gains from the futures market to reduce this cost to $984,375. The annualized yield on the T-bill is therefore 6.25 percent. The portfolio manager used the futures market to lock in the T-bill yield.

TABLE 9.4
Hedging the Future Purchase of Treasury Bills by a Portfolio Manager

Date	Cash Market	Futures Market
March	Anticipate purchasing T-bill with $1 million par value in June for $984,375. Expected yield of 6.25%.	Buy one June T-bill contract at 93.75 points. Implied yield is 6.25%.
June	Purchase T-Bill with $1 million par value for $988,125.	Sell 1 June T-Bill contract at 95.25 points.
	Loss of $3,750.	Gain of $3,750.
	Net change of $0. Actual T-Bill yield of 6.25%.	

The second hedge is for a securities trader that has agreed to sell T-bills in the future. On September 15, a securities trader contracts to deliver one 90-day T-bill with a par value of $1 million on December 15 to a customer for $984,375. This price allows the trader to make a small profit. The trader has contracted to deliver a T-bill with a 6.25 percent yield. The December futures contract is trading for 93.75 points or a yield of 6.25 percent. The trader takes a long position and buys one contract. This example is summarized in Table 9.5.

Suppose that on December 15, the December T-bill contract is trading at 92.75 points and the yield on T-bills is 7.25 percent. The yield implied by the futures price is also 7.25 percent. The trader offsets the long position and sells the futures contract. The net on the futures trade is –100 basis points or –$2,500. The manager pays $981,875 for a 90-day T-bill with a par value of $1 million, delivers the T-bill for $984,375, and uses the gains in the cash market to offset losses in the futures. The net cost of the delivered T-bill is $984,375. The annualized yield on the T-bill is therefore 6.25 percent. The trader makes the anticipated small profit.

In both of the examples, the traders took long positions in the futures market to protect their cash market positions. The traders will lose money, or will make less than expected, if the interest rates decrease. Thus, both traders are short the cash market—*the cash market here being the market for the actual debt instruments*—and therefore need to take long positions in the futures to protect that cash position.

Note that basis error and basis risk are not present in either example. We are not referring to the basis points relevant to interest rate changes. Rather, we are referring to the difference between the price of the debt instrument and the price of the futures contract for that instrument. Basis error and basis risk are generally not present when you are hedging the exact instrument that underlies the futures contract. Here we are hedging T-bill purchases and sales with T-bill futures. As with currencies and indexes, basis error is essentially zero in the futures contract expiration month. These markets for debt instruments are the most liquid of any market, and it is very inexpensive to deliver the debt instruments called for in the futures contract. These instruments are delivered through electronic communication, and the transactions costs are small. *This is quite different from delivery of physical commodities.* Two things, however, are more important than basis error and basis risk for interest rate contracts. These will be discussed following the next example.

The following is an example hedge in which the trader sells futures contracts. An agribusiness subsidiary of a large corporation has experienced cost overruns on a construction project. The firm will need to borrow $10 million in six months to complete the project. It is December 15, and the firm anticipates borrowing on June 15. The

TABLE 9.5
Hedging the Future Delivery of Treasury Bills by a Securities Trader

Date	Cash Market	Futures Market
September	Contract to deliver T-bill with $1 million par value in December for $984,375. Expected yield of 6.25%.	Buy one December T-bill contract at 93.75 points. Implied yield is 6.25%.
December	Purchase T-bill with $1 million par value for $981,875. Deliver T-bill. Receive $984,375.	Sell one December T-bill contract at 92.75 points.
	Gain of $2,500.	Loss of $2,500.
	Net change of $0. Actual profit equals expected profit.	

negotiated loan rate is the LIBOR plus 2 percent. However, the interest rate on the loan is not fixed until the funds are borrowed. It is anticipated that the construction will take place, and that this loan will be needed for three months after June 15. After construction is complete, this loan will be paid off and rolled into a larger loan package that is the responsibility of corporate headquarters. The financial officer for the agribusiness firm in charge of the project is concerned about interest rates increasing between December and June, increasing the cost of his part of the project. The June Eurodollar contract is trading at 94.50 points, or a yield of 5.5 percent. The financial officer should be able to lock in a loan rate of 7.5 percent and sells 10 June Eurodollar contracts. Table 9.6 provides the actions taken and the results in terms of the loan rate.

Suppose on June 15 the June contract is trading for 92.50 points. The loan is secured at 9.5 percent. The hedge is lifted, the net change in the futures price is +200 basis points, and the futures position returns $50,000. The present value of the $10 million loan is $9.775669 million.

$$\$10 \text{ million} / (1.095)^{0.25} = \$9.775669 \text{ million.}$$

The present value of the loan would be $9.820823

$$\$10 \text{ million} / (1.075)^{0.25} = \$9.820823 \text{ million}$$

if it could have been secured at 7.5 percent the previous December. The loss to the cash position is $45,154. Gains in the futures market are used to offset the increased costs of the loan.

This hedge presented three concepts related to interest rate hedging. The first is basis. Again, it is not the basis points that are relevant to interest rate changes. Rather, the loan rate is 2 percent over the Eurodollar rate. This is very familiar to traders of physical commodities and should pose no problem. In the example, the trader recognized that the expected interest rate for the loan is the rate implied by the futures contract price plus the basis. Further, the actual basis equaled the expected basis. This is because the 2 percent basis was contracted with the loaning bank. Suppose the financial officer calculated the 2 percent premium from historical data. If the actual basis in June is 2.5 percent over the LIBOR, then the loan will be secured at 10 percent. In this case, the present value of the loaned amount is $9.764541 million

$$\$10 \text{ million} / (1.10)^{0.25} = \$9.764541 \text{ million}$$

TABLE 9.6
Hedging a Future 90-Day Loan Based on the LIBOR Plus 2%

Date	Cash Market	Futures Market
December	Anticipate borrowing funds with $10 million par value in June for the LIBOR plus 2% or 7.5%. Market value of loan is $9,820,823.	Sell 10 June Eurodollar contracts at 94.50 points. Implied yield is 5.5%.
June	Borrow $10 million for LIBOR plus 2% or 9.5%. Market value of loan is $9,775,669.	Buy 10 June Eurodollar contracts at 92.50 points. Implied yield is 7.5%.
	Loss of $45,154.	Gain of $50,000.
	Net change of +$4,846. Actual loan rate of 7.29%.	

TABLE 9.7 Hedging a Future 90-Day Loan Based on the LIBOR

Date	Cash Market	Futures Market
December	Anticipate borrowing funds with $10 million par value in June for the LIBOR plus expected basis. Expected basis is 2% so expected loan rate is 7.5%. Market value of loan is $9,820,823.	Sell 10 June Eurodollar contracts at 94.50 points. Implied yield is 5.5%.
June	Borrow $10 million for LIBOR plus actual basis of 2.5% or 10%. Market value of loan is $9,764,541.	Buy 10 June Eurodollar contracts at 92.50 points. Implied yield is 7.5%.
	Loss of $56,282.	Gain of $50,000.
	Net change of –$6,282. Actual loan rate of 7.78%.	

and the loss in the cash market is –$56,282. This variation on the example is shown in Table 9.7.

Basis error and basis risk is present if you are not hedging the exact instrument underlying the futures contract or if the basis level is not contracted and it is calculated from historical data. Basis risk should be relatively small because the markets for debt instruments are well linked. But the premiums and discounts associated with debt instruments do change relative to the debt instrument interest rates traded on the futures market. In our earlier example, the premium over the LIBOR could have increased from 2 percent to 2.5 percent because business conditions changed during the six months over the hedge. In June it is perceived that short-term construction loans are more risky, and there is an increase in the basis premium.

As in markets for physical commodities, estimates of basis are essential for operating hedging program for debt instruments. Two things are also important for hedging with interest rate contracts. The first is the lumpiness of the contracts. Like exchange rate futures and stock index futures, these futures contracts involve large amounts of financial instruments, as shown in Table 9.1. However, the larger the debt instrument that needs to be hedged, the less this is an issue. But it can be important and almost always keeps gains or losses in the futures market from exactly offsetting losses or gains in the market for the debt instrument.

The second concern that arises through these examples is in matching the cash position with an appropriate position in the futures market. There are two pieces to the puzzle of defining an appropriate futures market position: (1) what particular contract and (2) how many contracts are necessary for an effective hedge? This concept will be covered extensively in Chapter 10, Index Futures Contracts. The same issue arises for hedging interest rate risk. The trader used 10 Eurodollar contracts in the preceding example. The loan was based off the LIBOR and the maturity is 90 days so the Eurodollar is the right contract. But is 10 the right number of contracts? The returns in the futures market offset losses in the cash when there was no basis risk, but not exactly. Did the financial officer hedge too much, or was the difference because the size of the cash position does not match up perfectly with a number of futures contracts?

The hedger can go about matching the cash position with the correct futures position four ways. First, at a minimum, the hedger should attempt to match the par value of the two positions. This is the *par value approach* and is what the financial officer did in the earlier examples. The number of contracts is the ratio of the par value of the debt instrument in the cash market to the par value of the futures contract.

$$\textit{Hedge Amount} = \frac{\textit{Par Value Cash Instrument}}{\textit{Par Value 1 Futures Contract}}$$

Further, the hedger should choose a futures contract in which the maturity of the underlying debt instrument is similar to the maturity of the debt instrument being hedged. But if there is no exact match, the approach does not provide a correction for this. The approach does not account for compounding differences or maturity differences between the instruments.

The second approach addresses another major weakness of the first approach. The par values of the debt instruments are not traded. It is the market values of the debt instruments that change with changing interest rates. The *market value approach* uses the ratio of the market value, or the price, of the cash debt instrument relative to the market value of the debt instrument underlying the futures contract.

$$\textit{Hedge Amount} = \frac{\textit{Market Value Cash Instrument}}{\textit{Market Value 1 Futures Contract}}$$

Returning to our example, the market value of the \$10 million loan for 90 days at 7.5 percent is \$9.820823 million.

$$\$10 \text{ million} / (1.075)^{0.25} = \$9.820823 \text{ million}$$

The market value of a Eurodollar contract trading at a 5.5 percent yield is

$$\$1 \text{ million } (1 - 0.055(90/360)) = \$0.98625 \text{ million.}$$

The ratio, which gives us the number of contracts, is

$$\$9{,}820{,}823 / \$986{,}250 = 9.96 \text{ contracts.}$$

Thus, we see that the financial officer was slightly overhedged in trading 10 contracts. This is one cause of the imbalance between gains and losses in the cash and futures market. But, because of the lumpiness of the contracts, the officer can do nothing about this.

The third possibility is the *basis point approach*. The hedger's real objective is likely to match gains and losses in the cash and futures markets, and not to match the par value or market value of the two positions. Matching the market values will come closer to matching gains and losses than matching the par values, but the hedge still may not be totally effective. The hedger can do a better job of matching gains and losses in the cash and futures markets by comparing the results of a one-basis-point change in the cash debt instrument to the futures contract. The hedge amount is

$$\textit{Hedge Amount} = \frac{\textit{Value 1 Basis Point Change Cash Instrument}}{\textit{Value 1 Basis Point Change Futures Contract}}$$

Again, using our example, a one-basis-point increase in the Eurodollar contract with a \$1 million par value is worth \$25. A one-basis-point increase in a \$10 million loan for 90 days at 7.5 percent is worth \$228. The ratio of the change in the value of the cash instrument to the change in the value of one futures contract is

$\$228 / \$25 = 9.12.$

This suggests that the officer needs to trade nine Eurodollar contracts. The officer is slightly overhedged using the market value approach because the loan rate is at a premium to the LIBOR. A one-basis-point change in the LIBOR has a slightly smaller impact on the loan market value than the Eurodollar market value.

While the financial officer in our example is overhedged according to the basis point approach, the magnitude is small and we could ignore it. However, if we modify the example slightly, the basis point approach shows its strength. Suppose the firm requires the loan for one year instead of 90 days, all of the other terms of the loan being identical. Now, a one-basis-point change in the interest rate costs the firm \$866. This is calculated from

$$\$10 \text{ million} / (1.0750)^1 = \$9.302326 \text{ million}$$

and

$$\$10 \text{ million} / (1.0751)^1 = \$9.301460 \text{ million}.$$

The ratio of the cash amount to the futures is 34.4 (\$866/\$25). The method suggests the officer needs to trade 34 Eurodollar contracts. You see that the officer needs to trade more contracts of a debt instrument with a 90-day maturity to hedge a cash debt instrument (i.e., the loan) with a one-year maturity. Table 9.8 presents an example outcome of a hedge following this recommendation.

Now, suppose the loan is not priced off the LIBOR but is priced off the prime rate. We can calculate the change in the market value of the loan given a one-basis-point change in the prime rate. It will be the same as a one-basis-point change in the LIBOR if the prime rate is equal to the LIBOR plus 2 percent. What we really need to know is how the prime rate changes relative to the LIBOR. The prime is more variable, and we will assume the prime rate is 25 percent more variable for the example. Thus, a 100-basis-point change in the LIBOR will be accompanied by a 125-basis-point change in the prime. The basis approach ratio can be augmented by a factor that captures the relative variability of the two interest rates.

TABLE 9.8
Hedging a Future One-Year Loan Based on the LIBOR Plus 2%

Date	Cash Market	Futures Market
December	Anticipate borrowing funds with \$10 million par value in June for the LIBOR plus 2% or 7.5%. Market value of loan is \$9,302,326.	Sell 34 June Eurodollar contracts at 94.50 points. Implied yield is 5.5%.
June	Borrow \$10 million for LIBOR plus 2% or 9.5%. Market value of loan is \$9,132,420.	Buy 34 June Eurodollar contracts at 92.50 points. Implied yield is 7.5%.
	Loss of \$169,906.	Gain of \$170,000.
	Net change of +\$94. Actual loan rate of 7.50%.	

$$Hedge\ Amount = \frac{Value\ 1\ Basis\ Point\ Change\ Cash\ Instrument}{Value\ 1\ Basis\ Point\ Change\ Futures\ Contract} \times Relative\ Variability.$$

Returning to our example, suppose the prime rate is about equal to the LIBOR plus 2 percent, so a one-basis-point change in the loan rate still costs the firm $866. Next, suppose the prime is 25 percent more variable than the LIBOR. The financial officer should trade

$$(\$866\ /\ \$25) \times 1.25 = 43.3$$

or 43 Eurodollar contracts. You also see that the financial officer needs to trade more contracts of a debt instrument that is less variable than the cash debt instrument (i.e., the loan) that you are trying to hedge. Table 9.9 presents an example outcome of a hedge following this recommendation.

How do you calculate the relative variability? The tool used is the same as that used to measure the relative variability in returns of different stock portfolios and is discussed in many finance texts. This tool will be discussed in more detail in Chapter 10. The relative variability between two interest rates is measured with the following linear model:

$$CMR_t = \alpha + \beta\, FMR_t + e_t$$

where CMR_t denotes the interest rate for the cash market instrument time t, FMR_t denotes the interest rate for which we have a futures contract during the same time period t, α and β are parameters to be estimated with the data, and e_t is the random unexplainable error. This model can be estimated with linear regression. All commercial spreadsheet software will perform this calculation. The parameter β measures the relative variability. If β equals 1, then the interest rate for the cash instrument moves one-for-one with the interest rate associated with the futures contract. A 1 percent increase or decrease in the futures market rate will, on average, be matched by a 1 percent increase or decrease in the cash instrument. If β is less than 1, the interest rate of the cash instrument is less variable than the futures market. If β is more than 1, the interest rate of the cash instrument is less variable than the futures market. In addition to the measure of relative variability, the model also provides an estimate of the average basis premium or discount through the estimate α.

TABLE 9.9
Hedging a Future One-Year Loan Based on the Prime Rate

Date	Cash Market	Futures Market
December	Anticipate borrowing funds with $10 million par value in June at prime. Expect prime to be LIBOR plus 2% or 7.5%. Market value of loan is $9,302,326.	Sell 43 June Eurodollar contracts at 94.50 points. Implied yield is 5.5%.
June	Borrow $10 million at prime. Prime is LIBOR plus 2.5% or 10%. Market value of loan is $9,090,909.	Buy 43 June Eurodollar contracts at 92.50 points. Implied yield is 7.5%.
	Loss of $211,417.	Gain of $215,000.
	Net change of +$3,583. Actual loan rate of 7.46%.	

The last procedure is the *duration approach*. This approach more completely accounts for differences in the maturity of the cash debt instrument and the futures contract used to hedge it. The hedge amount is calculated as the following

$$Hedge\ Amount = \frac{(1 + r_f) \times P_c \times T_c}{(1 + r_c) \times P_f \times T_f} \times Relative\ Variability$$

where P_c is the market value of the cash instrument, P_f is the market value of the underlying futures instrument, T_c is the maturity of the cash instrument, T_f is the maturity of the underlying futures instrument, r_c is the interest rate of the cash instrument, and r_f is the interest rate of the futures instrument.

Let's calculate the hedge amount for the 90-day $10 million loan.

$$Hedge\ Amount = \frac{(1 + 0.055) \times \$9,820,823 \times 0.25}{(1 + 0.075) \times \$986,250 \times 0.25} = 9.77$$

or the method suggests ten contracts. The hedge amount for the one-year $10 million loan is

$$Hedge\ Amount = \frac{(1 + 0.055) \times \$9,302,326 \times 1}{(1 + 0.075) \times \$986,250 \times 0.25} = 37.03$$

or the method suggests 37 contracts. And if the loan is priced off the prime rate, which is 25 percent more variable than the LIBOR,

$$Hedge\ Amount = \frac{(1 + 0.055) \times \$9,302,326 \times 1}{(1 + 0.075) \times \$986,250 \times 0.25} \times 1.25 = 46.28$$

or 46 contracts.

If you look closely at the formula you see the following: The greater the cash interest rate is relative to the futures interest rate, the fewer contracts are traded. The greater the cash instrument market value relative to the futures instrument market value, the more contracts are traded. And the greater the cash instrument maturity relative to the futures instrument maturity, the more contracts are traded. Thus, this approach corrects for differences in the market value between the cash and futures instruments, differences in maturity of the two instruments, and differences in the level of the underlying interest rates.

We will finish this section with two more example hedges. The first example is known as a strip hedge. The structure of the loan appears strange but the procedures of this hedge are useful in understanding the second example, converting a floating rate loan to a fixed rate loan. In the example strip hedge, a firm borrows $1 million at 3 percent over the LIBOR for three months, pays the interest after three months, rolls the loan over for a second three-month period, and then follows the same procedure for a third and fourth three-month period. Basically, the firm is borrowing the money for one year and making quarterly interest payments. The initial loan is taken out in December. The firm hedges the loan by selling one Eurodollar future for each one on the March, June, and September contracts. When the loan is refinanced in March, the firm buys back the March contract. When the loan is refinanced in June, the June con-

TABLE 9.10
Strip Hedge of a One-Year Balloon-Payment Loan Based on the LIBOR Plus 3%

Date	Cash Market	Futures Market
December	Borrow $10 million at 8.75%. Commit to roll over loan for three quarters at prevailing LIBOR plus 3%.	Sell 10 March Eurodollar contracts at 94.15 points. Sell 10 June Eurodollar contracts at 94.05 points.
	Expected interest cost is $888,750 based on expected yield of 8.8875% which is average of current rate and implied rates from futures.	Sell 10 September Eurodollar contracts at 94.00 points.
March	Pay interest of $218,750. Roll over $10 million loan for three months at 7.85%.	Buy 10 March Eurodollar contracts at 95.15 points.
		Loss of $25,000.
June	Pay interest of $196,250. Roll over $10 million loan for three months at 9.95%.	Buy 10 June Eurodollar contracts at 93.05 points.
		Gain of $25,000.
September	Pay interest of $248,750. Roll over $10 million loan for three months at 10.5%.	Buy 10 September Eurodollar contracts at 92.50 points.
		Gain of $37,500.
December	Pay interest of $262,500 and repay principal.	
	Total interest cost is $926,250.	Total gain of $37,500.
	Net interest cost is $888,750.	

tract is bought back, and likewise for the September contract. Currently, the March Eurodollar contract is trading at 94.15, June is trading at 94.05, and September is at 94.00. The firm should be able to lock in interest payments of 8.85 percent, 8.95 percent, and 8 percent. The interest payment for the first quarter is set when the loan was initiated. This example is summarized in Table 9.10.

The interest payment for the first quarter is set when the loan is initiated. However, the interest payment falls for the second quarter, and then rises for the next two. Through the hedge, the gains and losses in expected interest payments are offset by gains and losses in the futures market.

In the second example, a hedger converts a floating rate loan to a fixed rate loan. It is currently December, and an agribusiness exporter needs to borrow $100 million for one year. The firm will make quarterly payments to interest and principal, and a new interest rate is determined each quarter by LIBOR plus 3 percent. This hedge is different from the previous example, the strip hedge, in that the firm is less exposed to interest rate risk towards the end of the loan since principal is being repaid. The firm sells 74 Eurodollar futures in the March contract, 50 June contracts, and 25 September contracts. The firm lifts portions of the hedge components as the loan is repaid. The prices are 94.15, 94.05, and 94.00. The firm should be able to lock in interest payments of 8.85 percent, 8.95 percent, and 9 percent. Payments for the first quarter are based on 8.75 percent. This example is summarized in Table 9.11.

TABLE 9.11. Converting a Floating Rate Loan to a Fixed Rate Loan

Date	Cash Market	Futures Market
December	Borrow $100 million for one year at LIBOR plus 3%.	Sell 74 March Eurodollar contracts at 94.15 points.
	Current rate is 8.75%.	Sell 50 June Eurodollar contracts at 94.05 points.
	Expected interest cost is $4.8522 million based on expected yield of 8.8875% which is average of current rate and implied rates from futures.	Sell 25 September Eurodollar contracts at 94.00 points.
March	Interest paid is $2.012 million. Current rate is 9.85%.	Buy 74 March Eurodollar contracts at 93.15 points. (+100 basis points)
		Gain of $185,000.
June	Interest paid is $1.6656 million. Current rate is 10.45%.	Buy 50 June Eurodollar contracts at 92.55 points. (+150 basis points)
		Gain of $187,500.
September	Interest paid is $1.1165 million. Current rate is 7.5%.	Buy 25 September Eurodollar contracts at 95.50 points. (–150 basis points)
		Loss of $93,750.
December	Interest paid is $0.3236 million.	
	Total interest cost is $5.1177 million.	Total gain of $278,750.
	Net interest cost is $4.8390 million.	

The interest rate increases, effecting payments in the second and third quarter. The interest rate then declines in the fourth quarter. Increases in interest rates early in the repayment of the loan will create a problem for the firm. The amount of principal to be repaid on the loan is high at this point. Likewise, the decline in the interest rate later in the loan is comparatively not much of a benefit because most of the principal has been repaid. Thus, to hedge the loan, more nearby contracts are sold relative to deferred contracts. And, as the example shows, the gains and losses in the cash market are offset by gains and losses in the futures market.

The interest rate futures have a "matching" issue that is not always present in physical commodities. But the techniques are not difficult, and the basic idea behind the hedge is still the same: protect against risk.

INTEREST RATE FUNDAMENTALS

Interest rate fundamentals are probably the simplest of all those financial instruments for which futures contracts are traded. This is not to say that interest rates are easy to predict. Rather, the factors that influence the level and changes in interest rates are reasonably well known. While these factors remain difficult to predict, understanding what factors influence interest rates and where this information may be gathered will allow you to understand interest rate behavior.

Interest is the price paid by borrowers to lenders of money and capital. And like any price, the balance between supply and demand determines the equilibrium level, and changes in supply and demand factors will result in changes in the equilibrium price. If the supply of money and capital increases, the price will fall. If the demand for money and capital increases, the price will rise.

There is an anticipated long-run rate of return in the capital market. Borrowers of capital anticipate putting that capital to productive use and paying the lender principal plus interest. Lenders of capital are willing to delay current consumption for higher rates in the future, or are interested in spreading consumption of current wealth out over many periods. Money and capital markets allow both borrowers and lenders to pursue these goals. The long-run rate of return to money and capital depends on the productivity of borrowers and the preferences of lenders between consuming wealth now or some time in the future. This long-run rate of return is present in debt instruments with different maturities.

On a much less abstract level, the factors that determine the level and changes in interest rates come from four broad areas: business conditions, household conditions, fiscal policy, and monetary policy. The institutions in these areas make use of capital, have capital to loan, or influence the intermediaries in capital markets. We will address each of these areas, briefly, in turn. Within this framework, interest rates generally follow the broader business and economic cycle. There is a long-run rate of return paid and received on borrowed and loaned capital. Market-determined interest rates cycle around this rate of return as the economy moves from periods of excess demand to periods of excess supply. A strong and growing economy is usually constrained by production capacity and timing limitations. Expanding production requires money and capital and generally leads to increased interest rates. The central bank may also increase base interest rates to slow the economy and prevent inflation. Economic downturns always follow expansions. At some point, the economy turns from excess demand to excess supply. A weak economy has excess capacity and inventories. These conditions result in decreased demand for money and capital and lead to decreased interest rates. Likewise, the central bank will decrease base interest rates after inflation has been reduced to economic growth. Then the cycle repeats.

Businesses are the largest users of long-term capital, and are the largest providers and users of short-term capital. A strong and growing economy will lead to the increased need for business to borrow money and capital. If all else is constant in this market, then this demand will put upward pressure on interest rates. Specific components to watch within the area of business conditions include construction activity and inventories. Businesses will need to finance construction of plants and other facilities, and will need to finance the operating cost necessary to replenish shrinking inventories. Production, employment, overall price levels, and business profits are good indicators of business conditions. Increasing production, represented in Gross Domestic Product; increasing employment and decreasing unemployment; increasing retail, wholesale, and raw material prices; and increasing business profitability are all signals of a strong economy that must be matched with increasing demand for capital and the possibility of increasing interest rates. Statistics on these various factors are reported in publications released from the U.S. Department of Commerce and the Federal Reserve System.

Consumers are the other side of the economic coin. *Demand by consumers for goods and services is the primary factor leading to a strong economy*. The consumer spends the first dollar and then all businesses compete for a portion of that dollar. Initial spending by consumers leads to spending by retail firms, marketing firms, and initial producers. Strong consumer demand is synonymous with a strong economy. Likewise, weak economies are accompanied by weak demand.

Employment, personal income, and consumer credit conditions are indicators of consumer demand. Increasing employment, increasing personal income, and low consumer debt are precursors to expanded consumer spending. Likewise, increasing unemployment, decreasing real personal income, and high consumer debt will lead to decreases in consumer spending. Again, the U.S. Department of Commerce and Federal Reserve System compile these data and release publications with this information.

Consumers demand a significant portion of short-term credit, but they are the largest provides of long-term capital. This long-term capital is available through retirement savings. Short-term borrowing is usually largest during economic downturns, lessens at the bottom of the business cycle, and increases at the beginning of an economic upturn. Long-term capital is most available after sustained economic growth. The need for borrowing or provision of capital to loan by consumers generally accentuate conditions in the money and capital markets.

Government is one of the largest users of intermediate-term capital and is a significant borrower of short-term capital. Government capital needs are driven by fiscal policy. Fiscal policy is the combination of all spending on military, social programs and public works, net of revenue from tax collections. All else being constant, increases in government spending relative to tax receipts requires the financing of a deficit and will increase the interest rate, and decreases in spending relative to tax receipts will allow deficit reduction and will decrease the interest rate.

During the 1980s, the U.S. federal government borrowed heavily. The impact on interest rates was apparently rather limited because of the strong demand for U.S. government securities overseas. While current deficit and debt conditions appear to be improving, more deficit and debt financing appear to be needed in the future. Currently, the U.S. federal government has significant obligations in terms of future payments to the Social Security and Medicare programs.

We've saved the biggest and best for last. Monetary policy, as implemented by the U.S. Federal Reserve Bank, is one of the most important if not the most important factor influencing interest rates. The main objective of the Federal Reserve Bank is to promote stable economic growth. Economic growth can be increased or decreased by changing the money supply and by controlling the availability of credit. The Federal Reserve Bank (the Fed) carries out monetary policy through three main venues.

The Fed buys and sells U.S. government securities, treasury bonds, notes, and bills, through *open market operations*. When the Fed buys government securities, this increases the supply of money in bank reserves. Banks are able to loan this money. When the Fed sells government securities, this decreases the supply of money in bank reserves. Increases and decreases in the money supply through open market operations increase and decrease loanable funds in the banking system, and should result in decreases and increases in interest rates. Changes in the money supply in this fashion rapidly affect many financial institutions. The Fed used open market operations to increase the money supply following the stock market "corrections" of 1983, 1987, and 1997. The availability of these funds enabled many different financial institutions to meet their short-term obligations without excessive long-term problems associated with their limited cash reserves.

The Fed determines *reserves requirements* that must be kept by banks. Reserve requirements are the percent of deposits that every bank must keep to satisfy regulations and be insured through government sources. Reserve requirements can be changed to induce a change in interest rates. Increasing reserve requirements increases the reserve funds that each bank must hold, which decreases the money supply in capital and money markets and, in turn, will lead to an increase in the interest rate. This monetary tool is used infrequently when compared to open market operations.

There is also a market for these reserve funds or *federal funds*. Banks with reserves in excess of requirements loan these funds to banks with insufficient reserves. The federal funds rate is an important indicator of short-term interest rates and sustained changes in this rate often lead to changes in interest rates for longer-term debt instruments. The CBOT also trades a futures contract for 30-Day Federal Funds.

One method that banks use to increase their cash reserves is to borrow money from the Fed. Banks will borrow money from the Fed to loan in other markets if they have insufficient deposits to meet demands for funds. The interest rate on loans from the Fed is the *discount rate*. The discount rate is very much the interest rate floor in the market for short-term debt. Banks borrow money from the Fed and then mark up this rate to reflect the bank costs before loaning it again. Bank costs include the risk of default by borrowers. The discount rate is set by the Board of Governors of the Federal Reserve System. Lowering the discount rate will lower interest rates, particularly for instruments with short maturities, through the entire banking system. Increasing the discount rate will increase interest rates.

You will be helped if you remember that the interest rate is the price of money. Supply and demand interact to discover price, the same process you went through in earlier chapters dealing with agricultural commodities. Since the interest rate is a price, this helps you see why the interest rate changes when federal agencies release information showing strong demand for money (housing starts are up, for example) or when the Fed changes the supply of money available to the business world.

SUMMARY

Interest rate futures contracts are the most successful of all futures contracts. The volume of trade in futures contracts related to interest rates is the largest of all the categories of futures.

Interest rate futures contracts involve trading the market value of a debt instrument. Debt instruments are financial products that yield the holder either a one-time payment or a stream of interest payments. With short-term debt instruments, the investor buys the instrument at a market value which is less than the par value. At maturity, the investor redeems the instrument for the par value. The difference is the interest payment. With long-term debt instruments, the investor purchases the instrument at a market value. The market value may or may not be less than the par value. This is because the investor also receives a stream of interest payments between purchase and maturity. While the business or government that issues the instrument is responsible for paying the holder the par value at maturity, almost all debt instruments are traded more than once. As the market value, or price, of debt instruments with fixed par values (and possibly streams of interest payments) changes in this secondary market, the implied interest rate fluctuates. *Futures exchanges have developed contracts that allow traders in this market to manage their risk exposure.*

The different exchanges have developed futures contracts for different products. The CME focuses on short-term debt instruments; these include treasury bills and Eurodollar deposits. The interest rate underlying Eurodollar contracts is the London Interbank Offered Rate (LIBOR). The CBOT trades futures on long-term debt instruments—treasury bonds with 15 years to maturity and treasury notes with 10, 5, and 2 years to maturity.

Trading in these contracts is not a trivial matter. It requires some familiarity with bond pricing. Short-term instruments are the easiest to understand. The instrument is traded at some discount to the par value. The price of the futures contract is this discount. Thus, if the discount increases, the implied interest rate decreases. A trader that wants to sell the interest rate because it is expected to decline needs to buy the futures contract. Long-term instruments are more difficult. The instrument is also traded at a discount or premium to the par value. Thus, as with short-term instruments, the trader that wants to sell the interest needs to buy the futures contract. The difficulty is with the units. Long-term instruments are quoted in points and 32nds, and it is difficult to quickly convert this into an annual percent interest rate.

Hedging debt instruments are similar but also more difficult than other financial instruments. As with hedging stock portfolios, the trader needs to be careful to match the cash position with the proper futures position. There are a number of ways to do this. The more accurate methods entail increasing complexity. The procedures discussed in this chapter include the par value, market value, basis point, and duration approaches. The first two are the simplest and the least accurate. The third approach attempts to match changes in the value of the cash position with a futures position that offsets those changes. The fourth approach is similar to the third, but the approximation is better because the method accounts for the different maturities between the cash instrument and the instrument underlying the futures contract.

We reiterate, *a crucial part of hedging interest rates involves matching the cash and futures positions so that for changes in market interest rates, changes in the value of the cash position are offset by changes in the value of the futures position.* The hedger must account for differences in market value, changes in the market value with changes in different interest rates, and differences in the maturity of the cash and futures instruments.

Basis error is also present when hedging interest rates. Money and capital markets are as well integrated as exchange rate and stock index cash and futures markets, but interest rate markets must price risks not present in these other markets. Because of this, the premiums and discounts across different interest rates can change in unpredictable ways.

As with prices for other commodities, the price of money and capital is determined by the supply of and demand for money and capital. Interest rates, the price of capital, will increase if the demand for capital is high relative to its availability. Conversely, interest rates will fall if capital is readily available given the level of demand. In addition to market forces, the Federal Reserve Bank is an institution with instruments at its disposal that can influence the supply and demand for capital. The overall goal of the Federal Reserve Bank is stable economic growth, and to achieve this the bank will change the money supply and reserve requirements. Capital markets react accordingly.

With the continued industrialization, integration, and progress in global capital markets, the volume of trade in money and capital markets will continue to grow. *The need for futures markets with which to manage risk will likely grow as well.*

KEY POINTS

- *Interest rate futures contracts* are the most successful futures contract. Not all of the interest rate futures contracts that have been introduced have been successful, but there have been *tremendous successes*.
- Interest rate futures contracts involve *trading the market value of a debt instrument*.
- *Debt instruments are financial products* that yield the investor interest payments. *Short-term debt instruments* are purchased at market value, redeemed at par value, and the difference is the interest payment. *Long-term debt instruments* are purchased at a market value, redeemed for par, and receive coupon payments prior to maturity.
- Debt instruments are *traded in secondary markets*—not just between the issuer and first buyer. As the market value of debt instruments with fixed par value and interest returns changes, *the implied interest rate fluctuates*. Futures contracts allow traders to *manage this risk*.
- The CME trades futures on *short-term debt instruments*: treasury bills and Eurodollars. The CBOT trades futures on treasury bonds with *15 years to maturity*, and treasury notes with *10, 5, and 2 years to maturity*.
- Short-term instruments are traded at *discount to par value*. This is the *futures price*. If the discount increases, the *interest rate decreases*. Long-term instruments are also traded at a discount or premium to the par value. However, the units are points and 32nds of par. If the discount increases, the *interest rate decreases*.
- Hedging debt instruments is similar *but more difficult* than other financial instruments. The cash position needs to be *correctly matched with the proper futures position*. Par value, market value, basis point, and duration approaches were discussed. The hedger needs to account for *differences in market value*, changes in that value with *changes in different interest rates*, and *differences in the maturity of the cash and futures instruments*.
- *Basis error* is present when hedging interest rates. The premiums and discounts across different lending and borrowing rates *can change in unpredictable ways*. These changes reflect risks *not necessarily captured* in the price of the debt instrument traded in the futures contract.
- Interest rates are determined like any other *price*. Interest rates, *the price of capital*, will increase if the demand for capital is high relative to supply. And *interest rates will fall* if the demand for capital is low relative to supply. However, the Federal Reserve Bank plays a large role in *determining the interest rate*. The Federal Reserve Bank *controls money supply*, and capital markets react quickly to changes in money supply.

USEFUL REFERENCES

Steven C. Blank, Colin A. Carter, and Brian H. Schmiesing, *Futures and Options Markets: Trading in Financials and Commodities*, Prentice-Hall, Englewood Cliffs, NJ, 1991. Good intermediate text with coverage of financial futures. Good coverage of interest rate futures.

Robert T. Daigler, *Managing Risk with Financial Futures: Pricing, Hedging and Arbitrage*, Probus Publishing Company, Chicago, IL, 1993. A solid and comprehensive text with many examples. The author is presenting many research results. Very useful for interest rate, stock index, and exchange rate examples.

Robert E. Fink and Robert B. Feduniak, *Futures Trading: Concepts and Examples,* New York Institute of Finance, New York, 1988. A good basic introduction to financial futures.

John C. Hull, *Introduction to Futures and Options Markets*, 3rd ed., Prentice Hall, Upper Saddle River, NJ, 1998. An introduction to the mathematics of financial futures.

Robert W. Kolb, *The Financial Futures Primer*, Blackwell Publishers, Malden, MA,1997. An excellent introduction to financial futures. Many trading examples are presented. Some of the important subtleties of financial futures are also discussed.

Paul Krugman, *The Age of Diminished Expectations*, 3rd ed., The MIT Press, Cambridge, MA, 1994. An excellent book that discusses current topics of issue for the U.S. and world economy. The text is very useful for understanding the reactions of exchange rate, stock index, and interest rate markets.

Raymond M. Leuthold, Joan C. Junkus, and Jean E. Cordier, *The Theory and Practice of Futures Markets*, Lexington Books, Lexington, MA, 1989. A more advanced treatment of futures and hedging. Extensive focus on interest rate, stock index, and exchange rate futures.

Charles W. Smithson and Clifford W. Smith, Jr., with D. Sykes Wilford, *Managing Financial Risk: A Guide to Derivative Products, Financial Engineering, and Value Maximization*, Irwin Professional Publishing, Chicago, IL, 1995. A fairly simple treatment of more advanced financial management topics. The focus is mainly on options and derivative pricing with interest rate, stock index, and exchange rate examples.

CHAPTER 10

INDEX FUTURES CONTRACTS

INTRODUCTION

Following the successful introduction of exchange rate futures contracts, numerous futures exchanges began to develop and introduce other futures contracts for financial products. One of the more successful groups of these contracts are futures contracts based on stock indexes. The first stock index contract was traded at the Kansas City Board of Trade (KCBOT) in 1982 and is based on the Value Line Index. However, the volume of trade in this contract remains small. The most well-known and well-traded contract is that based on the Standard and Poor's 500 Index (S&P 500). This contract is traded at the Chicago Mercantile Exchange (CME). Contracts are also traded on the Japanese Nikkei 225 Index and the NASDAQ 100 Index at the CME.

Interestingly, there was no futures contract on the most well-known stock index, the Dow Jones Industrial Average (DJIA), until recently. Dow Jones blocked the introduction of a futures contract based on this index by the Chicago Board of Trade (CBOT) because of a copyright disagreement. In response, the American Stock Exchange created the Major Market Index (MMI) and licenced use of the index to the CBOT. Dow Jones and the CBOT recently overcame their differences, and the exchange introduced a futures contract based on the DJIA in late 1997. In a related development, the MMI has been licensed to the CME. There is also a futures contract of the second most well-known stock index, the New York Stock Exchange Composite Index. But this contract is traded at the New York Futures Exchange (NYFE) and the volume also is small. Futures contracts on the stock indexes reported for most western European countries and many Far East countries are traded most often on the native futures exchanges. However, no stock index futures contract is as successful at the CME's S&P 500.

In addition to being based on stock indexes, futures contracts are based on other types of indexes. Examples include the Goldman-Saches Commodity Index (GSCI), the Municipal Bond Index, and the U.S. Dollar Index. The GSCI measures inflation in a basket of world traded commodities, the Muni Bond Index captures variation in interest rates in the municipal bond market, and the U.S. Dollar Index measures the strength of the dollar relative to a basket of world currencies. These indexes are not as popular as the stock indexes, but the mechanics are similar and many other index contracts will likely be introduced in the future. You are encouraged to dig into the principles of index contracts. Index based contracts have very strong advantages over traditional futures contracts in that there is no delivery. The contract is typically cash settled, and the derivation, modification, and reporting of index values are the responsibility of an organization other than the futures exchange.

There promises to be many changes in the futures contracts offered on different indexes. This is one area on which the exchanges and markets are unsettled. New contracts are being developed and introduced, and older once very favored contracts are losing trading volume. The remainder of this chapter will work examples with stock index futures contracts. The specifics of the contracts are not as important as the underlying concepts. Basically, hedgers using index futures contracts have or will have assets in a cash position the value of which they are interested in protecting. This problem is identical to that of a trader with assets in agricultural commodities, other physical commodities such as metals, or currencies. In this case, the assets are stocks, and the futures contract is not priced in the same units as is the cash asset.

DEFINITION OF VARIOUS COMMON INDEXES

Understanding the definition of a specific stock index is critical for understanding how to use the index in a hedging program. It is also important for just being informed about the developed economies and financial markets. Many different stock indexes are reported in the financial press and broadcast news. However, few people are familiar with the actual construction of the indexes.

First, a stock index is a weighted average of stock prices. In all indexes, a sample of stock prices is summed after each is weighted and then the total is weighted by a divisor.

$$Index = \frac{\sum_{i=1}^{Sample} Price_i \times Weight_i}{Divisor}$$

The index is a simple average if each price is given equal weight of one over the sample size and the divisor is set equal to one. The simple average can be converted to an index of base 100 if the divisor is the resulting simple average divided by 100. This time period then becomes the base period. If prices change, the index changes, and the index measures change relative to the base period. For example, suppose the weighted average share price in 1997 is $35.00. We can create an index of 100 for that year by using 35/100 as the divisor.

$$Index_{1997} = (\$35/35) \times 100$$

Then, if in 1998 the weighted average share price is $38.15, the index becomes 109 ($\$38.15/35 \times 100$) and reflects the 9 percent appreciation in stock prices.

The factors that distinguish stock indexes from each other are the specific stocks in the sample and the choice of the weighting. The choice of stocks in an index is usually determined by location such as the country of origin or the exchange on which all of the stocks are traded. For example, the largest stock exchange in the world is the New

York Stock Exchange (NYSE), and the NYSE Composite Index summarizes the price of all stocks traded on that exchange. The index can then be further refined by grouping stocks of like industries, such as industrial, transportation, utility, or high-technology companies; by grouping stocks by the size of the underlying companies in the index, such as medium sized (midcaps) or small (small-caps) companies; or by grouping stocks by other company characteristics, such as growth or income-generating companies.

The second factor that determines what is measured by the index is the weighting procedure. There are in general two types of stock indexes, *value-weighted indexes and price-weighted indexes*. In a value-weighted index, the price of the stock affects the index based on the total amount of the stock issued. The more stock that a company has outstanding, the greater the impact the price of that stock has on the index. Changes in value-weighted indexes are largely due to price changes in stocks of large companies.

In a price-weighted index, the price of the stock affects the index based on the magnitude of that stock's price. The higher-priced the stock, the greater the impact the price of that stock has on the index. Changes in price-weighted indexes are largely due to price changes in stocks of companies with high prices.

Value-weighted indexes are most useful for traders with large stock portfolios. By definition, a large percentage of investor portfolios will hold stocks of companies that have issued a large amount of stock. Price-weighted indexes are useful to traders of more aggressive portfolios. High-performing stocks are always in demand and will usually be high priced. The S&P 500 and NYSE Composite indexes are value weighted. The DJIA, MMI, and Nikkei are price-weighted indexes.

The S&P 500 is the most widely used index in the U.S. finance industry. The index is constructed from the stock prices of 500 firms in a variety of industries. These stocks are some of the largest and commonly traded, and comprise about 75 percent of the volume of stocks traded on the NYSE. This is the reason for the popularity of the index. The contribution of each stock in the index is proportional to its value. The weight used for each stock price is the number of outstanding shares. Stocks with larger numbers of shares, and higher prices contribute more to the value of the index. The index is as follows:

$$S\&P\ 500\ Index_t = \frac{\sum_{i=1}^{500} Price_{it} \times Number_{it}}{Original\ Valuation} \times 10.$$

The Original Valuation was the average value of the price and share number combination of 500 stocks during the years 1941–43. The S&P 500 Index is based on that time period and would equal one without the additional multiplication by ten. The S&P 500 Index was close to 1,000 at the end of 1997. This implies that stock prices have appreciated 100-fold (1,000/10) since the early 1940s.

Every index must be adjusted from time to time. New companies are created, old companies go out of business, some companies merge with others, and single companies spin off divisions into separate companies. All of these may change the value of the index. Also, company actions such as stock splits, dividends, and the substitution of one type of stock for another may change the value of the index. The financial organizations that create and report various indexes modify the divisor following these changes to prevent the index from changing due to changes in the composition of stocks. The point of having the index is to communicate a composite of stock prices, not to have an index that is affected by the changing composition of firms in the index or by stock changes by individual firms. The method for doing this can be easily shown through an example.

Suppose Standard and Poors wants to remove a stock from the index that is trading at $25 per share and include a stock that is trading at $100 per share. An equal number of shares of both companies are being traded. Suppose the index is currently at 1,000 points. Removing a low-priced stock and substituting it with the same number of shares of a high-priced stock will increase the index. Suppose the new value would be 1,005. But the increase has nothing to do with changes in the market-value of stock prices, only the prices that S&P are reporting. To prevent this, S&P adjusts the divisor in the index. Suppose the divisor before the change was 0.955. The divisor is reduced the amount necessary so that the index is 1,000 before and after the change. The new divisor is derived from

$$\textit{Old Divisor} \times \textit{Old Index} \,/\, \textit{Index after Change} = \textit{New Divisor}$$

or for the example

$$0.955 \times 1000/1005 = 0.955 \times 0.995 = 0.950.$$

The new divisor is then used to calculate the index using current prices. The calculation will yield 1,000 based on current prices. The new divisor will also be used with prices observed in the future to calculate future values of the index. This will be done until the next change in the stocks in the index.

The New York Stock Exchange Composite Index is very similar to the S&P 500 Index. The NYSE Composite Index makes use of all of the stocks traded on the exchange. This is almost 1,800 stocks in late 1997. The index is also value weighted. We pose a question to you at this point. If the S&P 500 and the NYSE Composite Index are both value weighted and the S&P 500 are usually the 500 largest stocks on the NYSE, how correlated will percent changes in the two indexes be? Before we answer that, the following equation is used to construct the index:

$$\textit{NYSE Index}_t = \frac{\sum_{i=1}^{1800} \textit{Price}_{it} \times \textit{Number}_{it}}{\textit{Original Valuation}} \times 50.$$

The original valuation was based on all share prices as of 31 December 1965. The NYSE Composite Index was close to 500 at the end of 1997, so stock prices have appreciated 10-fold or 900 percent since December 31, 1965.

The answer: Percent changes in the NYSE Composite Index and the S&P 500 Index are very highly correlated. The correlation statistic is above 0.96. There are 1,200 more stocks in the NYSE Composite Index that are not in the S&P 500 Index, but these smaller stocks make up less than one-fifth of the value of stocks in the NYSE Composite Index. In fact, the value of stocks from less than 100 of the largest firms comprise more than 50 percent of the value of stocks traded on the NYSE. Any two value-weighted indexes that include the majority of the largest stocks will move together very closely.

The Dow Jones Industrial Average, Major Market Index, and Nikkei 225 Index are, as noted, price-weighted indexes. In a price-weighted index, the weight for each individual stock in the general formula above is set equal to one.

$$Index_t = \frac{\sum_{i=1}^{Sample} Price_{it}}{Divisor}$$

This implies that all stock prices are given equal weight in the index. The formulation makes the index easy to calculate but leads to potentially strange changes in the index. For example, a stock price that increases in price from $5 to $10 per share will have the same impact as an increase from $100 to $105 per share. Further, the $5 change in the price of a small company with a small number of stocks will have the same impact as that of a $5 change in the price of a large company with a large number of stocks.

As with the value-weighted index, the divisor is used to adjust the sum of the prices to account for stock splits and dividends. For example, suppose a company whose stock is trading at $200 per share offers a split to replace each share with four new shares. After the split, the stock should be valued at $50 per share. If this stock is in a price-weighted index, the index value will drop. The financial organization that calculates and reports the index will adjust the divisor upward so that the value of the index will be the same before and after the stock split. The same formula reported earlier for calculating a new divisor is used.

Price-weighted indexes such as the DJIA, MMI, and Nikkei are reported and used much like the value-weighted indexes such as the S&P 500 and NYSE Composite. However, you should remember the differences. The S&P 500 and the NYSE Composite Indexes are both broad-based indexes designed to capture overall movements in the stock market. Being value weighted, the two indexes mainly capture movements in the price of well-capitalized firms—those firms with large amounts of outstanding stocks. The DJIA is designed to capture price movements of stocks for firms categorized as industrial. MMI is designed to follow the DJIA, being constructed when Dow Jones was not interested in working with the CBOT in trading a futures contract based on the index. The MMI is comprised of 20 stocks whereas the DJIA is 30 stocks, and most of the stocks in the MMI are also in the DJIA. The Nikkei obviously captures conditions in the Japanese stock market. Although the Nikkei is not value-weighted, the index is comprised of 225 of the largest publicly traded companies in Japan. Stocks of smaller companies and stocks of specific industry segments tend to be more volatile than stocks of large well-capitalized companies and broader market segments. Therefore, indexes based on smaller companies and narrow market segments tend to be more volatile than more broad-based indexes. It is important to understand this idea when speculating or hedging with index futures. The next two sections will elaborate on this point.

It is easy to see why index futures are so widely watched and used. If you are holding a portfolio of stocks that at least roughly parallels the stocks in the Dow Jones Industrial Average, you can protect the value of that portfolio by going short in the futures index. Alternatively, you (or the manager of your retirement account) could spend some of the earnings to buy puts, perhaps even out-of-the money puts, to protect against a break in the stock market. Years ago, the only way to do that was to actually sell the stocks.

INDEX TRADING EXAMPLES

Index contracts are perhaps the easiest contracts for which to construct trading examples. In the previous section, we discussed what and how the different indexes were constructed. Pricing of these contracts through the futures market is very simple. The value of a futures contract is the index multiplied by a specified dollar amount. For example, the S&P 500 Index futures contract is $250 multiplied by the index value. The Nikkei 225 Index futures contract is $5 multiplied by the index. The net return to the person trading the index is the value of the index change between the sell and the buy multiplied by the index price weight or

$$(\text{Sell Index} - \text{Buy Index}) \times \text{Index Price Weight} = \text{Net Return.}$$

Table 10.1 presents an example of futures prices for different indexes as reported in the *Wall Street Journal* for 2 October 1997. And Table 10.2 presents a broader list of indexes traded at the exchanges in Chicago and New York. Let's work a couple of trading examples for clarity.

Suppose the current December S&P 500 contract is trading at 969.10. The trader thinks the contract is undervalued. He thinks that the actual index in December will be higher. The trader submits a buy order and takes a long position at 969.10. Suppose that after one month, the December contract is trading at 983.60. Believing that he has held the position long enough, he offsets the long position by selling the contract back. The net return for this S&P 500 example will be

$$(983.6 - 969.1) \times \$250 = \$3{,}625.$$

The following is a second example using the index that summarizes the Japanese stock market. Suppose the December Nikkei 225 contract is currently trading at 17570. The trader thinks the contract is overvalued and that the actual index will be lower in December. The trader submits a sell order and takes a short position at 17570. Suppose, after two weeks, the December contract is trading at 17050. The trader believes the risk of holding the contract longer is not worth the potential reward. The short position is bought back. The net return to the trader is the change in the index multiplied by the index price weight which for the Nikkei 225 is

$$(17570 - 17050) \times \$5 = \$2{,}600.$$

Index trading is very simple in terms of the mechanics of simple trades. You use direct long or short positions. *The subtlety in index trading is in understanding exactly what combination of stocks each index measures, how these stocks move together, and how the different indexes move together.* This is important for two reasons. First, for hedging, it is important to understand how changes in the value of a futures contract compare with changes in the value of different bundles of stocks. This is critical for determining how much to hedge and whether or not hedging will be effective. This is a lengthy topic and is discussed in the next section. Second, for speculators, the size of different index contracts are rather large and movements in the index can be substantial in terms of dollars given the size of the contract. Therefore, trading different spreads is an attractive alternative to a single long or short position. The remainder of this section presents some spread trade examples.

TABLE 10.1
Index Futures Contract Quotes on 2 October 1997

	Open	High	Low	Settle	Change
S&P 500 INDEX (CME) $500 × index					
Dec	964.75	969.70	961.10	969.10	+5.70
Mr98	973.00	979.50	971.80	979.45	+5.75
June	987.40	990.50	983.05	989.50	+5.75
Sept	—	—	—	1000.80	+5.75
Est vol 65,362; vol Wed 65,217; open int 188,136, -158					
Indx prelim High 960.46; Low 952.94; Close 960.46, +5.05					
NIKKEI 225 STOCK AVERAGE (CME) $5 × index					
Dec	17610	17610	17480	17570	–290
Est vol 601; vol Wed 857; open int 15,640, +178					
Indx prelim High 17875.52; Low 17415.18; Close 17455.04, +387.12					
NASDAQ 100 (CME) $100 × index					
Dec	1119.50	1127.00	1114.00	1126.50	+11.00
Mr98	1135.00	1142.25	1133.90	1142.25	+11.00
Est vol 1,863; vol Wed 3,111; open int 5,074, +179					
Indx prelim High 1113.51; Low 1102.05; Close 1112.89, +10.84					
GSCI (CME) $250 × nearby index					
Oct	200.60	204.00	199.50	203.70	+2.30
Nov	201.80	205.00	201.10	205.00	+2.50
Est vol 1,955; vol Wed 3,034; open int 23,364, +150					
Indx prelim High 204.24; Low 199.42; Close 203.91, +2.48					
U.S. DOLLAR INDEX (FINEX) 1,000 × USDX					
Dec	97.54	97.64	97.11	97.42	+.03
Mr98	—	—	—	97.28	+.03
Est vol 1,800; vol Wed 1,084; open int 9,709, +348					
Indx prelim High 97.80; Low 97.31; Close 97.64, +0.04					

Source: Wall Street Journal.

TABLE 10.2
Index Futures Contract Specifications

Index	Expiration Months	Size	Exchange
S&P 500 Index	Mar, June, Sep, Dec	$250 × index	CME
S&P 400 Midcap Index	Mar, June, Sep, Dec	$500 × index	CME
Major Market Index	Mar, June, Sep, Dec	$500 × index	CME
NASDAQ 100	Mar, June, Sep, Dec	$100 × index	CME
Dow Jones Industrial Avg	Mar, June, Sep, Dec	$10 × index	CBOT
Value Line Index	Mar, June, Sep, Dec	$500 × index	KCBOT
NYSE Composite	Mar, June, Sep, Dec	$500 × index	NYFE

Through 1997 much of the substantial gains in stock market indexes were realized due to increases in prices of large, well-capitalized firms. Small and medium-sized firms did not see the same gains. Suppose that early in the fall a speculator thinks that large stocks are overvalued relative to medium-sized firms. Therefore, the price of the mid-cap firms should gain relative to the large firms. *The trader buys three S&P Midcap 400 Index contracts and sells two S&P 500 Index contracts.* All three contracts expire the following March. The speculator trades three S&P Midcap 400 contracts for every two S&P 500 contracts because the first is roughly two-thirds the size of the second. The Midcap 400 index futures contract is valued at $500 multiplied by the index and the S&P 500 index futures contract is $250 multiplied by the index, but the Mid-cap 400 index is approximately one-third the number of the S&P index.

Suppose when the trades were initiated the S&P Midcap 400 futures contract was trading at 337.75 and the S&P 500 contract was at 979.45. Between the fall and spring of next year, the S&P Midcap 400 contract increases 3 percent to 347.88 and the S&P 500 contract increases 2.5 percent to 1003.94. The trader's intuition was correct, prices of midcap firms increased relative to large firms. Let's examine the trading returns. For the S&P Midcap 400, the return is

$$(347.88 - 337.75) \times \$500/\text{contract} \times 3 \text{ contracts} = \$15{,}195$$

and for the S&P 500 contract, the return is

$$(979.45 - 1003.94) \times \$250/\text{contract} \times 2 \text{ contracts} = -\$12{,}245$$

so the speculative strategy returns ($15,195 – $12,245) = $2,950. The two individual trades resulted in large returns and losses, but the combined strategy or spread had less risk.

For a second spread example, suppose a speculator examines the index prices for the S&P 500 contract associated with different expiration dates. The March contract is trading at 979.45 and the June contract is trading at 989.80. These index values imply that the market consensus is that the actual S&P 500 Index will appreciate slightly less than 1.1 percent between March and June. The speculator believes that the appreciation will be greater over this three-month period based on economic and business conditions that appear to be emerging and that the market consensus will change to reflect this before the March contract expires. The speculator sells the March contract and buys the June. Two months later, the March contract is trading at 924.45 and the June contract is at 945.26 when the speculator offsets the trades. The return to the March contract is

$$(979.45 - 924.45) \times \$250 = \$13{,}750$$

and the return to the June contract is

$$(945.26 - 989.80) \times \$250 = -\$11{,}135$$

so the net return is $2,615. Notice the market decreased substantially, but the differential between March and June increased. The change in the differential is what the speculator anticipated, and it does not matter that the overall market level also changed.

INDEX HEDGING EXAMPLES

The use of stock index futures contracts for hedging stock portfolios is not always as simple as the use of futures contracts for hedging grains, livestock, and other physical commodities. Rarely does a stock portfolio manager control a group of stocks that matches exactly with any reported index. This creates a type of basis risk, of course, not really different from that which users of grain sorghum face when they use corn to hedge sorghum or other types of feedgrain purchases. It is important that the hedger understand how changes in the value of different stock index futures contracts compare with changes in the value of stocks in the portfolio that the hedger manages. This is important for determining how many stock index futures contracts to trade and anticipating whether or not hedging will be effective in managing risk through protecting the value of the portfolio. This is the topic of this section, and we start with two examples to illustrate the problem.

Suppose there are two portfolio managers that work for a mutual fund company. Each person manages a portfolio of different types of stocks. One portfolio is comprised of high-technology computer-related firms and the other portfolio is comprised of utility and related companies. Both portfolios are valued at $50 million. In early October 1997, both managers are worried about the stock market being overvalued, stock prices falling, and the value of their portfolios decreasing. Both managers hedge their portfolios by selling S&P 500 Index futures. The December index is trading at 980. They each sell 204 contracts. This number of contracts protects each dollar of the cash position with a dollar in the futures position (980 × $250 per contract). Between early and late October, the December S&P 500 Index decreases 5 percent to 931 before the managers buy their contracts back and lift their hedges. However, the cash market position for the two portfolio managers is rather different. The high-technology portfolio is worth $46.875 million and the utility portfolio is worth $48.5 million. The futures position has made each manager $2.499 million. However, the high-technology portfolio manager has lost $3.125 million in value, while the utility portfolio manager has lost $1.5 million. The net position from the hedge of the high-technology portfolio manager is –$0.626 million and the net position of the utility portfolio manager is +$0.999 million. This hedge example is presented in Table 10.3.

The high-technology portfolio was underhedged and the utility portfolio was overhedged. The overall market index decreased 5 percent. However, the value of the high-technology portfolio decreased 6.25 percent, while the value of the utility portfolio decreased 3 percent. *All stocks and all portfolios do not mirror changes in the overall market index. Some stock values change more than the market index and some change less.* How do we construct a measure so that the proper level of hedging is followed?

The tool used is the same as that used to measure the variability in returns of different portfolios. Variability in returns is measured with a tool denoted as Beta. Beta (β) can be measured with the following linear model

$$IPPoR_t = \alpha + \beta\, MRoR_t + e_t$$

where $IPRoR_t$ denotes the individual portfolio rate of return for time t, $MRoR_t$ denotes the market rate of return for the same time period t, α and β are parameters to be estimated with the data, and e_t is the random unexplainable error. Data are selected to capture the desired time period for the rate of return. For example, we may be inter-

TABLE 10.3 Portfolio Hedging Example Where Managers Cover $1 in the Cash Market with $1 in the Futures Market

Date	Cash Market	Futures Market
	High-Technology Portfolio	
October	Stock portfolio worth $50 million.	Sell 204 December futures contracts at 980 covering $49.98 million.
December	Stock portfolio worth $46.875 million.	Buy 204 December futures contracts at 931.
	–$3.125 million loss.	+$2.499 million gain.
	Portfolio and cash worth $49.374 million.	
	Utility Portfolio	
October	Stock portfolio worth $50 million.	Sell 204 December futures contracts at 980 covering $49.98 million.
December	Stock portfolio worth $48.5 million.	Buy 102 December futures contracts at 931.
	–$1.5 million loss.	+$2.499 million gain.
	Portfolio and cash worth $50.999 million.	

ested in daily, monthly, or annual rates of return. This model is then estimated with linear regression. All commercial spreadsheet software will perform this exercise with the proper data.

The parameter β is the Beta of the portfolio. Beta measures the individual portfolio rate of return relative to the market rate of return. *If Beta equals 1 then the individual portfolio moves one-for-one with the market.* A 1 percent increase or decrease in the market rate of return will be matched by a 1 percent increase or decrease in the rate of return for the individual portfolio. In our earlier example, neither of the two portfolios had a Beta of 1. If Beta is less than 1, the rate of return for the individual portfolio is less variable than that of the market. For example, if the market return is +5 percent or –5 percent, the gain for the individual portfolio will be less than +5 percent or the loss will be less than –5 percent. If Beta is more than 1, the rate of return for the individual portfolio is greater than that of the market. For example, if the market return is +10 percent or –10 percent, the gain for the individual portfolio will be greater than +10 percent or the loss will be greater than –10 percent.

Different portfolios of stocks have different Betas. Some groups of stocks offer returns that are lower and more stable than the overall market—low-risk or low-Beta stocks, and some groups of stocks offer returns that are higher and more variable than the overall market—high-risk or high-Beta stocks. This is the problem experienced in earlier the hedge example. High-technology stocks are high-Beta stocks and utilities are low-Beta stocks. The high-technology portfolio decreased more than the S&P 500 future contract, and the utility portfolio decreased less than the futures contract. However, a second problem emerges in that we should not use the Beta from the model above to figure out how much to hedge. The Beta above measures the parallel movements between two *cash market* series. The individual portfolio cash value was compared to the overall spot or cash market index. Futures markets for different stock indexes are more variable than underlying spot market indexes. So, the portfolio manager needs to estimate Beta using futures prices of the market index. This model is

$$IPPoR_t = \alpha + \beta_{fm}\, FMRoR_t + e_t$$

where $FMRoR_t$ denotes the rate of return of the futures index for time period t, and α and β_{fm} are new parameters to be estimated with the data. The data must be more carefully selected. The time period should capture the length of the hedge. For example, the manager may be interested in hedging the portfolio for one day, one week, or one month. Again, this model is then estimated with linear regression.

Returning to our example, suppose the high-technology portfolio manager estimates the futures market Beta to be 1.25, and the utilities portfolio manager estimates the futures market Beta to be 0.6. These Beta estimates imply that the high-technology portfolio is 25 percent more variable than the futures market index and that the variability of the utility portfolio is only 60 percent of that of the futures index. The portfolio manager calculates the number of futures contracts necessary to cover the dollar value of the cash portfolio and multiplies that number by the futures market Beta or

$$\text{Number of Contracts} = (\text{Cash Portfolio Value / Futures Value Per Contract}) \times \beta_{fm}.$$

This adjusts the hedge position for the variability of the underlying cash portfolio. If the underlying portfolio is more variable than the futures index—Beta is greater than 1—the hedger need to trade more dollars of futures contracts than the value of the cash position. The high-technology portfolio manager will trade

$$255 \text{ contracts} = (\$50 \text{ million} / (980 \times \$250 \text{ per contract})) \times 1.25.$$

If the underlying portfolio is less variable than the futures index—Beta is less than 1—the hedger needs to trade fewer dollars of futures contracts than the value of the cash position. The utility portfolio manager will trade

$$122 \text{ contracts} = (\$50 \text{ million} / (980 \times \$250 \text{ per contract})) \times 0.6.$$

We can double-check these inferences with numbers from the example. Remember, the stock index futures contract decreases from 980 to 931. If the high-technology manager trades 255 contracts, this will generate a \$3.124 million gain in futures to offset the \$3.125 million loss in the value of the portfolio. Likewise, if the utility portfolio manager trades 122 contracts, this will generate a \$1.4945 million gain in futures to offset the \$1.5 million loss in the value of the portfolio. The hedge example is reworked in Table 10.4.

Now, the portfolios are hedged such that the changes in the value of groups of stocks more closely match the change in the value of the futures position. The futures gains and the cash market losses do not offset each other perfectly because each futures contract is worth approximately one-quarter million dollars of stocks. Thus, the gains and losses can easily be incorrect by this figure.

There is a second reason that the hedge examples may not work as perfectly in practice. *This is because of basis error and basis risk*. However, basis error and basis

TABLE 10.4 Portfolio Hedging Example Where Managers Cover a 1% Change in the Cash Position Value with Sufficient Contracts to Net a 1% Change in the Futures Position Value

Date	Cash Market	Futures Market
	High-Technology Portfolio	
October	Stock portfolio worth \$50 million.	Sell 255 December futures contracts at 980 covering \$62.495 million.
December	Stock portfolio worth \$46.875 million.	Buy 255 December futures contracts at 931.
	–\$3.125 million loss.	+\$3.136 million gain.
	Portfolio and cash worth \$49.999 million.	
	Utility Portfolio	
October	Stock portfolio worth \$50 million.	Sell 122 December futures contracts at 980 covering \$29.98 million.
December	Stock portfolio worth \$48.5 million.	Buy 122 December futures contracts at 931.
	–\$1.5 million loss.	+\$1.4945 million gain.
	Portfolio and cash worth \$49.9945 million.	

risk are measured differently for stock and other indexes. We used the tool Beta to measure the proper balance between cash and futures positions. For example, the high-technology portfolio is on average 25 percent more variable than the futures market index (average Beta of 1.25), and the utility portfolio is on average 40 percent less variable than the futures index (average Beta of 0.6). Thus, to protect their respective positions the two portfolio managers will hedge 125 percent and 60 percent of their cash positions. But the key thing to recognize is that the Beta measures *average variability*. There is no guarantee that the changes in the portfolio relative to the futures index will match the estimated Beta for *any one hedge*. Returning to our example, suppose the high-technology portfolio actually decreases 30 percent more than the index (actual Beta of 1.30), and the utility portfolio decreases 45 percent less than the index (actual Beta of 0.55). This example is presented in Table 10.5. This is the equivalent of basis risk for the index trader.

In addition to calculating the futures market Beta using the S&P 500 Index futures contract, the portfolio managers would be wise to compare the rates of return of their individual portfolios to other futures market indexes. For example, the manager may wish to compare returns to the individual portfolio to returns from the Value Line Index, the Nikkei, or the index from a European stock market. A model with a more precise estimate of Beta, one with a lower standard error, suggests that contract is a more effective hedging vehicle and that basis risk is lower. This would likely be the case for a portfolio that was comprised of small company stocks or stocks of Far East or European companies.

It may also make sense for the portfolio manager to make joint comparisons of the individual portfolio to several futures market indexes. Suppose the portfolio in question

TABLE 10.5
Portfolio Hedging Example with Basis Error and Where Managers Cover a 1% Change in the Cash Position Value with Sufficient Contracts to Net a 1% Change in the Futures Position Value

Date	Cash Market	Futures Market
	High-Technology Portfolio	
October	Stock portfolio worth $50 million.	Sell 255 December futures contracts at 980 covering $62.475 million.
December	Stock portfolio worth $46.75 million.	Buy 255 December futures contracts at 931.
	–$3.25 million loss.	+$3.124 million gain.
	Portfolio and cash worth $49.847 million.	
	Utility Portfolio	
October	Stock portfolio worth $50 million.	Sell 122 December futures contracts at 980 covering $29.98 million.
December	Stock portfolio worth $48.625 million.	Buy 122 December futures contracts at 931.
	–$1.375 million loss.	+$1.4945 million gain.
	Portfolio and cash worth $50.1195 million.	

is a combination of stocks from domestic and foreign companies. Then the most effective method of hedging may be to take positions in several futures contracts. The manager would examine the model

$$IPRoR_t = \alpha + \beta_{dfm}\, DFMRoR_t + \beta_{ffm}\, FFMRoR_t + e_t$$

where $DFMRoR_t$ denotes the rate of return of the domestic futures index for time period t, $FFMRoR_t$ denotes the rate of return of the foreign futures index for time period t, and α, β_{dfm}, and β_{ffm} are parameters to be estimated with the data. β_{dfm} and β_{ffm} identify the proportion of domestic and foreign futures contracts to use relative to the size of the cash position.

The concepts and methods discussed in this section so far have focused on general stock index trading and hedging, and at first they may appear to be out of context for producers, marketers, and users of agricultural products. However, these tools can be useful to a manager of a publicly traded agribusiness company. Further, the concepts and tools are useful to any business involved with hedging, including the farm firm.

Stock index futures can be used to hedge the value of a publicly trading agribusiness. To the extent that changes in the stock price of an agribusiness parallel changes in one of the stock index futures, the value of this stock can be protected. Hedging the value of an individual stock is the same type of exercise as hedging the value of a portfolio of stocks. The parallel movement between individual stocks and a broader market index is known as *systematic risk*. The basis risk associated with this hedge will be rather large. Further, hedging with futures will not protect the stock value from poor management, bad decision making, and low profits. This is *unsystematic risk* and is specific to the firm. In this latter case, the stock price of the individual firm can decrease irrespective of changes in the market.

The futures market Beta is also a useful concept and tool for any hedger. The idea it communicates is that the hedger is interested in protecting the potential change in the value of a cash position *with the proper proportion of futures contracts to the cash position so that losses in one market are offset by gains in the other*. It is unlikely that the manager of a stock portfolio can trade a futures contract that matches the content of his or her portfolio exactly. Rather, the manager trades a futures position that change in the dollar value comes close to matching the dollar value change in the cash position. The same concept can be applied to any commodity and is particularly useful for hedging commodities for which no futures markets exist. For example, no futures market exists for 450-pound calves, U.S. barley, grain sorghum, or feeder pigs. The CME feeder cattle contract, the corn or the Canadian barley contracts, and the live hog contract could be used, but it would require the hedger to know the proper futures-to-cash ratio. It requires knowledge of the proper number of futures contracts needed to protect the value of a cash position.

Let's work one more stock index hedge example. The example is summarized in Table 10.6. Suppose it is late October and the stock market decreased rather strongly the week before. However, unlike the decline in October 1987, many investors are viewing this as an opportunity to buy stocks. There are lines at many brokerage houses of customers wanting to place orders to buy actual stocks. In October 1987, the lines were of people wanting to sell. An astute portfolio manager would recognize that in a few days she will have new dollars with which to use to purchase stocks. However, the market seems to be rallying now and it may be bid up to precorrection levels before she has access to the new customer deposits. Anticipating the availability of $25 million in new funds within a week, the portfolio manager buys 105 December S&P 500 Index contracts at 950. One week later, the index and the December contract have increased 2.75 percent. The portfolio manager lifts the hedge, uses the $682,500 in futures gains to supplement the $25 million in new deposits, and makes stock purchases in the cash market which has continued to rally. This hedge essentially allows the portfolio manager to purchase stocks at prerally prices. It is exactly parallel to the anticipatory hedges used by grain exporting firms who expect China, for example, to buy wheat in the near future and the exporters want to "anticipate" that need by buying wheat futures now. The hedge as described would be effective if the manager is interested in purchasing a group of stocks with an overall Beta equal

TABLE 10.6
Anticipated Stock Purchase Hedge

Date	Cash Market	Futures Market
October	Stock portfolio cost $25 million.	Buy 105 December futures contracts at 950 covering $25 million.
November	Stock portfolio cost $25.6875 million.	Sell 105 December futures contracts at 976.
	–$0.6875 million loss.	+$0.6825 million gain.
	Portfolio cost $25.005 million.	

to 1. If the manager is interested in more volatile stocks, she will need to take a long position in the futures market greater than $25 million. If the manager is interested in less volatile stocks, she can take a long position in the futures market less than $25 million and still have an effective hedge. Note that the Beta level is important in each case.

A "matching" problem here is present in many hedge situations. The issue is definitely there in many agricultural applications. An example is the cattle feeder needing long hedges on lightweight feeder cattle and short hedges on the much heavier fed cattle he sells.

INDEX PRICING MODELS

This section presents a model that explains the rationale for different anticipated stock index values for contracts deliverable at different times in the future. For example, if it is currently October, what is the main economic reason for the futures contract that expires in December to be different from the contract that expires in March? This model examines the basis relationship that can be expected between the futures contract index value and the actual cash market index for hedges that are lifted prior to the expiration month. If a stock index hedge is lifted in October and the December contract is used in the hedge, what is the expected difference between futures and cash index?

The model is a slightly more complex version of that used to explain the difference in prices of two exchange rate contracts. The model is again based on the cost of carry. The futures price for any financial instrument will equal the spot price plus the cost of carrying the instrument to the delivery date. However, in this case, stocks pay dividends and the dividend plus interest on the dividend after its payment can be used to offset the carrying costs. In equation form,

$$F_{t,T} = S_1 (1 + C_{t,T}) - \sum_{i=1}^{N} D_i (1 + R_{s,T})$$

where $F_{t,T}$ is the futures price of the stock index at time t of a contract that expires at time T, S_t is the spot index in time t, $C_{t,T}$ is the cost-of-carry portfolio from t to T, D_i is the dividend on the ith stock of which there are N, and $R_{s,T}$ is the interest earned on the dividend from the time of its receipt s until T. The cost-of-carry equation can be rewritten as follows if the cost of carry is a percentage:

$$F_{t,T} = S_t (1 + r)^{T-t} - \sum_{i=1}^{N} D_i (1 + r)^{T-s}$$

where r is an annual interest rate, $T - t$ represents the length of time the instrument is held, and $T - s$ is the amount of time after the dividend is paid. For example, if the portfolio of 20 stocks costs the purchaser $1,000, the cost of carry and return on the dividend are 10 percent, the portfolio is held 12 months, and a $0.50 dividend is paid on all stocks after six months, then the future price of the portfolio will be

$$\$1,000 (1 + 0.10)^1 - \$10 (1 + 0.10)^{0.5} = \$1,089.51$$

in the long run. This long-run price of this portfolio will be factored into the futures price of the index.

Increases in the interest rate increase the opportunity cost of holding stocks and attracting investment capital away from the stock market, and current stock prices depreciate relative to future stock prices. Increases in dividends decrease the opportunity cost of holding stocks, thus attracting investment capital into the stock market, which causes current stock prices to appreciate relative to futures stock prices.

Further, as the futures contract approaches expiration, as t approaches T, the carry cost shrinks to zero, there is no opportunity for a dividend, and the spot price and the futures price converge.

$$F_{T,T} = S_T$$

The premium between the current value of the index and the futures price of the index, or between a nearby and a distant futures contract will reflect the anticipated growth in stock prices. In a properly functioning economy, we expect the economy to grow and stocks to appreciate in value. Thus, futures contracts for these future periods should reflect the anticipated appreciation and the anticipated stock index level at expiration. However, the link between the future value of the index and the current value is created by expected dividends and the opportunity cost of investment capital, and investors responding to these incentives will determine the premium over time. This is a very long-run concept, however. In the short run, stock prices appear to be influenced quite a bit by expectations and other factors. This is the topic of the next section.

INDEX FUNDAMENTALS

Stock index fundamentals are not as complex as those for currencies but are more complicated than fundamentals for some physical commodities. There are simply few good supply and demand models that will predict stock index levels. This is not to say there are no good methods for valuing stocks. There are disciplines in finance and accounting centered around stock and company valuation. However, the models appear to lose their explanatory power as stocks are aggregated into broader market or categorical indexes. In this section we do not address valuation of individual stocks. The reader interested in this topic should consult one of the numerous finance texts on the topic. Rather, we attempt to provide some insight into the behavior of broader market indexes.

Understanding stock index levels and changes in those levels requires some knowledge of macroeconomics and the operation of an economy. However, stocks are also speculative vehicles so that opportunity costs and expectations impact prices. Thus, stock prices, like currencies, behave much like any other commodity in that price is determined by the interaction and balance of supply and demand. If the demand for stocks increases, prices will increase. If the demand for stocks decrease, prices will decrease.

In the long run, stock price levels and the respective market indexes are determined by macroeconomic forces and equilibrium investment relationships. Economies are rather closed systems. Spending by consumers on goods and services goes to the firms providing those goods and services. That spending pays for the cost of resources used, production costs, management, and other services, and the remainder is returned to the owners of the firm or the stockholders. Wealth does not escape the system unless it is wasted, and supply and demand are linked in this system. Resources are the source of the wealth created, demand for goods and services creates spending, providers of the goods and services earn the spending, and the system grows through this multiplier effect. The economy grows through the consumption of resources, investment in provision of goods and services, and increases in productivity. In the long run, stock prices reflect the growth and productivity of the economy.

Stock prices should reflect the returns to investment in the provision of goods and services to the economy. Thus, stock prices reflect the sum of discounted future dividends paid by firms. The discount rate applied to the dividend stream is the opportunity cost of investing in businesses. In addition to purchasing stocks, investors could lend money directly to firms through the debt market. Investors can also lend money to the government. The profile of investment opportunities contains choices of various riskiness. There are high risk and low risk stock ventures, and high and low risk debt ventures with government debt offering a riskless rate of return. Investors evaluate this profile.

Productivity and growth in the economy will result in increased dividends. This will make stocks attractive relative to debt instruments and will increase the overall market index as investors make purchases. Statistics that capture dividend payments and that are frequently reported are the *price-to-earnings ratios*. This statistic is the individual stock price relative to previous dividends and is widely reported in the financial press. Increases in interest rates, primarily through actions by the central bank, will make debt instruments attractive relative to stocks but will require stock to earn higher rates of return in the long run. *Thus, changes in interest rates by the central bank often have profound effects on stock prices and stock market indexes.* Anticipating changes in interest rates by the Federal Reserve System is a major endeavor by financial institutions, and interest rate deliberations are kept secret because of this. Current monetary policy is to use interest rate control to maintain low inflation rates. Anticipating changes in the central bank rate thus involves anticipating changes in inflation.

An equation that is used as a base to stock price valuation is as follows

$$S_t = [E(S_T) + \sum_{t=1}^{T} D_t (1 + r)^t] \frac{1}{(1 + r)^T}$$

where S_t is the current stock price, $E(S_T)$ is the expected stock price T periods in the future, D_t is the dividend paid during time t, and r is the interest rate. We see that the current stock price is the expected future stock price plus discounted dividends, the total of which is discounted back to the current time period. If the investor does not intend to sell the stock in the future, then there is no expected future stock price and T gets very large in the formula. From this formula, we see the importance of dividends and the interest rate for long-term stock valuation. However, in the short term, expectations play a very important role.

If investors purchase stocks with the specific expectation of selling them at a higher price, or if investors short sell stocks with expectations of purchasing them back at lower prices, this can create speculative bubbles like those observed in the price history of a stock or index. Investors buy because they think stock prices are going higher. The act of purchasing pushes stock prices higher and the expectation is realized and reinforced. Investors usually continue to push stock prices higher. The follow-the-herd psychology so prevalent in the commodity markets is also present in the stock market and stock index futures market. *The conclusion that can be drawn here is that technical analysis should be an effective tool for timing purchases and sales.* Or, at least, technical analysis should be more effective than following emotional instincts.

The importance of expectations in determining trading patterns and influencing prices is recognized in the use of circuit breakers. If stock market indexes or futures on the indexes move down excessively, stock market and futures exchanges have policies in place to close trading for short periods of time. The Federal Reserve System can also suspend trading. Those familiar with futures markets will recognize these as limit moves and may remember trading being closed or sharply constrained following a limit move.

The performance of the stock market and various indexes through 1997 deserve discussion. The market has made substantial gains through this time. Some indexes are registering 20 percent to 30 percent returns over the previous year and price-to-earnings ratios are at or near all-time highs. High price-to-earnings ratios are not good. It implies investors have paid a lot for stocks and that dividends have not increased at the same pace. There has been considerable discussion about whether these growth rates, and in fact whether these price levels, are sustainable. The market corrected hard in October 1987, decreasing 15 percent to 18 percent, but the correction was very short lived. What do the fundamentals tell us? The picture is not clear. There are credible models and research that suggest the current price-to-earnings ratios cannot be maintained and that earnings will not increase. Rather, the future holds a real price depreciation—nominal stock prices will fall or will increase in price slower than the inflation rate. There is much discussion about a new economy, usually related to improved productivity through computers and management consolidation. The discussion also ties in an increasingly global economy and the benefits of international trade. However, while the discussion sounds good, the scientific evidence of even something that should be easy to measure, such as improved productivity through increased use of computers, is very hard to find. And yet the stock market continues to appreciate in the face of this evidence.

A point that does not receive as much discussion is the change in the investment behavior of the average household. Governments are trying to reduce their responsibility for providing retirement funds to citizens. In response, more households are investing in stocks—either directly or through organized mutual funds—in response to these incentives. This would improve demand for stocks and put buying pressure on the market. The improved demand for stocks is especially apparent given low interest rates and returns to debt instruments. This ties the stock market behavior back to the equations presented earlier that link stock index price through time to opportunity costs. And we should remember the discussion about speculative bubbles.

The long-term factors that influence stock prices and market indexes are productivity and economic growth. Changes in dividend returns and in long-term interest rates are the primary determinants of long-term changes in stock prices. Interest rate actions by central banks and investor expectations are the primary determinants of short-term changes in stock prices. The psychology of the market is there just as it is in commodity markets, and it would be hard to convince you that what happens in the stock market will not be a factor in what it does

tomorrow. If that day-to-day dependence is there, then all the technical tools discussed in prior chapters will help you in buying and selling stocks or in using stock index futures.

SUMMARY

Stock indexes seem to have become of a part of our daily lives. More indexes are watched and are widely reported. Their increasing importance is probably because we are taking more of a responsibility for providing for ourselves in retirement years. This increased interest could also be because of the recent astounding growth in the various indexes. In any case, indexes are used as an indicator of the well-being of the economy. *The futures contracts for the various indexes seem to be going through more of a period of change than growth.* There are booms and busts in popularity and contracts are rewritten to attract more volume. The lackluster interest may change if the market succumbs to a reversal in fortune—or at least a slower growth. These instruments can be important risk management tools for the growing wealth amassed in mutual funds.

In addition to stock indexes, several other indexes track inflation, bond prices, and the U.S. dollar. The concept of an index is very popular with futures exchange management, and more indexes will likely be traded in the future.

With the variety of indexes available, it is important for the trader to be aware of what the index measures. *All indexes are a composite measure of price level.* The prices are for publicly traded stocks in the various stock indexes. Stock indexes, and indexes in general, are value weighted or price weighted. Value weighting implies that the index is a weighted average of prices, where each price is weighted by the number of shares issued by the company. Price-weighted indexes are simple averages. Price-weighted indexes are most affected by changes in the price of high-priced stocks. The S&P 500 Index and NYSE Composite Index are value weighted. The DJIA and Nikkei 225 are price weighted.

The index also depends on the prices in the sample. A variety of indexes are available and being considered for trade, and they differ mainly by what stocks go into the sample. There are stock indexes futures now for midcap firms (medium sized), small firms, and growth firms. Indexes are also available for trade in most western European and Asian futures exchanges.

Indexes are simple to trade. The value of a position is the index multiplied by a dollar amount. The gain or loss on the trade is the difference between the level of the index at the sell point versus the buying point, times this amount. Trading these indexes could be very useful to people with portfolios of stocks, the value of which they wish to protect.

However, *hedging is not so simple. Any individual index will not likely match the combination of stocks in a specific trader's portfolio.* It is unlikely that a 1 percent change in the value of a trader's stock portfolio will be matched by a 1 percent change in the stock index or futures contracts on the index. There are portfolios that are more variable than the market index, higher risk portfolios, and portfolios that are less variable, low risk portfolios. The trader must understand the degree to which changes in the value of the portfolio move in parallel with changes in various indexes.

Currently, traders use the tool "Beta" to measure the degree with which their portfolios follow the market index. A similar futures market Beta must be calculated to measure the degree to which changes in the value of the portfolio follows changes in the value of the futures index. This futures market Beta measures the futures position as a percent of the cash position that should be traded to completely hedge the portfolio. For example, if the futures market Beta is 0.75, that suggests the trader should sell futures contracts worth 75 percent of the value of the cash portfolio. Likewise, a futures market Beta of 1.25 suggests establishing a futures position of 125 percent of the value of the cash position.

There is little basis risk in the stock index market. The buying and selling in the futures market pushes the value of the futures contract very close to the index in the expiration month. However, only four index futures contracts are traded for any particular calendar and there will be differences between the cash and futures index for nonexpiration months. This difference in predictable. The futures contract value is the value of the index that is anticipated in the expiration month. Therefore, a trader can work backwards in time to determine nonexpiration-month levels of the index given estimates of the rate of appreciation of the index.

The fundamental economic factors that affect the value of different stock indexes are well rehearsed. Stock prices reflect the market consensus of the well-being of the economy. A strong and growing economy creates wealth, and this prospect is priced into the value of capital assets. However, the stock market also reacts strongly to opportunity costs of investment. If returns to other capital assets are low, such as debt instruments and their respective interest rates, investors will tend to bid aggressively for stocks. This market is also subject to the psychological factors that overprice and underprice other commodities. Some market conditions are common regardless of the product traded.

Futures contracts for stock indexes, and indeed indexes for other assets as well, will likely continue to grow in importance. And these futures will also likely continue to change. *Contracts will be designed, written, and rewritten to help asset holders manage risk and attract speculative capital.*

KEY POINTS

- *Futures contracts on stock indexes* are some of the *most important financial futures contracts*. Further, there are futures contracts based on other indexes which measure other dimensions of risk and performance in the economy.
- The futures industry is on the verge of *a proliferation of stock index contracts*. New indexes are emerging that summarize the performance of different industry groups within the U.S. domestic stock markets and *indexes for stock markets in various foreign countries*.
- The contracts can be used by portfolio managers to *insulate the value of their portfolios* from changes in the market. This includes gains as well as losses.
- Indexes are *simple to trade*. The value of a position is the *index multiplied by a dollar amount*. The gain or loss on the trade is the difference between the level of the index at the sell point versus the buying point, times this dollar amount.
- Hedging is more complex. A *futures market Beta* must be calculated to measure the degree to which changes in the value of the portfolio *follow changes in the value of the index*. This futures market Beta measures the futures position as a percent of the cash position that should be traded to completely hedge the portfolio.

- *Little basis risk* exists in the stock index market in expiration months. Further, the difference between the cash and futures index for nonexpiration months is *relatively predictable*. The futures contract value is the value of the index that is anticipated in the expiration month.
- The *fundamental economic factors* that affect the value of different stock indexes reflect the well-being of the economy. However, the stock market also *reacts strongly to opportunity costs of investment in interest-bearing debt instruments*. If *interest rates are low*, then *stock prices tend* to *be high*. This market is also subject to the *psychological factors* that make technical analysis useful.
- Stock indexes futures contracts *will continue to grow in importance* and will *continue to change to attract traders*.

USEFUL REFERENCES

Steven C. Blank, Colin A. Carter, and Brian H. Schmiesing, *Futures and Options Markets: Trading in Financials and Commodities*, Prentice-Hall, Englewood Cliffs, NJ, 1991. Good intermediate text with coverage of financial futures. Good coverage of interest rate futures.

Robert T. Daigler, *Managing Risk with Financial Futures: Pricing, Hedging and Arbitrage*, Probus Publishing Company, Chicago, IL, 1993. A solid and comprehensive text with many examples. The author is presenting many research results. Very useful for interest rate, stock index, and exchange rate examples.

Robert E. Fink and Robert B. Feduniak, *Futures Trading: Concepts and Examples,* New York Institute of Finance, New York, 1988. A good basic introduction to financial futures.

John C. Hull, *Introduction to Futures and Options Markets*, 3rd ed., Prentice Hall, Upper Saddle River, NJ, 1998. An introduction to the mathematics of financial futures.

Robert W. Kolb, *The Financial Futures Primer*, Blackwell Publishers, Malden, MA, 1997. An excellent introduction to financial futures. Many trading examples are presented. Some of the important subtleties of financial futures are also discussed.

Paul Krugman, *The Age of Diminished Expectations*, 3rd ed., The MIT Press, Cambridge, MA, 1994. An excellent book that discusses current topics of issue for the U.S. and world economy. The text is very useful for understanding the reactions of exchange rate, stock index, and interest rate markets.

Raymond M. Leuthold, Joan C. Junkus, and Jean E. Cordier, *The Theory and Practice of Futures Markets*, Lexington Books, Lexington, MA, 1989. A more advanced treatment of futures and hedging. Extensive focus on interest rate, stock index, and exchange rate futures.

Charles W. Smithson and Clifford W. Smith, Jr., with D. Sykes Wilford, *Managing Financial Risk: A Guide to Derivative Products, Financial Engineering, and Value Maximization*, Irwin Professional Publishing, Chicago, 1995. A fairly simple treatment of more advanced financial management topics. The focus is mainly on options and derivative pricing with interest rate, stock index, and exchange rate examples.

CHAPTER 11

PRICE RISK MANAGEMENT STRATEGIES

INTRODUCTION

No single management strategy will prove superior for all decision makers. Which strategy is best will depend on many dimensions of the situation. Among the more important of those dimensions are the attitude of the decision maker toward risk, the financial position of the operation that influences the capacity to carry exposure to risk, and the abilities of the decision maker as a manager of the various risk management tools.

Before developing strategies, it is productive to step back and address just what is being discussed. The need is for effective management of price risk, but just what does that mean? How important is the price risk issue? Is it potentially devastating? Enough to run you as an agricultural producer out of business?

Table 11.1 provides a baseline for purposes of discussion. Corn is the key feedgrain in the U.S. and is a major export commodity. Hogs are an important livestock commodity and are widely produced throughout the United States. The table provides price ranges for nearby futures for the calendar years 1980 through October of 1997 and shows the general direction of price movement within the year ("D" for down, "U" for up).

Not surprisingly, there are about as many "up" years as "down" years. Does that mean just being a cash market speculator will work? The answer is no, not unless there are big financial reserves.

To cover all costs of production, corn prices of at least $2.00 are needed for most producers. Keep in mind that the prices in the table are *futures prices.* When the futures data in 1986 moved down to the $1.60 level, cash prices in some producing area dropped toward $1.00. There were reports of cash corn selling for $.96 per bushel in Des Moines, Iowa, in the fall of 1986.

Heavy subsidies from government farm programs have kept corn producers in business. In periods like that of 1986–87, 70–80 percent of the net farm income in some midwestern states came from the government programs. Without the subsidies, such low prices would have been financially ruinous to many producers. And there is another side to the story. The price rallies in 1980, 1983, and 1988 were all caused by droughts.

TABLE 11.1
Approximate Price Ranges for Corn and Hog Futures (Nearby Contracts) 1980–1997 with Indication of Direction of Price Movement within the Year ("U" = up, "D" = down)

Calendar Year	Price Range: Hogs (Direction) ($ per cwt.)	Price Range: Corn (Direction) ($ per bu.)
1980	$28–53 (U)	$2.60–4.00 (U)
1981	53–38 (D)	4.00–2.60 (D)
1982	40–68 (U)	2.80–2.20 (D)
1983	60–40 (D)	2.40–3.20 (U)
1984	40–58 (U)	3.60–2.70 (D)
1985	55–35 (D)	2.80–2.20 (D)
1986	38–64 (U)	2.60–1.60 (D)
1987	64–40 (D)	1.50–2.00 (U)
1988	45–40 (D)	2.00–3.60 (U)
1989	40–52 (U)	2.90–2.20 (D)
1990	47–67 (U)	2.40–2.90 (U)
1991	64–46 (D)	2.23–2.66 (U)
1992	38–52 (U)	2.74–2.18 (D)
1993	57–40 (D)	2.10–3.06 (U)
1994	30–53 (D)	3.12–2.10 (D)
1995	36–57 (U)	2.27–3.75 (U)
1996	44–67 (U)	5.54–2.56 (D)
1997	64–44 (D)	3.20–2.38 (D)

Many producers had little or no crop to sell at those high prices. Now, the 1996 farm bill legislation has removed the "safety net" for corn producers. Target prices and deficiency payments are gone, and producers are left to cope with what is likely to be an even more volatile market.

For hogs, a price of around $45.00 per hundredweight is needed to cover all costs for the typical producer. Some large operations have lower costs, but the data in Table 11.1 indicate frequent periods of major difficulty. The 1984 period was especially difficult for hog producers. Buffeted by higher corn prices and cash hog prices that dipped toward $30.00 in late 1984 and early 1985, many producers did not survive. In 1994, prices dipped briefly below $30.00 before the record high corn prices of 1995–96 reduced production and pushed prices higher. *The markets are very risky and the probability of a financially ruinous price move is not small. The capacity to manage that risk is very important.* The situation is not appreciably different for wheat, cattle, cotton, interest rates, exchange rates, and the other sources of risk that have surfaced or have increased in the 1970–1997 period.

In this chapter, price risk management strategies are discussed. Coverage cannot be exhaustive, of course, but an effort is made to provide an array of alternatives from which you as a potential decision maker can choose. Both the fundamental and technical approaches to analysis of the markets are employed and demonstrated.

ATTITUDE TOWARD RISK

A broad set of literature covers the importance of the decision maker's attitude toward risk. Here, the discussion will focus on a risk-averse decision maker and on a decision maker who seeks relatively high exposure to price risk when additional profit potential appears to be present. Most decision makers will fall on the continuum between the two, but this approach allows emphasis on how the correct strategy will vary with the attitude toward risk.

In most instances, an inverse relationship exists between the level of risk exposure and the potential returns. In an investment context, the more risky the venture, the greater the potential returns must be to attract investors. In the context of agricultural producers, the attitude toward risk often shows initially in the selection of enterprises. The risk-averse producer is more likely to be diversified as an attempt is made to spread production risk across several enterprises. This may prevent specialization in the crop or enterprise for which the production unit is best suited, but it may also reduce the possibility of a ruinous financial position in the event of extremely low prices on a particular crop or livestock enterprise when the producer is operating as a cash market speculator. The data in Table 11.1 suggest there is in fact reason to be concerned.

Typically, the same attitude will be transposed to the willingness to be exposed to price risk. *The risk-averse decision maker may be willing to suffer the opportunity costs associated with an unexpected price increase in the cash market in order to ensure there will be protection against falling prices.* A hedging or price risk management strategy will be adopted that fits that preference pattern.

Decision makers who are willing to accept exposure to price risk in the hopes of a higher return will opt for a different strategy. A producer may often choose to be a cash market speculator and look for the occasional strong move up in corn, wheat, cattle, hog, or cotton prices associated with unexpected developments in the supply-demand balance. These decision makers tend to prefer a selective approach to hedging. They want protection against potentially ruinous prices but are eager to be in a position to benefit from surges in cash prices.

The advent of options on the commodity futures in the early 1980s injected a new element into the situation. Risk-averse producers can buy put options to gain protection against declining prices and still be in a position to benefit from any unexpected surges in cash price. They may be willing to pay the option premium and tend to see the premium as the insurance premium for protection against lower prices. As noted in Chapter 7, options are often more flexible than the futures as price risk management tools.

Even with the options, the risk takers will often choose different approaches. They tend to be worried about paying the option premiums that they often argue are too high, and they get concerned about the fact that the cash market must rally enough to more than offset the premiums before they are in a position to benefit. This type of producer will tend to prefer some of the more sophisticated option strategies discussed in Chapter 7. They may be willing to carry exposure to risk for some range of lower prices in order to secure a net above the straight hedge for higher prices or for a preselected price range. In effect, the options are being converted or adapted to a strategy that has some of the characteristics of the selective hedge.

In very simple terms, the contrast can be shown as in Figure 11.1. The risk-averse producer and the risk-seeking producer have a different preference structure with regard to the level and variance of revenue. The variance is a statistical measure of variability, and the risk seeker is willing to accept exposure to more variability if the potential of higher revenue is present. Note, however, that the curve is increasing at an increasing rate, indicating that even the risk seeker will need to be compensated heavily as the measure of variability and therefore risk exposure grows.

The risk-averse producer operates with a different perspective. Such a producer will give up revenue to keep the risk exposure at tolerable levels. Note that the function

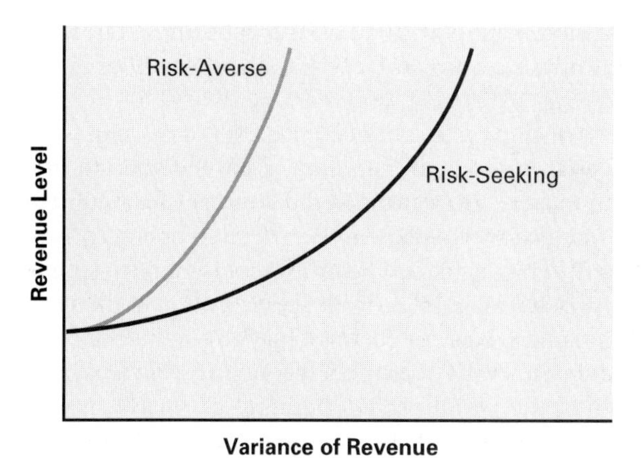

FIGURE 11.1
Simplistic Preference Functions for Risk-Averse and Risk-Seeking Decision Makers

is close to the vertical axis and also that it increases at an increasing rate, indicating that the rate of substitution between level and variability of revenue changes as the variability and related risk exposure grows.

Decision makers' preference functions can be developed into an analytical framework using the mean and variance of revenue or income streams. The book by Sharpe and Alexander listed in the references at the end of the chapter provides added coverage. *Here, the need is to recognize that different producers will make decisions that reflect differences in the underlying willingness to substitute level of income for variability of income.*

The attitude of the decision maker toward risk will be an important determinant of the strategies that will be employed. Risk-averse decision makers will opt for more conservative strategies and will be more nearly willing to pay the option premiums or to answer margin calls. Decision makers who are willing to accept risk for the chance of a higher return will opt for selective hedging approaches and the more sophisticated options strategies.

FINANCIAL POSITION OF THE FIRM

The financial position of the farm will influence the capacity to carry exposure to price risk. For the highly leveraged operation that is carrying a heavy debt load, a major and unexpected drop in price might put the firm out of business. So, in addition to being more likely to be diversified on the production side, the decision maker with limited finances will tend to opt for strategies that essentially guarantee protection against price declines. It is not unusual to find that the lending agency will require protection against price risk for the operations that are highly leveraged in financing the operation. These self-imposed and external requirements will often mean the producer is looking to guarantee at least a break-even price before the crop is planted or to cover the variable costs on at least a portion of the crop. In the livestock sector, producers may be required to forward-price cattle or hogs before they are even allowed by the lending agency to borrow funds to buy the feeder cattle or feeder pigs and place them in a feeding program.

There can be severe opportunity costs associated with such self-imposed or externally imposed requirements. At the producer level, the agricultural sector approaches the conditions of pure competition. The producer is a price taker and has little or no ability to influence price.

In an industry structure approaching the conditions of pure competition, there are few, if any, significant barriers to entry. New producers can come in or existing producers can expand by bringing in new capital. The result is that the market seldom discovers and offers prices for future delivery that open up a profit window over and above the price required to cover average total cost of production. *For the heavily indebted producer, this may mean having to pass the occasional opportunities that have a high probability of being profitable because of the lack of financial capacity.* The feeder pigs are not bought and placed into a feeding program that has a high probability of being profitable because the futures market never offers a guaranteed profit on the same day the pigs could be bought. Thus, the lack of financial capacity and the related conservative posture of the lender may place a serious constraint on what the firm can do and on the price risk management strategies that can be employed.

Later in the chapter we will come back to this point in discussing appropriate strategies for the producer with a heavy debt load. The complementary fundamental and technical analysis become extremely important in spotting opportunities that offer a high probability of profits. If the producer who is struggling with financing is denied these opportunities, both the producer and the lending agency are likely to suffer over time. After all, the need is for profitable ventures that contribute to the cash flow and enable the producer to service and reduce the debt load over time. The alternative is often a slow and painful exit from the industry, a process that was widely observed in agriculture during the early and mid-1980s and is starting to show up again in the dairy, swine, wheat and cotton sectors in the late 1990s.

Financial capacity comes into play in another important way in influencing strategies to be used in price risk management. A position directly in the futures always exposes the producer to the need for margin capital. The futures account must be margined initially and there must be adequate provision for funds to answer margin calls.

The lending agency is extremely important here. There must be recognition of the need for a separate credit line to handle margin needs. *As a producer, you should not be expected to handle margin-line needs from the production credit line or from personal funds.* There are countless examples of producers being forced to offset profitable hedges because of the inability to answer margin calls. If the price subsequently turns lower, as it often does, the producer who is in a poor financial position initially is then often denied the price protection that is so badly needed. A written agreement is needed that spells out the objective of the marketing plan. *The producer commits to the agreed-upon parameters in terms of how the futures or option transactions will be handled, and the bank commits to the needed margin funds.* The plan should be updated and modified based on experience and on an as-needed basis. Brokers should be brought into the planning process so that they know the objectives and the operating parameters as well.

If a credit line for margin needs is not available, the highly leveraged producer should select strategies that do not use the futures market directly. Cash contracts are a possibility. The margin requirements are transferred to the buyer. Options are another possibility if the producer has adequate capital to handle the initial option premiums. The cost of the pricc protection program is known up front when options are used and there is no exposure to the financial drain of margin calls.

The financial position of the firm will influence which price risk management strategy can be adopted. For the highly leveraged firm, it is important that the posture adopted not be so conservative that opportunities with a high probability of being profitable are ignored. Developing better understanding of price risk management strategies can help the highly leveraged firm take advantage of opportunities that offer a high probability of profit and can improve its financial position.

MANAGEMENT ABILITY

The abilities of the decision maker to manage exposure to price risk will clearly be an important criterion in selection of a strategy. A producer who is a skilled chart analyst or a producer who subscribes to the use of moving averages and understands their strengths and weaknesses is more likely to prefer a selective hedging strategy. Conversely, a producer who is poorly informed in charting techniques or does not feel that technical analysis is a valid analytical tool will make some other choice and be more likely to follow a conservative approach to hedging or to use options.

In either case, the ability of the decision maker as a fundamental analyst will be important. A selective hedging program, whether based on chart signals or moving averages, works best in a market that shows major and sustained price moves. If the underlying supply–demand situation is showing excessive ending stocks of corn, for example, major price increases in the coming crop year are not highly probable. Whatever the objectives of the producer, some approach other than a selective hedging approach might be in order.

To illustrate the impact of the fundamental outlook, it is useful to review its influence on the basic choice between direct use of the futures and options. In a market burdened by excessive stocks, the options can be a more expensive way to acquire price protection. Premiums do come down in the less volatile markets that emerge in the presence of burdensome stocks, but the time value of the options at planting time and early in the growing season keep the premiums relatively high. If no major price move develops because of the burdensome and excessive stocks and the markets trade sideways, the option value is likely to be at or near zero at harvest and the initial premium is forfeited. Price insurance was purchased via the options that was not, in an *ex post* context, needed. A position in the futures during that same type of market would cost commissions plus an interest charge on margin funds. The cost comparison could be between $.10 and $.20 per bushel of corn or more on the option premiums and $.02–$.03 per bushel when carrying protection directly in the futures.

An important thrust of this entire book is to describe and develop the abilities needed to effectively manage exposure to price risk. We have explained that fundamental and technical analysis are complementary. *That complementarity is apparent when it becomes clear that the management ability of the decision maker as a fundamental and technical analyst influences the selection of a pricing strategy.* As more specific strategies are illustrated in the following sections, the importance of both approaches to analysis will become even more apparent.

The ability of the decision maker as a technical and fundamental analyst will influence the choice of strategies. The more sophisticated strategies will be denied the manager who has few skills as an analyst of the fundamental price outlook, the technical picture in the market, or both.

MANAGEMENT STRATEGIES

There are numerous possible approaches to management of price risk.[1] To exercise a degree of control over the number of alternatives to be considered, the strategies developed will be related to the dimensions just introduced that would be expected to influence the choice of strategies. Strategies will be developed first for the risk-averse and conservative decision maker with limited financial capacity and limited analytical abilities. Strategies for the more nearly average decision maker will then be developed, leading up to coverage of strategies for the risk seeker who has substantial financial capacity, is a capable market analyst, and is a capable manager of exposure to price risk. You should be able to identify and adapt a strategy that fits your particular situation.

Conservative Strategies

For the risk-averse decision maker with limited financing and limited analytical abilities, a conservative approach will typically be best. Several alternatives are available, however.

Target Pricing *The primary objective is to ensure the economic viability of the operation.* Table 11.2 illustrates a reasonable approach for corn. Prorated per-bushel "charges" to allow servicing the annual debt and to cover living expenses for the family or to earn a preset return on the investment are added to the variable per unit costs of production. The result is a *target price,* the price needed to ensure economic viability. A strategy is then developed to secure the target price.

If a decision has already been made to plant a certain number of acres in corn, the immediate need is to ensure the target price. Given the limitations facing the producer, the most logical approach is to use a cash contract. Such an approach transfers the management of margins and the basis risks associated with a position in futures to the buyer. The coverage in earlier chapters indicates that it is the elevator, for example, who faces basis risk when a cash contract for harvest-period delivery is extended. To the producer, a contract price offer that incorporates a specific basis allowance means that they are apparently transferring both price and basis risk to someone else. It is important to keep in mind, however, that the elevator will tend to offer cash price contracts that, over time, transfer part or all of the basis risk to the producer. The cash bid will usually reflect some allowance for the elevator's exposure to basis risk, and that allowance gets imbedded in cash-futures basis patterns over time.

Not all the price protection has to be established at the same time, of course. Typically, producers will be willing to price some percentage of the expected production prior to planting and then add protection in increments until the desired level of protection is reached. Even though cash contracts are being used, *there are still ways to accomplish scale-up pricing once the target price is offered.*

[1]The strategies in this section are developed primarily from the viewpoint of the producer or holder of inventories who needs protection against declining prices. This is the more common need, but you are reminded that the strategies are appropriate for the buyer of grains, livestock, and so on, who needs protection against rising prices. Call options would be used instead of put options, and the buy signal from moving averages, for example, would be the signal on which long hedges are placed.

TABLE 11.2
Calculating a Target Price for Corn

Assigned Costs	Per Bushel
Variable costs, production	$1.60
Annual debt payment	.25
Family living*	.20
Target price	$2.05

*The "family living" prorated allocation could be based on a competitive return on the investment or a competitive salary for the manager instead of family expenses or living expenses.

The cash contracts offered by elevators for harvest-period delivery reflect an expected harvest-period basis adjustment based on historical data. There is no reason to expect that basis estimate to change significantly during the year, and that stability gives the producer an opportunity.

Elevator managers often keep a "wish book" with producers' names, number of bushels, and desired price levels. If the underlying futures market trades up, the cash contract offers by the elevator will trade up directly with the futures, reflecting a largely constant harvest-period basis allowance. *Without being involved directly in the futures, the producer is able to capture the benefits of a rallying futures market and price the product on a scale-up basis.* The elevators offer the cash contracts and then go short in the futures market when the cash contract is signed to protect the margin they have incorporated into the cash contract offer.

The simplicity of the target price strategy is appealing. With a target price established on something of a cost-plus basis, the first impression is that the producer is doing something about eliminating the problems of being a price taker. But that impression can be misleading, and there are in fact major shortcomings to this strategy.

Perhaps the most important disadvantage is the absence of any safety net to prevent a financially devastating dip in price. If the target price is never offered by the market, the producer is totally exposed to price risk and is caught in the role of a cash market speculator on 100 percent of the projected production. *The decision maker who is risk averse, in poor financial condition, and not a very effective market analyst faces the very real possibility of being totally exposed to price risk in the marketplace.* There is a need for a strategy that offers the appeal and simplicity of the target price approach but also offers protection against the major price break. Obviously, one approach would be to improve the analytical abilities of the decision maker and move to a more sophisticated and flexible strategy. But that improvement will take time, and not all producers will wish to become better analysts or be able to do so. Some other refinement is needed.

The simplicity of a target price strategy is appealing. The big shortcoming of the approach is the lack of any protection when the target price is never offered by the market. When that occurs, the producer who can least afford such an exposure may be totally exposed to price risk as a cash market speculator.

Target Pricing Plus What is needed is a safeguard against significant price breaks while the producer waits for the market to offer the target price. Virtually every refinement will require at least a marginal increase in access to financing and/or a marginal move toward more sophisticated analytical abilities. But some refinements do in fact require only an incremental change and are consistent with the producer's risk-averse orientation.

The use of moving averages can provide the refinement and the much-needed safety net. A crossover set of moving averages, such as those described in Chapter 5, are mechanistic and totally objective in nature. No particular analytical skill is required, but it is important that the decision maker understand what the moving averages can contribute. Figure 11.2 demonstrates the protection that the moving averages can bring with a realistic situation. For illustrative purposes, a simple set of 3-day and 10-day moving averages is used.

Feeder cattle were bought in late July with the December live cattle futures moving up. A target price of $75.00 per hundredweight is needed to cover the costs of feeder cattle, feed, interest, other variable costs plus an overhead or fixed cost assessment. As a cattle feeder, you might believe price will go above $75.00 but feel a need to get some protection if the $75.00 price is reached.

The chart shows what happened. A price of $73.02 was reached on July 29, and then the direction of the price trend turned negative. Reports were showing large numbers of cattle being placed on feed, and the long-standing demand weakness for beef was also a factor. The market dipped to the $66.00–67.00 level, price levels that would mean up to a $108.00 per-head loss on a 1,200-lb. steer.

The moving average safety net program would definitely help. A short hedge was placed in early August near $72.00 and then lifted a few days later at a loss of some $.50 per hundredweight. The second sell signal came near $71.00 and the short hedge was lifted near $69.75. A third sell signal came near $69.25 and was lifted near $66.50

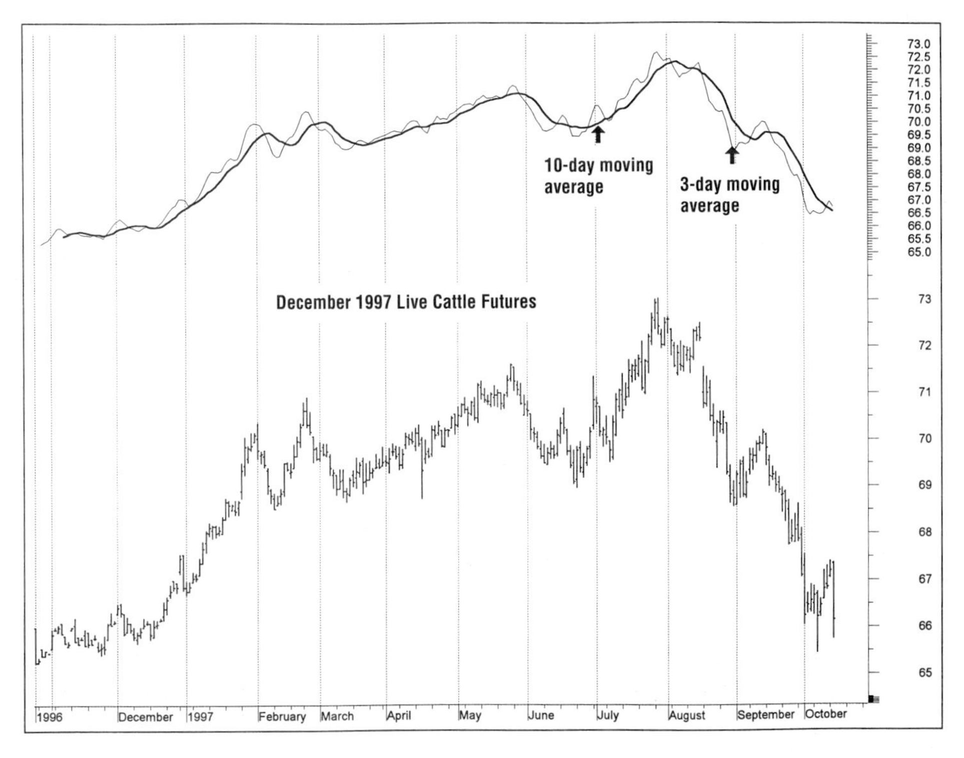

FIGURE 11.2
Demonstration of Moving Averages as a "Safety Net" in a Target Price Risk Management Program for Live Cattle

in early October. These three round turns generated a net of some $3.50 per hundredweight before commissions, and losses were reduced by some $42.00 per head.

In a market with sharp price movements as it trended lower, the moving averages helped to mitigate the financial burden on the cattle feeder. If the market moves still lower as December approaches, the protection in the form of short hedges would be reinstated on a sell signal from the moving averages.

With the grains and oilseeds, the decision on how much to price would still be a factor. Early in the year, the producer might opt to cash contract only 30–40 percent of expected production when the target price is offered the first time. If more is to be priced later *or* later pricing is to be at a higher price, the protection of the moving averages can be kept in place on the volume that has not been cash contracted. The procedure is thus a bit more complex with the grains, but is still quite manageable.

A simple program is to follow the moving-average signals on the number of bushels for which price protection will eventually be desired, say, 60 to 70 percent of normal production. As the initial target price objective is met and part of the crop is priced via a cash contract, the next position established via the moving-average signals would be on a smaller volume. Since the price target will be reached on a rallying market, it is unlikely that the moving averages will have short hedges in place when the initial cash contract is employed. If short hedges *are* in place, part of the short position in futures could be bought back to reflect volume now cash contracted.

The moving averages will not always be totally effective in providing backup protection, of course. The example with the cattle feeding program showed that only partial protection was present. If the market does work slowly higher to the target price, any hedges that are periodically placed and then lifted will probably be losing trades. *In an upward-trending market, it is virtually impossible for a selective hedging program that is periodically short in the market and then periodically out of the futures market to be profitable.* Any losses have to be viewed as a cost of having the safety net price protection in place. In markets that show major price breaks, *the moving averages will always have the producer short in the market* and provide the all-important protection without the extensive margin calls that can come when a short hedge is held in place while the market rallies.[2] In addition, the moving averages will never allow major accumulation of margin calls in an upward-trending market. Keep in mind that one reason this decision maker is using cash contracts is lack of financing for such outlays as margin calls.

Another simple refinement or extension that can provide a safety net for the target price strategy is to buy an out-of-the-money put option. By picking a strike price that would at least cover the variable costs of production, the producer has protection against a price break that might put the firm out of business.

The producer should buy put options to cover all of the expected production base that will eventually be cash contracted. In livestock programs, this would mean that essentially all the hogs or cattle would be priced if they are already in the feeding program. In grains, options should be bought relatively early on 60 to 80 percent of the

[2]It is important, remember, to use the correct set of moving averages for a particular commodity. The appendix to Chapter 5 provided suggestions based on our research and experience and shows extensions and refinements that could improve the performance of any selected set of averages. The potential user is encouraged to test several sets of moving averages to see which set appears to be most effective. It is a good idea to use the systems diligently "on paper" using an electronic market analysis package before actually taking market positions using the moving averages.

expected production or whatever level will eventually be cash contracted. One of the advantages of using put options is that in the event of a total or partial crop failure, there are no losses from short hedges in an upward-trending market. The only exposure is the premiums if the producer ends up pricing more via the options than is produced.

Buying an out-of-the-money put option keeps the cost of the program down and is consistent with the limited financial position. In addition, buying put options requires no sophisticated analytical ability. *The important point is that protection against bad price breaks is in place while waiting for the target price or prices to be offered.*

If the target price is reached and cash contracts are established, the option premiums will be forfeited if the market continues to trend up. If, after the cash contracts are set, the market turns lower, the put options have a chance to pick up value. It would be unusual to see the put option premiums at the close of the year exceed the premiums paid when the out-of-the-money put options were purchased, but that could certainly happen if the market declines sharply. There is thus always a chance that the put option will pick up at least some value and reduce the cost of the safety net protection if it is kept in place.

Another approach, of course, is to sell or offset the put options as the cash contracts are established. Premiums at the time the cash contracts are secured will typically be lower than when the options were bought since the price target will be above the levels offered by the market when the options are initially bought as a safety net measure. But the premiums will not necessarily be near zero. If it is early in the year, there will be significant time value still being reflected by the option premium. A $2.60 put option on December corn, bought at $.25 per bushel in late April, might still be trading at a premium of $.05 per bushel in early June when the December futures have traded up to $2.75 and the first price objective is met. If the options can be sold at $.05 per bushel, the effective cost of the safety net protection is reduced to $.20 per bushel plus commission costs.

Whether the options are sold or held should be determined largely by when the cash contracts are set (which will determine the time value of the options) and where the prices are in the projected price range for the year. If the options are bought as a safety net backup when the futures price is near the top end of the projected price range for the year, chances are they will later be in-the-money and provide a substantial premium if held. This can be more profitable than following a strategy in which the options are always sold when the cash contracts are established. Conversely, if the put options purchased are near the bottom end of the projected price range, the strategy of selling the options when cash contracts are set to realize any remaining premium tied to the time value would probably be more effective. Once again, we see the importance of fundamental analysis in projecting the probable price range for the year.

The target price strategy looks appealing, but it can be very dangerous. It is important that the potential user fully appreciate what will happen if the target price is never offered and there is no safety net in place. It could mean the demise of the business.

Using moving averages or a relatively inexpensive put option can provide a safety net of protection if the target price is never offered. You gain protection against a potentially ruinous price move, but the added feature will typically increase the cost of the program compared to just using cash contracts. The objective, of course, is protection against a major price break if the target price is never offered.

Trend Line Pricing An alternative to the target price approach is to establish price protection using the sell signals generated by trend lines. The trend line is one of the most simple chart techniques and can be used in a conservative hedge program by the risk-averse decision maker. Not much analytical ability is required, but the limited financial capacity could be strained unless the pricing is done via cash contracts.

If a trend line can be fitted to an upward-trending market early in the decision period, the decision maker can then use a close below the trend line as a signal to move aggressively in establishing cash contracts. Figure 11.3 illustrates with the March 1998 corn futures. A trend line connecting the lows in July and the lows in early October fits all the criteria of trend lines in Chapter 4 and allows the corn producer to relax. *There is no compelling reason to seek price protection until a change in trend is signaled via* a *close below the trend line.* The price risk associated with holding 1997 corn in storage is obviously still present, but it may be a type of risk exposure even the risk-averse producer will be willing to accept. Prices are in an uptrend, and as long as that positive direction is intact, the producer is benefiting. But you have to be prepared to extend that line and seek contract protection on a close below the line.

One additional appealing feature of this relatively conservative and simple strategy is the potential it offers to avoid suffering a major opportunity cost. If the market is in a major uptrend when the target price is reached initially, setting the price at that point via cash contracts can leave the producer open to the frustrations of watching the price levels move up after the price has been set in the cash contract.

If such a major uptrend does develop, the producer of a planned or growing crop still faces the tough decision of how much to price when the sell signal is generated at some level above the initial target price. The buyer in the cash contract will expect

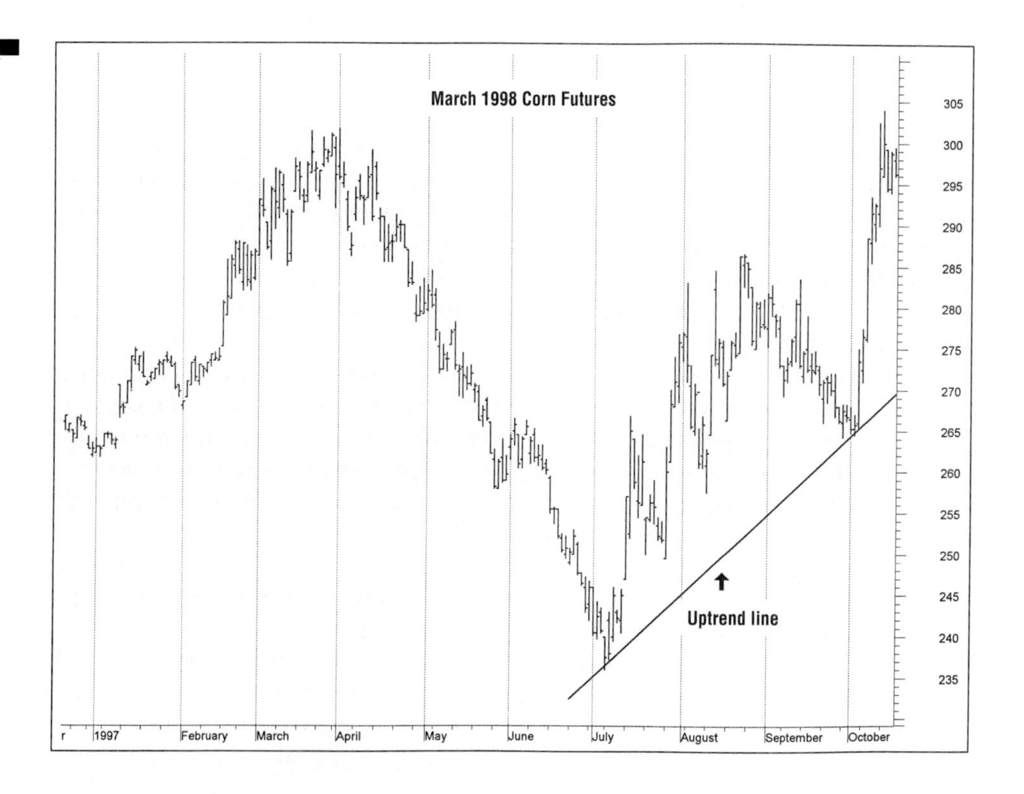

FIGURE 11.3
Demonstration of an Uptrend Line on the March 1998 Corn Futures Chart That Would Encourage Waiting on Placement of Short Hedges for a Storage Strategy

delivery of the physical product, and any "buyout" of the cash forward contract is usually expensive. *This is clearly a complication of cash contracts compared to a scale-up approach above a target price directly in the futures market or to a strategy using put options.* The short positions in futures can be bought back. For an options-based strategy, the only exposure in the event of a partial crop failure is the option premiums. Cash contracting more than is produced can be very expensive if the producer has to go into a high-priced cash market and buy grain to meet contractual commitments. Nonetheless, the appeal of the chart pattern shown in Figure 11.3 is apparent. As long as the uptrend holds and no close below the trend line is observed, the producer is in a position to benefit as the market moves above the target price. And for a stored crop, there is little worry about not having enough bushels.

There is the very real possibility, of course, that no easily recognized and sustained price uptrend will develop. If the market works higher without exhibiting a sustained trend, the producer can go straight to the target price approach. The objective is to get coverage at levels the firm can live with, and the target price approach can be used if the price targets are reached before a recognizable uptrend develops.

Clearly, there is the need for a safety net if the producer is monitoring the chart patterns and waiting for trend lines or target prices to appear. Either of the safety factors discussed in the target price section could be employed. Moving averages can provide protection, or the producer can buy an out-of-the-money put. Once a clear trend line is positioned by connecting two lows such as those in Figure 11.3, the need for a safety net is diminished. The producer now has "protection" via the anticipated sell signal. As soon as the prices rally from the lows in early October for three to five days, any short hedge position using moving averages (the averages should *not* be short, or would show a buy signal to lift short positions within a few days) could be removed or any put options sold at remaining premium value as the uptrend continues. *Then, it is very important that the markets be monitored carefully and that the hedges or cash contracts be set promptly when a close below the trend line generates a sell signal.*

A sell-stop-close-only order set on or just below the trend line will work on commodities where the exchanges accept the order. On the exchanges where the order will not be accepted, the producer will need to watch developments and place orders with the broker when the trend line is penetrated. The simple sell-stop order that is placed just below the trend line will also work and is accepted by all exchanges. Check with your broker to find out what orders you can use for your commodity. If needed, review the appendix to Chapter 4 which discusses types of orders in detail.

If the market is in an uptrend early in the production period, the sell signal generated by a close below the trend line can be an effective and conservative approach. A complication emerges in the form of having to decide how much to price when the sell signal is observed. If no clear trend line develops, the need for a safety net strategy is still present as the producer awaits development of a trend line or realization of target prices.

Intermediate Hedging Strategies

The decision maker described here is willing to accept some risk when the potential returns appear to justify the risk exposure. Financing is assumed adequate to allow the direct use of futures if that approach is preferred, and a separate credit line is available for margin financing. The margin line is presumed to be adequate to keep short hedges in place. The decision maker is presumed to understand moving averages and is comfortable with basic analyses in both the fundamental and technical dimensions. The analysis and related technique will be demonstrated across several commodities and alternative situations, but the focus then moves to a corn producer and the markets are followed through a production year. The objective is to make sure you recognize that the technique is broadly applicable and then to concentrate on a specific situation and follow it through.

Strategy 1: Moving Averages The use of moving averages to direct a selective hedging strategy was discussed in Chapter 5 and was discussed briefly in the previous section on conservative strategies. Figure 11.2 demonstrated the basic framework.

A short hedge is established early in the planning period each time a sell signal is generated by the moving averages. The producer would key off the sell signal to place hedges; the user of raw material would key off the buy signals to place long hedges. The effectiveness of the strategy prior to planting of the crop, or prior to the placing of livestock in a production program, may exert an influence on enterprise selection and on the level of production in that particular enterprise. If the use of moving averages to place long hedges on feeder cattle has already generated substantial profits before the cattle are placed, for example, this result could mean more cattle will be placed and fed than had originally been planned.

The overall objective of the strategy is to have protection in place against most of any major price break via short hedges. Alternatively, the objective is to gain protection against most of any price increase in the event of a long hedge. Since the moving averages are a trend-following system, protection will never be complete. The hedger's position will vary between being in the futures with a hedge in place and being a cash market speculator. The short hedger never has a long position in futures, and the long hedger is never short in the futures. If this point is not apparent, you should pause and go back over this in detail. *Remember, the potential short hedger places short hedges on sell signals and then buys them back or offsets on a buy signal.* If the market continues higher after the buy signal, the hedger is now a cash market speculator and the short hedge will not be replaced until the next sell signal occurs.

For the system to be effective, a set of moving averages must be selected that strikes an effective compromise between two somewhat competitive needs. First, *the averages must be sensitive enough to generate a signal in time to capture the bulk of the price moves*. Second, *the averages must not be so sensitive to market moves that too frequent signals are generated.*[3]

The importance of the correct set of averages was illustrated in Chapter 5. A quick-turning set of averages (such as 3–10) will perform differently than will a slow-turning set of averages (such as 9–18). The difference in performance comes at two primary points. When the market is staging a major move, there is a tendency for the market to either pause and consolidate or to make corrections of each price wave before moving on to higher or lower prices. If the consolidation patterns are treated as a market top or bottom by the short set of averages and a sell and subsequent buy signal is generated, that trade is likely to be a losing trade as the upward price trend resumes. The 9–18 combination is slower to turn, is less likely to incorrectly call a

[3]You might wish to review the discussion in Chapter 5 and the appendix to Chapter 5 that discusses refinements to the moving-average systems. The purpose of the refinements in the appendix is to generate a system that more nearly meets the need for a compromise discussed here.

consolidation pattern a top, and will keep the long hedge in place or keep the potential short hedger in the posture of a cash market speculator. This performance issue was discussed in Chapter 5 and in the appendix to Chapter 5.

The second visible difference in performance comes when the market is trading sideways and is exhibiting frequent and choppy price moves. Depending on the frequency and amplitude of the moves, the shorter set of averages that generates more signals may be able to avoid being wrong in the market and may even generate primarily neutral or profitable trades. A longer set of averages that is slower to generate signals can be caught generating a sell signal when the market is ready to turn higher for several days, or a buy signal when the market is ready to turn lower.

Critics of moving averages are quick to point to the tendency to generate frequent and losing trades in the choppy markets that are showing no major trends. Advocates will argue that any losses during such periods are nothing more than insurance premiums that have to be paid to ensure that the short hedge, for example, will be in place for the bulk of the major price break and have the producer off the short hedge for the bulk of the major price rally.

It is easy to see both the strong and weak points of the moving averages using the December 1997 lean hog futures shown in Figure 11.4. Picking up in early February, the 3–10 set of moving averages would have generated 10 sell signals by mid-October. Examination of the performance of the moving averages shows several sell signals that were, in hindsight, wrong. The brief rally in early June, for example, generated a buy signal and removed the short hedges—only to see them replaced a few days later. Essentially the same thing happened in early September. But the other side of the issue, the advantages of moving averages, is also clearly present.

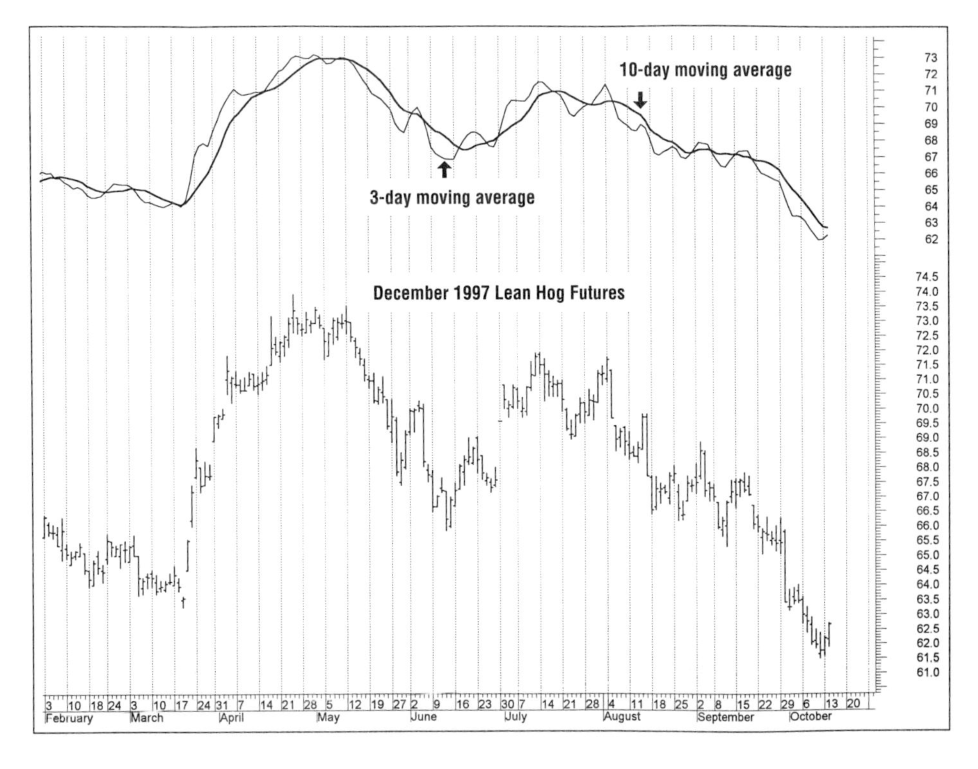

FIGURE 11.4
Performance of Moving Averages in a Major Bull Market in Hogs

Early short hedges would have been lifted by a buy signal on March 20 when the market closed at $65.50. The benefits of much of the rally that reached $73.90 on April 24 were realized. The next sell signal occurred on May 5 at a close of $72.30. Those short positions were quickly bought back and then the system turned short again at $72.45 on May 13. Those hedges were lifted on June 3 at $69.95, and then replaced again a few days later. During August and September, the volatile market brought an "on again, off again" status that you would prefer not to see, but since the market was trending lower, most of the transactions would have been profitable after commissions.

The key is that the positive features of moving averages are present. During the major price rally in March and April, the benefits of a rising cash market were there. No margin calls accumulated. Then, protection was in place for much of the price break during May and June. A safety net feature is clearly present. And if you want to eliminate those "mistakes," you can do it by going to a longer set of moving averages such as the 9 and 18. But you pay in other ways for the reduction in trades. With the 9 and 18, the buy signal to lift short hedges would not have been seen until March 24—and the closing price is up to $68.25. The subsequent sell signal to replace the short hedges would not have occurred until May 14 at $72.10. These hedges were lifted on June 26 at $67.32, an impressive performance during the downtrend.

Overall performance of such a moving average strategy will tend to smooth the net return flow over time. Whether the average profits will be above or below the average for the cash speculative position will depend on the market patterns (frequency and amplitude of price moves) for the particular commodity and how effective the decision maker is in selecting the correct set of moving averages.

Figure 11.5 documents the advantages of moving averages as discussed in Chapter 5 and shows why many users swear by moving averages. The average monthly per head return to the computer-simulated cattle feeding operation in the research on which this chart is based is increased *and* there is a major improvement in the stability of the monthly cash flow over time. The distribution of the monthly net returns is altered. There are fewer large losses and a higher frequency around the average level of net returns. The large windfall gain to the cash market speculator is also eliminated by the losses that a moving average strategy tend to register during an upward trending market, but the elimination of the large losses is the key. *For a decision maker who wishes to avoid major exposure to price risk and who does not have unlimited capital, the moving averages may in fact be a preferred strategy*. Profit performance can be improved and/or stabilized with a strategy that does not require exceptional capacity as a fundamental or technical analyst by simply following a workable set of moving averages.

Moving averages assist the decision maker in efforts to gain protection against major price moves and put the selective hedger in the position of being a cash market speculator when the price trend is in his or her favor. The big disadvantage of moving averages is the losses that tend to accrue in choppy, sideways markets. It is important to select the correct set of moving averages for a particular commodity, and the system must be employed in a disciplined way. The signals must be followed and not overruled in a subjective or judgmental way.

Strategy II: Moving Averages Plus As referenced in Chapter 5, a number of analysts have conceptualized and tested refinements that they argue will improve the performance of even the correct set of moving averages. Coverage here cannot be exhaustive, but selected refinements can be introduced fairly easily. The references at the end of the chapter provide more detail. You are reminded that the appendix to Chapter 5 covers these refinements in a procedural context.

One approach is to use a penetration rule. When a set of moving averages makes a mistake, such as prematurely calling a top in the market, the decision maker usually sees the short average drop only slightly below the longer average and then quickly move back above the longer average. This scenario developed on the hog futures shown in Figure 11.4 during mid-April. Note the 3-day average equaled the 10-day, but did not move significantly through it.

The initial long positions were still correct, however, and there is no top in the market as yet. To avoid this type of mistake, and it is evident on the hog chart in Figure 11.4 on other occasions, an added requirement is imposed. The shorter average must penetrate the longer average by a preset amount or the signal is ignored.

Exactly what penetration requirement to impose is a researchable issue. Available research referenced in Chapter 5 suggests increments around $.15 per hundredweight for livestock and $.01, $.02, and $.03 for corn, wheat, and soybeans, respectively. Historical data sets can be analyzed by computer programs designed to isolate the optimum penetration increment, and that type of analysis is now within your reach with spreadsheets or computer software packages readily available in the market.

A second refinement is to use a third or leading moving average. If the base set of averages is a 9–18, for example, a shorter average can be used to confirm the buy or sell signal. A widely used and widely available set (on many electronic systems) is the 4–9–18, in which the sell or buy signal is confirmed by the 4-day moving average. Once again, you will find discussion in Chapter 5 and detailed discussion of this refinement in the appendix to Chapter 5.

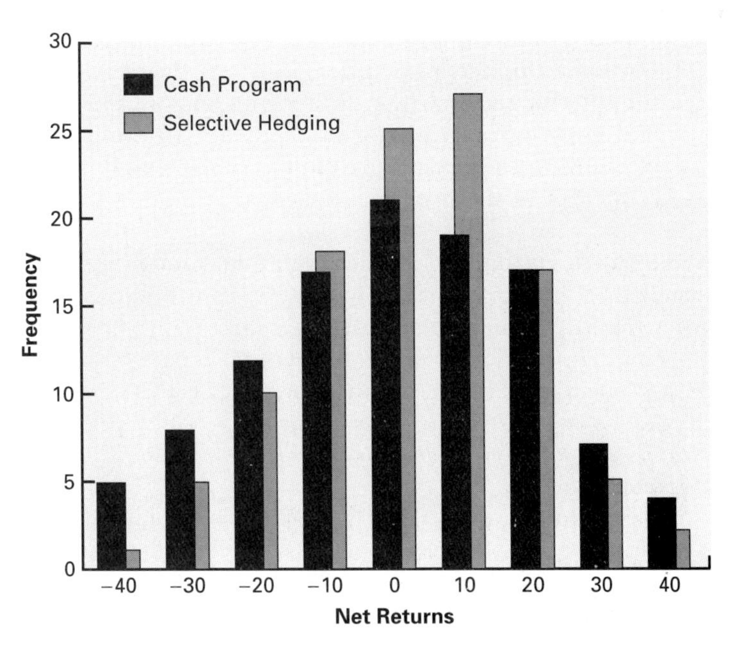

FIGURE 11.5
Impact of a Moving Average Selective Hedging Program on the Mean and Dispersion of Net Returns to Cattle Feeding in the Texas-Oklahoma Area

Source: Don Riffe and W. D. Purcell, *Hedging Strategies to Protect the Financial Positions of Cattle Feeders and Lenders*, Okla. Ag. Exp. Sta. Bul. B-743, Stillwater, Oklahoma, June 1979.

If the 9-day average crosses the 18-day average from above, the sell signal is confirmed if the 4-day average leads the 9-day average and is below the 9-day average when the sell signal is generated. The logic to the approach is appealing. If the signal generated by the 9-18 is one that, *ex post*, will be seen as a premature calling of a top in the market, then the use of the confirmed average can help eliminate the mistakes. In an upward-trending market that consolidates briefly and then resumes the upward trend, the 4-day will turn up more quickly than the longer moving averages as the consolidation pattern is completed, and the premature sell signal may not be confirmed.

Both the penetration rule and the leading moving average have the potential to reduce the number of trades and improve the net performance of the selected set of moving averages. Performance will be improved in instances in which premature tops and bottoms would otherwise be signaled and in the choppy sideways pattern which frequent trades would otherwise be generated.

A third way to improve the performance of moving averages, also introduced in Chapter 5, is to use the relative strength index as a safeguard against incorrect signals near the tops and bottoms in the market. If a buy signal is generated when the RSI is above 70, that signal could prompt the buying back of short hedges in a market that is overbought and ready to turn lower. Buy signals that occur when the RSI is above 70 could be ignored.

Near the bottoms in the market, sell signals that occur when the RSI is below 30 would be ignored. Thus, the RSI provides protection against the incorrect signals that may occur near tops and bottoms. It is mistakes at those levels that detract from the performance of the moving averages.

In the choppy and sideways markets, the RSI will not offer significant help. It is seldom that the markets are overbought or oversold, based on typically used RSI levels, in this type of market pattern. Some other safeguard must be applied. The penetration rule or use of a third or leading moving average can be employed.

The use of penetration rules, leading (shorter) moving averages, and the RSI can reduce the number of trades signaled by moving averages and improve their performance. Much of the improvement comes via preventing the premature calling of tops or bottoms and the related reduction in the number of trades.

Strategy III: Fundamental and Technical Analyses Combined The use of the standard bar chart buy and sell indicators is one of the most widely used approaches to a selective hedging program. Hedgers must be in a position to finance margin calls, but they will seek to lift or offset the hedge when the appropriate buy signal emerges. That should preclude the extended margin calls that come with a more conservative approach that does not allow the lifting and possible replacement of hedges. *The decision maker who selects this strategy must have both fundamental and technical analytical skills.*

On the fundamental side, the first need is to project the likely direction of any major price trends and/or to establish the probable price range within which trade will occur. It does no good to watch for a chart signal to suggest a short hedge be placed around $4.00 in corn or around $80.00 in fed cattle if those prices have little or no chance of being reached. Before turning to a demonstration of the use of the bar chart signals, therefore, it is important to look at fundamental analysis at a level the average decision maker being discussed here could be expected to handle.

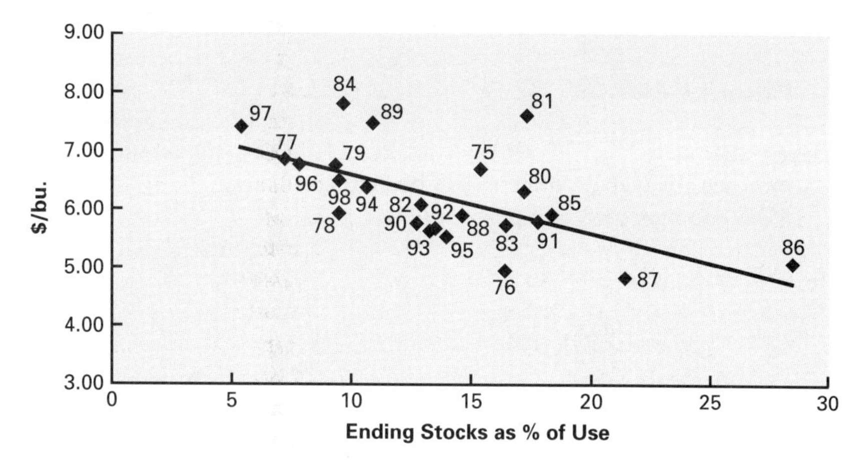

FIGURE 11.6 The Relationship Between Price and Ending Stocks as Percent of Use, Soybeans, 1975–1998

For grains, the relationships between price and ending stocks as a percent of total use for the crop year provides a simple but effective way to analyze the supply–demand fundamentals. Developments build on the discussions and presentations of Chapter 3 and extend the analysis to a particular decision situation. Soybeans will be used to illustrate.

The World Agricultural Supply and Demand Estimates reports first introduced in Chapter 3 provide the data needed to estimate a relationship between ending stocks and prices. Figure 11.6 shows the fitted relationship for soybeans that was developed in Chapter 3.[4] The years shown in the plot are the second year in the crop year, so 97 refers to the 1996–97 crop year that ended August 31, 1997.

Early in the year, an estimate of the yearly average price received by farmers can be generated by using the estimates of ending stocks and total use from the most recent supply–demand report released by the USDA. Calculate ending stocks as a percentage of total use and locate that point on the horizontal axis. Then, move up vertically to locate the corresponding point on the function. By coming across horizontally to the vertical axis, a price estimate is generated.

Table 11.3 illustrates the calculations from the algebraic equation with a final price estimate of $6.84 per bushel for soybeans. This approach looks more sophisticated, but reading the price off the chart will be very effective. There is too much variation in this relationship to worry a great deal about precision in numbers.

The 5.4 used in the equation in Table 11.3 reflects the last available estimate of total usage (2.444 billion bushels) and ending stocks (132 million bushels) for the 1996–97 crop year. The $6.84 estimate is for the average cash price to farmers for the 1996–97 year. At the time of the update, the November 1997 soybean futures were trading near $7.00 with a life-of-contract trading range of $5.97 to $7.17. The midpoint of that range is $6.57. Both the $6.57 and the current trading level of $7.00 are generally consistent, after basis allowances, with a $6.84 cash price.

[4]The equation is in Appendix 3B of Chapter 3. For the decision maker without access to a fitted equation, a useful approximation can be generated by simply looking at the scatter plot. Sketch in a curvilinear or linear function that is placed so that deviations or "misses" by the points above the line offset or match the deviations below the line. In other words, the deviations from the line that is being sketched should sum to zero. The result will be a quite useful approximation of a fitted mathematical function.

TABLE 11.3
Calculating an Estimate of the Season Average Price for Soybeans Using the Price–Ending Stocks Relationship

The general form of the model is:

$$PR = a + b_1 ES$$

where

PR = season average price ($ per bushel) and
ES = ending stocks as a percent of use.

The fitted model was:

$$PR = \$7.38 - \$0.0995\ (ES).$$

If the estimate of ES is 5.4, the projected price will be

$$PR = \$7.38 - \$0.0995\ (5.4) = \$6.84 \text{ per bushel}$$

Given the estimate of the general price level for the year, attention then turns to the need for some idea of the price range that will be needed to capture the average price for the year. A very simple approach is to examine the deviations above and below the line and convert those to price equivalents. Most crop years in Figure 11.6 would be within $ 1.00 per bushel of the fitted line. A range of $6.84 plus and minus $1.00, or $5.84 to $7.84, would contain the average price with a very high degree of probability.

If the decision maker understands basic statistics, a confidence interval can be calculated around the estimated price using the standard error of the regression. A 95 percent confidence interval, the interval within which 95 of 100 repeated observations would be expected to fall, is calculated as price plus and minus 2 standard deviations.

Readers with awareness of statistical measure of dispersion, such as standard deviation will have no problem with this approach. Others might prefer to just use the deviations from the fitted function based on visual inspection and not get involved in the statistics.

Whichever the approach taken, it is important to keep in mind that the relationship is between ending stocks as a percent of use and the average cash price for the crop year. Day-to-day or any short-run prices would be expected to vary over a wider range than the range that is likely to contain the average price for the year.

A simple but effective alternative to the single-equation model approach involves the use of an *elasticity framework*. For many users, this may be the preferred approach because the only data required are estimates of the period-to-period changes in quantity produced. This approach was covered in general terms in the discussion of fundamental analysis in Chapter 3.

Table 11.4 expands the presentation to another commodity and demonstrates an approach to estimate slaughter hog prices. A demand elasticity of –0.6 is employed and the expected price for year 2 is generated. The usefulness of the approach depends on the accuracy of the elasticity coefficient, and it requires that the user be fully aware of the implicit assumptions being employed. Keep in mind that demand elasticity is simply percent change in quantity divided by percent change in price. In Table 11.4, the expected 5 percent increase in production translates to an 8.3 percent projected decrease in hog prices.

Using projected quantity changes and the own-price demand elasticity of –0.6 to generate a price estimate *assumes that the demand for hogs is constant on a period-*

TABLE 11.4
Demonstration of the Elasticity Framework in Projecting Year-to-Year Hog Prices

Year t, quarter 1: Price = \$53.00 per hundredweight

Based on projections from the USDA quarterly *Hogs and Pigs* reports, production for year t+1, quarter 1 is expected to increase by 5 percent. Using a demand elasticity for hogs at the farm level of –0.6, the expected price change for year t+1, quarter 1, would be calculated as follows:

$$-0.06 = \frac{+0.05}{X} \qquad X = -0.083$$

where X is the expected change in price given the elasticity of –0.6 and assuming that demand for hogs does not shift.

Therefore:

Price year t + 1, quarter 1
= \$53.00 – \$.083 (\$53.00)
= \$53.00 – \$4.40
= \$48.60

---

to-period basis. If the demand function itself is shifting, then the price estimate will be too high or too low, depending on the direction of the shift in demand. The price estimate generated under the initial assumption that the level of demand is constant is an excellent place to start, however, because demand does not typically change a great deal from quarter to quarter or even year to year. *It is the price–quantity relationship captured by the demand curve that dominates in importance in determining price, and shifts in supply along that demand curve can be converted to price changes using the elasticity framework.*

An obvious shortcoming of the elasticity framework approach is the absence of any formal way to place a confidence band around the estimate to help in generating expectations of the probable price range for the year. By going back and applying the procedure across several historical years, you could generate estimates of the errors in the estimation process and the variation of monthly or quarterly prices around the estimate. This process was mentioned in Chapter 3, and the results, in terms of accuracy, appeared reasonable and useful. It is important that you keep in mind that it is just the expected changes in quantity that are being used to estimate price through the elasticity framework. Demand is being held constant. A model this simple, it is really just $P = f(Q)$, is not going to be completely accurate in a complex marketplace.

Any possible shifts in demand can be incorporated, of course. Demand will shift due to changes in tastes and preferences, changes in consumer incomes, and changes in prices of other products. On a year-to-year basis, tastes and preferences are not likely to vary a great deal, and consumers' incomes will not change enough to make a major difference. In the short run, changes in the prices of important substitute products can influence demand.

A published demand analysis[5] estimated the cross-elasticity between beef and pork to be 0.369. This suggests that a 1.0 percent change in beef prices would prompt a 0.369 percent change in pork consumption in the same direction, all other factors equal. This cross-elasticity measure of the relationship between beef and pork provides a mechanism to account for demand shifts due to changes in the price of substitute products.

[5]Reference is to Wayne D. Purcell, *Analysis of Demand for Beef, Pork, Lamb, and Broilers*, Res. Bul. 1-89, Research Institute on Livestock Pricing, Agricultural Economics, Virginia Tech, Blacksburg, VA, July 1989.

Incorporating this into the illustration on year-to-year changes in hog prices is a bit complicated, but it is worth the effort. *At a minimum, the decision maker should know the direction in which hog prices would move for given changes in the cattle sector.*

In the example in Table 11.4, the quantity of pork for quarter 1 of year $t + 1$ has been predicted and set. If cattle prices are projected by the USDA or a private-sector advisory firm to increase by 5 percent, what will this mean to the hog prices? To convert to price equivalents, we can look at it as if the expected 1.85 percent ($0.05 \times .369$ gives 0.0185) increase in quantity of pork is being denied the consumer and convert it to price implications as follows:

$$-0.6 = \frac{-0.0185}{X}.$$

Here, X is the percentage change in price that is associated with the inability to increase quantity—that is why the negative sign is shown on the –0.0185. Solving for X, we get

$$X = 0.031.$$

The revised hog price for quarter 1 of year $t + 1$ is therefore

$$\$48.60 + .031\ (\$48.60) = \$50.11.$$

This analysis takes liberties with the underlying theoretical framework. Technically, the cross-elasticity of 0.369 is applicable only at the retail level and should not be applied directly to the live hog price calculations. *But the primary need here is to make sure you appreciate the importance of our decision maker understanding the direction of the expected price impact.* Beef is a substitute for pork. If beef prices are expected to go up, the demand for pork—and the derived demand for hogs—will tend to increase as well. The final price of $50.11 shown in Figure 11.7 comes from (1) a price decrease from $53.00 to $48.60 if the only change was in pork supplies, and (2) a shift in demand for pork due to higher cattle prices that pushes the price back up to $50.11.

This conceptual framework as a way of thinking is what is important—and it has broad applications. Without knowing the detail of how soybean meal and corn substitute for each other in livestock ratios, it is important to understand that increases in the price of corn will increase the demand for soybean meal and therefore increase the price of soybean meal, other things being equal. A decision maker of intermediate ability needs this framework.

Given an initial projection of price for a livestock or a grain commodity[6] and the establishment of a possible price range, the decision maker is then in a position to use

[6]Keep in mind that the USDA is offering the results of this type of analysis in its projections on production and prices in the *Livestock, Dairy, and Poultry Situation and Outlook* report and situation and outlook reports for feed, wheat, oilseeds, cotton, and wool, and so on, available by subscription. It is thus less important that you be able to do the analysis than it is that you at least understand what is done by the USDA or the state-level extension specialist. The available forecasts will then be used with more confidence. These outlook reports are available at http://www.mannlib.cornell.edu/usda/ (select ERS button) on the Internet, and the supply–demand reports are at the same address (but select the WAOB button).

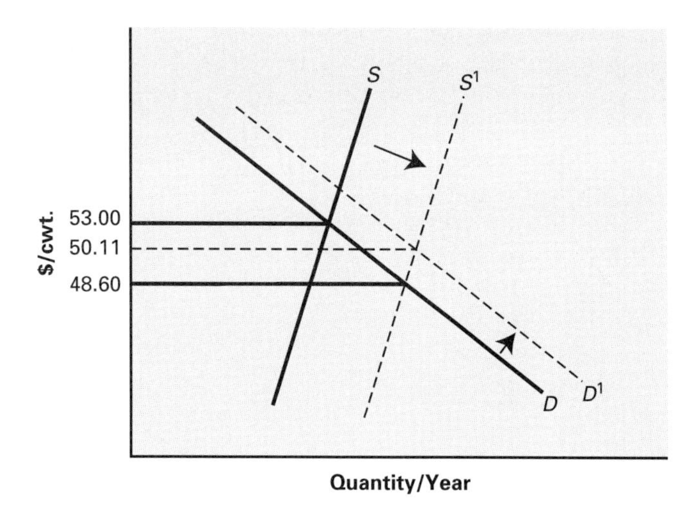

FIGURE 11.7
Demonstration of Price Implications of a Change in Supply and Demand Shift Due to Change in Price of a Substitute Product

the chart signals. The bar chart patterns likely to be employed by the user of intermediate strategies are the same patterns that were discussed at some length in Chapter 3. In this chapter, the use of the chart will be demonstrated via tracing through the array of decisions that would face a corn producer starting in the spring of the year after the decision has been made as to how much acreage to plant in corn. *Keep in mind that once an estimate of the direction of year-to-year changes in price and a probable price range have been established, the procedures will then be essentially the same whether it is a growing corn crop, hogs in a feeding program, or cattle that have just been placed in the feedlot.*

The assumptions being employed are the following:

1. The producer will follow a selective hedging strategy;
2. There are adequate funds in a margin line to allow forward-pricing directly in the futures market;
3. Historical basis data for the fall or harvest period are available;
4. The producer understands the use of options; and
5. Historical data indicate the producer will harvest at least 75 percent of his or her normal yields in any given year.

Prior to employing chart signals, fundamental analysis is needed to determine the general level of price expectations. Alternatives vary, ranging from formal price forecasting models to simply plotting basic price-quantity relationships. The need is an estimate of the probable price range within which bar chart signals can be employed in a selective hedging program. The user must either do the analysis or have some understanding of the process behind forecasts available from the USDA, university extension specialists, or private consultants and advisory firms.

FIGURE 11.8
Bar Chart for December Corn Futures

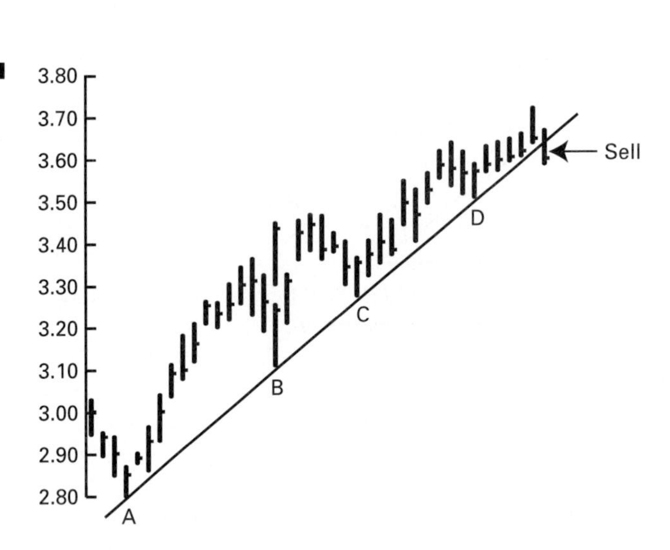

Pricing Decision Process: Bar Chart Signals In late March, the producer starts monitoring the December corn futures. The contract has been trading since the previous August and shows the pattern exhibited in Figure 11.8.[7]

An initial sell signal is generated in mid-April when an uptrend line is penetrated. Figure 11.8 demonstrates, with the close below the trend line confirming a reversal in the direction of price trend. The producer notes that the trading volume is relatively high the day the trend line is penetrated, and he also observes that the RSI had been approaching an overbought condition. Both of these developments reaffirm the sell signal, but the close below the trend line is sufficient. The producer sells enough 5,000-bushel contracts on the CBOT to forward-price or hedge 30 percent of expected production at $3.50 per bushel by placing a limit-price order with the broker the next morning.[8] The order is placed at the closing price for the prior day, the last day shown in Figure 11.8, and is filled shortly after the opening of trade. The producer knows that the trading range for any particular day will include the previous day's close most of the time.

With short hedge protection in place on 30 percent of the expected crop, the producer now faces a decision of either (1) when to lift the hedges, or (2) when to add to the hedges if no buy signal is generated before another sell signal emerges. Corollary with these issues is the perpetual question of which trend lines to draw, a question that is always there for the chart analyst. Given the price scale on the charts, the producer tries to stick with trend lines that (1) are not more than 45 degrees in slope and (2) have the two connecting points at least 10 trading days apart. The idea, as was discussed in

[7]For ease of exposition, the constructed chart patterns shown in this chapter will not represent as many trading days as would be shown during an August-to-March period. This abstraction should not in any way detract from the usefulness of the developments in the chapter. The objective here is to demonstrate, not to try to convince the reader that the chart patterns appear on real-world charts. Those illustrations have been shown throughout the book.

[8]The 30 percent is a judgment call that may be influenced by how high in the projected price range for the year the market is trading when the sell signal occurs. If the $3.50 price is near the 50th percentile in the price range, for example, the producer may be less aggressive than if the sell order had occurred near the high end of the expected price range.

Chapter 4, is to try to stick with the major trends and avoid getting caught up in the short-run gyrations in the market.

After the short hedges are established, the market works lower for several days. A correction of the dip in prices then emerges, and the rally retraces roughly 38 percent of the down move in price. Open interest declines on the price rally, and trading volume is relatively light. Both patterns suggest a short-covering rally and the producer anticipates that prices will turn lower again.

A relatively weak close on the day the 38 percent correction is completed suggests that the rally has run its course. The next day the market gaps down and the downward move is resumed. After three days of lower prices, a downtrend line could be drawn across the price high for the day the 38 percent corrective rally was recorded and the producer starts to wish he had more than 30 percent of the crop hedged. But it is still early and the planting process is just starting to move into full swing. Remembering the need to be disciplined, the producer refuses to rush in and sell in the now-lower market. A decision is made to monitor the chart actions and wait to see what develops. In actuality, the producer is speculating in the cash market on the 70 percent of the crop that is not priced.

Later in the planting period, the weather patterns turn dry and there is talk about the possibility of germination problems. The market responds to the concerns about the weather, and prices start to move higher. In early June, the market closes above a downtrend line that could now be shown on the chart, and the short hedges are lifted by a buy-stop order that had been placed about 1 cent above the trend line and lowered every day by the producer's broker. The hedges are lifted at $3.20, giving a net profit before commissions of $.30 per bushel. *The producer is now a speculator in the cash market, has no hedges in* place, *and is elated with the success to date* with *the selective hedging efforts.*

The markets have a way of restoring humility, however. A resistance plane is drawn across the high recorded on the earlier 38 percent correction, and the short hedges are replaced by a limit-price sell order at $3.29 as the market rallies toward that plane. Consistent with accepted techniques, the producer places the order just under the resistance plane that crosses near $3.30. After all, prices were as low as $2.80 earlier, and the producer remembers wishing that more than 30 percent of the expected crop had been hedged. It is early June and most of the corn is up and growing nicely. This time 50 percent of the crop is hedged.

But the dry weather patterns persist, bringing changes in the USDA's estimates of supply for the year and the market pushes up toward the life-of-contract high at $3.60. Several margin calls are answered and the producer watches as the price levels climb quickly higher. After the second consecutive close above the contract high at $3.60, the producer follows through on preestablished plans and lifts the short hedges that were placed earlier at $3.29. The broker, following an agreed-on procedure in the marketing plan involving producer, banker, and broker, had entered a market order to buy near the close as the second close in new, higher prices emerged. The order is filled at $3.65, yielding a net loss of $.36 per bushel before commissions. Figure 11.9 illustrates this. *The producer now has a net per bushel loss in the futures account, but is back to a cash market speculator status and is in position to benefit if the market continues to rally.*

A few days later, in June, widespread rains come to the Cornbelt, and the market moves rapidly lower. A key-reversal top is recorded. A limit-price sell order at the closing price of the key-reversal day is not filled on the following trading day. The market moves lower, and the producer hedges 60 percent of the expected crop via a sell-stop

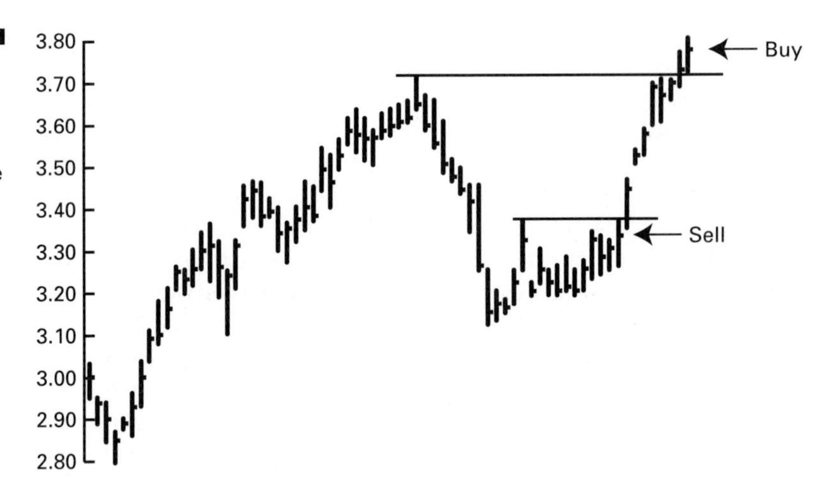

FIGURE 11.9 Demonstration of Covering Short Hedges at Contract Highs on the December Corn Chart

order placed under the latest uptrend line. A sell-stop order is placed at $3.45 but is filled at $3.42 as the market moves quickly lower. Figure 11.10 illustrates this action.

The next day, the market gaps lower. The producer places a limit-price sell order near the bottom of the gap, and hedges an added 15 percent of his expected crop at $3.34 four days later. At this point, 75 percent of the crop is hedged at a weighted average futures price of $3.40 (60 percent at $3.42, 15 percent at $3.34). With an expected harvest-period basis of –$.40, the corn is forward-priced at $3.00.

The downtrend then continues and the producer monitors the situation carefully. The producer can now reasonably expect 85–90 percent of normal yields, perhaps more, and decides to add price protection on another 15 percent of expected production using a put option if an attractive selling opportunity emerges. Watching the chart patterns, the producer decides to add the last increment of protection if a 50 percent correction to the last price high in the bottom of the chart gap develops.

The 50 percent correction does not materialize, but the producer's vigilance is rewarded when a consolidation pattern that looks like a bear flag starts to develop. After about seven trading days, the flag formation is complete. After a close below the flag portion of the formation that has developed near the $3.10 level, a $3.10 December put is purchased at a premium cost of $.15 per bushel. It is now mid-August. Figure 11.11 records the chart patterns and the latest action to buy the $3.10 put.

FIGURE 11.10 Replacing Short Hedges on December Corn After a Break of a Trend Line

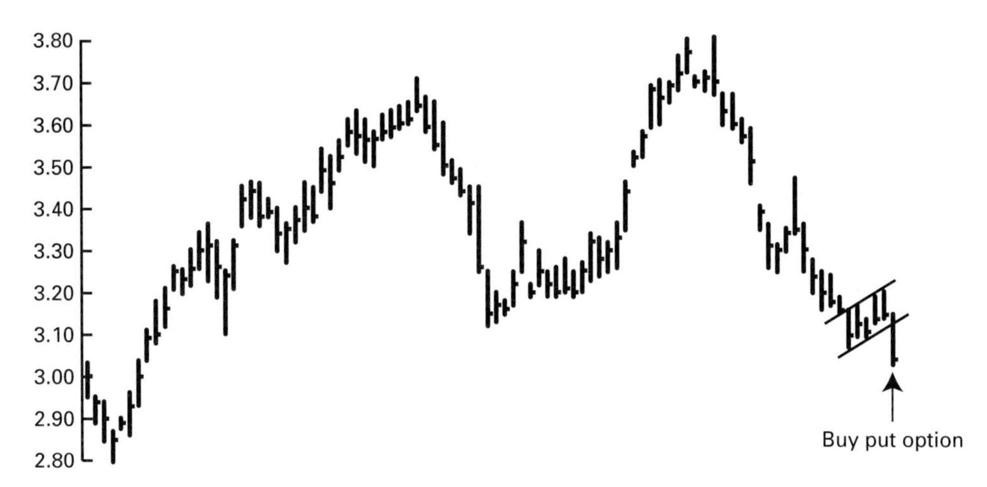

FIGURE 11.11
Adding Price Protection by Buying a Put Option via Sell Signals on the December Corn Chart

Using the put option on the last 15 percent of the crop reflects an important advantage of options to the producer. If 90 percent of a normal crop were hedged in the futures market, the producer faces potential problems if the August period brings a return to dry and hot weather. A drop in yields below 90 percent of normal would mean any rally in the futures market is not completely covered in the cash market. *Losses in the futures account would then be at least partly out-of-the-pocket losses, not just opportunity costs.* In the event of a partial crop failure, the options have an advantage. Losses in an upward-trending market are restricted to the $.15 option premium. In a higher priced market, the $3.10 put option will be worthless and the producer would just allow the option to expire. This part of the pricing program, you might decide, will not be treated as a selective hedge but as an approach that makes you comfortable with pushing pricing protection up to 90 percent of normal yields.

The downtrend in prices continues. As the market approaches life-of-contract lows at $2.80, you must, as a producer, monitor the situation carefully. Buy orders are placed at $2.81 to lift the hedges for 75 percent of the expected crop in the futures market, and the order is filled. A few days later, however, the market records two consecutive closes below the $2.80 support plane, and the short hedges are replaced at $2.75. Then, two days later, a key-reversal bottom is recorded and the market closes back above the old $2.80 life-of-contract low. On the next day, you have to conjure up the discipline to lift the short hedges again at $2.83. Figure 11.12 highlights the action around the $2.80 support plane during September. The market then works higher as the October *Crop Production* report estimates yields slightly below prereport expectations. No hedges are in place as the crop is sold in late October. The $3.10 put is sold at a premium value of $.20 with the December futures trading at $2.89.

Table 11.5 summarizes the results of the trades during the year. Before commissions and any assessment of interest on margin funds, the trades net $30,300 assuming the normal crop would be 100,000 bushels. That result translates into an improvement of $.303 per bushel when prorated across the entire crop, an improvement of $.404 if prorated across the 75,000 bushels priced in the futures market, or $.337 per bushel when considered across the 90,000 bushels priced either in the futures or in the options.

Clearly, you would have fared well as a producer during this particular year. If the harvest-period basis turns out to be around –$.40, then the cash corn would be

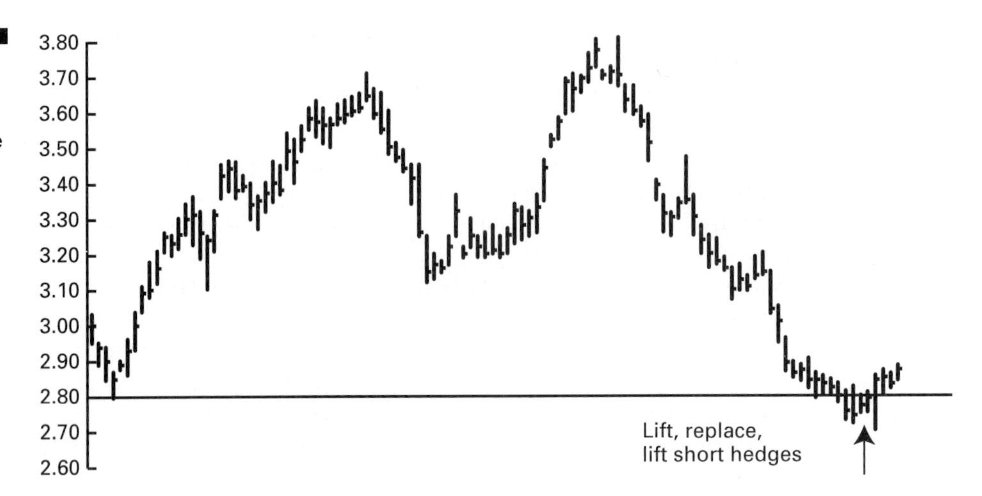

FIGURE 11.12 Actions in a Selective Hedging Program at the Support Plane on the December Corn Chart

sold around $2.50. After allowing for commissions and interest on margin funds, the net from the futures and options program would be nearly $.30 per bushel. Add that to the $2.50 cash market price, and the effective price is $2.80 for a normal crop of 100,000 bushels.

Examined from another viewpoint, the effective or realized price of $2.80 would require a hedge at $3.20 given the –$.40 harvest-period basis expectation. That means any producer following a conservative one-time approach to hedging would have to place the hedge by selling December futures at $3.20. *It is difficult to conceive of a situation in which any conservative producer would have priced 90 percent of expected production at $3.20 or higher*. Producers who are interested in a more flexible posture and willing to be selective hedgers thus have a chance to improve their final positions.

It is important to remember that during some crop years, it would be virtually impossible for a producer following a selective hedging program to improve over a conservative hedge, or to improve the final average price in comparison to a cash market speculative position. *But a better average price is not the only, or even the most important, objective of the selective hedger.* Compared to the cash market speculator, the disciplined selective hedger will have the significant added advantage of being

TABLE 11.5 Summary of Trades During the Year in the Selective Hedging Program

Hedge Placed	Hedge Lifted	Quantity (1,000 bu.)	Net
$3.50	$3.20	30	$9,000
3.29	3.65	50	–18,000
3.42	2.81	60	36,600
3.34	2.81	15	7,950
2.75	2.83	75	–6,000
Put option bought @ $.15, sold @ $.20.		15	750
		TOTAL	$30,300

Net before trading costs for the entire 100,000 bushel crop = $.303 per bushel.

hedged when a sell signal raises the possibility of significantly lower prices. That protection against a ruinous price drop may be worth the possibility of a lower average price during some years when the market trends are not as pronounced and sustained.

You might voice two questions at this point. First, there is the appearance of the transactions being dominated by technical analysis. If that is the case, why worry about the fundamentals? The second question is almost inevitable: Why produce corn? Why not just speculate?

The technical side only appears to be dominating actions of the producers. The technical analysis, after all, generates the timing of actions. The actions and exactly when they are taken get a lot of attention. *But the fundamental side is present.* Decisions on how much to price, how urgent the need for protection is perceived to be, and when you will be willing to tolerate functioning as a cash market speculator are all tied to the fundamental picture. This leads to the oft-repeated conclusion: The two approaches are complementary.

Why not just speculate? First, it is obviously that this individual has chosen to be a farmer and not a commodity speculator. But the issue runs much deeper than that. *It is much easier to be a successful selective hedger in the futures markets than it is to be a successful speculator.* There is a very, very important difference: *The selective hedger has the cash product.* To be successful, the selective hedger must realize profits over time and stabilize revenue flows so that financing needs can be met. Within a 5-year or a 10-year period, the selective hedger can suffer losses in the futures account in an upward-trending market that is difficult to manage. The money in the futures account is lost, but it is matched by gains in the cash market. So, the futures account loss is an opportunity loss, and the business is viable and thriving. Across the same period and in the same markets, the speculator who takes similar positions could go broke. *There is a basic and fundamental difference between selective hedging and speculating.*

The bar chart signals can be effective guides to a selective hedging program for the decision maker with an average financial position, some skill as an analyst, and an ability to manage moderate exposure to risk. When major trends develop, a selective hedging program based on the bar chart signals has the potential to both provide protection and raise the average per-bushel or per-unit return.

Sophisticated Hedging Strategies

In general, the decision maker who is willing to accept more exposure to price risk will be the decision maker who feels comfortable with managing exposure to price risk. Financing must be adequate, of course, if the wide array of available tools are to have a chance of being employed. In many respects, the sophisticated approach will parallel that discussed in the previous section for the more nearly intermediate approaches. Fundamental analysis will still be employed and the bar chart will still be a favorite tool. Option strategies will be employed, but in such a way as to increase the chances of revenue while making a conscious decision on when exposure to price risk will be tolerated.

Fundamental analysis must be completed to establish a price forecast and a probable price range for the year. Techniques were established in the previous section. *The more knowledgeable trader is likely to select the more sophisticated tools such as the regression models to forecast prices.* Any application of the elasticity framework is likely to also include the capacity to handle the cross-elasticity issues. Analysis of the meat sector, for example, may be conducted at the retail level and then the concept of derived demand employed to generate a price estimate at the live animal level. In the process, price spreads or marketing margins must be analyzed for seasonal or other patterns that would influence what live animal price will be associated with a particular retail price. The analysis is more complex but deals with the issue that prices are set at the consumer level and that most demand analyses are conducted at the retail consumer level.

Table 11.6 provides an example of how the analysis of the hog market might be extended to a more sophisticated plane than that covered earlier in the chapter. Retail pork prices are deflated to remove the influence of changes in the overall price level. The period-to-period change in quantity is either forecasted or estimates of the projected quantity change are pulled from publications such as the USDA's *Livestock, Dairy, and Poultry Situation and Outlook* reports. In Table 11.7, the per-capita supply of pork for the upcoming quarter is projected to be up by 6 percent. Using a retail-level demand elasticity of –0.67, price of pork at retail would be expected to be down 9 percent.

An initial price of pork is estimated and then adjusted for the crossover influence of an expected 5 percent increase in beef prices (cross-elasticity coefficient = 0.35), an expected 8 percent increase in poultry prices (cross-elasticity coefficient = 0.25), and for an expected 1 percent increase in real per capita disposable income (income elasticity coefficient = 0.5). The final price estimate is then inflated by multiplying by the projected level of the consumer price index. The farm–retail price spread is examined across recent years for average, minimum, and maximum levels, and those numbers are used to generate an expected live hog price and a range within which live hog prices might vary.

It is readily apparent that this approach is for the relatively sophisticated decision maker. You are encouraged to spend time on Table 11.6, however. For those who can use this approach, the rewards in terms of quality of the fundamental analysis are apparent. Starting at the retail level and incorporating the demand shifters via the cross-elasticity and income-elasticity measures is a more rigorous approach than just looking at the price impacts of a quantity change at the producer level. The process also focuses attention on variation in the farm–retail price spread as a source of variability in hog prices. For those who might be more comfortable with a more simplistic approach, it is important to understand that many participants in the futures markets do employ fundamental analysis at this level of sophistication or greater. The consensus price being discovered in the futures reflects this effort and analysis.

Given the initial price projections based on fundamental analysis, the decision maker can then proceed to apply numerous technical strategies. Moving averages might be employed, but they are likely to be sophisticated extensions of the moving averages. Bar chart analysis will surely be considered, but in a sophisticated way.

Not all of the strategies will be treated here in great detail. It would be redundant to go through a complete "decision scenario" comparable to that detailed above in dealing with intermediate strategies. Selected extensions will be offered to ensure adequate coverage and to better illustrate the vast array of technical tools from which the decision maker can choose.

The big need insofar as moving averages are concerned is a means of improving the performances of the moving averages in the sideways and choppy markets. The use of penetration rules and/or leading indicators was discussed in Chapter 5, the appendix

TABLE 11.6
Generating an Estimate of Live Hog Prices Based on Retail Prices of Pork

You have:

1. The year-to-year change in per-capita supplies of pork for the upcoming quarter is 0.06;
2. Retail-level demand elasticity for pork is -0.67 and current deflated price is $1.565;
3. Real beef prices are expected to be up 0.05 and cross-elasticity coefficient with pork is .35;
4. Real poultry prices are expected to be up 0.08 and cross-elasticity coefficient with pork is 0.25; and
5. Real per-capita disposable income is expected to be up 0.01 and the income elasticity for pork is 5.

New retail price if only supply changes:

$$-0.67 = \frac{0.06}{X} \quad X = -0.09$$

$1.565 – 0.09 (1.565) = $1.424

Impact of higher beef prices:

$$0.35 = \frac{X}{0.05} \quad X = 0.0175$$

$$-0.67 = \frac{-0.0175}{X} \quad X = 0.0261$$

$1.424 + .0261 (1.424) = $1.461

Added impact of higher poultry prices:

$$0.25 = \frac{X}{0.08} \quad X = 0.02$$

$$-0.67 = \frac{-0.02}{X} \quad X = 0.0299$$

$1.461 + 0.0299 (1.461) = $1.505

Added impact of higher income:

$$0.5 = \frac{X}{0.01} \quad X = 0.005$$

$$-0.67 = \frac{-0.005}{X} \quad X = 0.0075$$

$1.505 + 0.0075 (1.505) = $1.516

Once the per-capita supplies are projected and "fixed" at 1.06 of the previous year, then the influence of demand shifters such as the projected change in beef prices must be converted to price implications *at a fixed quantity of pork.* Therefore, the impact of a 0.05 projected increase in beef prices is carried back through the demand-elasticity framework for pork. The *inability* to take a 0.0175 increase in pork supplies is converted, via the cross-elasticity framework, to a 0.0261 increase in pork prices. Comparable adjustments are made for the impact of the higher poultry prices and the increase in income.

To inflate:

$1.516 (1.569) = $2.378
where the 1.569 reflects a CPI (198284 = 100) estimate for the upcoming quarter.

To convert to live-hog price at the farm level:

Retail price – (farm to retail price spread) = net farm value
$2.378 – 1.630 (average spread) = $.748

Net farm value ÷ 1.6 = live hog price
$.748 ÷ 1.6 = $.4675 or $46.75 per hundredweight

The average projected quarterly price is therefore $46.75. If the range in the farm–retail spread across recent quarters has been 1.598 to 1.712, the range in projected hog prices would be

($2.378 – 1.598) ÷ 1.6 = $46.87
($2.378 – 1.712) ÷ 1.6 = $41.62

The price range is $41.62 to $46.87, and this demonstrates the importance of the magnitude of the farm-retail price spread as one of the major contributors to change in farm-level hog prices.

to Chapter 5, and again in Chapter 6 on the psychology of the market. In that discussion, selected prices of research were referenced. The work by Riffe and Purcell referenced in this chapter used a 3-day moving average to confirm the signals of the 5- and 15-day combinations for cattle feeders. The leading average reduced the number of trades and increased the net per trade.

Many analysts have adapted measures of market volatility and market oscillators from the developments by Wilder. Conceptually, the idea is to use a measure of volatility to determine whether the trader should be using moving averages or whether they should go to some other approach. In choppy and sideways markets, the level of volatility is high, and moving averages will not be the best choice for a selective hedging program. A major book by Wilder was mentioned in Chapter 4 and is listed in the references at the end of that chapter. The reader interested in this type of system is encouraged to take a look at the book. It is one of the best references on advanced technical trading systems currently available. The book by Schwager is also excellent, is relatively new, and covers a broad array of the technical tools used in the markets.

In discussing the intermediate-level strategies, the use of market corrections was demonstrated. The more sophisticated trader would use the concept of a correction, but might reinforce that approach via familiarity with the Elliott wave theory. Entire books have been written on this approach and references are shown at the end of the chapter. Knowledge of the tendency for bull markets to come in five waves with waves 2 and 4 the correction waves can be important to the decision maker. In bear markets, there tends to be three waves with wave 2 the corrective wave for wave 1. Some analysts have developed projection techniques using the length of the moves in the developing market moves.

The December 1997 soybean oil chart shown in Figure 11.13 demonstrates. A soybean producer who is still a cash market speculator and who is waiting to set price protection on stored soybeans would feel more comfortable after examining the oil chart. The first rally that developed in September appears to be wave 1. The corrective wave 2 lasted into early October, followed by wave 3 to the upside during the month of October. The brief correction that followed would be the corrective wave 4, and the last thrust on the price chart would then be wave 5. *Given this interpretation, the producer will expect topping action to start soon. The five-wave bull market may be over. If the soybean oil futures top, then the soybeans might top too—and the producer is more vigilant in watching for pricing opportunities in soybeans.*

All this is not as easy as it might appear on an *ex post* basis. As the "waves" develop, it is sometimes very difficult to separate moves within a wave from the moves that will ultimately be the primary market waves. Experience and sophistication as a market analyst are needed here.

Sophisticated applications of options continue to emerge as users probe possibilities and adapt the relatively new tools to their needs. In the context of a relatively sophisticated and well-financed decision maker who is willing to accept and manage exposure to price risk, the option strategies selected will typically be strategies that allow controlled or managed exposure to price risk in order to open up the possibility of a larger return. A demonstration of such strategies was offered in Chapter 7.

The importance of effective fundamental analysis is especially apparent here. If the decision maker is to deliberately accept exposure to the risk of lower prices or to the risk of higher costs of inputs in order to open up the possibility of better prices or lower costs across selected price ranges, then *there has to be a degree of confidence in the analysis of expected price patterns.* Technical analysis can help, of course. Assume, for example, that fundamental-based pricing models suggest that the prices

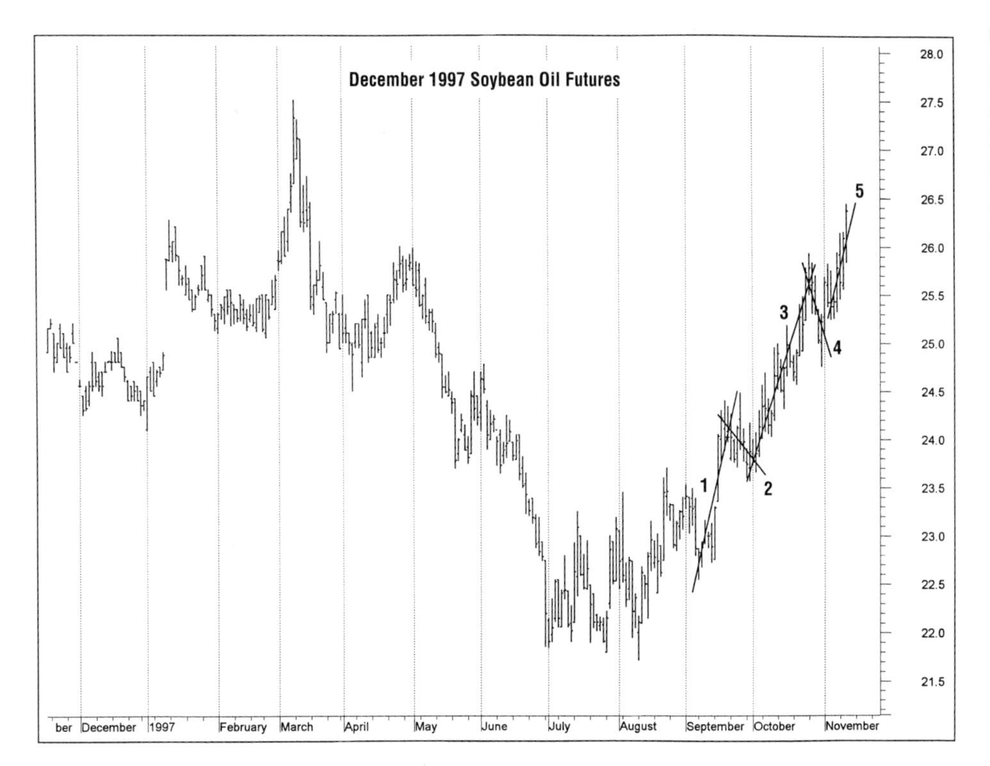

FIGURE 11.13
A Possible Five-Wave Bull Market on the December 1997 Soybean Oil Futures Chart

of corn, soybeans, wheat, cattle, or hogs are likely to trend higher. If the fundamental analysis is calling for higher priced soybeans, for example, support comes from the technical dimensions of the market if the new crop of November soybeans is near the contract low with the RSI showing the market to be approaching an oversold status. Under such circumstances, the producer might seek to protect against falling prices while leaving open the potential of higher prices, and would tend to use options. But the more sophisticated user might also prefer a better net price than is available from the simple buying of a put option.

Table 11.7 provides an example that provides a higher price floor for soybeans than just buying a put option and leaves open the potential of higher prices. In the illustration, the cash futures basis is assumed to be zero for ease of exposition.

The numbers in the body of the table are the net prices for the various strategies (down the left side) given various closing prices for the November futures (across the top). For example, the "Buy $7.50 put" strategy yields a net of $8.02 per bushel when the November futures are at $8.50 at the end of the production period. It is apparent that strategies 3 and 4 will be superior *if the price trend is up.* If the market drops, protection is less effective than it would be by just buying the put option. The risk exposure is substantial for strategies 3 and 4 and net prices are worse than being a cash market speculator at significantly lower prices.[9] *Such strategies place emphasis*

[9]To illustrate for strategy 4, when November futures go to $7.00, selling a $7.50 put generates revenue of $.48, but is now losing $.50. Before commissions, the net is –$.02 per bushel. Buying an $8.00 call involves an outlay of $.39 in option premiums, and the $8.00 call is worthless with the market at $7.00. Therefore, the options strategies show a net –$.41 (–$.02 and –$.39), and the net price is the $7.00 cash less $.41 or $6.59.

TABLE 11.7
Net Prices for Straight Hedge and Alternative Option Strategies in a Soybean Market Expected to Trend Higher: Expected Basis = 0

	Closing Price November Futures						
Strategy	**$6.50**	**$7.00**	**$7.50**	**$8.00**	**$8.50**	**$9.00**	**$9.50**
1. Sell November futures @ $7.50.	7.50	7.50	7.50	7.50	7.50	7.50	7.50
2. Buy $7.50 put.	7.02	7.02	7.02	7.52	8.02	8.52	9.02
3. Sell $8.00 put.	5.80	6.80	7.80	8.80	9.30	9.80	10.30
4. Sell $7.50 put, buy $8.00 call.	5.59	6.59	7.59	8.09	9.09	10.09	11.09

Closing price November futures = $7.57.
Option premiums are as follows:

Strike Price	Puts	Calls
$7.00	$.25	$.84
7.25	.35	.69
7.50	.48	.57
7.75	.62	.48
8.00	.80	.39
8.25	.97	.32

If the expected cash futures basis were –$.50, the "nets" in the table would all be $.50 lower, but the comparisons of the strategies would not be affected.

on fundamental analysis, and should be employed—even by the sophisticated user—only when there is a reasonable degree of confidence that the price trend will be up.

The oft-repeated need for a well-defined plan of action is relevant here. If the supply–demand balance starts to change to the bearish side, adjustments may be needed. Both strategies 3 and 4 involve selling put options, and those positions must be covered if lower prices loom imminent. *The contingency plan should be prepared in advance.*

At many points in the book, it has been assumed that you can convert the reasoning behind the development of a short hedge strategy to a long hedge strategy. But that assumption might not always be correct, and that is especially the case with a sophisticated option strategy. Table 11.8 records the actions taken and Figure 11.14 pictures the performance of a mixed trade strategy relative to simply buying a call option to protect against higher corn costs. Note that the more sophisticated strategy (strategy 3) will be effective if the price of corn is in the projected price range. If prices move out of the range to the high side, the corn user is exposed to the risks associated with higher costs.[10] *The ability to predict the direction of price movement and a price range using fundamental and/or technical analysis is again very important to the success of the strategy, and such a strategy should be employed only when there is a high degree of confidence in that analysis.*

[10] If futures prices go to $3.50, to illustrate, the net cost of the corn will be $3.55. Selling the $2.50 put gives premium revenue of $.08. Selling the $3.00 call yields a premium revenue of $.17, but that position is losing $.50 when futures are at $3.50 for a net of –.33. The net cost before commissions would be $3.55, well above the strategy of buying a $2.70 call that places a cost ceiling at $2.67. At higher prices, the net from the mixed options strategy yields a higher cost than would be yielded by being a cash market speculator.

Situation: (1) December corn futures @ $2.73
(2) Option premiums are as follows:

Strike Price	Puts	Calls
$2.50	$.08	$.37
2.60	.14	.34
2.70	.23	.27
2.80	.27	.23
2.90	.35	.20
3.00	.42	.17

(3) Harvest period basis projected to be –$.30

Strategies: 1. Buy in cash market.
2. Buy $2.70 call.
3. Sell $2.50 put, sell $3.00 call.

TABLE 11.8
Demonstration of Option Strategies to Place a Ceiling on Feed Costs

In some instances, there may be no strong evidence that the price will trend in either direction. In this setting, the more sophisticated decision maker might be interested in accepting the risk associated with a move outside a price range. Table 11.9 demonstrates for hogs. For prices above $60, the producer is willing to carry the opportunity costs associated with higher prices. Selling the call option prevents benefits of cash prices above $60. For prices below $50, the producer is exposed to the problems associated with lower prices. *The producer feels comfortable with his ability to predict that prices will vary within a $10 range between $50 and $60, and he seeks improved returns from the adopted strategies within that price range.* Figure 11.15 demonstrates performance of the mixed option strategy relative to simply selling the futures. As long as prices stay above $48.10, the mixed options strategy of selling a $60.00 call and a $50.00 put yields a superior return to the cash market. Above $53.10, the mixed options strategy is superior to selling the futures at $55.00. Obviously, the returns to the strategy will improve if the price range can be narrowed. If added analysis indicates prices are likely to stay in a $52.00 to $58.00 range, selling

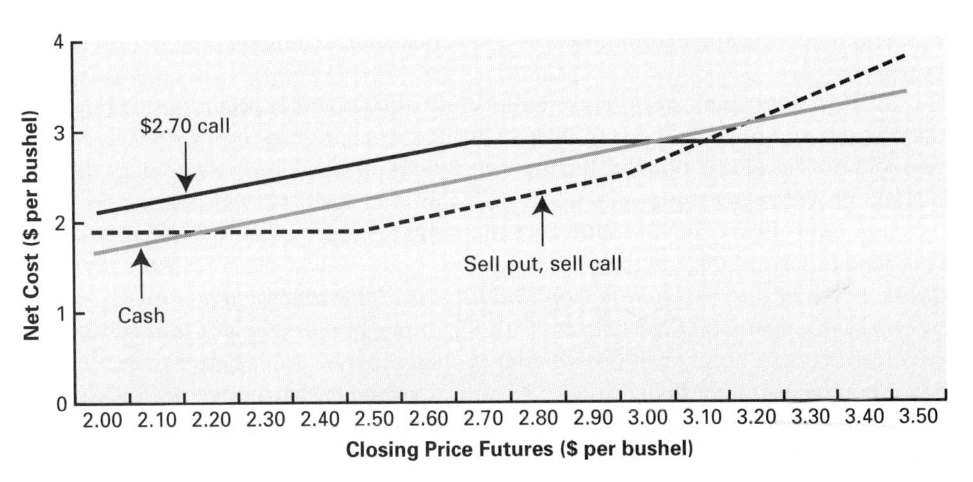

FIGURE 11.14
Comparison of Buying in Cash, a Call Option, and a Mixed Option Strategy: Corn Costs

TABLE 11.9
Demonstration of an Options-Based Strategy Designed to Yield Superior Returns to a Hog Operation in a Specific Price Range

Situation: (1) December Hog Futures @ $55.00.
(2) Option Premiums are as follows:

Strike Price	Puts	Calls
50	$ .90	$5.10
52	1.45	4.10
54	2.25	3.25
56	3.25	2.30
58	4.65	1.65
60	5.50	1.00

(3) Expected cash futures basis when hogs are sold is zero.

	Closing Price December Futures							
Strategy	**$48**	**$50**	**$52**	**$54**	**$56**	**$58**	**$60**	**$62**
1. Sell December futures @ $55.00.	55	55	55	55	55	55	55	55
2. Sell $60.00 call; sell $50.00 put.	47.90	51.90	53.90	55.90	57.90	59.90	61.90	61.90

the $52.00 put and the $58.00 call will increase the net by $3.10 ($1.45 from selling the $52.00 put plus $1.65 from selling the $58.00 call) compared to the cash market.

Consistent with the more sophisticated strategies, there must be a plan to protect the program if the analysis turns out to be wrong. Let's assume a hog producer adopts a strategy of selling the $52.00 put and selling the $58.00 call. These actions would contribute $3.10 per hundredweight in premium revenue before commissions.

The need for a contingency plan is apparent if we assume the market gets shocked with a *Hogs and Pigs* report that is unexpectedly bearish. The futures market starts to move lower and there is a very real possibility that the $52.00 level, the bottom end of the price range, will be taken out. Below $52.00, the producer is facing both a declining cash market and the exposure of having sold the $52.00 put. *The risk exposure is twice that of a cash market speculator.*

One straightforward approach is to buy the $52.00 put back. The premium will be increasing as the market moves lower, but the producer has the $3.10 with which to work. In a declining market, the $1.65 revenue from selling the $58.00 call should be secure.

If the outlook has deteriorated significantly, the producer might opt to buy back the $52 put and buy a $50 put to protect against a major drop in price. Alternatively, two $52 puts could be bought. Buying two would offset the previous sell of the $52 put and provide a net position of a $52 put to protect against lower prices.

Numerous plans can be followed which will be appropriate will depend on the emerging price outlook and the probability that prices will go sharply lower. *As a producer, you will have to update fundamental analyses and keep a close eye on the technical dimensions of the charts.* With $3.10 per hundredweight in revenue with which to work, a number of possibilities would generate a net better than buying a $52.00 put or a straight hedge at $52.00. Since extremely sharp price declines are the unexpected scenario, the results in most of the instances would be better than the straight hedge.

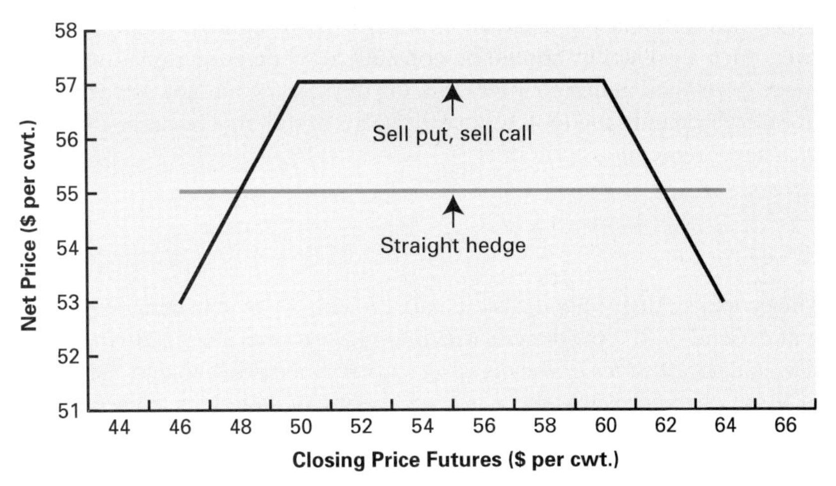

FIGURE 11.15
Comparison of a Straight Hedge and Options Strategies: Hog Prices

The more sophisticated strategies will tend to be selected by decision makers with above-average analytical skills, the financial prowess to use any strategy, and the orientation toward accepting some price risk when the potential reward is significant. These strategies will involve the sophisticated approaches to technical analysis and will often involve options strategies that deliberately leave open exposure to cash price risk in price ranges that are outside the expected or projected price range.

HEDGING STRATEGIES IN PERSPECTIVE

Strategies in this chapter and throughout the book have been developed with an eye toward effectiveness. That is, developments have been oriented toward strategies that not only reduce the variability of revenue flows but also raise their average level. That is the motivation behind selective hedging and behind the attention to fundamental analysis to help the decision maker decide when the role of cash market speculator is the correct role.

It would be inappropriate to imply, however, that you will have to be able to forecast prices accurately and be excellent at reading the charts before hedging will work. Numerous studies have confirmed that hedging on a routine basis or hedging using a simple indicator will in fact significantly reduce the variability of revenue flows and, in some instances, raise the average level of the revenue flow. A study by Schroeder and Hayenga, listed in the references at the end of the chapter, found that a strategy of hedging cattle in the feedlot whenever a $4-per-hundredweight margin was offered was very effective. *Developments in this chapter accept the fact that use of the futures and options can help and attempts to move you toward even more effective strategies.*

It is appropriate to remind you once again that the tax implications of strategies that involve selling call or put options have not always been completely clear. Prior to a Treasury Department administration action in the summer of 1994, losses incurred due to selling a call, for example, were usually treated as speculative losses. Selling a call above the market and buying a put below the market, the "fence" in Chapter 7, *appears* to be okay in light of the Treasury Department action. *It is not clear in late*

1997 that selling a put will not be treated as speculative activity by the IRS, however. Such a possibility should be considered when contemplating strategies such as those developed in this chapter. Talk to your tax accountant. We will all need to monitor developments and look for clarifications in this area because some of these strategies have tremendous potential.

SUMMARY

The proper strategy for a decision maker to employ in managing exposure to price risk will depend on the producer's *attitude toward risk, the financial capacity* to carry risk, and the *abilities of the decision maker* as a market analyst. Given an assessment of those characteristics, there is a wide array of strategies from which any decision maker may choose.

Both fundamental and technical analysis will prove to be important. Early in the relevant decision period, it will be important to project the price level for the period and the range within which the price is likely to vary. *Fundamental analysis is essential.* The choice of strategies will be influenced by the expected amplitude of price swings and the likelihood of major price trends.

More conservative hedging strategies, such as *target price hedging,* need a backup strategy in the event the target price is never offered. A mechanical system such as *moving averages* has potential to protect against a ruinous price dip. Out-of-the-money *put options* can be used to provide a price floor in some circumstances.

Decision makers with a better financial position and who can accept and manage some exposure to price risk will tend to select strategies designed to improve net returns compared to a conservative hedging approach. A selective hedging strategy can be based on *moving averages* or it can be managed using widely recognized *bar chart patterns.*

Strategies based on *options* can allow the decision maker to *choose price ranges* within which exposure to risk will be accepted in *search of a higher price floor* or net return within the expected price range.

The more sophisticated strategies should be based on *more sophisticated efforts in fundamental analysis.* The capacity to *anticipate price direction* and to *identify a price range within which price is likely to remain* is closely related to fundamental analysis and can be reinforced by such technical dimensions as the life-of-contract trading range.

KEY POINTS

- It is important that decision makers select a strategy consistent with their *attitude toward risk, the financial capacity* to carry exposure to price risk, and their ability to *manage exposure to price risk effectively.*
- *Conservative strategies* that involve a one-time placing of a hedge need a *safety net* to block potentially ruinous price moves in the event target price objectives are not offered by the markets.
- Selective hedging strategies have the potential to *decrease exposure to price risk* and to *increase the net returns,* but such a set of outcomes will not always be possible. The main objective is to *decrease exposure to price risk,* and strategies should be employed with that objective in mind.

- More sophisticated strategies that are designed to provide protection over specific price ranges require effective *fundamental and technical analysis. Fundamental analysis* is especially important to guide selection of strategies that work best for a *certain direction of price trend* or are effective *within a selected price range.*
- The relatively new options may be the most important tool for the capable manager who seeks returns above a straight hedge and *who is willing to accept exposure to price risk within a certain price range. Net prices or net costs* for mixed option strategies can be superior to *straight hedges* or direct use of *put and call options* if the probable price range can be predicted with confidence.

USEFUL REFERENCES

Chicago Mercantile Exchange, *Trading Tactics: A Livestock Futures Anthology,* CME, Chicago, 1986. This is a very useful reference that discusses and applies many of the fundamental and technical approaches to market analysis discussed in earlier chapters.

Robert R. Prechter, Jr., D. Weis, and D. Allman, "Forecasting Prices with the Elliott Wave Principle," *Trading Tactics: A Livestock Futures Anthology,* Chicago Mercantile Exchange, Chicago, 1986. A demonstration of Elliott wave theory and related procedures.

Ted C. Schroeder and Marvin L. Hayenga, "Comparisons of Selected Hedging and Option Strategies in Cattle Feedlot Risk Management," *Journal of Futures Markets,* Vol. 8, No. 2, April 1988, pp. 141–156. The study found that a simple profit margin target approach and an options-based strategy performed best in reducing revenue variability and raising the average level of revenue.

Jack D. Schwager. *Technical Analysis,* John Wiley & Sons, New York, 1996. The author presents a brief treatment of a single moving average in Chapters 3 and 8 and a more sophisticated coverage in combination with oscillators in Chapter 15.

William F. Sharpe and G. J. Alexander, *Investments,* 4th ed., Prentice-Hall, Englewood Cliffs, NJ, 1990. This reference provides detailed coverage of portfolio theory and the choices between level and variability of revenue flows in marketing strategies.

Phillip W. Sronce and J. R. Franzmann, *Hedging Slaughter Hogs with Moving Averages,* Bulletin B-768, Oklahoma Ag. Exp. Station, July 1983. This work was referenced in Chapter 5 but is repeated here as an example of the research on more sophisticated uses of moving-average systems.

CHAPTER 12

CASH-FUTURES RELATIONSHIPS: CAUSES AND CRITICISMS

The nature of the relationship between cash and futures prices has long been a subject of discussion and even debate. Earlier chapters have touched on the source of some of the issues in discussing the basic concepts of hedging and in price discovery. There is clearly a relationship between price of the cash commodity and trade in contracts that call for future delivery of that some commodity. *That relationship is formalized in terms of the cash-futures basis, and we have found that the behavior of that basis is critical to the success of hedging or risk-transfer programs in all commodities. But it is not the basis or the observed basis patterns that are the primary source of the controversy.*

The problems arise with the tendency to assume or assign casualty. To the uninitiated, the tendency for the cash and futures markets to move together is enough to argue that the move in futures *caused* the move in cash. Add the tendency for futures to occasionally react more dramatically than cash to a new and unexpected bit of information and to show a more pronounced psychological dimension, and some individuals and trade groups are ready to demand legislation to ban trade in futures. There is a presumption that the cash prices would somehow be higher and/or less variable if there were no trade in futures. An example of the thinking is apparent in an April 1990 report released by the Center for Rural Affairs in Nebraska. The report states, "The futures market is now being used in a way it was not intended to be used—as a price discovery mechanism for the cash market."[1] Reference was to the live cattle futures traded at the CME. *Contributing to price discovery has always been one economic justification for trade in futures contracts,* and there are informed observers who would argue that it is the most important function of trade in futures and options.

Lack of information and misinformation hinders the progressive use of futures markets. The techniques, strategies, and approaches presented in earlier chapters will not be adopted or adapted by the decision maker who is not quite sure how the futures market impacts the cash markets. Use of the futures markets or options will not even be

[1]*Competition and the Livestock Market*, Report of a Task Force Commission by the Center for Rural Affairs, Walthill, NE, April 1990, p. vi.

considered if there is a perception that the markets are not needed or are a negative influence. *You should be aware of the fact that there exists continuing controversy, and you are entitled to know something about the issues that prompt continued dialogue about the role and influence of the futures markets.* The controversy was fueled again by the FBI "sting" operation at the Chicago Mercantile Exchange and the Chicago Board of Trade in 1988–1989 that uncovered rules violations and questionable tactics by traders at the exchanges. The criticism will not be quick to go away. During 1997, the issue of lifting a long-standing ban on off-exchange agricultural options is being debated—and is very controversial. The off-exchange option-type investments may or may not be tied to futures and options, but all are often seen as in the same area. If the ban is lifted and abuses do develop, any criticism is sure to spill over to the organized futures exchanges.

In this chapter, the relationship between cash and futures prices is explored, explained, and placed in perspective. The objective is to move you as a student of the markets and as a decision maker to a plane of understanding that will encourage intelligent decisions on whether, when, and how to use the futures markets, which has been described in some detail throughout the book.

SOME COMMON CONCEPTIONS AND MISCONCEPTIONS

Every individual and every trade group associated with a commodity has an opinion about the role of futures and whether that role is positive or negative. To cover all the many points of view would be impossible and is unnecessary. Discussion of the more prevalent points of view should suffice.

Futures Trade Is Not Needed

Why do we need trade in futures? Why is the cash market not sufficient? These questions are commonplace among the critics of trade in commodity futures.

Following are some of the many possible illustrations of the ways trade in futures makes an economic contribution.

> A bank commits itself to a significant outlay of loans to agricultural producers over the next six months to a year at or near a particular interest rate. To protect against the possibility of interest rates rising and creating a situation in which outstanding loans are earning less than it costs the bank to borrow its own funds, the bank hedges its position by taking an appropriate position in financial futures such as T-bill, T-bond, or Eurodollar futures. *If the possibility of hedging did not exist, the bank would be exposed to the risk of rising interest rates, and the interest rate charged the borrower would surely increase.*
>
> In Iowa, a hog farmer places feeder pigs on a feeding floor. The farmer has budgeted the feeding operation and, with corn at $2.75 per bushel, estimates a profit of $10 per head. Selling prices for the hogs have been set via a contract with the local packer. If corn prices increase significantly during the feeding period, the $10-per-head projected profit can be wiped out. *Without the opportunity for protection on the corn prices via a long hedge using the futures or by buying call options, the farmer might experience difficulty in financing the operation or be hesitant to accept the risk.*

Throughout the grain-producing regions of the U.S., commercial elevators buy grain from producers and place it in storage. When grain is not sold to processors or exporters immediately, the elevator manager is faced with an inventory risk of staggering proportions. Consider, for example, a facility holding several million bushels of wheat bought at $4 per bushel. The manager hedges that risk by selling wheat futures or buying put options. *If there were no protection against the inventory risk, the elevator operation would be forced to protect itself by reducing its bids to producers to cover the cost of risk exposure over time or to find some other way to protect the value of the inventory. Virtually 100 percent of wheat, corn, soybeans, cotton, and other storable products held in inventory are hedged.*

Multinational grain firms and banks active in the world market were seriously impacted during the 1970s, 1980s, and 1990s by fluctuating exchange rates. Carefully conceived investment plans were wiped out or threatened by dramatic changes in the value of the U.S. dollar relative to other currencies. In response to an obvious and emerging need, trade in foreign currency futures was introduced. *Without the protection of the currency futures, U.S. firms and financial institutions would be seriously constrained as to the role they can play in a dynamic world market.*

In many years, only 10 to 15 percent of corn, wheat, and soybeans is hedged directly by the producer. The level of cash contracting varies but often surges to 50 percent or more in some years when producers are worried about prices. When cash contracts are used, the elevator does the hedging. *Directly or indirectly, the costs of exposure to price risk in a significant percentage of our storable commodities is passed to the speculator outside of the agricultural sector.*

The 1996 farm bill legislation continued the setting of the support price for milk well below the market price. Dairy farmers face a growing level of price risk. In the complex dairy industry, prices in different producing regions have been based at least partly on cheese prices discovered at the Wisconsin Cheese Exchange. During 1997, facing growing concern about its effectiveness, the Wisconsin Cheese Exchange disbanded. There was an immediate need for some new approach to discovering a "base" price for milk. Both the Coffee and Cocoa Exchange in New York and the Chicago Mercantile Exchange are starting trade in milk futures. It appears the responsibility for establishing price levels for the massive dairy industry in the U.S. is shifting to one or both of these exchanges, and the dairy farmer will now have a way to manage volatile milk prices. *You would agree that trade in milk futures could prove to be very important to the dairy industry and dairy farmers across the U.S.*

It appears the futures markets are needed. The markets are a part of the institutional framework that finances economic activity and stores, handles, and transports the product of that economic activity. In many of these areas, we could argue that if the futures markets did not exist, something—some other type of institution—would have to be developed to allow the transfer of price risk and to perform the functions of the existing futures markets. If there are no mechanisms in place to transfer the costs of exposure to price risk, then prices to producers would be lower and/or prices at the consumer level would be higher over time, and society in general would be the loser.

Futures Prices Cause Breaks in the Cash Market

The issue of causality comes up more directly here and it is an important issue. Investigation of causality between cash and futures markets can be treated along a continuum from the very simple to the very complex. At a simplistic level, the argument typically heard runs something like this:

> *The futures market caused the break in cash prices. Everything was going fine until the futures prices dropped and that caused the cash prices to fall.*

This "conclusion" crops up in the slaughter cattle market, for example, when a dip in live cattle futures precedes or parallels a drop in cash prices. We see the same thing in the cotton market, in the soybean market, and in the many other areas in which parallel futures and cash markets exist. There is the presumption that the futures market is the culprit when the cash market moves to the disadvantage of cash market participants. And the appearance of causality *is* there. *What's wrong with concluding, when a drop in futures prices precedes a drop in cash prices, that the change in futures caused the change in cash?*

To get at this question, it is important to remember that both markets are discovering price for the same commodity. Both markets react to information coming out of essentially the same supply–demand setting. The only difference is the time period for which prices are being discovered. Even that difference disappears as the maturity date for the futures contract approaches. In this environment, if one market has the capacity to register the impact of a new piece of information more quickly, then it will react before the other market reacts.

It appears the capacity of the centralized futures market to react quickly to new information is the root of many of the charges of "causality." Most research suggests that the futures market is an efficient market in that it responds, on any particular day, to all the publicly available information on that day. The key question is: Is it logical to assign causality to the futures market because it registers first the impact of a change in the information on supply and/or demand? Is it not logical to assume the cash market would have reacted in the absence of a futures market once the new information filtered into the much more decentralized cash market? And is it not possible that the cash market may in fact react more quickly to certain types of information?

The participants in the two markets *are* different. Much of the direction in the futures market comes from the buying and selling actions of skilled analysts in the large firms who are also active in the cash market. Other impact comes from buy or sell recommendations of large brokerage firms who employ analysts to appraise both the fundamental and the technical aspects of the market. It *is* true that some of the trading in the futures markets is by the often poorly informed speculator, but this is not the major "market moving" part of the total trade. Some of these participants are small speculators, but small hedgers can and often do fit the same profile of not being well informed. The small traders tend to jump on the bandwagon after a change in price direction has been initiated primarily by the actions of the large traders who are generally skilled market analysts involved in the daily market.

Participants in the cash market are often dramatically different in terms of the frequency of their exposure to the market, their ability as market analysts, and their access to a broad base of information. The small investor may be in the treasury bill market only once a year when he puts together $10,000 to invest in the cash market. Portfolio managers with the large banks are in the futures daily. A Cornbelt cattle feeder may sell cattle only once a year. Market analysts with the packing firms, the large feedlots, and brokerage houses try to stay on top of cash and futures markets for cattle every day. The Midwest corn farmer will be prone to base his expectations of the corn market on what he sees around his area in terms of crop potential. The analyst with the major grain exporter or the large brokerage firm will make an attempt to know what the weather and crop conditions are all over the country and around the world and often employ their own meteorologists. Often, they will travel to other producing countries or through the producing regions of the U.S. to get first-hand information on crop prospects. It is not surprising, given the makeup of the markets, that the futures market sometimes reacts more quickly to new information.

Research efforts are continuing to emerge, but the consensus appears to be toward the presence of major components of interaction between the cash and futures markets.[2] Complex analyses show that for some commodities, the futures market is more efficient than the disaggregated and geographically dispersed cash markets and does register the influence of action types of information change first. Those same analyses, however, also often show a lagged response in the futures to earlier developments in cash, suggesting that the cash markets do react first to certain types of information. Overall, the two markets tend to interact and work together.

Simply observing that futures markets react or change before cash markets do is not sufficient grounds to conclude the move in the futures market caused the subsequent move in the cash market or brought a price move that would not have been eventually realized in the cash market. The quicker move in futures, when it develops, may be evidence of a highly efficient, effective price discovery process in the futures market.

Any causality that does consistently flow from futures to cash will often be across a longer time period than the day-to-day variations in the market. Over time, activity in the futures market does get involved in the price discovery process in the cash market by exerting an influence on the level of supplies. *Under these conditions, a causal flow from futures to cash would be expected.* Further, this type of causal flow is very important because it tends to moderate the supply–demand imbalances that would otherwise tend to evolve.

Consider, for example, the situation facing corn farmers who are trying to decide whether to hold their corn in on-farm bins or to sell at harvest. Assume it is November and they are considering selling in November or holding in storage until the following May. As a farmer, you will need to consider the following:

1. The current cash price: assume it is $3.00.
2. All costs of holding until the following May—interest on the money tied up in the stored crop, shrink, spoilage, and any other variable costs of storage: assume these costs total $.30 per bushel from November to May.
3. The expected change in price between November and May.

[2]This is especially true in the livestock commodities. In the grains and oilseeds, it is more nearly accepted that the futures markets provide much of the price discovery activity. Among the studies that show interaction across the markets in the livestock markets are Charles E. Oellermann, B. Brorsen, and P. Farris, "Price Discovery for Feeder Cattle," *The Journal of Future Markets*, Vol. 9, No. 2, April 1989, pp. 113–121, and Michael A. Hudson and W. Purcell, *Price Discovery Processes in the Cattle Complex: An Investigation of Cash-Futures Interaction*, Va. Ag. Exp. Sta. Bul. 85–12, Blacksburg, VA, Fall 1985.

To make storage equally profitable to selling in November, producers facing these conditions must receive a price of at least $3.30 in May. One alternative is to look at the May futures for corn to see whether you can hedge a profit to your storage operation. Let's assume there is evidence to show the local cash market tends to run $.20 per bushel under May futures around May 1, reflecting a closing basis expectation of –$.20. If May futures are trading at $3.50 on this particular day in November, the situation is:

$3.50	May futures
−.20	to convert futures to local cash equivalent
= $3.30	forward price for May
−.30	costs of storing
= $3.00	net price to the hedged storage program.

This store–sell decision was discussed in some detail in Chapter 2, and the rule developed there was to store if the projected basis improvement exceeds the cost of storage. When that rule is met, the stored product *can* be hedged or forward-priced at a profit. But a review of that detailed decision criterion is not needed here. *The objective here is to show the interaction between the two markets and to demonstrate how the futures market does exert influence on the cash market.*

At a $3.50 trading level for May futures, you face a break-even position. You would probably sell the cash grain in November because there are still some uncontrollable risks, such as the basis risk associated with storage introduced in earlier chapters. But if May prices move well above $3.50, you and other farmers would be encouraged to store. This will decrease sales in the November period and tend to boost cash prices. It would be logical to argue, then, that higher prices in the May futures tend to *cause* higher cash prices in the November period because the level of the May futures prices, and changes in those prices, will influence storage decisions. Note that to the extent farmers store because the May futures price is high enough to allow a profit to storage, the product is being distributed across the crop year and made available throughout the year. There *is* a causal influence here, and we would expect the markets to work this way. Storing and selling May futures to hedge the inventory will boost cash prices at harvest, depress the May futures as they are sold to place short hedges, and generate basis levels during November at which the very efficient holder of corn would face essentially a break-even situation. The market relationships would then be back in equilibrium. There is clearly a parallel in the spring months when the farmer is attempting to decide whether to plant soybeans or corn. The decision might be influenced by the springtime trading levels of November soybeans and December corn futures. There are other similar settings including the banker trying to decide which way interest rates will move and the cattle feeder who watches the relationship between live cattle futures and costs of producing a slaughter steer in deciding how many cattle to feed.

Futures markets expand the set of alternative courses of action open to decision makers and provide an input to many basic economic decisions such as the storage decision, decisions on what to produce and how much, decisions on whether to seek protection against rising interest rates, and decisions on placing feeder cattle or feeder pigs into feeding programs. These types of decisions will influence cash prices because they influence the supply of product being offered or produced both now and later.

FIGURE 12.1
Price Impacts of Storing Grain from November to May

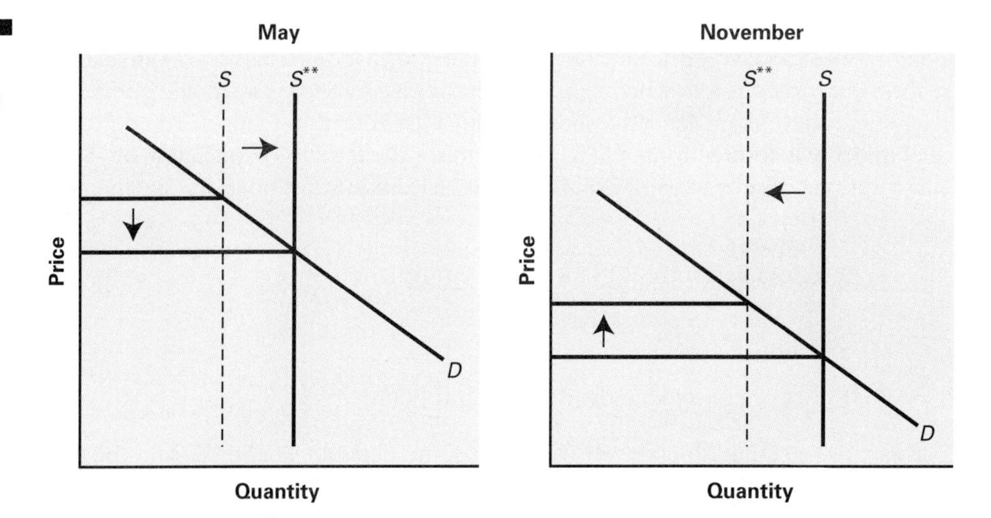

Before leaving this point, it is important to recognize the possible implications of the futures market being brought into decision processes. In the preceding illustration, it is clear that a November trading level above $3.50 in May futures will encourage storage. The outcome will be fine for the producer who decides to store and forward-prices or hedges by selling May futures. But there is another potential use—a misuse—of the futures markets. *Those producers who view May futures prices as a prediction of cash price and base their November decisions to sell or store on that "prediction" are asking for trouble if they do not follow through and hedge the grain.* If enough producers respond to this type of stimulus and misuse the futures in this way, it is possible that cash prices in May will be down. Supply is being shifted from November out toward May, and if this shifting is overdone and exceeds expectations of those trading the futures, price expectations for the following May will surely fall as the information filters into and through the markets. Figure 12.1 shows the graphics. Price will tend to be pushed up in November, down in May.[3]

The futures markets are not, as a rule, perfectly accurate predictors of later cash prices.[4] The futures markets, we have found, do not have to be good predictors of cash prices to provide an effective hedging mechanism. *It is the behavior of the basis that determines the effectiveness of hedging programs.* Much of the negative attitude toward futures among producers and producer groups can be traced back to those who have misused the markets as cash price predictors and then argue the futures market caused their problems. At any point in time, the futures quote is the consensus of what

[3]In the graphical presentation, there is no need to worry about the *magnitude* of the price changes. As producers and/or elevators decide to store because the May futures are high enough to justify storage, these actions do in fact increase the cash price in November and push the May futures lower if the cash grain is not sold, is stored, and is hedged by selling May futures. The different between November cash and May futures prices *during November* will be decreased and the stimulus for storage will start to disappear. That is what Figure 12.1 is designed to demonstrate.

[4]Later in the chapter, evidence will be provided that suggests the futures are typically as effective as our best efforts in terms of complex econometric models. It is in fact a difficult area in which to be accurate because so much of what happens comes from a largely unpredictable behavioral response by all the decision makers involved.

the price will be at a later time period. But that consensus is based on available information, and we have seen over and over that the information base is constantly changing. Discovered prices for later time periods must change accordingly.

Futures markets should not be used solely as predictors of cash prices. Decision makers who have no intention of using the futures market to forward-price via a hedge should be careful not to bring futures prices into their decision models as predictors of cash prices in making hold–sell storage decisions or in deciding, for example, which crop to plant.

At the other end of the continuum, the issue of causality is anything but simple. Analysis to measure the direction and magnitude of causality is difficult. The references at the end of this chapter will provide detail if you want to pursue this area.

In general, the research shows futures-to-cash causality on a day-to-day basis in those commodities in which the pricing function is essentially committed to the futures market. This includes many of the storable commodities such as corn, wheat, and soybeans. At a central point in Illinois, for example, the cash bid available to the corn producer for immediate delivery of corn is determined by

Nearby Futures[5] + Basis = Cash Bid.

The price discovery action occurs primarily in the futures market. Cash bids tend to move with the futures quotes. The cash bid for any particular day is tied to the closing price for futures on the previous day and then adjusted, on some occasions, during the day.

After dramatic moves in futures due to new information, such as a crop production estimate that is unexpectedly high or low, a manager in the cash market is sometimes observed to adjust early-morning bids to what he or she *expects* to happen in the futures market later in the day. A manager may *take protection* by adjusting cash bids.

Assume, for example, that the futures market for corn drops the daily allowable limit of 12 cents per bushel on a Tuesday. Early Wednesday morning, the manager of the cash elevator might take protection by lowering cash bids in anticipation of a further drop in futures when the corn futures market opens at 9:30 central time. With at least some "overnight" trade now being conducted for buyers and sellers around the world, there will even be some information on how much futures will change when the trading switches to full blast at 9:30. The task of discovering price is thus being essentially left to the futures market and there does appear to be causality from futures to cash. But we need to keep in mind that there *is* interaction. *What the elevator managers do in the cash market can influence trading levels of the futures*. If producers' selling of corn essentially stops after the elevators drop their bids, analysts trading the futures will start to register the resistance of producers to lower prices. *The future prices can be supported by those producers' reactions as those reactions are regis-*

[5]Corn futures, the reader should recall, are traded for the months of March, May, July, September, and December. The "nearby futures" rule is applied to the closest contract up to the first day of the month. For example, the December contract will be used during October and November until November 30. Then, a switch is often made to the March contract on December 1, to the May contract on March 1, and so on.

tered in the cash market. Later in the trading day, any early decline in futures may disappear.

The elevators may remove the "protection" as it turns out it was not needed. Thus, actions in the cash market can and do feed influence into the futures market.

The much-discussed live cattle futures fall in a category of commodities in which the direction of causality is less apparent. Analysis to determine the direction of causality is difficult in terms of both methodology and analytical techniques. The reference by Hudson and Purcell cited in footnote 2 in this chapter, and listed at the end of the chapter, examines the issue for live cattle. The researchers found no strong evidence of unidirectional causality between daily futures and cash prices. There *was* evidence of bidirectional causality, suggesting that the two prices are in fact discovered in the same supply–demand setting and receive impact from the same sets of information. In the live or fed cattle market, there is no strong evidence that the futures market is always more efficient in receiving, and registering the impact, of new information. The futures market does not dominate the cash market. In 1987, a study by the General Accounting Office, in response to a major controversy concerning the role of trade in live cattle futures, found no evidence that the futures market was dominating cash markets in a detrimental way. In a February 1990 study, Weaver and Banerjee reached the same conclusion with regard to live cattle futures. But we still hear, in the late 1990s, strong criticism of trade in live cattle futures.

For nonstorable commodities, the question of causality requires research and analysis. Examination of the limited studies reveals no strong evidence that futures prices unilaterally cause cash price movement or vice versa. The two prices appear to be discovered simultaneously with each market "feeding" information over to the other, often with time lags of one or more days.

Futures Trade Increases Variability in Cash Prices

This charge led to legislation in 1958 to stop trade in onion futures. Similar changes were leveled against futures trade in many commodities throughout the 1970s, 1980s, and into the 1990s. The situation is difficult to assess for the following reasons:

1. *Price variability in virtually every commodity—food, fiber, or any other—increased dramatically during the last few decades*. Increased exposure to the world market, especially recently via NAFTA and GATT,[6] the removal of strict production control programs in many of the agricultural commodities, and increasing levels of inflation during the 1970s and early 1980s were among forces ushering in the price variability. The 1996 farm bill legislation removed acreage controls on most of our food and fiber crops, and most analysts expect to see still further increases in price volatility. To observe that this variability occurred and is occurring during the period that has seen a dramatic increase in trade in commodity futures is not a sufficient reason to argue that trade in futures caused increased price variability. Correlation or association does not mean causality.
2. *Analysis to test a hypothesis in this area is complex*. Comparing price variability in the cash market before the advent of futures trade with variability after the

[6]Reference is to the North American Free Trade Agreement (NAFTA) and the broader General Agreement on Tariffs and Trade (GATT).

advent of futures trade is difficult. Comparisons are not appropriate unless all other significant causes of cash price variability can be controlled or eliminated. Such control of other forces is very difficult to accomplish in economic analyses.

3. *The level of price variability is a function of which prices are measured.* In general, daily prices will be more variable than weekly, weekly more variable than monthly, and so on, when variability is measured by statistical measures such as variance or the coefficient of variation. If the futures market is an efficient market on a day-to-day basis, then the daily prices in cash might be more variable than weekly or monthly cash prices because they are being discovered in the presence of the highly responsive futures market. But this does not mean the variability in monthly or quarterly prices will also increase. The short-term response to daily changes in supply–demand information that is registered in the futures markets might eliminate the emergence of a supply–demand imbalance and major changes in monthly or quarterly prices at a later date. For many commodities, the monthly or quarterly prices will be more important to the economic well-being of the firms involved.

The research by Hudson and Purcell attempted to get at this issue by dividing the variability in daily cash and futures prices for live cattle into systematic and random components. The systematic component of day-to-day price changes is that portion attributable to changes in information in the underlying supply–demand situation. The random component is the portion attributable to imperfections in the pricing process. During the 1970s and into the 1980s, as the volume of trade in live cattle futures increased dramatically, the variance or variability of the systematic component of the cash price series decreased and the variability of the random component increased. Those results suggest that the presence of trade in futures tended to *decrease* the variability in the cash market. As noted earlier, Weaver and Banerjee reached a similar conclusion. The presence of trade in live cattle futures does not increase, in a statistically significant way, variability in cash cattle prices.

Trying to determine whether the variability of cash prices is significantly influenced by the existence of futures trade is a difficult research task. More work is needed. To date, there is no strong evidence to support the claims of critics who argue futures trade increases variability in cash prices. Most of the available research, in fact, concludes that the presence of futures markets decreases variability in the related cash market across a number of commodities by increasing the effectiveness and efficiency of supply response to price changes.

THE SUPPLY RESPONSE ISSUE

Evidence on the accuracy of futures quotes as predictors of cash prices is mixed, but that should be no big surprise. The potential for a short-run supply response, especially in the agricultural commodities, can mean futures quotes will turn out to be inaccurate predictors. But the same difficulty faces the econometric models designed to predict prices. An early study by Just and Rausser[7] indicated that the future markets

[7]Richard E. Just and G. Rausser, "Commodity Price Forecasting with Large Scale Econometric Models and the Futures Market," *American Journal of Agricultural Economics*, Vol. 63, No. 2, May 1981, pp. 197–208.

for many commodities are just as accurate in predictions as the complex econometric models. The potential of an unexpected supply response is a primary difficulty in any attempt to forecast prices, and this area needs further discussion.

To illustrate, let's assume that it is October 1 and a midwestern hog-corn farmer is trying to decide whether to buy gilts for breeding to expand the breeding herd. From the date of breeding, it will take about nine months to have added slaughter hogs ready to sell. What information does the decision maker use in making the decision?

The current prices for hogs, corn costs, interest rates—all these and other traditional economic variables will be brought in. *But some type of price expectation will have to be included, and marketing economists are increasingly recognizing that the distant futures quotes are being widely used as price expectations.* In this illustration, our midwestern producer might look at the October 1 trading levels for the June or July hog futures for the next year.

What if, on October 1, the distant July lean hog futures contract is trading near $75? At current corn costs, the producer estimates his break-even cost for lean hogs on a carcass basis at $60 per hundredweight. A $75 selling price would mean profits of nearly $28 per hog for a 250-pound slaughter hog that produces a 185-pound carcass.

If many producers look at the same price expectation and decide to expand, the increased supply of hogs the following summer could push the price well below $75. But the marketplace will have a hard time deciphering just what is going on in the hog sector until the December and the March quarterly *Hogs and Pigs* reports are released by the USDA. Then, if the expansion is greater than has been expected, the futures prices will decline—and the current cash market could also drop as expansion plans are aborted and some gilts are sold in the cash market.

Analysts who examine the efficiency of the futures markets would argue that the futures market should be able to incorporate the expected expansion and not get caught showing distant futures prices that are, in an *ex post* context, too high. But there are two difficulties with this argument.

First, it is always difficult to anticipate how strongly decision makers will react. It is easy to find periods in which profit indicators were at only moderate levels and a major expansion developed. In other periods, those same profit indicators were higher for several calendar quarters before even a modest expansion was launched. Most analysts would agree that 1986 was the most profitable year for hog producers in the 1970s and 1980s, but there was a net liquidation of hog numbers during the year in many of the producing states. Hog producers were using their profits to pay off debt.

There were good profits in 1987 and parts of 1988 and 1989, and cash prices pushed well above $60 on a liveweight basis in early 1990. Still there was no expansion. Due to their financial position, producers may periodically use the better hog prices to reduce debt versus launching an expansion. In addition, banks or other lending agencies are notorious for being conservative about expansion if they have seen any recent period of forced herd liquidation due to low hog prices. A similar pattern developed in 1996 and 1997. Lean hog futures prices for the summer months of 1997 traded above $80 per hundredweight—but there was little or no expansion. Hog prices had dipped below $30 on a liveweight basis (below $40–41 on a lean weight basis) in late 1994. Record high corn prices developed late in 1995 and through the summer months of 1997. Bankers, and producers, remembered and were reluctant to expand until late in 1997 when corn was much cheaper and some of the losses, and debts, of 1994 had been covered. But when expansion came, it came with gusto. Pork production in 1998 may be up as much as 10 percent compared to 1997. *The behavioral reactions of producers are very difficult to anticipate correctly in terms of magnitude and timing.*

The second difficulty comes in the form of concern over the adequacy of the available set of information. It is difficult to discover the correct price for some future time period when the widely used information (*Hogs and Pigs* reports, in this example) is released only quarterly and is subject to a sampling error of up to 3 percent in either direction. Research by such authors as Colling and Irwin[8] often concludes that the sustained moves up or down in the futures markets are the results of "information shocks" when periodic reports are released and are not the result of any inefficiency in the futures markets.

The debate will continue, but it is important to remember that a short-run supply response within the year is possible in hogs and in the cattle-feeding sector. Even in the crops, acreage can be switched at the last minute from corn to soybeans, from soybeans to cotton, and so on, in response to changes in price expectations. During the period of heavy reliance on the subsidies from governmental programs (prior to 1996), price expectations influenced program participation and therefore exerted a significant influence on planted acreage. Each time prices trend higher, there was always talk about what price would be required to cause producers to abandon the programs and plant all their acreage versus meeting the set-aside requirements to be eligible for program benefits. If producers' reactions are bigger or smaller than the marketplace expected, the futures quotes for the fall contracts, prior to planting time in the spring months, will invariably be too high or too low when the harvest period arrives. And since 1996, even the modest attempts by the USDA to manage the supply side of the price equation in key crops are gone.

The possibility of a supply response will exert significant influence on the cash futures relationships and on how accurately futures prices can predict final cash prices. In assessing the efficiency of the futures markets, we need to keep in mind the related issues of behavioral responses by producers and the adequacy of the information base in the areas the future markets are helping to discover prices. At best, it will be difficult for the futures markets to anticipate correctly the magnitude of the behavioral response of producers of agricultural commodities, especially in commodities in which publicly available supply–demand information is infrequent and subject to sampling error.

THE CONVERGENCE ISSUE

In earlier chapters, there was reference to the threat of delivery that forces convergence in the cash and futures markets. This is an extremely important determinant of the relationship between cash and future prices. *If the cash and futures markets do not converge to some expected level of basis with a useful degree of reliability, the hedging process will not work effectively*. It would do no good for you as a decision maker to expose yourself to a level of basis risk that parallels the level of cash price risk you would face if no hedging is done. Critics of trade in futures often argue the convergence is not reliable and that the futures markets are therefore not an effective risk-transfer mechanism.

[8]Phil L. Colling and S. Irwin, "On the Reaction of Live Hog Futures Prices to Informational Components in Quarterly USDA *Hogs and Pigs* Reports," *Proceedings*, NCR-134 Conference on Applied Commodity Price Analysis, Forecasting, and Market Risk Management, Chicago, April 20–21, 1989, pp. 17–35.

Whether we are talking about corn, hogs, cotton, cattle, or financial instruments, it is the possibility of delivery that forces the needed convergence. Both hedgers and speculators get involved in making the system work.

Let's look at fed or slaughter cattle to illustrate. There is a provision in the live cattle futures contract that allows for delivery at several designated points and, since June 1996, delivery can be either on a live or carcass-evaluation basis.

The delivery period for any commodity starts at a previously scheduled date around the first of the month for which a futures contract is being traded. As the delivery period approaches, we will find that futures prices will not be allowed to stay above cash prices by more than the costs of delivery.

Assume it is June 1 and the June live cattle futures are trading at $75. Cash (deliverable) cattle near a delivery point are selling around $71. A feedlot with a short position in futures could announce its intention to deliver its cattle under the futures contract by tendering a certificate for delivery. The procedure is a bit more detailed than this, but the cattle feeder is looking at a $75 futures price with roughly a $1.50 cost of delivery and a net price for cattle delivered under the provisions of the futures contract of $73.50.

The estimated net price from delivery is $2.50 per hundredweight above the cash prices. Cattle feeders and other holders of short futures position would continue to deliver as long as this situation prevailed. The delivery process means short futures positions held by the cattle feeder as hedger (or others) do not have to be bought back and that tends to let the futures market decline. Persons holding long positions in futures, especially speculators who do not want to receive delivery of cattle, rush to sell futures to offset their long positions before they face the risk of being assigned to take delivery. The futures are forced lower by the selling actions of these holders of long futures positions. Convergence to a basis level approximating the costs of delivery during the delivery month is assured.

Traders or speculators, acting as arbitrageurs, also help to assure convergence. A knowledgeable trader on the cash market could sell the futures at $75, buy deliverable cattle, and announce intent to delivery points is very important and helps to ensure complete convergence. *The markets will converge to a reasonable approximation of the delivery costs and that is all that is required for the hedge to work.*[9]

If the cash market is above the futures, a less frequent scenario, another set of actions is required. Long hedgers who can kill and process the cattle hold their long futures positions rather than selling futures to offset. If, for example, the futures were $68 and the cash market $71, the long hedger (a packer, for example) could just wait to be assigned delivery.[10] Since the short hedger or other holder of short positions

[9]Analysis shows that the cash-futures basis in all delivery points tended to average –$1.00 to –$1.50 over time. The average basis for a particular month may occasionally widen toward –$2.00, but these variations are infrequent, and the basis risk or possible basis variability was still far less than the $5.00 to $10.00 moves in the cash market that are fairly common occurrences. The change in June 1996 moved the futures contract specifications to a higher-quality level, and producers of lower-quality cattle may now face an average basis of –$2.00 or even –$3.00. But the hedging process is still effective if convergence to the expected basis levels occurs.

[10]Under the certificate of delivery systems used in live cattle, the packer or other long hedger can specify the delivery location at which they are willing to accept cattle. This provision protects them against the possibility of being assigned delivery at points not convenient to their plant locations. The article by Purcell and Hudson in the America Enterprise Institute series cited at the end of the chapter provides more detail.

who delivers pays delivery costs, the packer could be facing a $3 per hundredweight lower cost of cattle via delivery than via buying in the cash market.

The examples are a bit extreme but the procedure illustrated is valid. In both cases, we should recognize that some exposure to uncertainty is involved. The cattle feeder, or other trader seeking profits by delivering cattle, faces some uncertainty about whether the cattle will be graded at the needed level and will in fact be deliverable without penalties. The process can take more than one day to complete, so the spread between futures and cash that prompted delivery can change while the delivery process is being completed.

When the cash market is above futures, the long hedgers have to hold their positions and wait for cattle to be assigned. The holder of long positions cannot initiate the delivery process. That initiative is always with the traders holding short positions. The advantage associated with being willing to accept delivery can start to disappear as the last delivery date approaches and the packers face the risk that holding the long position in futures will not always work to lower cattle costs.

In any other commodity, any financial instrument, or anything for which there is futures trade, the process is similar. *The possibility of delivery forces the needed convergence*. For future contracts that use the cash settlement process instead of physical delivery, the economic forces that prompt cash futures convergence are the same. Arbitrage between futures and cash by knowledgeable trades will force cash-futures coverage.

Logical economic forces ensure convergence. For commodities requiring physical delivery (fed cattle, corn, etc.) the process of convergence is slower when the cash prices exceed futures prices during the delivery period. Users holding long positions have to wait until essentially the last delivery day to force convergence.[11] As long as there is economic advantage to be gained by delivering or accepting delivery, the hedger is protected from exposure to a level of basis risk that would approach the level of price risk in the cash markets. There is reason, therefore, to be a hedger versus a cash market speculator.

Basis Problems

Problems of unpredictable basis behavior or lack of the expected futures convergence can and do arise. Generally, it is argued that the producer or other potential hedger who is located at some distance from a delivery point will not be able to hedge effectively because of unpredictable variations in the basis—what we are calling basis risk. One result is a persistent request to the CME, CBOT, or other exchanges for more delivery points. But those requests are usually resisted by the exchanges.

Adding more delivery points does not necessarily improve the effectiveness of hedging programs. No product or commodity will command the same price in all geographical locations. Corn, for example is worth more in New Orleans where it is accessible to the ships involved in world transit than it is in a farmer's storage bins in central Iowa. In the livestock sector, the geographical price spreads can be very large.

[11]The perceptive reader will ask the question, Why not let the trader holding the long position initiate the delivery process? This possibility is being discussed. The contracts and delivery procedures are adjusted over time to make sure trade in futures offers an effective and viable hedging or price risk transfer mechanism.

Feeder cattle (600–800-pound steers) will command a higher price in the Amarillo, Texas, area than in Alabama because the feedlots that need those steers are concentrated in the Southwest. When Montgomery, Alabama, was made a delivery point for feeder cattle a number of years back, delivery there was discounted significantly compared to the par delivery points in Oklahoma City, Kansas City, and so on. But the discount was set at a constant $6 per hundredweight, and that created an added problem: the difference between cash prices in Montgomery and in Amarillo was not constant throughout the year. (This and other problems helped prompt the move to cash settlement in feeder cattle.)

If the difference were a constant $6, then we could argue that hedges in Alabama would be just as effective as those in Texas without the Montgomery delivery point. The Alabama producer would face the following:

Futures Price + Basis = Forward Price.

The forward price in Alabama would in fact be lower than that in Texas because the basis adjustment is not only negative but is larger in absolute terms. But that would not contribute to basis risk if that adjustment was just as reliable and predictable as the smaller adjustment in Texas. We could suggest then, that the effectiveness of hedges in Alabama will be related to the stability of the intermarket differences in the cash markets. If the relationship between the cash markets is stable, the effectiveness of hedges is not influenced by access to a nearby delivery point.

The magnitude of the basis adjustment is not a determinant of the effectiveness of the hedging program. It is the stability or predictability of the basis that is important. It follows, therefore, that if intermarket cash differences are stable, the hedge in the area distant to the delivery point will be effective. And it is important to recognize that where cash settlement is feasible, the hedge can be even more effective.

CASH SETTLEMENT ISSUES, CONCERNS

In recent years, there has been strong interest in moving to cash settlement as opposed to physical delivery of the product. The move started in the financial futures and stock index futures where "cash settlement" has facilitated a huge expansion in activity. Interest in cash settlement has spread to the agricultural commodities as well. It is a controversial issue.

There are apparent advantages to cash settlement. Delivering the actual commodity always requires costly transporting and handling of the commodity. In the livestock sector, long hedgers who accepted delivery of cattle have long complained about the poor condition of the livestock after they go through the delivery procedure.

The advantages to cash settlement are

1. Reduced costs of "settling" futures positions,
2. Elimination of the need to transport the product,
3. Possible improvement in performance of the basis, and
4. Alignment of the expiration of the futures contract and the maturity date of the options on the futures.

The reasons for advantages 1 and 2 are apparent. Advantage 3 is a researchable issue, but research evidence does suggest improved basis performance in commodities such as feeder cattle after the change to cash settlement starting with the September 1986 futures contract. Basis appears to be less variable since the advent of cash settlement.

Advantage 4 is important. With physical delivery, the announced expiration dates for options are well in advance of the expiration date for the underlying futures contract. The August live cattle futures contract, for example, will always expire on August 20 (if not a holiday or weekend), but the expiration date for the options on the August contract will be set during the last half of July.[12] The time is needed for the logistics of the delivery process to be handled.

The holder of a put option has the right to a short position in the underlying futures at a prescribed strike price. If the futures prices fall below the strike price, the right to a short position will increase in value as the futures market declines. Normally, the buyer or holder of the put option just sells the option at its increased value (higher premium) and thus gains the needed protection against declining cash prices. But the holder of the put also has the right to exercise the option and request a position in the futures market.

The buyer of the put option might elect to exercise and demand a short position in the futures if the premium on the option is not increasing enough to fully reflect the decline in the underlying futures market. Alternatively, the holder of the put may feel the cash futures basis is too wide and would want to be assigned a short position in futures so delivery would be possible. Clearly, there are problems that emerge from the early expiration of the option. *With the option expiring before the delivery period for the futures begins, the producer choosing to use options to forward-price is denied the possibility of getting involved in the delivery process even when the basis is not at expected levels.*

In terms of procedure, cash settlement is simple. As the expiration date, holders of long and short positions in the futures simply have their accounts settled by using the current level of the cash price series. For example, if the cash price series for feeder cattle is $81.25, all futures positions that have not been offset will be settled using the $81.25. A short hedger who sold the futures at $84.50 would be credited with $3.25 per hundredweight. A short hedger who sold at $78.00 would see his futures account settled via a debit of $3.25 per hundredweight. The holder of a long position at $80.00 would be credited with $1.25 per hundredweight.

Cash settlement is not a cure-all, however. Some of the issues were discussed briefly in earlier chapters. *It is extremely important that the cash series used for settlement be representative of trade in the cash market, be competitively determined, and be free from potential manipulation.* If the cash series can be influenced by a single buyer or seller in the cash market, that firm can enhance the value of futures positions to its advantage by exerting pressures designed to change the cash-price series.

For example, a holder of short positions at $84 in feeder cattle will clearly benefit if the cash price series can be forced down to, say, $79, before cash settlement of futures positions is completed. Holders of long positions in the futures would want to see the cash series forced up as much as possible to enhance the profitability of their futures positions.

[12]Early in 1990, the CME proposed changes that would move the maturity date of the options on live cattle futures to a date early in the delivery month for the maturing futures contract. This was done, and in 1997 discussion continues about moving the maturity date of the options to still later dates.

Attempts are made to minimize the possible problems of cash price manipulation by using a broadbased index. In feeder cattle, a weighted average of cash prices spread across numerous markets in 27 states was initially used. Later, this was changed to markets across 12 midwestern states. It is difficult to conceive of any one firm being able to influence such a broad cash market.

In other commodities, such as the fed cattle traded via the live cattle futures, the solution is not so apparent. There are far fewer markets or geographical areas with active fed cattle trade in the cash market. And since the mergers and acquisitions of the late 1980s, as few as four firms slaughter 80 percent or more of the steers and heifers that go into the boxed beef trade. In this type of situation, there is usually an attempt to broaden the base for the cash-price series or index by using cash prices for heifers as well as steers or by bringing in a related series such as the boxed-beef-value series that involves retail-level activity. Still, there is and should be concerns about moving to cash settlement procedures when there is any chance the cash-price index could be manipulated. A study by Kahl, Hudson, and Ward[13] recommended against moving to cash settlement in the live cattle futures because of the concern that the cash-price index could not be constructed such that it would be free from potential manipulation.

Cash settlement moves are being discussed again in the late 1990s. The USDA is collecting cash prices more broadly, and some feel the move to cash settlement is feasible. In hogs, the move has already been made. The CME is shifting to a "lean-hog" futures contract that discovers price on a carcass basis and is cash settled to a cash price index collected from a number of markets. There is also always the possibility that moving to cash settlement would damage the viability of markets in which physical delivery has been allowed. Part of the downward-trending volume in fed cattle at Omaha and Sioux City, for example, is the delivered cattle under futures contracts. If either of these markets were to disappear because of a move to cash settlement for live cattle, producers selling cattle in those markets will have lost their access to one of the few remaining terminal markets.

Cash settlement of futures positions has obvious advantages in that it reduces the costs of delivery and eliminates the time difference between maturity of the futures and options on futures. If it is to work effectively, however, the cash series must be free of potential manipulation by large firms who deal in the cash market. In highly concentrated markets such as the fed cattle markets, there are reasons to be concerned about the immunity of cash-price series from possible manipulation.

AN OVERALL OBSERVATION

You might reasonably observe that the futures markets have not been effectively criticized at all to this point in this chapter. Are there no negatives? There are but the purpose here was to deal with the long-standing areas of controversy.

There are bad players in the futures markets as in any sector. The FBI sting made public in 1989 revealed some of the problems that can come up on a day-to-day basis.

[13]Kandice H. Kahl, M. Hudson, and C. Ward, "Cash Settlement Issues for Live Cattle Futures Contracts," *Journal of Futures Markets*, Vol. 9, No. 3, June 1989, pp. 237–248.n

Too often, brokers are paid strictly on commission, and that has undoubtedly contributed to their tendency to pressure the user to trade too often. *Under that type of pressure, what are hedging programs can turn into speculative programs in the futures when the original idea was to avoid speculation in the cash market.*

Trade is too thin in the distant contracts in some commodities for effective use of the markets. Trade in options in the distant contracts is often especially thin, but the options are still relatively new. It is not always apparent that the exchanges are doing all they can to encourage trade in the distant contracts. Trade in the distant contracts was encouraged back when long-term capital gains were treated differently. There has been no obvious effort to look at ways of restoring trade in those sometimes thin contracts that are six months or more in the future.

The price discovery component of the feeder cattle market appears to be weak. Since the move to cash settlement in September 1986, all the feeder cattle futures prices appear to be "tracking" the cash index very closely. On some days, there is no more than a $.35 per hundredweight spread in the settlement prices of all contracts being traded. This area needs more investigation. The CME has made changes in the feeder cattle contract, but its price discovery function could still be criticized.

In recent years, trading funds have moved into the agricultural commodities, especially in corn, soybeans, wheat, live cattle, and lean hogs where futures trading volume is significant. Many of these funds trade strictly on technical indicators with little regard to supply–demand balances or imbalances. Research by Murphy indicates these funds may hurt the effectiveness of price discovery in live cattle futures, and many observers would argue that the same is true for other commodities. The CFTC continues to look at their monitoring and enforcement policies with regard to trading funds.

Certainly some areas could be criticized. *There is an ever present need to make the futures as useful and productive as possible*. The exchanges, the federal regulatory agencies, and users all have a responsibility here.

SUMMARY

Misconceptions and misunderstanding of the relationship between the cash and futures markets can discourage use of the futures markets in areas of economic activity in which the price risk transfer and price discovery dimensions of the markets are needed. This chapter discusses a limited number of those issues in lay terminology.

It does often appear that moves in the futures market cause changes in cash prices, but this interpretation is too simplistic. *There are supply-response issues to be considered, and the issue of whether the underlying base of information is adequate for efficient price discovery processes deserves attention*. In general, the research findings suggest the two markets interact and work together in the pricing process. Since the futures markets are highly visible, it may be that the futures markets become the "messenger" for bad news to come in terms of pending changes in cash price. There is then a tendency to want to "kill the messenger."

You are is encouraged to complete and review this book with an open mind about the futures markets and the use of futures markets. There is no substantive body of evidence to suggest the futures markets do anything other than their prescribed tasks of contributing to price discovery and providing effective price risk transfer mechanisms. Since the exchanges are motivated to offer a contract that works for potential hedgers, there is constant attention to possible improvements. In recent years, moving to cash settlement is receiving attention, and the willingness of the exchanges to examine alternatives is tangible evidence of their interest in offering futures contracts and related trading procedures that fit the needs of the decision maker.

The most compelling evidence of the importance of the economic functions performed by trade in futures is the rapid growth in the number and types of contracts offered and the record levels of trade at all exchanges in recent years. When risk exposure became pervasive in the 1970s and 1980s, trade in futures contracts emerged to contribute to price discovery processes and, most importantly, provide a risk-transfer mechanism. If the need were not present, the markets would not receive the widespread use we have seen in many commodities, and this growth in use levels has continued through the late 1990s for many of the futures instruments.

KEY POINTS

- Increasing exposure to price risk across a wide spectrum of economic activity suggests that a *mechanism such as futures markets is needed to allow the transfer of price risk.*
- *Trade in futures contracts does not appear to dominate cash prices and pricing in the cash market.* For nonstorable commodities such as slaughter cattle, the markets are found to interact and work together in the price discovery process.
- *Variability in cash prices is not increased when trade in futures contracted is started.* Though a difficult issue to research, the consensus of the available literature is that *variability in cash prices is decreased by trade in futures.*
- Futures prices will not typically be any more accurate predictors of cash prices than will econometric models or other analytical attempts to predict prices because *a supply response can evolve that changes the level of supply for the later time period.* Since the supply response is a function of decision makers' behavioral responses, *the futures markets cannot always correctly anticipate the magnitude of the supply response.*
- Logical economic forces involved in the delivery process *help to ensure cash-futures convergence and a predictable pattern of behavior for the basis.*
- *Cash settlement of commodity futures has significant advantages* in the form of reduced costs and alignment of futures and options expiration, *but a cash-price series that is free from potential manipulation is critically important.*
- *Widespread adoption and use* of futures contracts for hedging purposes is perhaps the most compelling and tangible evidence that *trade in futures contracts serves an economic purpose.*

USEFUL REFERENCES

Phil L. Colling and S. Irwin, "On the Reaction of Live Hog Futures Prices to Informational Components in Quarterly USDA *Hogs and Pigs* Reports," *Proceedings, NCR-134* Conference on Applied Commodity Price Analysis, Forecasting, and Market Risk Management, Chicago, April 20–21, 1989, pp. 17–35. The authors identify the problems in the available information for analysts of the hog markets.

Competition and the Livestock Market, Report of a Task Force Commissioned by the Center for Rural Affairs, Walthill, NE, April 1990. The study focuses on structural change in the cattle sector in the late 1980s and looks at the role of live cattle futures in the consolidated industry of the 1990s.

Michael A. Hudson and Wayne D. Purcell, *Price Discovery Processes in the Cattle Complex: An Investigation of Cash-Futures Price Interaction*, Bulletin 85–12, Virginia Ag. Exp. Station, Blacksburg, VA, 1985. The study analyzes the time-related relationship between day-to-day changes in cash and futures for live cattle.

Richard E. Just and G. Rausser, "Commodity Price Forecasting with Large Scale Econometric Models and the Futures Market," *American Journal of Agricultural Economics*, Vol. 63, No. 2, May 1981, pp. 197–208. The authors test the predictive accuracy of the futures markets versus sophisticated econometric models and conclude that the futures markets are just as accurate for many commodities.

Kandice H. Kahl, M. Hudson, and C. Ward, "Cash Settlement Issues for Live Cattle Futures Contracts," *Journal of Futures Markets*, Vol. 9, No. 3, June 1989, pp. 237–248. After investigating, the authors recommend against cash settlement in live cattle futures because of concerns that no acceptable cash price or cash-price index can be developed.

Robert David Murphy, *The Influence of Specific Trader Groups on Price Discovery in the Live Cattle Futures Market*, unpublished Ph.D. dissertation, Virginia Tech, January 1995.

Charles E. Oellermann, B. Brorsen, and P. Farris, "Price Discovery for Feeder Cattle," *The Journal of Futures Markets*, Vol. 9, No. 2, April 1989, pp. 113–121. A test of the causal flows between cash and futures, the study finds that the futures market does respond first to new information and finds interaction between cash and futures markets.

Wayne D. Purcell and Michael A. Hudson, "The Economic Roles and Implications of Trade in Livestock Futures," in *Futures Markets: Regulatory Issues*, American Enterprise Institute, Washington, D.C., 1985. A general treatment of the role of livestock futures including a discussion of the certificate delivery system for live cattle.

Robert D. Weaver and A. Banerjee, "Does Further Trading Destabilize Cash Prices? Evidence for U.S. Beef Cattle," *Journal of Futures Markets*, Vol. 10, No. 1, February 1990, pp. 41–60. The authors test a hypothesis that trade in futures destabilizes cash prices but find no evidence that such is the case.

CHAPTER 13

AN OVERALL PERSPECTIVE

BACKGROUND

The future markets can be powerful price and cost risk management tools. Advocates would argue that the still relatively minor direct use of the markets by participants at the producer level in the agricultural sector suggests a major need for extended and more effective educational programs. Skeptics would counter with the charge that the agricultural futures contracts do not fit the needs of many relatively small producers, that the contracts and trading procedures are not always appropriate, and that the futures markets either are not needed or should be changed in a significant way. In the dairy sector, where government programs have historically minimized producer-level exposure to price risk, new futures instruments are being offered and the dairy producer will need to learn to manage price risk. And in this and other sectors, critics argue that we do not need trade in futures.

A conclusion that embraces the postures of both the advocates and the skeptics is most likely the correct one. Most of our land-grant universities have conducted formal classes and extension educational programs during the 1970s, 1980, and 1990s on the use and application of commodity futures in managing exposure to price risk. Those educational programs have been supplemented by programs in the private sector coordinated and often subsidized by the commodity exchanges and by private advisory groups. *Still, surveys indicate that only a small percentage of agricultural producers use the futures markets or options on futures directly.* In the grains and oilseeds, the futures markets are important indirectly in making possible the extension of cash contracts for future delivery by elevators, exporters, processors, and other buyers. A survey of a sample of midwestern farmers by Shapiro and Brorsen (the study is listed in the references) estimated 11.4 percent of grain and oilseed production was being hedged directly in the futures and that an added 20.5 percent was cash contracted, with producers thus using the futures markets indirectly.

The indirect or secondary extension of the futures markets to producers is less widespread in the livestock commodities. It is only in recent years that significant use of cash contracts that have a price provision has come into procurement programs for slaughter hogs. In many instances, however, only the cash futures basis is included, and the producer is still fully exposed to price risk and is, therefore, still a cash market speculator. At the cow-calf and stocker operator level, a vast majority of the stocker and feeder cattle are produced and sold with the producer operating totally as a cash market speculator and totally exposed to the risk of dramatic price moves. And as noted, the exposure to price risk in dairy products is just starting.

In a special March 13, 1987, survey, the Commodity Futures Trading Commission attempted to identify who held positions in cattle, hogs, and feeder cattle. Roughly one-half of the positions in live cattle, hogs, and feeder cattle futures were held by hedgers. With average monthly open interest at 82,773; 29,305; and 17,923 contracts for live cattle, hogs, and feeder cattle, respectively, in 1987, it is clear that only a small percentage of the livestock is hedged. If 41,500 of the live cattle contracts were held by hedgers, for example, that would amount to about 1,577 million cattle (38 head per contract). The cattle on feed in the 13 major feeding states totaled 9.24 million head on January 1, 1987, and was at 9.77 million head on December 31, 1987. At most, it would appear, 17 percent of the cattle on feed were hedged. The percentage would be less for hogs and for feeder cattle, and the situation is much the same in the late 1990s. Trading volume and open interest are still small compared to the size of the industry, and *Position of Traders* reports show only a modest hedger presence. *It appears most producers of livestock operate as cash market speculators.*

The rapid growth of financial futures raises the possibility of a second area in which indirect use of futures markets could be very important to producers, processors, and other business entities involved in the production, processing, and distribution of food and fiber products. In most instances, the individual entrepreneur will not use the markets directly to hedge against rising interest rates or the implications of changing exchange rates because of the size of the future contracts. The financial institutions will be looked to for action in the futures so that loans, for example, could be extended to borrowing clients that have a fixed rather than a variable interest rate. But very little is being done, and the agricultural sector—and especially producers—continues to absorb most of the costs associated with exposure to the risk of fluctuating interest rates.

It may be that there is too little competition to force bankers to change their policies. As long as they can operate on a margin and pass the risk back to borrowers, there is no major incentive to change. But another possible reason applies to producers as well. Neither the agricultural loan officers in the bank nor the producer is very comfortable with marketing and marketing issues. By choice they are production oriented—and this is especially true of the producer. *Perhaps it will continue to be the choice of many producers to sit on the tractor rather than at a desk in front of a microcomputer analyzing the markets and financial issues.* Against that possibility is the concern that it is not an informed choice but an unnecessary barrier thrown up against use of the markets that stops many producers—and their bankers. This book has tried to deal with that issue by showing that it is relatively easy to use the markets effectively and by pointing to the potential that is not being fully realized. In an era when the Internet brings information to our fingertips and when the "information revolution" is the only revolution we have, you would expect the users and potential users of the markets to be more amenable to their application.

The revitalization of options for the agricultural commodities in the 1980s and the rapid growth in options trade during the 1990s has the potential to change the situation. Using put options to establish a price floor on crops or livestock to be sold, or call options to place a ceiling on raw material costs, eliminates some of the major barriers to the direct producer-level use of the futures. *The ever present concern about the opportunity costs of pegging prices that turn out to be too low or pegging costs that turn out to be higher than were later offered is eliminated. In addition, producers' concerns and perpetual problems with arranging for and managing margin accounts are eliminated.*

Use levels suggest that the options are not yet proving to be very attractive to producers, however. In recent years, during the early summer months, trade in feeder cattle options on the fall feeder cattle futures has been so thin and so sparse that using the options has still been difficult. Across a time horizon of four to eight months, it will be the producer and potential hedger of stocker and feeder cattle who will be interested in the put options on the distant futures. With an annual calf crop exceeding 30 million head across the past decade, it is clear that not many of the feeder cattle are being floor-priced using the options. Much of the open interest in the distant feeder cattle futures and options appears to be held by the feedlot complex looking to gain some protection against rising costs of feeder cattle, their most costly input. Speculators and traders arbitraging between the feeder cattle and live cattle options are involved, but their activity tends to be focused in the nearby contracts and the trade in the distant contracts is thin and often difficult to manage. The use of options by producers is more widespread in the grains and oilseeds, but would still account for only a small percentage of total production.

In terms of perspective, then, it appears that significant parts of agricultural production is still completed with the producers operating as cash market speculators. Much of the hedging that is done is by the minority of the very large producers. *That fact argues in support of continued progress in education, in understanding, in awareness of the potentials the markets offer, and in support of constant monitoring of any problems and needed changes in the contracts and in trading procedure.* In the remainder of this final chapter, the objective is to pull these needs and issues together into a total picture of the requisites of an effective trading program.

Much of agricultural production is still completed with the producer operating as a cash market speculator. The potential of the markets is not being realized, perhaps because there are still problems of understanding and negative attitudes on the part of potential users, or problems with the applicability and relevancy of the futures and options that are being traded.

THE TOTAL PICTURE

A recurring theme in this book has been the importance of both fundamental and technical analysis of the commodity markets and the complementarity of the two approaches. The successful commodity trader, whether speculator or hedger, must come to recognize the need for a dual approach to analysis of the markets. *The failure to do so contributes to the sparse use of the markets, to the frequent misuse of the markets, and to the inclination to misinterpret what is being accomplished as a mistake when the futures side of a hedge results in significant opportunity costs.* As the development proceeds, you are encouraged to keep in mind the treatment of the psychology of the market developed in Chapter 6 in detail.

In terms of specific dimensions of an overall orientation to the markets, perhaps the most important is the contrast between the infusion of new information in the fundamental and technical dimensions of the markets. *On the fundamental side, significant changes in the base of information, and certainly in the publicly available base of information, tend to come infrequently.* During the growing season for corn, the crop production estimates come monthly and the first estimation based on a survey of producers does not usually come until August.

Cattle on Feed reports for the seven major feeding states are released monthly. In the hog sector, the reports on the supply-side numbers are available only on a quarterly basis. There are no significant privately produced reports for hogs comparable to *Cattle-Fax* in the cattle sector which provide updates on information such as weekly placements and shipments from *Cattle-Fax* member feedlots.

From the perspective of an individual and relatively small producer, therefore, a formulated perception of a significant change in the underlying supply–demand or fundamental picture will often be slow to change. If a position—short hedge, long hedge, purchase of a put option—has been established in the market, it can take several days, weeks, or even months for individual producers to change their perceptions of the expected price direction based strictly on their access to and appraisal of the fundamental information.

In sharp contrast, information is constantly flowing into the technical side of the market. Not only do prices change and adjust constantly, and the price action is the technical dimension of the market, but price action also reflects the injection of both publicly available *and* privately held information. Earlier in the book, there was a discussion of the efficiency of the futures markets. Most research efforts support the idea that the futures markets typically reflect not only the publicly available information, but also the impact of much of the privately held information as the large commercial firms, who have their own information network, act in the markets on the basis of their privately generated intelligence. Further, the point has been made many times in the book that the large firms do attempt to hire and use capable analysts and market technicians.

What we have, then, is the specter of individual producers holding firm to their fundamentally based biases and perceptions while the market proceeds to make new price highs or price lows. By the time the perceptions change, a significant opportunity cost has been incurred as the market makes, for example, new life-of-contract highs and presents a technical pattern that confirms that a significant change in price direction has occurred. If short hedges are held firmly in place while the market adjusts and moves to new life-of-contract highs, margin calls accumulate, the producer has suffered a substantial opportunity cost, and the hedge is inevitably viewed as a mistake. *There can be little questions that the typical agricultural producer makes no distinction between the opportunity cost associated with a hedge that,* ex post, *was placed at prices too low and an actual loss such as selling cash product below the costs of production.* The producer becomes frustrated and tends to back away from the markets.

The essence of the argument here is that the user must chart the markets. In the many educational programs we have conducted across the years, it has been argued that recognition of a change in the direction of price trend is too late if one waits for it to "trickle down" and emerge in the form of a coffee shop consensus. Fast-moving markets can impose a heavy opportunity cost or move quickly away from a pricing opportunity. The decision maker cannot afford to let several days or even weeks slide past before recognizing what is happening.

A minority of decision makers involved in agricultural production have come to understand the importance of monitoring the markets and being prepared to act when the need is there or when an opportunity presents itself. But still too often, the response by the majority of producers to the need to understand both the fundamental and technical dimensions of the markets and to monitor daily price action is "I'm too busy, I don't have time." That attitude is one of the reasons many still act as strict cash market speculators. There is the concern that understanding the markets is too difficult—and it is not. There is the argument that it takes too much time—and it does not. *What a bit of time and persistent monitoring and charting of the markets can do is offer more potential to improve the economic viability of the operation than a like amount of time and energy spent on further refinement of production technique and refinement of production-oriented skills.* If a producer reflects on this, takes a look, and still refuses to "get off the tractor" and manage the business, then that choice has to be honored, and that particular individual is expressing a preference for the cash market speculator role.

Without question, one dimension of the total picture being discussed here is the need to understand the importance of both the fundamental and technical dimensions of the market and how to manage exposure to price-risk management based on that understanding. There are times when being a cash market speculator is the right choice, times when being heavily hedged or protected via options is the right choice, and even times when price protection is taken at a loss relative to costs of the operation. The key is knowing when to adopt those various postures; both approaches to market analysis are needed in making those very important decisions.

In forming a total picture of what is needed to be effective in the markets, it is important to include both the fundamental and technical dimensions of the market. Neither, used alone, can be totally effective in a price-risk management program because the two approaches to analysis are complementary in their application.

Related to the capacity to recognize the importance of a broad approach to analysis, the issue of misuse of the markets emerges. Here, reference is not to the tendency for users to start as legitimate hedgers and then allow themselves to slide into speculation in the futures markets. That tendency is a problem and it is a misuse of the markets. *The objective here is to highlight the tendency to put reliance on the futures markets as a predictor of cash prices and to fail to recognize the importance of the constantly changing supply–demand balance*.

In earlier chapters, there is discussion of the supply response that can develop, even within the year, in most agricultural commodities. The possibility is especially important in the livestock commodities. Cattle can be placed on feed and moved to slaughter weights in as few as 80 days. With modern technology, improved feeding techniques, and superior genetic potential, a producer can have increased numbers of hogs to slaughter weights in less than nine months from the day the gilt is moved into the breeding herd. In terms of the biological dimensions, therefore, it is clearly possible to change the expectations of supply and the actual supply of cattle or hogs within the year in response to a price stimulus.

In the grains, oilseeds, cotton, and so on, the intrayear supply response will be less significant, but it is still important. Prior to planting, acreage can be switched from one crop to another if the producer's price expectations are more favorable for one crop versus another. Since one source of price expectations is the preplanting quotes on the harvest-period futures contracts, the possibility of responding to those futures-based price expectations is clearly present, and a change in supply is the result.

After planting, the capacity to respond is diminished. Fertilization and herbicide rates can be adjusted, however, and tillage practices can be designed to enhance yields. On harvested acreages near 60 million acres in soybeans and well above 70 million acres in corn, a yield increase of a few bushels per acre can change the total supply significantly.

Before extending discussion of the implicit argument about misuse of the futures price quotes, it is productive to review the discussion of Chapter 3 and the coverage of the psychological dimensions of the markets and of decision makers' behavior in Chapter 6. In Chapter 3, the concept of a *micro–macro paradox* was introduced. *At the micro or individual firm level, no single decision maker is able to exert enough influence on either the supply or demand side to influence price. But the combined or aggregate (macro) influence of all the individual decisions can bring a major change in the supply of corn, soybeans, or cattle.*

In Chapter 6, the tendency for individuals to follow the crowd—"the herd tendency"—was introduced. What looks good to several producers often looks good to their neighbor and there is a tendency for everyone to go along. Countless examples of this type of behavior can be documented. Surges in placements of cattle into feedlots are common. Hog producers continue to expand in response to earlier inflated price expectations even after a significant increase in supply is assured or has even been documented. When not blocked by government program restrictions, producers have been observed to make a huge switch from corn to soybeans or vice versa. The harvest-period price relationships are then effectively reversed in terms of per-acre profits relative to preplanting expectations.

The net result of all this is that any profit window that opens and offers the possibility of attractive profits to agricultural producers or processors will be quickly closed by the aggregate response of many relatively small producers. For the producer who responds to the price incentive and forward-prices the expanded production, there is no problem. If the supply response is larger than the market had anticipated, and the magnitude of the behavioral response is *very* difficult to anticipate, the price in the future time period will be driven down relative to the price that prompted the expansion. If price protection is set by hedges, options, or cash contracts, the producer is protected, and much of the economic pain of the price decline is transferred outside the agricultural sector to the speculator who is willing to accept the risk.

For producers who act on early price expectations and who do not hedge or contract and establish price protection, the outcome is very different. The aggregate response, if large enough eventually to drive cash price sharply lower, means the producer will sell any expanded production *and the original base of production at a lower price*. Herein lies the misuse of the markets. *Futures prices should never be used as price expectations and the production program expanded if much of the total production is not going to be forward-priced in either the futures market or via cash contracts.* Decision makers *must* understand the need for protection. If such is not the case, it is more appropriate to criticize the handling and management of the pricing program than to criticize the futures market for its periodic inability to accurately predict the magnitude of decision makers' response to a price incentive.

It is imperative that decision makers be aware of the price implications of an aggregate supply response in a setting where individual firms actions cannot influence price. Any response to a futures market price

incentive must be made with understanding of those aggregate influences and a willingness to get the prices established via hedging, options, or cash contracts.

An added and essential ingredient to the overall orientation being developed here is the much-discussed discipline. It helps little to understand the fundamental and technical dimensions of the markets, to appreciate the price implications of an aggregate supply response, and to adapt the price-risk management program to the risk preferences and abilities of the decision maker and to the financial capacity of the firm if there is no discipline in the application of the program.

In practice, traders use numerous rules in an attempt to bring discipline to their programs. Speculators keep reminding themselves and each other that "the trend is my friend." That simple rule amounts to the recognition that it is very difficult to trade successfully and profitably if the entry position is short in an upward-trending market or long in a downward-trending market. Following that simple rule in combination with a rule such as "look for as 3:1 advantage," which means never enter the market unless the potential gain is projected to be at least three times the apparent risk of loss potential, can help guard against undisciplined trading.

For potential hedgers, some of the technically oriented trading programs discussed in Chapters 4 and 5 will help. Moving averages take the subjective dimension out of the trading program. Point-and-figure charting techniques provide the same type of objectivity. Trying to make decisions on trades to be made and orders to be placed when the market is not open can protect against the temptation to get caught up in the emotions of the market. But an added step, one that has been mentioned several times in earlier chapters and one that should not be overlooked, is to "write it down."

A written plan should always be used. *A written three-party agreement that lays out the objectives of the program and the responsibilities of the producer, the financial institution, and the broker is needed.* When positions in the futures will be employed, provisions for credit for margins and who is to answer margin calls must be established. It is especially important that the role of the broker be delineated. With rare exceptions, it is preferred that the broker's role be restricted to the effective execution of orders—and most brokers that deal with hedging programs would prefer that approach.

The plan must encompass an adequate level of detail. In Chapter 4 and in Chapter 11 there was considerable discussion of selective hedging programs using bar chart signals. In Chapter 7 and again in Chapter 11, discussion of options strategies was extended to more sophisticated approaches that seek to improve, across some preselected price range, over a straight hedge or just buying a put. *The detail in the written plan must include a planned course of action in anticipation of market patterns that will require hard decisions and quick actions.* This is especially important if the price trend proves to be counter to that expected or the price ranges observed in the markets turn out to be wider than those anticipated in the marketing plan because the supply–demand balance registers unexpected changes.

On approaches to life-of-contract highs, for example, the written plan should include what will be done in the event of two consecutive closes at new highs. If short hedges are to be lifted, both the producer and the lender must understand the implications of being back in the posture of a cash market speculator, and the broker must faithfully execute the plan. And there must be agreement on what will be done if the market later turns lower. Making all participants party to the contract will help ensure consistent and disciplined execution of the plan. *It is relatively easy to follow through on a preestablished plan as to what will be done in the event of new life-of-contract highs. It is next to impossible to handle that same development on a spur-of-the-moment basis.*

In the option-based strategies, one approach examined in Chapter 7 involved selecting a price range across which a particular option strategy would be applied. Outside that preselected range, the decision was to carry the risk associated with price moves outside the identified price range. But that decision should be backed up by a safety net provision if the financial viability of the firm could be threatened. A written plan is essential.

Selling a $60 put on hog futures when the price is expected to be well above $60 can generate premium income if prices do move higher. This is one approach to the use of options that has the potential to improve the net price compared to a straight hedge. But if an unexpected surge in supply pushes prices lower, the firm is doubly vulnerable below $60. Not only is there no protection against falling cash prices, but selling the $60 put will bring futures account losses and margin calls if the price drops into the mid-$50s and lower. The producer is losing in the cash market as prices drop, and selling the $60 put will bring losses of a comparable magnitude in the futures. *It is imperative that the price-risk management plan lay out in advance of the developments what will be done if futures prices drop below $60.* The put could be bought back, a cheaper put (such as $55) could be bought, or futures could be sold. As discussed in Chapter 11, there are many possibilities. *The important thing is to pick the alternative that fits the firm best and follow through with discipline.* The same need holds for the grains or oilseeds or any position in options on futures contracts.

In an effective and comprehensive marketing program, there must be a written plan that clearly delineates what is to be done and who carries the responsibility for action. Since the program is based on expectations of the direction of price trend and/or an expected price range, it is important to include provisions for action if the market does not behave as expected due to a supply response, a surge in demand, or some other unexpected development.

A FINAL WORD

Recently, a speaker suggested that the potential applications of options on agricultural commodities are limited only by our imagination. He was right, but there is a parallel.

The potential and applications of the futures markets (and indeed the options) are limited only by our lack of understanding—and imagination. *If we take the time to understand the fundamental and technical dimensions of the market, formulate designated plans to fit the needs of the firm, and follow through with discipline, that potential can be realized.* And, concurrently, there will always be a need to critique the contracts and exchange trading rules to make sure the protections are indeed present, the contracts will be effective, and trade is conducted with adequate protection for all potential users.

If you get to that point, the economic health of the agricultural sector can be ensured to a degree, and you have a better chance of being profitable. And on that optimistic note, we will close this edition.

USEFUL REFERENCES

B.I. Shapiro and B. Wade Brorsen, *Factors Influencing Farmers' Decisions of Whether or Not to Hedge*, Purdue University, Lafayette, IN, April 1987. The study provided information on the number of producers using the futures markets and how they were being used in the mid-1980s.

Gregory J. Kuserk, *Trading in Livestock Futures and Options Markets: A Survey of Traders with Open Positions on March 13, 1987*. Commodities Futures Trading Commission, Washington, D.C., February 1988. The survey provides detail on the type of trader holding positions and on the number of positions for a specific day.

GLOSSARY

Anticipatory Hedge: Hedge placed in anticipation of future action in the cash market and which serves as a temporary or short-run replacement for the cash position.

Arbitrage: Actions taken in response to an apparent profit opportunity due to market imbalances. An example would be buying in the cash market and delivering in the futures market because the futures market appears to be unusually high relative to the cash market.

At-the-Market: An order to buy or sell at the first price obtainable when the order reaches the trading floor.

At-the-Money: Option with a strike price equal the trading level of the underlying futures.

Bar Chart: A daily record of futures market activity which shows the high price, the low price, and the closing or settlement price for a particular futures contract.

Basis: The difference between cash and futures prices for a particular commodity. It is defined as cash minus futures for a specific location and for a specific point in time.

Basis Allowance: The expected basis at the end of a production or planning period for a particular commodity for which futures contracts are trading. The "allowance" adjusts the price for a particular futures contract to a particular market location.

Basis Contract: A contract between buyer and seller of grain, livestock, or other commodity that specifies a level of the cash-futures basis that will be used in final settlement of the contract.

Basis Risk: The possibility that the cash-futures basis will not move to expected or predicted levels.

Bearish: The belief that the market will go to lower price levels.

Beginning Basis: The cash-futures basis at the beginning of a decision period such as a storage period in grains.

Break: A rapid and significant decline in price often associated with prices moving below a support level.

Breakaway Gap: A gap on a bar chart that occurs when the market moves rapidly from a topping or bottoming formation or from a consolidation pattern.

Broker: A person paid a commission for acting as an agent in buying or selling futures or options on futures.

Bullish: The belief that the market will go to higher price levels.

Buy Order: An order to buy a commodity futures contract or option on a futures contract.

Buy Signal: A chart development, chart pattern, or related action that indicates the futures market is likely to move higher and therefore should be bought.

Buy-Stop Order: An order that becomes a market order when the price specified in the order is touched from below.

Buy-Stop-Close-Only Order: An order that is executed by buying futures contracts if the market closes above the price level specified in the order.

Call Option: The right, but not the obligation, to a long position in futures.

Cash Bid: A price bid by a buyer seeking to buy the cash product immediately or for later delivery.

Cash Contract: A contract, usually for later delivery, that specifies price, quality, and other conditions for delivery of a cash commodity.

Cash-Futures Basis: The same as "basis," it is the difference between cash prices in a specific market and for a specific time period and a specific commodity futures contract.

Cash Settlement: A process whereby open positions, long and short in futures contracts, are settled or "accounted" using a measure of the cash price level versus delivery of the physical commodity under the provisions of the futures contract.

Causality: The notion that one event causes another related event, as in the futures market "causing" a change in the cash market, or vice versa.

CBOT: Chicago Board of Trade.

Cell: A cell or block in a point-and-figure chart that records a price increase or a price decrease.

Cell Size: The size of the cells in cents per bushel, dollars per hundredweight, etc., in point-and-figure charting of commodity futures action.

Close: A term used to refer to the "closing" or "settlement" price of a futures contract at the end of the daily trading period.

Close-Only Order: An order that will be activated only if the closing price will exceed (buy order) or be below (sell order) a price level specified in the order.

Closing Basis: The cash-futures basis at the end of the decision period such as the end of a storage or production period.

CME: Chicago Mercantile Exchange.

Commissions: The charges by brokerage firms for completing transactions in the futures market.

Commodity Futures Exchange Commission: Federal commission with authority to oversee and regulate trade in commodity futures.

Congestion Area: A pattern on the bar chart of futures trading activity that is characterized by a prolonged period of trade in a relatively narrow price range.

Conservative Hedger: Hedger who ascribes to the philosophy that once a hedge is placed, it will not be lifted or offset until the end of the production, storage, or other decision period.

Consolidation Pattern: A pattern on the bar charts that is characterized by recognizable shapes as the market consolidates or "rests" in the middle of a major price move.

Corrective Rally: A price move up that "corrects" or retraces part of a recent decrease in price.

Cost of Carry: Storage, interest, and other variable costs of holding a product in storage.

Cover: The term often used to refer to the action of completing a round turn, as in "covering short positions" when short or sell positions are closed out or offset by buying back.

CPI: Consumer price index, a widely used measure of overall price inflation.

Crossover Action: Term used to refer to a short moving average penetrating and moving above or below a longer moving average.

Day Trader: Trader who focuses on intraday trading versus longer run positions that are carried from one trading day to the next.

Deficiency Payments: Payments to producers eligible for the subsidies allowed for in the 1985 and 1990 farm bills when cash prices fall below a target price prescribed in the farm bill legislation.

Deflated Price: A price that is adjusted for the influence of overall price inflation as measured by such indices as the consumer price index.

Delta Factor: Used in the context of options, it is the probability of a price move from a specific price level.

Delta Neutral: A marketing or trading program that compensates for the delta factor in options by adjusting the number of options so that the moves in the underlying futures are "matched."

Demand Curve: A schedule of the quantities consumers will take at alternative prices.

Demand Elasticity: Defined as percent change in quantity/percent change in price, elasticity is a property of a demand curve that measures quantitative responsiveness to a price change.

Disappearance: Refers to the demand or use components in the USDA supply-demand tables.

Double Bottom: Pattern on a bar chart that shows two price lows at or near the same price level.

Double Top: Pattern on a bar chart that shows two price highs at or near the same price level.

Ending Basis: The cash-futures basis at the end of a decision period such as the end of a storage period or the end of a production period. The same as closing basis.

Ending Stocks: The stocks of corn, wheat, etc., that are left over at the end of a crop year and must be carried forward to the next crop year.

Equilibrium Price: A market-clearing price, and the only price at which the quantity suppliers are willing to offer equals the quantity buyers are willing to take.

Exchange Member: Firm that owns a "seat" or the right to trade on an organized futures exchange.

Exchange Rates: The rate at which one currency, such as the Canadian dollar, can be exchanged for another currency, such as the U.S. dollar.

Exercise: Formal requesting of the right to a short or long position in the underlying futures contract for put and call options respectively.

Exhaustion Gap: A chart gap that appears near the completion of a major and sustained move in price of a commodity futures contract.

Federal Reserve (Fed): The Federal Reserve System that has the potential, through committee selected actions, to influence the money supply and interest rates by open market operations, changing the discount rate at which member banks borrow funds, etc.

Fill: Act of fulfilling an order to buy or sell commodity futures.

Fiscal Policy: Tax, spending, and related federal policy decisions that have the potential to influence the nature and level of economic activity.

Flag: Consolidation pattern that takes on the distinctive shape of a flag in the midst of a major price move in commodity futures.

Floor Broker: Brokers who trade in the futures or options pits on an exchange and fill orders for one or more brokerage firms and/or trade their own account.

Forced Sale: Sale that is forced due to margin calls or other financial limitations and which may be inconsistent with orderly management of market positions.

Forward Price: The price being offered by the futures market for future delivery, defined as the futures price plus an adjustment for the cash-futures basis.

Fundamentals: The supply and demand forces that ultimately determine the direction of price movement and the level of price.

Futures Contract: Contract for future delivery traded on organized futures exchange and that specifies quality standards, delivery specifications, delivery locations, etc.

Fundamental Analysis: Analyses of the underlying supply-demand forces in an attempt to project direction, probable price range, etc., of the prices of a particular futures commodity.

Gap: Price range on a bar chart of daily price action in which no trade occurred.

GNP: Gross national product, a measure of economic activity in the U.S. Government Loan Program: Program that offers legislatively determined "loan prices" for eligible producers of corn, wheat, etc. with the loan price becoming effectively a minimum price guarantee.

Good 'Til Canceled (GTC): An order for a futures market position that will remain in place until it is canceled by the commodity broker.

Head and Shoulders: A topping or bottoming formation on commodity futures charts that takes on the distinctive shape of a person's head and shoulders.

Hedger: Futures or options trader who has a position, or will have a position, in the cash market and wishes to transfer the risk of cash-market price fluctuations to someone else.

Hedging: The establishing of a position in the futures market that is equal and opposite the position, or intended position, in the cash market with an objective of transferring cash price risk to someone else.

High: The highest price recorded for a particular commodity futures contract in a particular trading session or time period.

High-Volume Day: A day during which trading volume for a particular commodity future is unusually high relative to trading volumes for recent days.

IMM: International Monetary Market, located in Chicago, offering trade in futures contracts for financial futures and foreign currencies.

In-The-Money: Put option at a strike price above the underlying futures price or a call option below the underlying futures price.

Inelastic: Refers to a demand elasticity coefficient of less than 1.0 in absolute value, indicating quantitative response to a price stimulus will be relatively small.

Inflation Adjusted: A price or income series that has been adjusted for the influence of overall price inflation by dividing the series by a measure of price inflation such as the consumer price index (CPI).

Initial Margin: Margin required, per contract, before a potential trader can be involved in buying and selling futures contracts for a particular commodity.

Intrinsic Value: The difference between the strike price for a commodity futures option and the trading level of the underlying futures contract.

Island Reversal: Chart pattern that shows chart gaps before and after price activity for one or more trading days.

Key Reversal: A bar chart pattern characterized by a new contract high or low, a trading range that exceeds that of the previous day, and a lower close (for a topping pattern) or a higher close (for a bottoming pattern).

Life-of-Contract: Used to refer to the entire period for which a particular commodity futures contract has been trading.

Limit-Down: Refers to a commodity futures contract that trades down the allowable daily move from the previous day's close.

Limit Move: Largest move allowed, whether up or down, from the closing price of the previous day for a futures contract.

Limit-Price Order: An order to buy or sell a commodity futures contract that specifies a particular price level at which the trader is willing to enter the market.

Limit-Price-Buy Order: An order to buy a commodity futures contract that specifies the maximum price at which the trader is willing to buy.

Limit-Price-Sell Order: An order to sell a commodity futures contract that specifies the minimum price at which the trader is willing to sell.

Limit-Up: Refers to a commodity futures contract that trades up the allowable daily move from the previous day's close.

Liquidity: The presence of active trade in a futures market to ensure an order can be quickly filled.

Loan Rate: The specific price at which a producer could enter corn, for example, into the government loan program as an alternative to selling the crop in the cash market.

Long: Refers to a "buy" position in the market—to buy is to go "long."

Low: The lowest price recorded for a particular commodity futures contract in a particular trading session or time period.

Maintenance Margin: The level of margin funds that precipitates a margin call if the account balance falls below the specified maintenance level.

Margin Call: Monies that must be sent to the brokerage firm to maintain a futures market position when the market is moving against the trader's position.

Margin Liquidation: Refers to traders being forced out of their futures positions because they are unable to meet margin calls.

Margin Requirement: Monies that must be on deposit with the broker before futures can be sold.

Marginal Cost: The change in cost associated with generating an added unit of product or service.

Marginal Revenue: The change in revenue associated with generating an added unit of product or service.

Marginal Value Product: A measure of marginal revenue, it refers to the revenue stream or the change in revenue associated with using an added unit of input.

Market-Clearing Price: Equivalent to equilibrium price, and is the only price at which the quantity suppliers are willing to offer equals the quantity buyers are willing to take.

Market Order: Order to buy or sell commodity futures at the first available price.

Micro-Macro Trap: Tendency for aggregate actions of many small producers to influence price in a way that no single producer's actions could influence price. The individual decision maker is therefore exposed to the aggregate or "macro" influences on quantity and price that are beyond the control of the individual or "micro" firm.

Market-if-Touched (MIT): An order to buy or sell commodity futures that becomes a market order if the specified price level is reached or "touched."

Measuring Gap: A chart gap that may appear near the 50 percent point in a sustained price move that can be used in conjunction with a breakaway gap to project the price move.

Monetary Policy: Actions by the Federal Reserve System designed to control money supply and interest rates and in other ways influence, via policy related actions, the levels of activity in the U.S. economy.

Moving Averages: An average of a specified length that "moves" through a data set as N consecutive settlement prices are added and divided by N.

Nearby Futures Contract: Closest of the futures contracts that are being traded in a time context.

Neckline: A line connecting the price lows in a head-and-shoulders top or the price highs in a head-and-shoulders bottom.

Nominal: Prices or costs that have not been adjusted for the influence of overall price inflation.

Offset: Buying back after selling futures, or selling after buying futures, to cancel previously established futures positions.

Open: Refers to the price recorded for a particular commodity futures contract when the trading session begins or opens.

Open Interest: The total number of contracts that are outstanding and have not been offset by either delivery or an opposite buy-sell transaction.

Open Outcry: A required procedure in futures pits, referring to the "crying out" of a willingness to sell or buy at a certain price.

Opening Basis: The cash-futures basis at the end of a decision period such as a storage period in grains. The same as beginning basis.

Operating Margin: Margin a business entity seeks or employs, as in the operating margin per bushel for an elevator buying and selling wheat or soybeans.

Opportunity Cost: The potential benefits of a particular course of action that are missed because some other course of action is employed.

Option: Right to a position in an underlying futures contract.

Order: Request to have a buy or sell action performed by a broker.

One Cancels the Other (OTO): Refers to a constraint placed on two orders so that one is automatically canceled if the other is filled.

Out-of-the-Money: Put option at a strike price below the underlying futures or a call option above the underlying futures.

Overbought (oversold): An expression used when there is belief the market will not go higher (lower) because almost every trader interested in the market has already bought (sold).

Overnight Position: The positions in futures or options that are not canceled or offset before trade closes and thus are carried forward to the next trading day.

Own-Price Elasticity: The elasticity of demand, where reference is to the price-quantity relationships for a particular commodity as compared to the relationships between two commodities.

Penetration Rule: A requirement that a particular level or some measure will be penetrated by a preset amount before any action will be taken.

Pennants: Consolidation pattern that takes on the shape of a small, symmetrical triangle.

Point-and-Figure Chart: A technique for plotting price action for a commodity futures contract that employs only price action and ignores the time dimension.

Premium: The market-determined value of an option for a particular futures contract and for a particular price level or "strike price."

Price Discovery: The dynamic process by which buyers and sellers interpret available information and seek to generate a price that balances the available supply and demand and clears the market.

Price Expectation: Price anticipated for some future time period that gets brought into production, storage, and related decision processes.

Price Floor: Minimum price established by adjusting a particular option strike price for premium costs and a projected cash-futures basis.

Price Risk: The risk associated with unpredictable fluctuations in price.

Price Taker: Firm that sells products in an economic structure that is characterized by many small producers where no one seller can influence price.

Production Hedge: Act of hedging a growing crop or hedging livestock that are in the feeding phase.

Profit Margin: A margin reflecting a difference between costs and a price or between costs and a price the decision maker would like to realize.

Profit-Taking Rally: A period of rising prices associated with buying actions by holders of short positions in futures. The same as a short-covering rally.

Profit Window: A profitable opportunity given costs of providing a product and service and immediately available prices, or prices for a future time period, that would allow a profit to be secured.

Put Option: The right, but not the obligation, to a short position in futures.

Relative Strength Index (RSI): One of many measures of market momentum designed to help determine when a market is overbought or oversold. The Index is based on moving averages of closing futures prices.

Resistance Area: Range of prices in which the market is expected to encounter resistance to higher prices.

Resistance Plane: A specific resistance point or plane across a price high at which the market is expected to encounter resistance to higher prices.

Reversal Requirements: The required price move on point-and-figure charts for a change in price direction to be recorded.

Round Turn: The completion of a "sell and buy back" or of a "buy and then sell" set of transactions.

Selective Hedger: Hedger whose objective is to be hedged when price moves are disadvantageous to the operation and to be a cash market speculator when price moves are advantageous.

Sell Order: Order to sell commodity futures.

Sell Signal: A chart development pattern or related development that indicates the futures market is likely to move lower and therefore should be sold.

Sell-Stop-Close-Only Order: A sell order that will be filled only if the market closes below the price level specified in the order.

Sell-Stop Order: A sell order that becomes a market order if touched from above.

Settlement: Comparable to closing price, the settlement price is the price at the end of the closing session as determined by exchange officials within any range of recorded prices at the close of trade.

Short: Refers to a "sell" position in the market—to sell is to "go short."

Short-Covering Rally: Price surge that is associated with holders of short positions buying back or offsetting those short positions.

Speculator: Futures or option trader who has no position in the cash market and is attempting to earn profits as an investor in commodity futures or options.

Stops: Orders designed to limit the exposure if the market is moving against the trader's position.

Storage Hedge: Selling commodity futures to forward price a commodity being held in storage.

Strike Price: Designated price levels for which put and call options are traded.

Supply Curve: Schedule of the quantities that will be offered by producers at alternative prices.

Support Area: Range of prices in which the market is expected to encounter support against lower prices.

Support Plane: A specific support point or plane across a price low at which the market is expected to encounter support and protection against lower prices.

T-Bill: Treasury bill, a short-term or 90-day financial instrument for which futures are traded on the International Monetary Market.

Target Price: A price specified in farm bill legislation that is set at a level matching estimates of costs of production. The term may also be used as the price desired by a producer to cover costs, living expenses, etc., and thus becomes a trigger for hedging or pricing action.

Technicals: Chart and related considerations used by traders to predict the direction of price movement based on the past history of price movements.

Technical Analysis: An approach to the market which relies on the belief that the best information on where the market is going is the past history of the price itself.

Thin: A market condition in which high levels of liquidity are not present,

Time Stamp: The stamping of the time at which an order to buy or sell is received, or when it is filled.

Time Value: That portion of the premium or an option that is associated with the time left before the option matures.

Trading Volume: Number of contracts or other measure of trade activity for a particular commodity futures contract or contracts.

Trend Line: Line on a bar chart that connects two price lows or two price highs.

Triangle: Consolidation pattern on a bar chart that takes the distinctive shape of a triangle.

Variable Costs: Costs associated with the level of production and are therefore costs that could be eliminated if operations cease.

Volatility: Fluctuations in price or other measure of economic activity.

Volume: The number of contracts traded in a particular trading session. Volume data are recorded for each futures month traded, but the reference to trading volume usually refers to the volume of trade in all the months being traded. The same as trading volume.

Worst-Case Basis: A basis that shows the cash price at its lowest level relative to a futures contract in a historical context.

INDEX